Free e-book

For access to a free PDF of *Just Enough* SAS®: *A Quick-Start Guide to* SAS® *for Engineers*, go to the following URL:

support.sas.com/justenoughsas

Just Enough SAS®
A Quick-Start Guide to SAS® for Engineers

Robert A. Rutledge

The correct bibliographic citation for this manual is as follows: Rutledge, Robert A. 2009. *Just Enough SAS®: A Quick-Start Guide to SAS® for Engineers.* Cary, NC: SAS Institute Inc.

Just Enough SAS®: A Quick-Start Guide to SAS® for Engineers

Copyright © 2009, SAS Institute Inc., Cary, NC, USA

ISBN 978-1-59994-649-8

All rights reserved. Produced in the United States of America.

For a hard-copy book: No part of this publication may be reproduced, stored in a retrieval system, or transmitted, in any form or by any means, electronic, mechanical, photocopying, or otherwise, without the prior written permission of the publisher, SAS Institute Inc.

For a Web download or e-book: Your use of this publication shall be governed by the terms established by the vendor at the time you acquire this publication.

U.S. Government Restricted Rights Notice: Use, duplication, or disclosure of this software and related documentation by the U.S. government is subject to the Agreement with SAS Institute and the restrictions set forth in FAR 52.227-19, Commercial Computer Software-Restricted Rights (June 1987).

SAS Institute Inc., SAS Campus Drive, Cary, North Carolina 27513.

1st printing, April 2009

SAS® Publishing provides a complete selection of books and electronic products to help customers use SAS software to its fullest potential. For more information about our e-books, e-learning products, CDs, and hard-copy books, visit the SAS Publishing Web site at **support.sas.com/publishing** or call 1-800-727-3228.

SAS® and all other SAS Institute Inc. product or service names are registered trademarks or trademarks of SAS Institute Inc. in the USA and other countries. ® indicates USA registration.

Other brand and product names are registered trademarks or trademarks of their respective companies.

Contents

About This Book vii

Chapter 1 Getting Started 1

1.1   Introduction 2
1.2   Installing the Sample Code 2
1.3   Starting a SAS Session 4
1.4   Writing, Submitting, and Saving SAS Code 6
1.5   Using autoexec.sas 8
1.6   Just Enough Syntax 9
1.7   SAS Programming 10
1.8   SAS Procedures 12
1.9   Debugging Basics 14
1.10 Chapter Summary 16

Chapter 2 DATA Step Programming 19

2.1   Introduction 20
2.2   Creating SAS Data Sets 22
2.3   Saving SAS Data Sets 30
2.4   SAS Functions and CALL Routines 31
2.5   The RETAIN Statement 41
2.6   Selecting Subsets of Data Sets 42
2.7   Sorting Data Sets 44
2.8   Merging Data Sets 47
2.9   More Than Enough 50
2.10 Chapter Summary 54

Chapter 3 Data Out, Data In—Spreadsheets 57

3.1   Introduction 58
3.2   Exporting a SAS Data Set to a Spreadsheet 58
3.3   Importing Spreadsheet Data to SAS 62
3.4   More Than Enough 70
3.5   Chapter Summary 74

Chapter 4 Data Out, Data In—Relational Databases 77

- 4.1 Introduction 78
- 4.2 Using PROC SQL with SAS Data Sets 79
- 4.3 Using SAS with a Database Management System 84
- 4.4 More Than Enough 94
- 4.5 Chapter Summary 96

Chapter 5 Summarizing Your Data 99

- 5.1 Introduction 100
- 5.2 PROC MEANS 102
- 5.3 PROC TABULATE 112
- 5.4 PROC REPORT 114
- 5.5 PROC BOXPLOT 116
- 5.6 PROC ANOM 120
- 5.7 PROC UNIVARIATE 124
- 5.8 PROC FREQ 130
- 5.9 More Than Enough 132
- 5.10 Chapter Summary 134

Chapter 6 Plotting Your Data with SAS/GRAPH Software 137

- 6.1 Introduction 138
- 6.2 Viewing, Saving, and Naming Your Graphic Ouput 142
- 6.3 PROC GPLOT 146
- 6.4 PROC GCHART 180
- 6.5 More Than Enough 198
- 6.6 Chapter Summary 208

Chapter 7 The Output Delivery System 211

- 7.1 Introduction 212
- 7.2 Publishing Your Report in RTF 216
- 7.3 Publishing Your Report in PDF 218
- 7.4 Publishing Your Report to the Web 220
- 7.5 Using ODS to Save and Select Procedure Output 232
- 7.6 More Than Enough 238
- 7.7 Chapter Summary 244

Chapter 8 Plotting Your Data with ODS Graphics 247

- 8.1 Introduction 248
- 8.2 ODS Statistical Graphics 250
- 8.3 PROC SGPLOT 252
- 8.4 PROC SGPANEL 296
- 8.5 PROC SGSCATTER 298
- 8.6 More Than Enough 304
- 8.7 Chapter Summary 306

Chapter 9 Analyzing Quality Data with SAS 309

- 9.1 Introduction 310
- 9.2 Deciding Whether the Process Is In Control 314
- 9.3 Measuring Process Capability 318
- 9.4 Monitoring the Process 320
- 9.5 P-Charts for Fraction Failing 322
- 9.6 U-Charts for Defects per Unit 324
- 9.7 More Than Enough 326
- 9.8 Chapter Summary 328

Chapter 10 Analyzing Reliability Data with SAS 331

- 10.1 Introduction 332
- 10.2 PROC RELIABILITY 338
- 10.3 PROC LIFEREG 346
- 10.4 PROC LIFETEST 352
- 10.5 PROC PHREG 354
- 10.6 The Reliability of Repairable Units 358
- 10.7 More Than Enough 360
- 10.8 Chapter Summary 362

Chapter 11 SAS Macro Programming 365

- 11.1 Introduction 366
- 11.2 Macro Variables 368
- 11.3 Macro Programs 378
- 11.4 Utility Macros 390
- 11.5 More Than Enough 404
- 11.6 Chapter Summary 406

Bibliography 409

Index 419

About This Book

Purpose

SAS is not difficult to learn, but it is a lot to learn. There are so many different SAS products, procedures, and options that it's difficult to know where to start, and it's easy to invest a lot of time in learning before you reach the point of being able to use SAS effectively to do really useful work.

The purpose of this book is to provide "just enough" instruction on a broad variety of topics so that a new SAS user can become productive very quickly. Because the coverage is very broad, it is necessarily not as deep as the coverage in other sources. So there are many references to books and papers that you can use to better understand the topics introduced here, and most of the papers are included for download as PDF documents.

Much of the material in the book is general enough to be useful in many fields, but the examples are geared toward the analysis of quality and reliability data, and Chapters 9 and 10 are specifically aimed at those topics.

Is This Book for You?

If you are an engineer or statistician doing a serious amount of data manipulation and analysis with spreadsheets and other interactive tools, you could really benefit from using SAS to:

- Extract data from spreadsheets or relational databases.
- Summarize and analyze your data using a wide variety of statistical methods.
- Create plots and graphs that help you to understand your data and explain it to others.
- Publish your results to text documents and Web sites.
- Automate the entire process, from data extraction to report publication.

SAS provides an end-to-end solution that enables you to do all of these things. And it is very powerful because it is a high-level language that enables you to do a lot with just a few procedure calls.

Typographical Conventions

SAS code is shown indented, using the Courier font, with SAS keywords in uppercase. For example:

```
DATA Temp; Item="Hammer"; Price=9.99; RUN;
PROC PRINT DATA=Temp; RUN;
```

- DATA, RUN, PROC and PRINT are SAS keywords in uppercase.
- Temp, Item, and Price are user-supplied names in mixed uppercase and lowercase.

Placeholders for user-supplied names or options will be shown in italics. For example:

```
SYMBOL1 VALUE=value COLOR=color;
```

- VALUE and COLOR are SAS keywords.
- *value* and *color* are placeholders for user-supplied option values.

The contents of the Log window are shown in a box with dashed gray borders, like this:

```
DATA Temp;Item ="Hammer"; Price=9.99;  RUN;

NOTE: The data set WORK.TEMP has 1 observations and 2 variables.
NOTE: DATA statement used (Total process time):

      real time        0.01 seconds
      cpu time         0.00 seconds
```

The contents of the Output window are shown in a box with solid borders, like this:

```
Obs       Item       Price
 1        Hammer     9.99
```

Companion Web Site

You can access the companion Web site for this book from:

http://support.sas.com/companionsites

The companion Web site includes several features that relate to this specific book, including more information about the book and author, book reviews, and book updates; book extras such as example code and data; and contact information for the author and for SAS Press.

Example Code—Examples from This Book at Your Fingertips

You can access the example programs from this book by linking to its companion Web site at support.sas.com/companionsites. Select the book title to display its companion Web site, and select Example Code and Data to display the SAS programs that are included in the book.

For an alphabetical listing of all books for which example code is available, see support.sas.com/bookcode. Select a title to display the book's example code.

If you are unable to access the code through the Web site, send e-mail to saspress@sas.com.

PDF of This Book

The full text of this book is available for free in PDF format. See the URL on the first page of this book to access the PDF. Save the PDF on your computer so that you can access the content at any time without lugging the book around, and print just the pages you need on a particular topic if you find that easier than reading from the screen.

Getting Help

You can learn a lot about SAS from this book, but the subject is enormous, so by necessity, a lot is left out. The following sections include recommendations for learning more.

Books

There are several SAS books listed in the Bibliography, and throughout this text there are suggestions on books to read for more detailed information on particular topics. If you want to create your own SAS library, I would recommend starting with the following books, which cover topics all users will need to be familiar with. Then look to the books listed in the Bibliography, or look for the full list of books available from SAS on the SAS Web site, http://support.sas.com/publishing/, to learn more about areas of interest to you.

Aster, Rick. 2005. *Professional SAS® Programming Shortcuts, Second Edition.* Paoli, PA: Breakfast Communications Corp.

Burlew, Michele M. 2006. *SAS® Macro Programming Made Easy, Second Edition.* Cary NC: SAS Institute Inc.

Carpenter, Art. 2004. *Carpenter's Complete Guide to the SAS® Macro Language, Second Edition.* Cary NC: SAS Institute Inc.

Delwiche, Lora D., and Susan J. Slaughter. 2003. *The Little SAS® Book: A Primer, Third Edition.* Cary, NC: SAS Institute Inc.

Miron, Thomas. 1995. *The How-To Book for SAS/GRAPH® Software.* Cary, NC: SAS Institute Inc.

SAS Web Site

The main SAS Web site is:

http://www.sas.com/

From there you can navigate to the support page that has links to information on upcoming conferences, SAS code samples, SAS books, and a complete set of online documentation in both HTML and PDF format. The HTML version is very convenient for quick searches, but the PDF version is preferable if you want to print out several pages or a whole section. The SAS support site is here:

http://support.sas.com/

The online documentation in both HTML and PDF format is here:

http://support.sas.com/documentation/onlinedoc/index.html

Google

It is sometimes quicker and easier to use Google than to search through the SAS documents. For example, to find out how to include page numbers on an RTF document, just Google [SAS RTF "Page Numbers"]. Even if the answer is on the SAS Web site, you can probably find it quicker using the Google site: option, for example [site:support.sas.com SAS RTF "Page Numbers"].

SAS_L

"SAS-L is an electronic mail discussion group that was developed to allow SAS users worldwide the potential to communicate with some the best and most powerful SAS users in the world virtually instantaneously via discussion groups or LISTSERV e-mail servers." -- A direct quote from the paper "SAS-L – A VERY POWERFUL RESOURCE FOR SAS USERS WORLDWIDE." 247-28.pdf.

Check it out at http://listserv.uga.edu/archives/sas-l.html and read the paper (included in your \JES\docs\ folder) for tips on how to use SAS-L.

SAS Technical Support

If all else fails, try SAS Technical Support. They can be reached in the US at 919-677-8008. For other countries, you can find SAS Technical Support contact information here:

   http://support.sas.com/techsup/contact/index.html

SAS Technical Support is outstanding. You shouldn't call for help with every missing semicolon, but I suggest you make a point of calling at least once early in your SAS career just so you understand the quality of support that's only a phone call away. Be sure to have your site license available (see Section 1.3.3).

SAS Training

The Web site below describes the many courses offered by SAS Institute in instructor-led and self-paced e-learning formats. These courses are a good way to get a more in-depth understanding of SAS programming in general, and of the topics that are of particular interest to you.

   http://support.sas.com/training/index.html

SAS Global Forum (Formerly SUGI)

SAS Global Forum is the new name (starting in 2007) for the yearly SAS conference formerly known as the SAS Users Group International (SUGI) Conference. The conference is held each spring at various locations in the US and includes:

- Hundreds of contributed papers covering all areas of SAS programming.
- A SAS demo area where SAS developers demonstrate their latest innovations.
- Hands-on workshops teaching elementary and advanced programming techniques.
- The SAS Publishing booth where you can review and purchase SAS programming books.

These conferences provide a great opportunity to learn a lot about SAS in just a few days' time. I strongly recommend you attend if at all possible. If you don't get to attend the conference, you can still access the full program at the SAS Web site. Go to http://www.sas.com/ and search on "SAS Global Forum" to find the page for the latest conference. From there you can download the individual papers in PDF format. These papers are a great place to find concise descriptions of new features and creative methods that you can use to enhance your SAS programming projects.

SAS Conference Papers

Several papers from past SAS conferences are included with this book. See the companion Web site for a link to the conference papers. Once you have copied the ZIP files, the ~JES\docs folder includes:

- A subfolder titled "pdfs" that contains a selection of papers from past SAS conferences.
- A CSV file titled "SUGI_Papers.csv" that contains a list of all the papers.
- An HTML page titled "papers.html" that provides a simple way to browse the papers.

If you open the papers.html file in a browser, you will see a page similar to Display 0-1.

Display 0-1: The papers.html File Opened in a Web Browser

The papers are grouped according to the chapter of this book that they are related to. Click on a link in the left frame to select the chapter you are interested in. Then click on a link in the right frame to select a paper. The paper will open with Adobe Reader in another tab in the same window, in another browser window, or in a separate Adobe Reader window, depending on your browser settings. For example, if you click on the link 255-30.pdf in the right frame, the corresponding paper will open, as shown in Display 0-2.

Note: You might need to adjust your browser settings to allow PDF documents to be displayed.

Display 0-2: Paper 255-30.pdf Opened with Adobe Reader

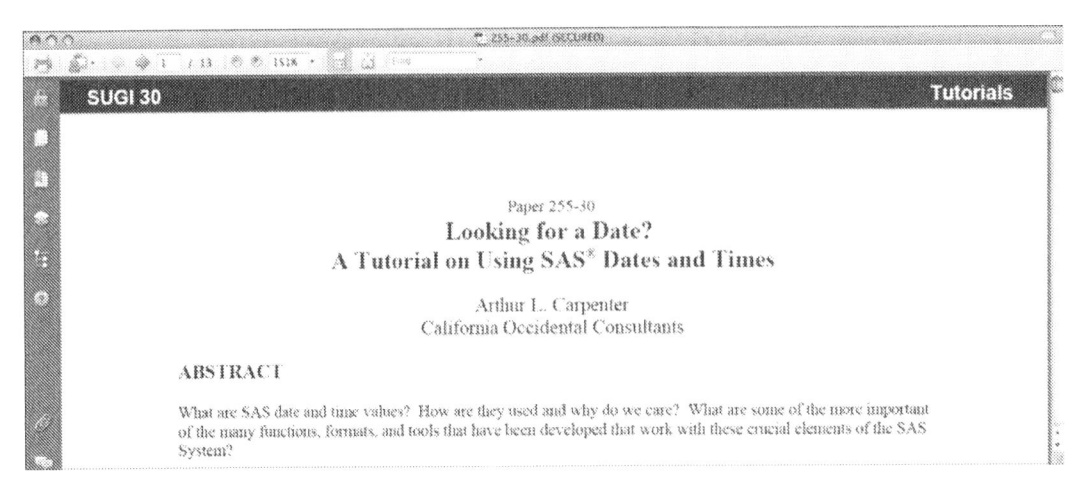

The SUGI_Papers.csv file is shown (in part) in Display 0-3. The papers are arranged by chapter, and there is a 12th "chapter," which is not in the book, but provides a place to add additional papers that are not related to a specific chapter. In Section 11.4.13 of Chapter 11 you will learn how to run the SAS code that reads the CSV file and creates the papers.html page. This allows you to add new papers to the Web page by saving the PDFs to the pdfs folder, adding the paper names to the CSV file, and then rerunning the code. In this way you can keep the page up to date with the latest papers from SAS Global Forum that you find interesting. And you can of course add papers from any other source, such as technical journals, as long as they are in PDF.

Display 0-3: The SUGI_Papers.csv File Opened in MS Excel

Additional Resources

SAS offers you a rich variety of resources to help build your SAS skills and explore and apply the full power of SAS software. Whether you are in a professional or academic setting, we have learning products that can help you maximize your investment in SAS.

SAS Community	http://support.sas.com/community/
SAS Global Certification	http://support.sas.com/certify/
SAS Knowledge Base	http://support.sas.com/resources/
SAS Learning Center	http://support.sas.com/learn/
SAS OnDemand for Academics	http://support.sas.com/ondemand/
SAS Publishing Bookstore	http://support.sas.com/publishing/
SAS Technical Support	http://support.sas.com/techsup/
SAS Training	http://support.sas.com/training/

Comments or Questions?

If you have comments or questions about this book, you may contact the author through SAS as follows:

Mail:

SAS Institute Inc.
SAS Press
Attn: Robert A. Rutledge
SAS Campus Drive
Cary, NC 27513

E-mail: saspress@sas.com

Fax: (919) 677-4444

Please include the title of the book in your correspondence.

For a complete list of books available through SAS Press, visit support.sas.com/publishing.

SAS Publishing News: Receive up-to-date information about all new SAS publications via e-mail by subscribing to the SAS Publishing News monthly eNewsletter. Visit support.sas.com/subscribe.

Acknowledgments

I would like first of all to thank Julie Platt, Editor-in-Chief of SAS Press, for her support of this project, and John West, my editor, for his guidance, patience, and understanding in getting me through the process and making it all come together.

I owe a special debt of gratitude to the technical reviewers: Garth Helf, Bob Rodriguez, Dan Heath, Jim Seabolt, Sue Walsh, Kevin Hobbs, Marcia Surratt, Mike Patetta, Catherine Truxillo, Liz Edwards, Charlie Mullin, Sally Walczak, Kathryn McLawhorn, Chevell Parker, Bari Lawhorn, and Lelia McConnell. They took the time to not only correct my mistakes, but also to offer very valuable comments and suggestions for improving the organization and flow of the content. Thanks to their efforts, I believe that the final result is much better than my original draft. I would also like to thank the writers and editors of the excellent SAS online documentation, which I relied on every step of the way to get the facts as complete and accurate as possible, and the heroes at SAS Technical Support, who got me quickly and easily through the parts that I couldn't find in the documentation.

A lot of the tips and tricks included in the book were acquired from my SAS programming colleagues at IBM and Sun Microsystems. I would never have gotten started as a SAS user without the patient and generous mentoring of Garth Helf at IBM. I also appreciate the numerous techniques and pointers provided over the years by my SAS colleagues at IBM: Linda Tai, Qin Pan, Jason Wang, Mutsumi Asahi, and Miwa Wasa (Matsuo), and at Sun: Coleen MacMillan, Trang Nguyen, Emma Gottesman, Tom Salvesen, and Ian Petrie. Their contributions are not explicitly pointed out, but their fingerprints can be found throughout the text. I would also like to thank Peter Carr, my manager at Sun, for his encouragement and support.

Finally, I would like to thank my wife, Irene, for her generous and enthusiastic support for a project that consumed so much of my time.

Chapter 1

Getting Started

1.1 Introduction 2
1.2 Installing the Sample Code 2
1.3 Starting a SAS Session 4
1.4 Writing, Submitting, and Saving SAS Code 6
1.5 Using autoexec.sas 8
1.6 Just Enough Syntax 9
1.7 SAS Programming 10
1.8 SAS Procedures 12
1.9 Debugging Basics 14
1.10 Chapter Summary 16

1.1 Introduction

This chapter shows you how to get started using SAS.

- Sections 1.2 and 1.3 show you how to install the files and folders included on the companion Web site and how to start a SAS session.
- Sections 1.4, 1.5, and 1.6 cover some basic SAS syntax, including how to write, submit, and save a simple program, and how to customize your SAS environment with autoexec.sas.
- Sections 1.7 and 1.8 give a brief overview of the key elements of SAS programming.
- Section 1.9 contains some basic advice on detecting and recovering from coding errors.

Even if you already know the basics, please read Section 1.2 on installing the JES files and Section 1.5 on configuring your autoexec.sas file.

1.2 Installing the Sample Code

The whole point of this book is to help you to do useful work with SAS as quickly as possible. And the easiest way to become effective with SAS, or any other programming language, is to begin with a working example of code that does something similar to what you need, and then to modify that code to do exactly what you need. Therefore, this book relies heavily on sample code to illustrate various aspects of SAS programming, and to provide templates for your use in future projects. So, before you go any further, please download the JES sample code and other documents from the companion Web site to the hard drive on the computer where you will be running SAS.

1.2.1 SAS for Windows

If you are running SAS for Windows:

1. Copy the JES.zip and JES_Docs.zip files from the companion Web site to any convenient folder on your hard drive.

2. Unzip each file as follows:

    - In Windows XP, right-click the file and **Extract All**.
    - For earlier versions of Windows, use a ZIP file utility like PKZip or WinZip.

3. Place the docs folder from the second ZIP file into the JES folder from the first ZIP file.

4. You will then have a JES folder similar to the one shown in Display 1-1.

5. Note the location of your JES folder, which will be referred to as ~\JES\ in this book. In the example shown in Display 1-1, ~\JES\ = c:\JES\.

Display 1-1: Contents of the JES Folder

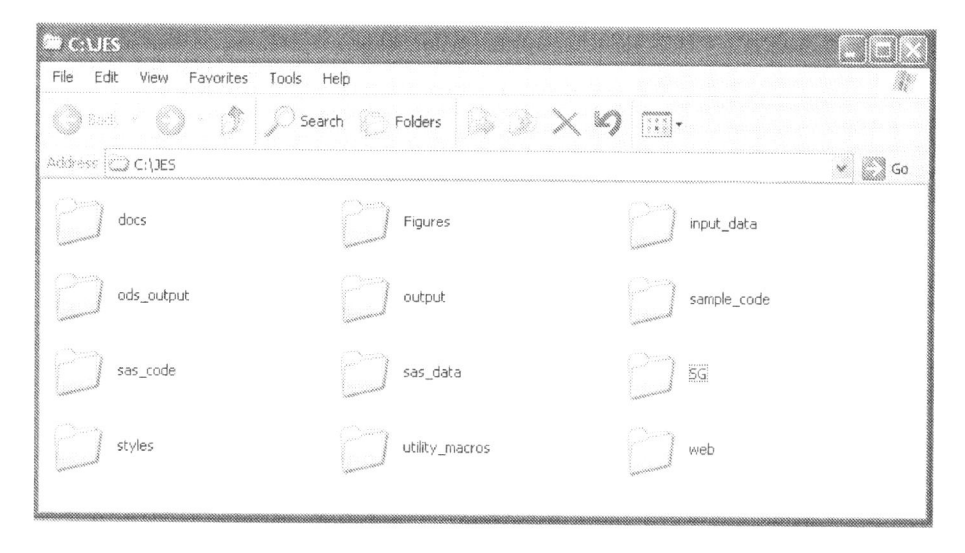

Several of these folders are empty and will be used to store the data and results that you create when running the examples. Other folders contain SAS code and other files for use in the examples.

- **sample_code** contains the SAS code for all of the examples in this book.
- **utility_macros** contains a starter set of SAS macros that you can use in your own programs.
- **input_data** contains a sample CSV (comma-separated values) file which will be used in Chapter 3.

1.2.2 SAS for UNIX

If you are running SAS for UNIX:

1. Copy the JES.tar.gz and JES_Docs.tar.gz files from the companion Web site to any convenient folder in your home directory.

2. Unzip and un-tar each file as follows: In a terminal window:
   - cd (your path)
   - gunzip JES.tar.gz
   - tar –xvf JES.tar

3. Place the docs folder from the second TAR file into the JES folder from the first TAR file.

4. You will then have a JES folder similar to the one shown in Display 1-1.

5. Note the location of your JES folder, which will be referred to as ~\JES\ in this book.

1.2.3 Windows vs. UNIX

For the most part, SAS code written for Windows works equally well under UNIX and vice versa. The exceptions mostly have to do with code that speaks to the operating system in its own language. For example, if you want to tell SAS where to find a folder, Windows expects to see backward slashes (\) while UNIX expects to see forward slashes (/) in the path. This book uses the Windows version for all path references, so UNIX users will have to adjust path references as required. For example, ~\JES\ is used to refer to the location of your JES folder, even though ~/JES/ would be more appropriate for UNIX users.

4 *Just Enough SAS: A Quick-Start Guide to SAS for Engineers*

1.3 Starting a SAS Session

1.3.1 SAS for Windows

When you start SAS for Windows, five windows open: three visible and two minimized.

- **Program Editor:** The window where you write and submit SAS code.
- **Log:** The session Log, Notes, and Messages are written here when you submit code.
- **Explorer:** This is like the Windows Explorer, but is used to explore various SAS objects.
- **Results:** This window contains results, such as charts and tables, generated by your code.
- **Output:** This window contains text-only output from various SAS procedures.

Display 1-2: The SAS for Windows Programming Interface

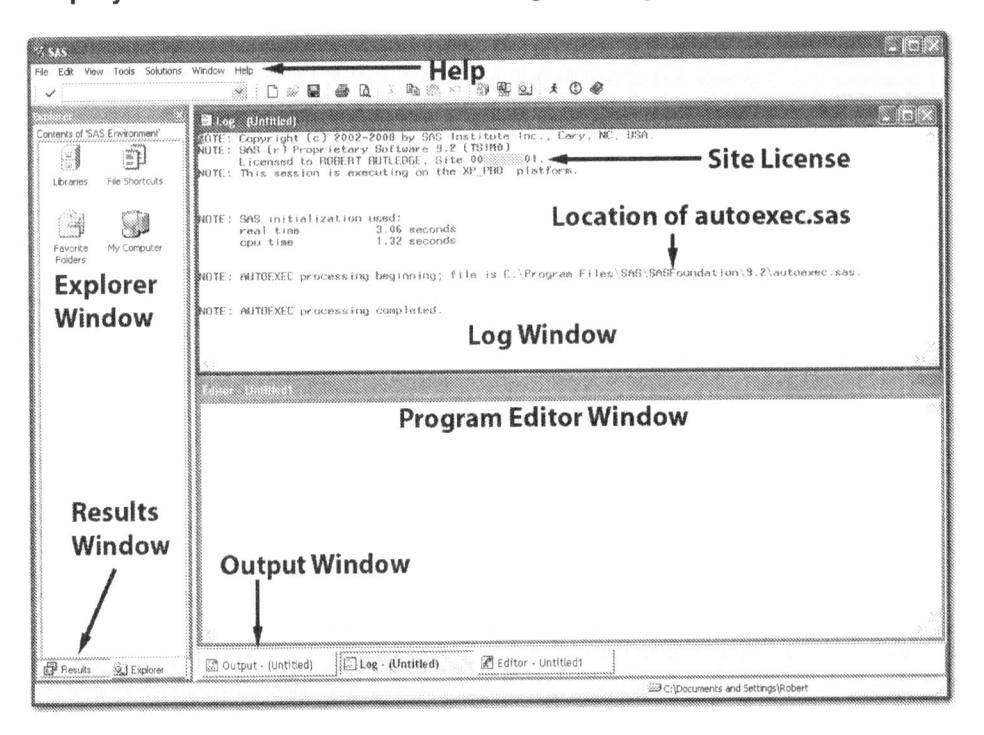

1.3.2 SAS for UNIX

When you start SAS for UNIX, the same five windows open, as well as a toolbox, which changes depending on which of the other windows is active. In Display 1-3, the Results and Output windows have been closed. You can reopen any closed window by selecting it from the View menu in any open window.

Use the Toolbox to change the Explorer window to a split-screen as indicated in Display 1-3.

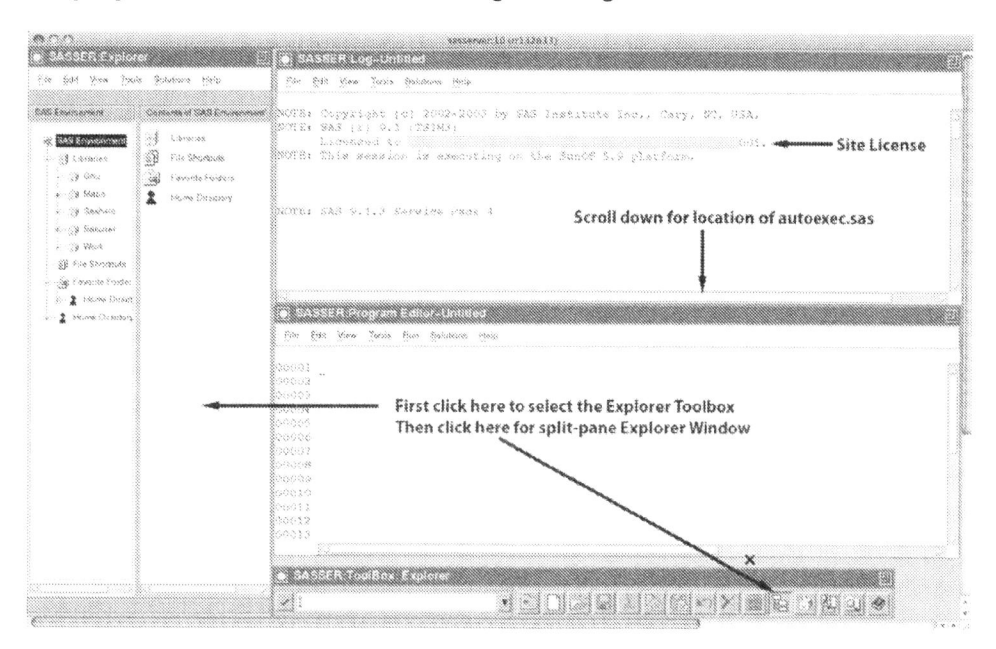

Display 1-3: The SAS for UNIX Programming Interface

In the Program Editor window, select **Tools → Options → Preferences → Editing**, and make sure that **Automatically store selection** is unchecked, and Cursor is set to **Insert** rather than **Overtype**.

Display 1-4: The Preferences Window

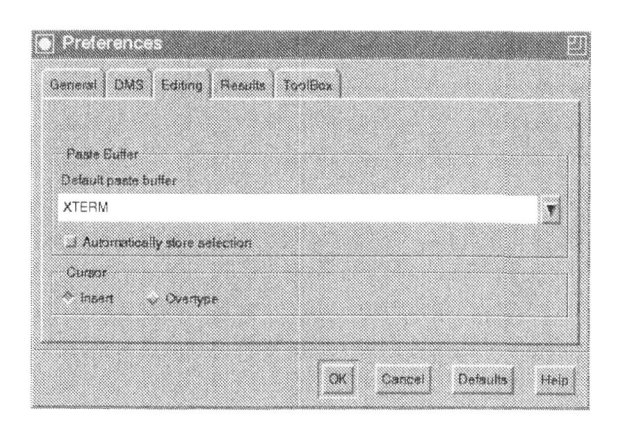

1.3.3 Site License and Path to autoexec.sas

When you start a SAS session, a note is written to the Log window with your site license, as shown in Displays 1-2 and 1-3. Make a note of this number as you will need it when you call SAS Technical Support.

When SAS starts up, it first tries to run the code in a file called autoexec.sas if it can find such a file. You might or might not also see a note indicating that the autoexec.sas code file was run. If you don't see the AUTOEXEC note, that's OK; you will create your own autoexec.sas file later. If you do see the AUTOEXEC note, write down the path to the autoexec.sas file so you can edit this file, as shown in Section 1.5.

1.4 Writing, Submitting, and Saving SAS Code

1.4.1 Macro Functions and Variables

Macro programming is discussed in Chapter 11, but a few simple ideas are introduced here as they will greatly simplify the process of running the examples in this book.

- %PUT is a macro function which writes a text string to the Log window.
  For example: **%PUT Some text;** writes "Some text" to the Log window.
- %LET is a macro function which assigns a value to a macro variable.
  For example: **%LET myName = Bob;** assigns the value "Bob" to the macro variable &myName.
- %INCLUDE is a Base SAS function, not a macro function; it runs a file containing SAS code.
  For example: **%INCLUDE "c:\mySASfolder\test.sas"** runs the code in test.sas.
- Macro function names have a prefix of "%" and macro variable names have a prefix of "&".

1.4.2 Writing and Submitting Code

All of the code used in this book is included in the ~\JES\sample_code folder, which you downloaded following the instructions in Section 1.2. The code used in each section is identified with a highlighted statement like this: The code for this section is in ~\JES\sample_code\ch_1\hello_world.sas. For this section, however, there is so little code to enter that it will be just as easy to type it in yourself. It is customary to introduce a new programming language or environment with a program that simply writes "Hello World!" to some appropriate output location. To create a "Hello World" program in SAS, type these lines in the Program Editor window, replacing "(your name)" with your own name.

```
%LET myName = (your name);
%PUT Hello World! My name is &myName;
```

There are several different ways to run this program:

- **Run → Submit** (i.e., select **Submit** from the Run menu).
- Click the Running Man icon at the top of the screen (Windows only).
- Press the F3 key.

When you run the program, the code and then the results are written to the Log window, as shown here.

```
%LET myName = (your name);
%PUT Hello World! My name is &myName;
Hello World! My name is (your name)
```

If you are running SAS for Windows, the code in the Program Editor window is unchanged. If you are running SAS for UNIX, however, you will notice that the code has vanished from the Program Editor window. You can bring back the lost code in either of two ways:

- **Run → Recall Last Submit** (i.e., select **Recall Last Submit** from the Run menu).
- Press the F4 key.

UNIX users can avoid having to recall the missing code by using the mouse to highlight the code you want to run, and then submitting the code in either of two ways:

- **Run→Submit**.
- Press the F3 key.

The result is the same except that the code does not disappear from the Program Editor window.

1.4.3 Saving Code

The process for saving SAS code is just the usual process for saving a file in any text editor. And, you can in fact use any text editor to edit your SAS code, although the Program Editor is usually more convenient. To save the "Hello World" code that you wrote in the previous section, select **File→Save As** in the Program Editor window, navigate to the ~\JES\sas_code folder, and save the file as hello_world.sas. It is conventional, but not necessary, to use a .sas extension for a SAS program. (The rules for constructing valid SAS names are given in Section 1.6.2.) Note that the name of the new file, "hello_world.sas", now appears at the top of the Program Editor window. To save an updated version of your code with the same name, just select **File→Save**.

To open SAS code that you have saved, select **File → Open Program** (or **File→Open** in UNIX) in the Program Editor, navigate to the folder where you saved the code, and open the file.

Windows users: You can have several Program Editor windows open at the same time. When you open a SAS program, it appears in a new window of its own.

UNIX users: You have only one Program Editor window to work with. When you open a SAS program, it is appended at the bottom of the window, below any code that was already there. This is almost always *not* what you want to happen, so before you open a program, you should clear out the Program Editor window by selecting **Edit→Clear All**.

1.4.4 Using the %INCLUDE Function to Run SAS Code

You can use the %INCLUDE function to run code without bringing it into the Program Editor.

1. In the Program Editor window, select **Edit→Clear All** to get rid of whatever code is there.
2. Type in the statement shown below, replacing ~\JES\ with the path to your JES folder.
3. Submit the code using any of the methods described in the previous section.
4. Check the Log window to verify that the code you saved in hello_world.sas was run correctly.

```
%INCLUDE "~\JES\sas_code\hello_world.sas";
```

You can avoid having to type in your JES path repeatedly by storing the value in a macro variable.

```
%LET JES = ~\JES\;
%INCLUDE "&JES.sas_code\hello_world.sas";
```

- Type in the code above, replacing ~\JES\ with the path to your JES folder.
- The first line creates a macro variable, &JES, which contains the path to your JES folder.
- The second line replaces &JES with whatever you typed in between = and ; in the first line, and then runs the %INCLUDE statement.
  - Note that the macro variable &JES is resolved only within double quotes. This code does not work if single quotes are used.
  - The period after &JES is a delimiter indicating the end of the variable name. Without the period, SAS would look for a variable named &JESsas_code.

The value of &JES will be lost when you close your SAS session, but you can make sure that it is always available by including it in your autoexec.sas file, as shown in Section 1.5.

1.5 Using autoexec.sas

Each time SAS starts up, it looks for a file named autoexec.sas and, if it finds one, runs the code in that file before doing anything else. On a Windows system, the location where SAS looks for autoexec.sas is the same folder where the SAS executable is located; for example:

C:\Program Files\SAS\SASFoundation\9.2\autoexec.sas

On a UNIX system, the typical location where SAS looks for autoexec.sas is in your home directory; for example:

/home/your_name/autoexec.sas

Depending on how SAS was installed, you might or might not already have an autoexec.sas file. If you do, you would have seen the path to your autoexec.sas file in the Log window when you started your SAS session (see Display 1-2 or 1-3 in Section 1.3). If you already have an autoexec.sas file, open it up with the Program Editor and add the following lines at the bottom, replacing ~\JES\ with the actual path to your JES folder.

```
%LET JES = ~\JES\ ;   ❶
%PUT JES=&JES;   ❷
%INCLUDE "&JES.sas_code\hello_world.sas";   ❸
```

❶ The %LET function creates a macro variable, JES, which contains the path to your ~\JES\ folder.

❷ The %PUT function prints the value of &JES to the Log window.

❸ The %INCLUDE function runs your hello_world.sas code, but only if you actually saved it (see Section 1.4), and got the path right.

If you don't already have an autoexec.sas file, create one in the Windows or UNIX location indicated above, enter the above lines, and save the file. Run the code to verify that you got the path right. If you did not, the third line won't run and will generate an error message in the Log window.

Then, shut down your SAS session by selecting **File→Exit** from the Program Editor window, and open a new SAS session. If all goes well, you should see something like this in the Log window when the new session opens.

```
NOTE: AUTOEXEC processing beginning; file is C:\Program
    Files\SAS\SASFoundation\9.2\autoexec.sas.

JES=c:\JES\
Hello World! My name is (your name)

NOTE: AUTOEXEC processing completed.
```

If you don't see the AUTOEXEC processing lines in the Log window, check to see that the name, autoexec.sas, and location of the file that you saved is correct. If the path is not correct, fix it and try again. If the path is correct, but hello_world.sas doesn't run, make sure you've saved it to the right place. The path to your JES folder, &JES, is needed for most examples in this book to tell SAS where to find code and data files and where to send results, so please ensure that this path is correct before continuing.

1.6 Just Enough Syntax

You will of course be learning SAS syntax throughout the book. The point of this section is to bootstrap the process by giving you "just enough" syntax to get started. The code for this section is in ~\JES\sample_code\ch_1\just_enuf_syntax.sas.

1.6.1 SAS Statements

The code below, and the results in the Log window, illustrate a few basic rules of SAS syntax. Note that the Program Editor automatically color codes your SAS code, making it easier to understand. For example, macro function names, such as %LET, are in blue, and comments are in green. Other uses of color coding will become clear as you look at the other code samples.

```
/* This is a comment  */ ❶
* And this is another comment; ❷
%LET First=SAS; %LET Second = Program; ❸
%LET both
= &first &SECOND; ❹
%PUT The results are: &First, &Second, --- &Both.ming; ❺
```

❶ Anything between /* and the next */ is treated as a comment and is not executed.

❷ Anything from * to the next ; is treated as a comment and is not executed.

❸ You can have multiple statements on a single line. Each statement ends with a semicolon.

❹ A single statement can span multiple lines.

❺ A period is used as a delimiter to tell SAS where a macro variable name ends.

When you run the code, the lines you submitted, and the resulting output, are written to the Log window.

```
/* This is a comment  */
* And this is another comment;
%LET First=SAS; %LET Second = Program;
%LET both
= &first &SECOND;
%PUT The results are: &First, &Second, --- &Both.ming;
The results are: SAS, Program, --- SAS Programming
```

1.6.2 SAS Names

The names of SAS data sets and variables:

- Are case insensitive. For example, %LET, %let, and %LeT are considered equivalent. In this book, however, SAS keywords appear in uppercase letters.
- Contain at most 32 characters. The characters must be letters, numbers, or the underscore character.
- Must begin with a letter or the underscore character.

1.7 SAS Programming

The key elements you will use to construct your SAS programs are DATA steps, procedures, macros, and the Output Delivery System. These are all discussed in detail in the following chapters, but some simple examples are included here to give you an idea of where you're going. The code for this section is in ~\JES\sample_code\ch_1\sas_programming.sas.

1.7.1 DATA Steps (Chapter 2)

DATA step programming is used to create, merge, and transform SAS data sets. The code below uses the SAS function RANPOI to create the data set, New, containing 2500 random values from a Poisson distribution with mean 100.

```
DATA New;
  DO i = 1 TO 2500;
      X=RANPOI(0, 100);
      OUTPUT New;
  END;
RUN;
```

1.7.2 Procedures (Various Chapters)

SAS has hundreds of built-in procedures that you can invoke to automatically process your data and produce results such as tables, graphs, parameter estimates, etc. A partial list of these procedures is given in Section 1.8. The next example uses the UNIVARIATE procedure to fit a normal distribution to the random numbers in the New data set generated in the DATA step, and performs the Anderson-Darling test for goodness-of-fit. The tabular results are sent to the Output window, and a graph is sent to the Graph window.

```
GOPTIONS RESET=ALL GUNIT=PCT HTEXT=3 FTEXT=SWISSB;
PROC UNIVARIATE DATA=New;
  VAR X;
  HISTOGRAM X / Normal(MU=EST SIGMA=EST);
  INSET N MEAN STD NORMAL(AD ADPVAL) / HEIGHT=2.5 FORMAT=5.2
POSITION=NE;
RUN; QUIT;
```

1.7.3 Macro Programs (Chapter 11)

The SAS macro language allows you to wrap up useful chunks of code that can be customized with parameters and used throughout your projects. A macro definition begins with %MACRO followed by the name of the macro to be defined, and ends with a %MEND statement. After the macro has been defined, it is invoked by entering the macro name, preceded by a percent sign. The macro below runs the same code as in the previous example for the data set specified in the macro call, which in this case is the same data set, New.

```
%MACRO simpleMacro(dataSetName);
   TITLE "Distribution of 100 Random Poisson Variables";
    PROC UNIVARIATE DATA=&dataSetName;
    VAR X;
    HISTOGRAM X / Normal(MU=EST SIGMA=EST);
    INSET N MEAN STD NORMAL(AD ADPVAL) /
    HEIGHT=2.5 FORMAT=5.2 POSITION=NE;
   RUN; QUIT;
%MEND simpleMacro;
%simpleMacro(New)
```

1.7.4 The Output Delivery System (Chapter 7)

The Output Delivery System provides a very simple method to direct the output from your SAS programs to various destinations, including PDF, RTF, HTML, XML, and CSV files. The final example reruns the simplePlot macro and sends the resulting chart to the HTML page ~/JES/web/test.html.

```
ODS HTML PATH="&jes.web" (URL=NONE) BODY = "test.html";
   %simpleMacro(New)
ODS HTML CLOSE;
```

Note: This code depends on the macro variable &JES that you defined in Section 1.5, and saved to autoexec.sas. If the variable &JES is not defined, or if the path does not exist, you will see an error message in the Log window. If so, go back to Section 1.5 and make sure that &JES is defined correctly.

When the code has run successfully, open the new file, ~\JES\web\test.html, in a Web browser, as shown in Display 1-5.

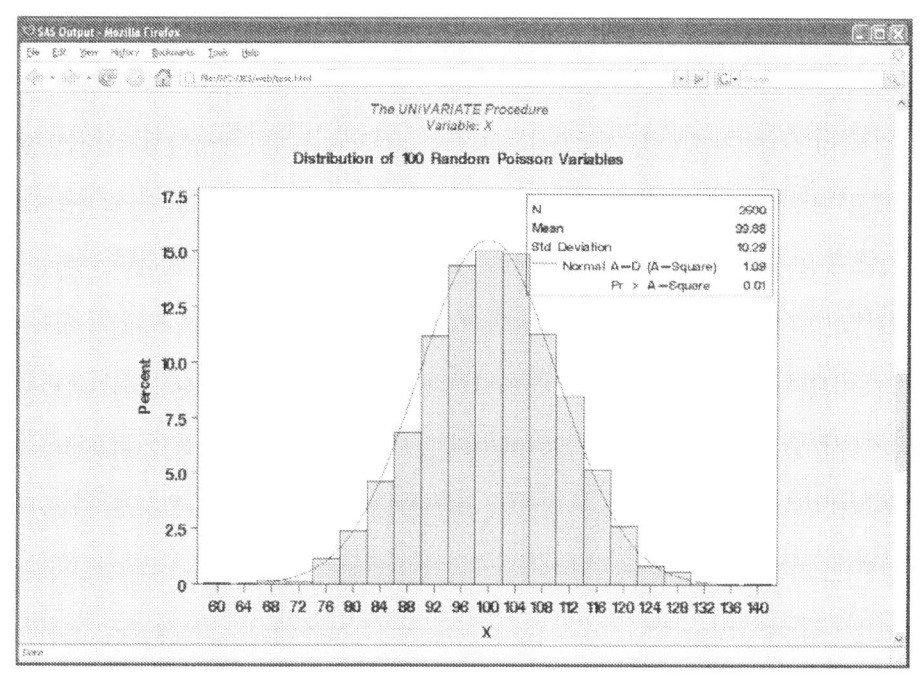

Display 1-5: The test.html Web Page Created by ODS HTML

You have published a sophisticated data analysis to your own Web site, using only a few lines of SAS code!

1.8 SAS Procedures

Tables 1-1 to 1-4 list all of the procedures in four SAS products: Base SAS, SAS/GRAPH, SAS/STAT, and SAS/QC. The list is complete for SAS 9.2, but procedures are added with each new release of SAS, so you should check the SAS online documentation for new procedures.

Some of the procedure names are easy to decipher (e.g., TTEST) while others are decidedly cryptic; e.g., MI, and others such as PRINCOMP will be recognized only by those with specialized subject knowledge, which would be principal components analysis in this case. In any case, the best way to know what a procedure actually does is to consult the documentation. You can open the SAS Help and Documentation in the Program Editor by using the Help pull-down menu (see Display 1-2) and selecting SAS Help and Documentation. A separate window opens that has a left pane with four tabs (Contents, Index, Search, and Favorites) and a right pane that displays the selected topic. The Contents tab has an expandable tree, and the SAS product documentation contains a wealth of information about the procedures, their specific syntax, and details about the various algorithms. The Search tab has a field where you can enter keywords, click **List Topics**, and the documentation will be searched for the entered text. If there are specific sections that you refer to often, you can add them to the list on the Favorites tab. To quickly find documentation on any of the procedures in Tables 1-1 to1-4, or to look for newly added procedures, select

SAS Products → SAS Procedures → SAS Procedures by Product

Or, follow the path for each product shown at the bottom of each table. You will be able to use only the SAS procedures in the products that are included in your SAS license. You can see which products you have by running PROC SETINIT. (You might note that PROC SETINIT is nowhere to be found in the tables on the facing page, and in fact I can't find it anywhere in the online documentation, but it works.)

```
PROC SETINIT NOALIAS; RUN;
```

When you run this line, the products included in your SAS license and the expiration dates are written to the Log window.

```
Product expiration dates:
---Base Product                       14DEC2008
---SAS/STAT                           14DEC2008
---SAS/GRAPH                          14DEC2008
---SAS/QC                             14DEC2008
---SAS/INSIGHT                        14DEC2008
---SAS/ACCESS Interface to ORACLE     14DEC2008
```

Every SAS license includes Base SAS. This book assumes that you also have at least SAS/GRAPH and SAS/STAT. If you will be working with quality and/or reliability data, I would highly recommend that you also have SAS/QC. Some of the examples require SAS/QC, and these will be noted in the text.

If you need to access a relational database, you will also need an appropriate SAS product such as

- SAS/ACCESS Interface to ORACLE
- SAS/ACCESS Interface to DB2
- SAS/ACCESS Interface to ODBC

The procedures discussed in this book are highlighted in Tables 1-1 through 1-4.

Table 1-1: Base SAS Procedures

APPEND	BMDP	CALENDAR	CATALOG	CHART	CIMPORT
COMPARE	CONTENTS	CONVERT	COPY	CORR	CPORT
DATASETS	DBCSTAB	DISPLAY	DOCUMENT	EXPLODE	EXPORT
FCMP	FONTREG	FORMAT	FORMS	**FREQ**	**IMPORT**
ITEMS	JAVAINFO	**MEANS**	MIGRATE	OPTIONS	OPTLOAD
OPTSAVE	PDS	PDSCOPY	PLOT	PMENU	**PRINT**
PRINTTO	PROTO	PRTDEF	PRTEXP	PWENCODE	RANK
REGISTRY	RELEASE	**REPORT**	SCAPROC	SOAP	**SORT**
SOURCE	**SQL**	STANDARD	SUMMARY	**TABULATE**	TAPECOPY
TAPELABEL	**TEMPLATE**	TIMEPLOT	**TRANSPOSE**	TRANTAB	**UNIVARIATE**
VAXTOINTEG	WEBMDDB				
Paths to documentation in SAS Help: Base SAS → Base SAS 9.2 Procedures Guide→Procedures Base SAS → Base SAS Procedures Guide:Statistical Procedures					

Table 1-2: SAS/GRAPH Procedures

G3D	G3GRID	GANNO	GAREABAR	GBARLINE	**GCHART**
GCONTOUR	GDEVICE	GEOCODE	GFONT	GIMPORT	GINSIDE
GKEYMAP	GKPI	**GMAP**	**GOPTIONS**	**GPLOT**	GPROJECT
GRADAR	GREDUCE	GREMOVE	GREPLAY	GSLIDE	GTILE
MAPIMPORT	**SGPANEL**	**SGPLOT**	**SGRENDER**	**SGSCATTER**	
Path to documentation in SAS Help: SAS/GRAPH → Procedures and Statements → All Procedures					

Table 1-3: SAS/STAT Procedures

ACECLUS	ANOVA	**BOXPLOT**	CALIS	CANCORR	CANDISC
CATMOD	CLUSTER	CORRESP	DISCRIM	DISTANCE	FACTOR
FASTCLUS	**FREQ**	GAM	GENMOD	GLIMMIX	GLM
GLMMOD	GLMPOWER	GLMSELECT	HPMIXED	INBREED	KDE
KRIGE2D	LATTICE	**LIFEREG**	**LIFETEST**	LOESS	LOGISTIC
MCMC	MDS	MI	MIANALYZE	MIXED	MODECLUS
MULTTEST	NESTED	NLIN	NLMIXED	NPAR1WAY	ORTHOREG
PHREG	PLAN	PLS	POWER	PRINCOMP	PRINQUAL
PROBIT	QUANTREG	REG	ROBUSTREG	RSREG	SCORE
SEQDESIGN	SEQTEST	SIM2D	SIMNORMAL	STDIZE	STEPDISC
SURVEYFREQ	SURVEYLOGISTIC		SURVEYMEANS		SURVEYREG
SURVEYSELECT		TCALIS	TPSPLINE	TRANSREG	TREE
TTEST	VARCLUS	VARCOMP	VARIOGRAM		
Path to documentation in SAS Help: SAS/STAT→SAS/STAT User's Guide					

Table 1-4: SAS/QC Procedures

ANOM	**CAPABILITY**	**CUSUM**	FACTEX	ISHIKAWA	MACONTROL
OPTEX	PARETO	**RELIABILITY**	**SHEWHART**		
Path to documentation in SAS Help: SAS/QC→SAS/QC User's Guide					

1.9 Debugging Basics

When you begin to write SAS programs, you will surely make some mistakes. Just as beginning skiers are first taught how to fall without getting hurt, it's a good idea to begin your SAS programming with a few intentional mistakes, so that you learn how to recognize and recover from them. If you are just starting out, you will not yet understand the syntax of the sample code used here. Just run the code to understand the basics of the error detection and correction process. The sample code for this section is in ~\JES\sample_code\ch_1\bugs.sas.

1.9.1 Detecting Problems

The first step in recovering from a mistake is to notice that you made one. SAS helps out with this by writing WARNING and ERROR messages to the Log window. An ERROR message points to something which is definitely wrong, while a WARNING message points to something irregular, but which might or might not be a bug. When you run a new program, you should review the SAS log for ERROR or WARNING messages. Errors must be fixed, and warnings should at least be understood. The code below contains a bug: you are trying to reference a data set, Old, which does not exist.

```
DATA New; SET Old; RUN;
```

When you run the code, an error message is written to the Log window.

```
DATA New; SET Old; RUN;
ERROR: File WORK.OLD.DATA does not exist.
NOTE: The SAS System stopped processing this step because of errors.
WARNING: The data set WORK.NEW may be incomplete.  When this step was stopped there were 0
     observations and 0 variables.
WARNING: Data set WORK.NEW was not replaced because this step was stopped.
```

The next example includes the same bug, but when you run this one, you might not notice the error message in the Log window because only the last few lines will be visible. The only way to be sure your program ran without any ERROR messages is to scroll to the very top of the Log window, and then use **Edit→Find** to search for the word "ERROR". Be sure to clear the Log window (**Edit→Clear All**) before you run your code. Otherwise, you will just rediscover errors you already fixed but which are still in the Log window from a previous run.

```
%MACRO sas_bugs;
%DO n = 1 %TO 1000;
  %PUT n=&n;
  %IF &n=666 %THEN %DO; DATA new; SET old; RUN; %END;
%END;
%MEND sas_bugs; %sas_bugs
```

1.9.2 Correcting Problems

Most errors will be simple mistakes in the code that, hopefully, you will figure out by reading your code and the ERROR message carefully. However, there are a few common bugs which effectively hang the SAS interface, and which you should learn to escape from. These usually happen when SAS is expecting to see something in your code that never comes; for example:

- a semicolon at the end to complete a SAS statement
- a RUN; statement to complete a DATA step or procedure
- an ending quote symbol to balance a beginning quote symbol
- a %MEND statement to complete a %MACRO definition

While SAS waits for the other shoe to drop, it misinterprets everything else you enter—which can be a very frustrating experience. The code below is a typical example. When you run this code, SAS waits for the quote to be completed.

```
DATA a;
   X= "some text with unbalanced quotes'; OUTPUT a;
RUN;
```

In this example, it is obvious what went wrong: The double quotes at the beginning of the line are not closed by the single quote at the end, so SAS thinks the quoted text is continuing. You can tell from the color coding that SAS thinks that OUTPUT a; is part of the quoted text. But if this occurred in the middle of a 1,000-line program, it could be very hard to find. Fortunately, you can escape from this and other similar situations by running the following line of code:

```
*'; *"; RUN; %MEND;
```

Depending on how deep a hole you have dug for yourself, you might need to run this line several times. You'll know you're done when you see an error message that says "No matching %MACRO statement for this %MEND statement." The line works on unbalanced quotes because anything between an asterisk and the next semicolon is treated as a comment by SAS, *unless* there is an open quote, in which case it is just part of the quote. For example, *"; will close a double quote if one is open, but will not start a new quote.

1.9.3 Escaping from a Runaway Program

In some cases, your code will go off into never-never land, and you will lose control of the interface. You have no chance of fixing this with a line of code because in this situation you can't enter code, or do anything else. The code below causes this problem by trying to execute the nonsensical statement X=1/X.

```
DATA a;
   DO x=1 TO 100;
      x = 1/x;
   END;
RUN;
```

To escape from this one, you will have to interrupt the program as follows:

Windows: Select **Break** (Exclamation inside circle) from the top of the SAS window, and then select **Cancel Submitted Statements**.

UNIX: Open the Session Management window. This is located in different places for different versions of UNIX, so you might have to hunt around a bit. Best to find it now before you get into real trouble. When you get there, select **Interrupt** and then **Cancel Submitted Statements**.

Display 1-6: The SAS for UNIX Session Management Window

1.10 Chapter Summary

1.10.1 Recap

After finishing this chapter, you should know how to

- Find SAS documentation online.
- Find the site number for your SAS license.
- Contact SAS Technical Support.
- Submit SAS code from the Program Editor window.
- Write text to the Log window with the %PUT statement.
- Create a macro variable with a %LET statement.
- Save your code and data.
- Edit your autoexec.sas file.
- Look for ERRORS and WARNINGS in the Log window.
- Escape from some common programming problems that cause the interface to hang.

And you should have edited your autoexec.sas file as instructed in Section 1.5. This is important because from here on the book will assume that the variable &JES is correctly defined.

1.10.2 For More Information

SAS Online Documentation
The online documents in both HTML and PDF format are available at

http://support.sas.com/documentation/onlinedoc/sas9doc.html

Books
Aster, Rick. 2005. *Professional SAS® Programming Shortcuts, Second Edition.* Paoli, PA: Breakfast Communications Corp. Chapters 1–4 give more detailed information on the log, program files, startup, and system options.

Delwiche, Lora D., and Susan J. Slaughter. 2003. *The Little SAS® Book: A Primer, Third Edition.* Cary, NC: SAS Institute Inc. Chapter 1 covers most of what's in this chapter, and more, and in greater detail. Chapter 10 contains a lot more good advice on debugging your SAS programs.

SAS Conference Papers
These papers give a good introduction to the SAS-L Web site that was mentioned in Section 1.1. For a link to the papers see the Companion Web site for this book. The filenames are shown in bold below—e.g., **[247-28.pdf]**.

Matthews, JoAnn, and Doug Zirbel. 2003. "SAS-L—A Very Powerful Resource for SAS Users Worldwide." *Proceedings of the Twenty-Eighth Annual SAS Users Group International Conference.* Cary, NC: SAS Institute Inc. Paper 247-28. **[247-28.pdf]**

Whitlock, Ian. 2008. "The Art of Debugging." *Proceedings of the 2008 SAS Global Forum Conference.* Cary, NC: SAS Institute Inc. Paper 165-2008. **[165-2008.pdf]**

1.10.3 Exercises

1. Write a SAS program to write your name and address to the SAS Log window.

2. Save your program in a file called myname.sas in the ~\JES\sas_code directory.

3. Arrange for SAS to write "Welcome Back (your name)!" to the Log window automatically each time you start SAS.

4. Open your favorite browser and bookmark both the PDF and HTML versions of the SAS online documentation.

5. Go to the SAS-L Web site to better understand the kinds of help you might find there:

    http://listserv.uga.edu/archives/sas-l.html

6. View the documentation on the APPEND procedure in both HTML and PDF formats. Hint: Use the tables in Section 1.8 to help you find out where to look.

7. Get prepared to use SAS Technical Support:

    - What is your site license number? _____
    - What version of SAS are you running? _____
    - What is the phone number for SAS Technical Support? _____

Chapter 2

DATA Step Programming

2.1 Introduction 20
2.2 Creating SAS Data Sets 22
2.3 Saving SAS Data Sets 30
2.4 SAS Functions and CALL Routines 31
2.5 The RETAIN Statement 41
2.6 Selecting Subsets of Data Sets 42
2.7 Sorting Data Sets 44
2.8 Merging Data Sets 47
2.9 More Than Enough 50
2.10 Chapter Summary 54

2.1 Introduction

The data you analyze with SAS is typically stored in *SAS data sets*. DATA step programming is used to create and manipulate SAS data sets. This section shows how to view SAS data sets, and the remainder of the chapter shows how to create, transform, and save your data sets. The code for this section is in ~\JES\sample_code\ch_2\intro.sas.

This is the first code sample in intro.sas. The syntax is explained in Section 2.2.3. For now, just run the code to create a SAS data set so that you can learn how to view it. Open the intro.sas file in the Program Editor window and run these lines of code using one of the methods described in Section 1.4.

```
DATA XSquare;
   DO X = 0 TO 5;
       Y=X*X;
       OUTPUT;
   END;
RUN;
```

When you run the code, the XSquare data set is created, and a message is written to the Log window.

NOTE: The data set WORK.XSQUARE has 6 observations and 2 variables.

In SAS terminology, XSquare is a *data set* with two *variables* (X and Y) and five *observations*, but you can also think of XSquare as a *table* with two *columns* and five *rows*. These terms are used interchangeably in this book.

SAS Terminology	Equivalent Terms
Data set	Table
Variable	Column
Observation	Row

You can view the new XSquare data set in a Viewtable window as follows:

- If you are using SAS for Windows: In the Explorer window, double-click **Libraries**, double-click **Work**, and then double-click **XSquare**. The result is shown in Display 2-1.
- If you are using SAS for UNIX: In the Explorer window, select **Work** in the left pane, and then right-click on **XSquare** and select **Open**.

Display 2-1: The XSquare Data Set in a Viewtable Window

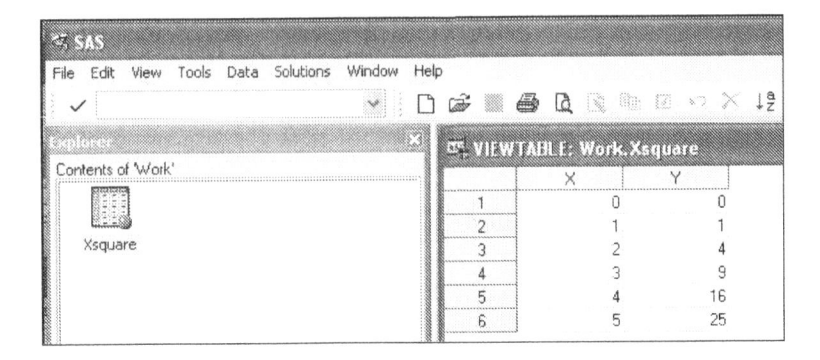

In Display 2-1, you can see that the new data set is called WORK.Xsquare rather than just Xsquare. SAS stores each data set in a *library*, and every SAS data set has a two-level name in the form *<Library Name>.<Member Name>*. If you don't supply a library name, SAS assumes that you want to create the data set in a temporary library called WORK. Section 2.3 shows how to store SAS data sets in permanent libraries.

You can also view the contents of a data set by using the PRINT procedure to print the contents to the Output window, as illustrated in the next code sample.

```
PROC PRINT DATA=XSquare; RUN;
```

When you run this code, the XSquare data set is written to the output window. If the Output window does not open automatically, you can open it by double-clicking **Data Set WORK.XSQUARE** in the Results window.

Display 2-2: The XSquare Data Set Written to the Output Window

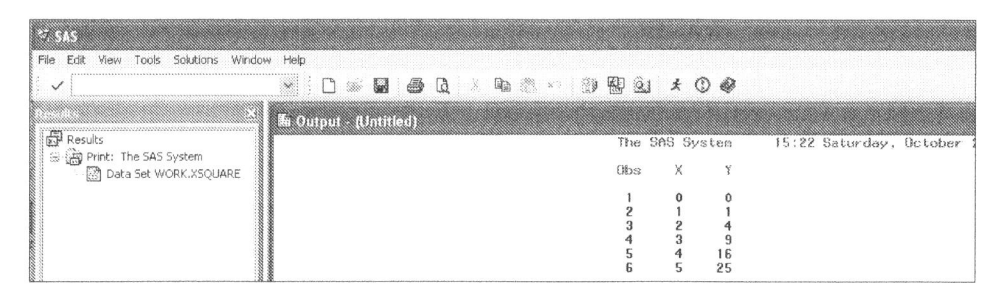

Display 2-2 shows the Results and Output windows after the code is run. Note that, by default, the observation number (Obs) is printed along with the variables. Obs is not a variable in the data set.

For the examples in this book, data sets are shown as PROC PRINT output like this:

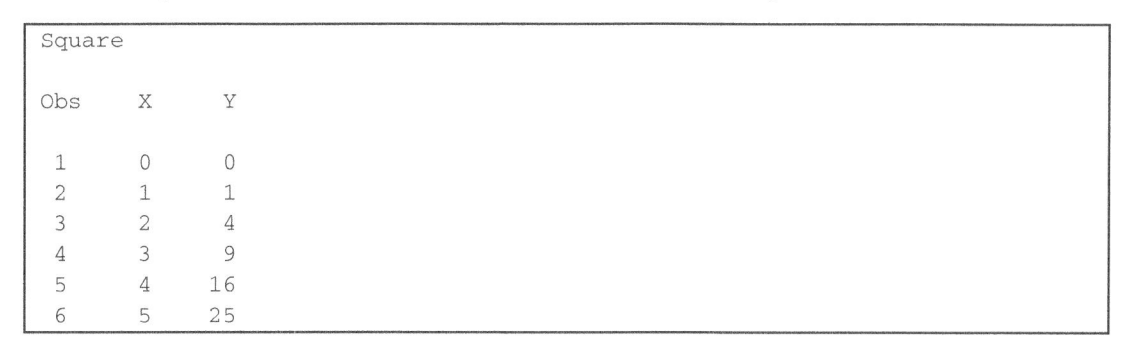

```
Square

Obs    X    Y

 1     0    0
 2     1    1
 3     2    4
 4     3    9
 5     4   16
 6     5   25
```

The timestamp shown in Display 2-2 is removed, the data is left-justified, and the data set name is printed at the top of the display using OPTIONS and TITLE statements as described in the last example in Section 2.2.5.

2.2 Creating SAS Data Sets

Chapters 3 and 4 show how to create SAS data sets by bringing in data from spreadsheets and relational databases. This section shows you how to create data sets by using a DATA step to enter each row of data or to copy data from other SAS data sets. The code for this section is in ~\JES\sample_code \ch_2\create_datasets.sas.

2.2.1 Creating a Data Set by Entering Each Line of Data

One way to create a SAS data set is to type in each row of data. The first example uses a DATA step to create a data set with three columns (or variables) and two rows (or observations) by manually entering each variable for each observation. SAS data sets have only two types of variables:

Numeric variables
　Any kind of number, including integers, rational numbers, dates, and times.

Character variables
　Any string of at most 32,767 characters.

This example creates a data set with one character variable and two numeric variables. Open create_datasets.sas in the Program Editor window, select these lines, and submit the code.

```
DATA Temp;   ❶
  Vendor = "ChiTronix Components";   ❷
  Test = 5000;   ❸
  Fail =  25;
  OUTPUT;   ❹
  Vendor = "Duality Logic"; Test = 1000; Fail = 7; OUTPUT;   ❺
RUN;   ❻
```

❶ The DATA step begins with the keyword DATA followed by the name of the data set to be created, in this case Temp, and ends with a semicolon.

❷ The Vendor = "Chitronix Components"; statement creates a character variable named Vendor and assigns the value "Chitronix Components" (without the quotation marks) to the variable.

❸ The Test = 5000; statement creates a numeric variable, Test, and assigns the value 5000 to the Test variable. The Fail =25; statement creates a numeric variable, Fail, and assigns the value 25 to the Fail variable.

❹ The OUTPUT; statement tells SAS to write a row (or observation) including the Vendor, Test, and Fail variables, into the Temp data set.

❺ This line includes four statements similar to the preceding four statements, adding another row (or observation) to the data set. You can put multiple SAS statements on a single line, using a semicolon to indicate the end of each statement.

❻ The DATA step ends with a RUN; statement.

```
Temp

Obs    Vendor                  Test      Fail

 1     ChiTronix Components    5000      25
 2     Duality Logic           1000       7
```

2.2.2 Using INPUT and DATALINES Statements

Another way to enter the same data is to use INPUT and DATALINES statements. The code below uses these statements to create the Temp2 data set, which is identical to Temp.

```
DATA Temp2;
   INPUT Vendor $20. Test Fail;  ❶
   DATALINES;  ❷
ChiTronix Components  5000  25
Duality Logic         1000   7
   ;  ❸
RUN;
```

❶ The INPUT statement tells SAS to create three variables: Vendor, Test, and Fail. The $20. after Vendor tells SAS that Vendor should be a character variable of length 20. Note that a period is required after the $20.

❷ The DATALINES statement tells SAS that the data is in the lines to follow. The rows (or observations) are entered on the next two lines.

❸ The semicolon tells SAS not to expect any more lines of data, and the RUN; statement ends the DATA step.

This method of creating a SAS data set is more efficient because the variable names are defined in the INPUT statement and don't have to be repeated for each observation. Also, there is no need for an OUTPUT statement as SAS knows to write each line to the data set that is being created. However, if you use the INPUT and DATALINES statements, you need to be careful to ensure that the data lines are properly aligned. The next example uses $25. instead of $20. in the INPUT statement, with unintended consequences.

```
DATA Temp2X;
   INPUT Vendor $25. Test Fail;
   DATALINES;
ChiTronix Components  5000  25
Duality Logic         1000   7
   ;
RUN;
```

When you run this code, SAS includes the first 25 characters on each line in the Vendor variable, and then looks for the numeric variables Test and Fail starting at the 26th character, leading to incorrect values for both Vendor and Test.

```
Temp2X

Obs           Vendor                Test    Fail

 1     ChiTronix Components  500      0      25
 2     Duality Logic         100      0       7
```

Of course, for large data sets, it is usually much more efficient to import the data from a spreadsheet or relational database, as shown in Chapters 3 and 4.

2.2.3 Using DO Loops, Conditional Logic, and SAS Functions

The code below uses a DO loop, conditional logic, and two SAS functions to create a table of Poisson probabilities. SAS provides hundreds of functions that you can use in creating data sets. Many of these are described in Section 2.4.

```
DATA Poisson_Table;
   DO K = 0 TO 10 BY 1;  ❶
       F = CDF('POISSON', K, 5);  ❷
       P = PDF('POISSON', K, 5);
       IF P < .05 THEN Chance = 'Unlikely';  ❸
              ELSE Chance = 'Likely  ';
       IF F > .95 THEN DO;  ❹
              F_GT_95 = "YES";
       END;  ❺
       OUTPUT;  ❻
   END;  ❼
RUN;
```

❶ The DO statement tells SAS to execute the following lines with K set equal to 0, 1, …10.

❷ The next two lines use the SAS functions CDF and PDF to compute the Cumulative Distribution function (CDF) and probability mass function (PDF) of the Poisson distribution with mean 5. SAS functions are discussed in Section 2.4.

❸ The IF–THEN–ELSE statements set the value of Chance to "Unlikely" or "Likely" depending on the value of P.

❹ The IF *(condition)* THEN DO; statement causes all of the following lines, up until the END statement (❺), to be run if and only if the condition, in this case F>.95, is true.

❻ The OUTPUT statement writes a row to the data set.

❼ The END; statement ends the DO loop.

When you run this code, the Poisson_Table data set is created.

```
Poisson_Table

Obs      K         F           P          Chance      F_GT_95

  1      0      0.00674     0.00674      Unlikely
  2      1      0.04043     0.03369      Unlikely
  3      2      0.12465     0.08422      Likely
  4      3      0.26503     0.14037      Likely
  5      4      0.44049     0.17547      Likely
  6      5      0.61596     0.17547      Likely
  7      6      0.76218     0.14622      Likely
  8      7      0.86663     0.10444      Likely
  9      8      0.93191     0.06528      Likely
 10      9      0.96817     0.03627      Unlikely     YES
 11     10      0.98630     0.01813      Unlikely     YES
```

Note that the value of F_GT_95 was not defined for cases where F <= .95; therefore, its value is missing.

2.2.4 Creating a New Data Set from Existing Data Sets

Another way to create a SAS data set is to copy data from one or more existing data sets by using a SET statement in a DATA step. The next code sample creates a new data set, Temp3, which is an exact copy of the Temp data set. This code will work only if the Temp data set already exists. If necessary, go back and re-create Temp using the code in Section 2.2.1.

```
DATA Temp3; SET Temp;
RUN;
```

DATA Temp3; tells SAS to create a new data set named Temp3, and SET Temp; tells SAS to copy in all the variables and observations from Temp. Note that an OUTPUT statement is not required here. All rows of Temp are written to Temp 3.

You can also create a new data set by concatenating two or more existing data sets. The code below creates a new data set, More, to be added to the Temp data set.

```
DATA More;
   Vendor = "Empirical Engineering"; Test = 7000; Fail = 100; OUTPUT;
RUN;
```

The OUTPUT statement is optional here because there is only one row to be written to the More data set. When you run this code, the More data set is created.

```
More

Obs          Vendor              Test     Fail

 1     Empirical Engineering     7000     100
```

Next, a DATA step is used to concatenate More to the previously created Temp data set.

```
DATA New; SET Temp More;
RUN;
```

The SET statement tells SAS to create the New data set by reading in all the rows of Temp and then all the rows of More. When you run this code, the New data set is created. Note that the final "g" has been dropped from "Empirical Engineering". Section 2.2.5 shows how to use a LENGTH statement to fix this problem.

```
New

Obs     Vendor                  Test     Fail

 1      ChiTronix Components    5000     25
 2      Duality Logic           1000      7
 3      Empirical Engineerin    7000     100
```

2.2.5 LENGTH, INFORMAT, FORMAT, and LABEL Statements

Every variable in a SAS data set has five attributes.

- TYPE = "Text" for character variables or "Number" for numeric variables.
- LENGTH = Number of bytes used to store the variable.
- INFORMAT = The SAS informat used to read in the variable.
- FORMAT = The SAS format used to display the variable.
- LABEL = A label that can be used in place of the variable name in output.

TYPE and LENGTH must be defined for every variable, either explicitly or by default. INFORMAT, FORMAT, and LABEL are blank unless explicitly defined. To see the attributes of a data set, right-click the data set in the Explorer window, and select **View Columns** from the drop-down menu. If you do this for the Temp data set, the properties window shown below will open.

Display 2-3: View Columns View of the Temp Data Set

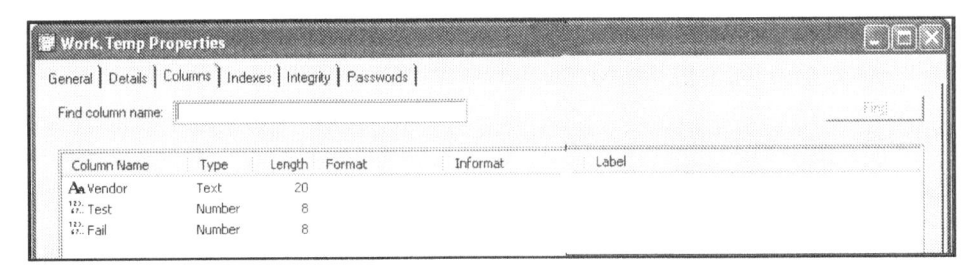

When the Temp data set was created, the length of the variable Vendor was set to 20 because the first value read in, "ChiTronix Components", has 20 characters. When "Empirical Engineering" was read in, only the first 20 characters were kept, causing the missing "g" noted in the example in Section 2.2.4. The lengths of the numeric variables were set to 8 bytes by default. The default length for numeric variables seldom causes problems, but the default length for character variables often causes the kind of truncation problems seen in this example.

LENGTH Statements

If you are creating a data set with a character variable of varying length, it is a good idea to use a LENGTH statement to explicitly set the length of the variable in order to avoid truncation errors.

```
DATA Good; LENGTH Vendor $50;  ❶
  SET Temp More;
  IF Test>0 THEN Rate = Fail/Test;  ❷
RUN;
```

❶ The LENGTH Vendor $50; statement tells SAS that Vendor should be a character variable ($ denotes character) of length 50 characters, which is more than enough for "Empirical Engineering".

❷ This statement creates a new variable, Rate, defined when Test>0, which is used in the next example to illustrate formatting of numeric variables. The Rate variable is computed as Fail/Test. You can use the symbols +, -, *, / to add, subtract, multiply, or divide two variables. Use ** for exponentiation. For example, 5**2 is 5 squared or 25.

When you run this code, the Good data set is created, and the third observation includes the full name of the Vendor, "Empirical Engineering".

```
Good

Obs    Vendor                   Test    Fail      Rate

 1     ChiTronix Components     5000      25    0.005000
 2     Duality Logic            1000       7    0.007000
 3     Empirical Engineering    7000     100    0.014286
```

Note that the Rate variable includes more decimal places than you might want to display. This can be corrected with a FORMAT statement, as illustrated in the next example.

INFORMAT and FORMAT Statements

INFORMAT and FORMAT statements are used to define the SAS formats that are used when reading in and writing out data, respectively. INFORMAT statements are discussed in Chapter 3. The next code sample shows how to use a FORMAT statement to control the appearance of the variables in your output.

```
DATA Better; LENGTH Vendor $50;  SET Good;
   FORMAT Rate 6.4 Test Fail 8.0 Vendor $25.;
RUN;
```

The FORMAT statement tells SAS how to format the output of variables. Rate 6.4 means that Rate will be written as 8 characters with 4 places after the decimal point. Test Fail 8.0 means that both Test and Fail will be written as 8 characters with no decimal point. Vendor $25. means that Vendor will be written as 25 characters, even though 50 are available. See *The Little SAS Book* for a comprehensive table of commonly used SAS formats.

```
Better

Obs    Vendor                   Test    Fail      Rate

 1     ChiTronix Components     5000      25    0.0050
 2     Duality Logic            1000       7    0.0070
 3     Empirical Engineering    7000     100    0.0143
```

The names that you choose for your variables might not be meaningful to the persons reading your reports. In this example, someone reading the table might not know that "Test" means "Number of Units Tested" and "Fail" means "Number of Units Failed." SAS provides the ability to assign a label to each variable, and you can use this capability to give more understandable labels to the column headings, as shown in the next example.

LABEL Statements

You can use a LABEL statement to create a variable label, which can be used in place of the variable name in printed and graphical output. The next code sample creates a new data set, Best, from the Better data set, adding labels for the Test, Fail, and Rate variables. Labels can include characters, for example, spaces, which are not valid in a SAS variable name, and so can be used to enhance the readability of your output.

```
DATA Best; SET Better;
   LABEL   Test = "Number of Units Tested"
           Fail = "Number of Units Failed"
           Rate = "Fraction Failed";
RUN;
```

Display 2-4 shows the new data set, Best, opened in a Viewtable window.

Display 2-4: The Best Data Set

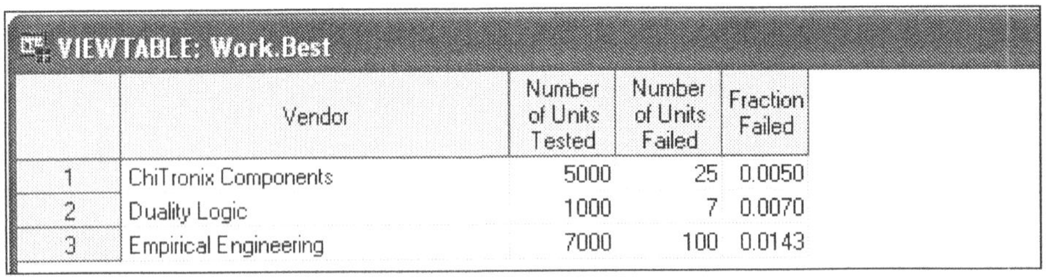

	Vendor	Number of Units Tested	Number of Units Failed	Fraction Failed
1	ChiTronix Components	5000	25	0.0050
2	Duality Logic	1000	7	0.0070
3	Empirical Engineering	7000	100	0.0143

Note that you now see the variable labels, but not the variable names in Viewtable. You can see the names instead by selecting **View→Column Names**, as shown in Display 2-5.

Display 2-5: The Best Data Set After Selecting View→Column Names

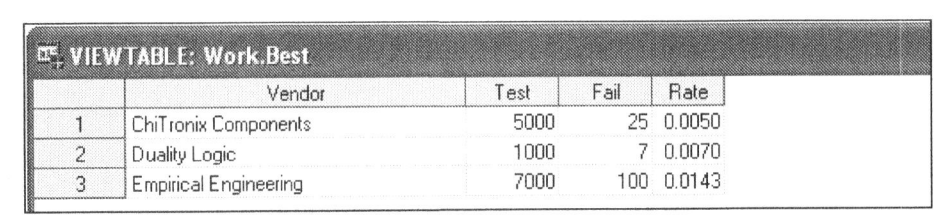

	Vendor	Test	Fail	Rate
1	ChiTronix Components	5000	25	0.0050
2	Duality Logic	1000	7	0.0070
3	Empirical Engineering	7000	100	0.0143

You can also right-click the data set and select **View Columns** to see the name, type, length, format, informat, and label of each variable.

Display 2-6: The Properties View of the Best Data Set

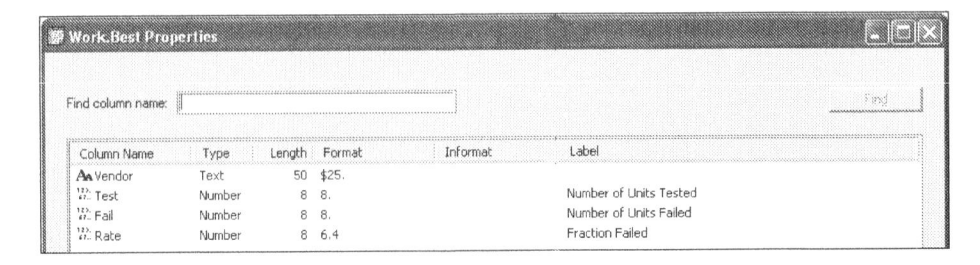

Column Name	Type	Length	Format	Informat	Label
Vendor	Text	50	$25.		
Test	Number	8	8.		Number of Units Tested
Fail	Number	8	8.		Number of Units Failed
Rate	Number	8	6.4		Fraction Failed

As noted in Section 2.1, another way to view your data sets is to print them to the Output window using PROC PRINT. The next line of code uses PROC PRINT to send the Best data set to the Output window.

```
PROC PRINT DATA=Best; RUN;
```

Display 2-7 shows the contents of the Output window after the PROC PRINT is run.

Display 2-7: The Best Data Set Written to the Output Window by PROC PRINT

```
                        The SAS System     14:34 Saturday, October 25, 2008    1
            Obs    Vendor                   Test      Fail      Rate

             1     ChiTronix Components     5000        25      0.0050
             2     Duality Logic            1000         7      0.0070
             3     Empirical Engineering    7000       100      0.0143
```

You can customize this view of the data set using some global options, a TITLE statement, and some PROC PRINT options, as illustrated in the next code sample.

```
OPTIONS NOCENTER NODATE LINESIZE=80 NONUMBER;   ❶
TITLE "Best";   ❷
PROC PRINT DATA=Best NOOBS LABEL;   ❸
  VAR Rate Fail Vendor;   ❹
RUN;
TITLE;   ❺
```

❶ The OPTIONS statement is used to set the value of the named options. NOCENTER prevents centering of the output. NODATE and NONUMBER prevent printing of the date and page numbers. LINESIZE controls the width (number of characters) of the output.

❷ The TITLE statement replaces the default title, "The SAS System", with whatever text you specify.

❸ The NOOBS option prevents the printing of the "Obs" column. The LABEL option forces the labels to be printed instead of the variable names.

❹ The VAR statement controls which variables are printed and the order of printing.

❺ The second TITLE; statement is not required, but it cancels the first TITLE statement and prevents the "Best" title from being inadvertently used with subsequent output.

When you run this code, the results in the Output window will be as shown in Display 2-8. This is the method used to display data sets in this book, except that the LABEL option is usually not used, and the data sets are shown as text rather than as screen shots.

Display 2-8: The Best Data Set in the Output Window, Using Options and a TITLE Statement

```
Best

           Number of
Fraction    Units
Failed     Failed      Vendor

0.0050        25       ChiTronix Components
0.0070         7       Duality Logic
0.0143       100       Empirical Engineering
```

2.3 Saving SAS Data Sets

SAS stores each data set in a *library*, and every SAS data set has a two-level name in the form *<Library Name>.<Member Name>*. If you don't supply a library name, SAS assumes that you want to create the data set in a temporary library called WORK. Any data sets stored in WORK will disappear as soon as you close your SAS session. To save a permanent copy of a data set, you first use a LIBNAME statement to associate a library name with the folder where you want to store the data.

```
LIBNAME JES "&JES.sas_data";
```

This line tells SAS that you will use the library name JES to refer to the folder sas_data at the path you assigned to the macro variable &JES (see Section 1.5). When you submit this line, a note is written to the Log Window saying that "Libref JES was successfully assigned…." If you see an error message instead, go back to Section 1.5 to make sure that &JES was defined correctly.

```
LIBNAME JES "&JES.sas_data";
NOTE: Libref JES was successfully assigned as follows:
      Engine:        V9
      Physical Name: c:\JES\sas_data
```

When you get a successful response to the LIBNAME statement, check the Explorer window to see that a new library, JES, has been created, and then double-click the JES library to see the contents. Display 2-9 shows the first 10 data sets in JES. These data sets are used in the examples in this book.

Display 2-9: The JES Library

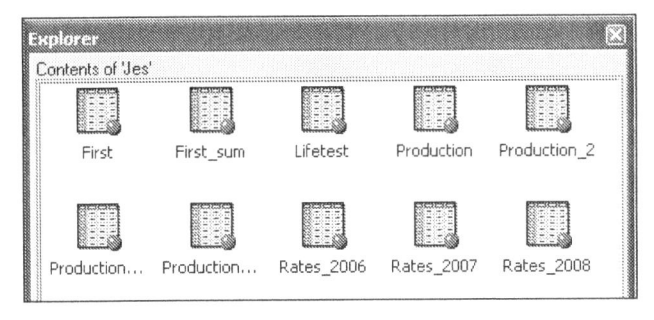

If the Best data set, which was created in Section 2.2.5, is still defined, you can copy it to the JES library by running this line of code.

```
DATA JES.Best; SET Best; RUN;
```

Then use the *Windows* (not SAS) Explorer or the UNIX file manager to verify that a new file, best.sas7bdat, was created in your ~\JES\sas_data folder. This is your new permanent data set.

Many examples in this book depend on the JES library being defined so that the data sets can be accessed. Add the following line to your autoexec.sas file (see Section 1.5) so that the JES library will be defined each time you start a SAS session.

```
LIBNAME JES "&JES.sas_data";
```

2.4 SAS Functions and CALL Routines

You can create new variables in a SAS data set using the symbols for addition, subtraction, multiplication, division and exponentiation (+, -, *, /, **). For example, in Section 2.2.5, the Rate variable in the Good data set is defined as Fail/Test. In addition, SAS provides a number of functions and CALL routines that you can use for more complex computations.

A SAS *function*

performs a computation, usually on one or more arguments, and returns a value.

A SAS *CALL routine*

can create new variables or alter variable values, but is not used with an assignment statement.

The code for this section is in ~\JES\sample_code\ch_2\functions.sas. The first code sample illustrates the difference between a function and a call routine.

```
DATA Rand;
  Seed=12345;
  DO I = 1 TO 5;
    CALL RANNOR(Seed,  X);  ❶
    Y =  RANNOR(Seed);  ❷
    OUTPUT;
  END;
RUN;
```

❶ The CALL RANNOR statement uses the value of Seed to generate a random variable from the normal distribution, and stores the result in variable X, without an assignment statement, "X = ." The value of Seed is changed by each invocation of CALL RANNOR.

❷ The RANNOR function also generates a normal variable and stores the result in variable Y.

Using a positive integer as the Seed value ensures that the results are the same each time the code is run. To get a different result each time, set Seed to zero.

```
RAND
Obs        SEED          I         X            Y
 1      1600293460       1     -0.04298     -0.09999
 2       593300711       2     -0.09999     -0.24349
 3      1565263655       3     -0.24349     -0.22226
 4      1576936339       4     -0.22226      0.07353
 5      1640848258       5      0.07353      0.49937
```

Most of the examples in this book use functions rather than CALL routines. The main exception is that CALL routines are used for random number generation, because they provide better control over the Seed values. Some of the more useful functions and CALL routines are listed in the following section:

- Section 2.4.1 Numeric Functions and CALL Routines
- Section 2.4.2 Date, Time, and Datetime Functions
- Section 2.4.3 Character Functions
- Section 2.4.4 The LAG*n*() and DIF*n*() Functions

2.4.1 Numeric Functions and CALL Routines

Table 2-1 lists some of the more useful numeric functions and CALL routines. The list is heavily weighted towards the functions used in this book, especially functions related to probability distributions. Be sure to check out the full list to see what other functions are available.

Base SAS → SAS Language Reference: Dictionary
   →Dictionary of Language Elements→Functions and CALL Routines

Optional parameters are indicated by the use of angular brackets. For example:

CDF('NORM', $x <,\mu, \sigma>$)

This means that when the first parameter of CDF is 'NORM', the parameter x is required, and the parameters μ and σ are optional. If μ and σ are not specified, then a standard normal distribution ($\mu=0$, $\sigma=1$) will be assumed.

In addition to basic math functions, LOG, SQRT, ROUND, etc., the probability functions are particularly useful for quality and reliability analysis. The code for this section is in ~\JES\sample_code \ch_2\numeric.sas. The code sample shows how to compute the CDF and inverse CDF of a normal distribution.

```
DATA Stats;
   DO x = -2 TO 2 BY 1;
      F1 = CDF('NORM', x);      ❶
      F2 = CDF('NORM', x, 1, 1);  ❷
      P = PROBIT(F1);           ❸
      OUTPUT;
   END;
RUN;
```

❶ The function CDF can be used to compute the cumulative distribution function for any of the distributions listed in Table 2-1. The first parameter, 'NORM', specifies the distribution to be used. For the normal distribution, there is one required parameter, x, and two optional parameters, μ and σ, which are the mean and standard deviation of the normal distribution respectively. In this case, the optional parameters are omitted, so a standard normal distribution is assumed.

❷ The second use of the CDF function specifies a normal distribution with both mean and standard deviation equal to 1.

❸ The function PROBIT(F) returns the inverse of the standard normal distribution.

```
STATS

Obs    x       F1         F2         P

 1    -2     0.02275    0.00135    -2
 2    -1     0.15866    0.02275    -1
 3     0     0.50000    0.15866    -0
 4     1     0.84134    0.50000     1
 5     2     0.97725    0.84134     2
```

The inverse probability functions BETAINV and CINV are used in Section 5.2.7 to compute confidence intervals for the parameters of the binomial and Poisson distribution.

Table 2-1: Selected Numeric Functions and CALL Routines

Function or CALL Routine	Meaning
(Italicized variables, *X*, *Y*, etc., represent any numeric constant or the name of any variable in the data set.)	
Elementary Math	
ABS(*X*)	Absolute value of *X*
EXP(*X*)	*X* raised to the power **e**
LOG(*X*)	Natural (base **e**) logarithm of *X*
LOG10(*X*)	Base 10 logarithm of *X*
LOG2(*X*)	Base 2 logarithm of *X*
MOD(*X, D*)	Remainder from the division of *X* by *D*
SQRT(*X*)	Square root of *X*
Logic [The number 0 is FALSE. Any non-zero number is TRUE.]	
X AND *Y*	= 0 if *X* and *Y* are TRUE, = 0 otherwise
X OR *Y*	= 1 if either *X* or *Y* is TRUE, =0 otherwise
NOT *X*	= 1 if *X* is FALSE, =0 if *X* is TRUE
Truncation and Rounding	
CEIL(*X*)	Smallest integer >= *X* e.g., CEIL(1.5) = 2 CEIL(-1.5) = -1
FLOOR(*X*)	Largest integer <= *X* e.g., FLOOR(1.5) = 1 FLOOR(-1.5) = -2
INT(*X*)	The integer portion of *X* e.g., INT(1.5) = 1 INT(-1.5) = -1
ROUND(*X*)	*X* rounded to the nearest integer
ROUND(*X, u*)	*X* rounded to the nearest multiple of *u*
Cumulative Distribution Functions	
CDF('BINO', *m,p,n*)	Prob[X<=m] if X has the binomial distribution with parameters (*n,p*)
CDF('CHIS', *x, df <,nc>*)	Prob[X<=x] if X has the chi-square distribution with *df* degrees of freedom and optional non-centrality parameter *nc*
CDF('LOGN', *x, τ, λ*)	Prob[X<=x] if X has the lognormal distribution with location *τ* and scale *λ*
CDF('NORM',*x <, μ, σ>*)	Prob [X<=x] if X has the normal distribution with mean *μ* and standard deviation *σ*. Default is *μ* = 0, *σ* = 1
CDF('POIS', *n, m*)	Prob[X<=n] if X has the Poisson distribution with mean *m*
CDF('WEIB', *x, a <,λ>*)	Prob[X<=x] where X has the Weibull distribution with shape parameter *a* and scale parameter *λ*. Default is *λ* = 1
See the SAS Help and Documentation for the syntax of the other distributions which can be used with the CDF function, including Bernoulli, Beta, Cauchy, Exponential, F, Gamma, Geometric, Hypergeometric, Laplace, Logistic, Negative Binomial, Normal Mixture, Pareto, Student's T, Uniform, and Wald (Inverse Gaussian).	
Probability Density or Mass Functions	
The probability density for continuous distributions, or probability mass for discrete distributions, for the same distributions used with the CDF function, using the same syntax. For example:	
PDF('BINO', *m,p,n*)	Prob[X=m] if X has the binomial distribution with parameters (*n,p*)
PDF('NORM',*x <, μ, σ>*)	Probability density at *x*, if X is normal with mean *μ* and standard deviation *σ*
Inverse Probability Functions	
BETAINV(*p, a, b*)	The *p*-th quantile of the Beta distribution with parameters *a* and *b*
CINV(*p, df, <nc>*)	The *p*-th quantile of the chi-squared distribution with *df* degrees of freedom and optional non-centrality parameter *nc*
PROBIT(*p*)	The *p*-th quantile of the standard normal distribution
Random Number Functions and CALL Routines	
RANNOR(*Seed*)	Generates a random variable from the standard normal distribution
CALL RANNOR(*Seed, X*)	Generates a random variable from the standard normal distribution

2.4.2 Date, Time, and Datetime Functions

A *SAS date* is stored as an integer equal to the number of days since January 1, 1960. A *SAS datetime* is stored as a decimal equal to the number of seconds, to the nearest millisecond, since midnight of January 1, 1960. A *SAS time* is the number of seconds since midnight, to the nearest millisecond. Table 2-3 shows some of the more useful SAS date, time, and datetime functions. The code for this section is in ~\JES\sample_code\ch_2\date_time.sas. The first code sample shows how to enter a fixed date or datetime (or the current date, time, or datetime) into a SAS data set.

```
DATA Temp;
   aDate    = '13AUG1954'd;              ❶
   aDateTime = '13AUG1954:08:30:25'dt;   ❷
   Today=TODAY();                        ❸
   Time = TIME();
   Now = DATETIME();
RUN;
```

❶ Enter a specific date as "DDMMMYYYY"d. The d at the end tells SAS that this is a date.

❷ Enter a specific time as "DDMMMYYYY:HH:MM:SS"dt. The dt tells SAS that this is a datetime.

❸ The function TODAY() returns the current date. The function TIME() returns the current time to the nearest millisecond. The function DATETIME() returns the current date and time to the nearest millisecond.

When you run this code, your Temp data set will look something like this. Of course, your values of Today, Time, and Now will correspond to the date and time when you run the code.

```
Temp
aDate      aDateTime      Today     Time         Now
-1967      -169918175     17244     78648.50     1489960248.5
```

You can use FORMAT statements to make dates, times, and datetimes easier to read. Table 2-2 shows some commonly used formats, and the appearance of the output when these are applied.

Table 2-2: Selected Date, Time, and Datetime Formats

FORMAT	Example	FORMAT	Example
DATE7.	15MAR07	DATETIME13.	15MAR07:08:30
DATE9.	15MAR2007	DATETIME16.	15MAR07:08:30:26
MMDDYY10.	03/15/2007	DATETIME20.	15MAR2007:08:30:26
WEEKDATE15.	Thu, Mar 15, 07	TIME8.	08:30:26

The next bit of code applies appropriate formats to make the variables in Temp easier to understand.

```
DATA Better; SET Temp;
   FORMAT aDate Today DATE9. aDateTime Now DATETIME20. Time TIME8.;
RUN;
```

Note that a single FORMAT statement can apply formats to several variables.

```
Better
   aDate         aDateTime            Today        Time            Now
13AUG1954    13AUG1954:08:30:25     19MAR2007    21:50:49    19MAR2007:21:50:49
```

Working with SAS dates and datetimes can be confusing. For more information, see the SAS conference papers and the SAS Technical Support document listed in Section 2.10.

Table 2-3: Selected SAS Date, Time, and Datetime Functions

Function	Meaning
TODAY()	Today's date [Number of days since Jan 1, 1960].
DATETIME()	The current date and time [Number of seconds since Jan 1, 1960].
DATEPART(*datetime*)	The date part of *datetime*.
TIMEPART(*datetime*)	The time part of *datetime*.
MONTH(*theDate*)	The month (as an integer from 1–12) corresponding to *theDate*.
DAY(*theDate*)	The day of the month (from 1 to 31) corresponding to *theDate*.
YEAR(*theDate*)	The year (e.g., 2006) corresponding to *theDate*.
WEEKDAY(*theDate*)	The day of the week (1–7, with 1=Sunday) corresponding to *theDate*.
QTR(*theDate*)	The quarter (1–4, with 1=JAN–MAR) corresponding to *theDate*.
JULDATE7(*theDate*)	The Julian version *theDate* in the form YYYYDDD.
MDY(*month,day,year*)	The date corresponding to *month, day, year*.

INTNX(*interval, start_from, N, \<alignment\>*)	Increments the value *start_from* by *N* *interval*s, where *start_from* can be a date, time, or datetime variable, *N* is an integer, and *interval* can be WEEK, MONTH, HOUR, YEAR, QTR, etc. For example:
	    INTNX('WEEK', '13JUN2005'd, 1) = 20JUN2005
	The optional *alignment* parameter controls the position of the returned date within the interval. The choices are BEGINNING, MIDDLE, or END, or simply B, M, or E. For example:
	    INTNX('WEEK', '13JUN2005'd, 1, 'E')= 25JUN2005
	because 25JUN2005 is the first Saturday after 20JUN2005.

INTCK(*interval, from, to*)	The number of *interval*s between *from* and *to*. The value of *interval* can be WEEK, MONTH, HOUR, YEAR, QTR, etc. The values of *from* and *to* can be dates, times, or datetimes. The type of *interval* (date, datetime, or time) must match the type of value in *from*. For example: INTCK('DAY', '13JUN2005'd, '25JUN2005'd) = 12

The next code sample illustrates the use of several of these functions.

```
DATA OneWeek;
   FORMAT theDate WEEKDATE29.  weekEnd DATE7. nextMonth MMDDYY10.;
   DO I=0 TO 6;
      theDate = '28MAR2007'd + I;
      M = MONTH(theDate); D = DAY(theDate); Y = YEAR(theDate);
      Q = QTR(theDate);
      DOW = WEEKDAY(theDate);
      nextMonth = MDY(M+1, D, Y);
      NDays    = INTCK('DAY', theDate, nextMonth);
      weekEnd = INTNX('WEEK', theDate, 0, 'E');
      OUTPUT;
   END;
RUN;
```

Note the missing values for nextMonth and NDays because 04/31/2007 is not a valid date.

```
OneWeek
                   theDate weekEnd   nextMonth  I  M  D    Y   Q DOW NDays
      Wednesday, March 28, 2007 31MAR07 04/28/2007 0  3 28 2007  1   4    31
       Thursday, March 29, 2007 31MAR07 04/29/2007 1  3 29 2007  1   5    31
         Friday, March 30, 2007 31MAR07 04/30/2007 2  3 30 2007  1   6    31
       Saturday, March 31, 2007 31MAR07            .  3 31 2007  1   7     .
         Sunday, April  1, 2007 07APR07 05/01/2007 4  4  1 2007  2   1    30
         Monday, April  2, 2007 07APR07 05/02/2007 5  4  2 2007  2   2    30
```

2.4.3 Character Functions

Tables 2-4 and 2-5 list some of the SAS character functions that are very useful for manipulating and extracting information from character variables. The examples in this section illustrate the use of some of these functions. The code for this section is in ~\JES\sample_code\ch_2\text.sas.

This first example uses the JES.Contacts data set, so be sure that you have followed the instruction in Section 2.3 to define the JES library. JES.Contacts contains two lines of messy data, which are typical of what you might find in manually entered records.

```
JES.Contacts

  Name          City         State       Number

John X. Doe    Lodi           nj         201-555-0123
Mary Murphy    San   Jose     CA         408.555.678
```

The first code sample uses SAS character functions to improve the formatting of the City, State, and Number variables, and to extract three new variables: First, Last, and City_State.

```
DATA Contacts_1; SET JES.Contacts;
   City       = COMPBL(City);    ❶
   State      = UPCASE(State);   ❷
   Number     = TRANSLATE(Number, '-', '.');  ❸
   Area       = SUBSTR(Number,1,3);  ❹
   First      = SCAN(Name, 1, ' ');  ❺
   Last       = SCAN(Name,-1,' ');
   City_State = City||", "||State;   ❻
RUN;
```

❶ The COMPBL function is used to remove the extra spaces between "San" and "Jose".

❷ The UPCASE function ensures that the State variable is uppercase.

❸ The TRANSLATE function converts the phone numbers to a standard format.

❹ The SUBSTR function is used to extract the area code from the phone number.

❺ The SCAN function is used to extract the first name from the Name variable, and then again to extract the last name from the Name variable.

❻ The concatenation operator, ||, is used to create the City_State variable.

```
Contacts_1

  Name         City       State   Number         Area   First   Last      City_State

John X. Doe   Lodi         NJ     201-555-0123   201    John    Doe       Lodi      , NJ
Mary Murphy   San Jose     CA     408-555-678    408    Mary    Murphy    San Jose  , CA
```

Table 2-4: Selected SAS Character Functions

Function	Meaning
Functions that change or rearrange the characters in a string	
UPCASE(*string*)	Convert all characters in *string* to uppercase.
LOWCASE(*string*)	Convert all characters in *string* to lowercase.
LEFT(*string*)	Move leading spaces in *string* to the end of *string*.
RIGHT(*string*)	Move trailing spaces in *string* to the beginning of *string*.
COMPRESS(*string, char, modifier*)	Remove all occurrences of the characters in *char* from *string*. The optional *modifier* is a character string in which each character modifies the action of the function. For example: "N" adds all numerals, letters, and the underscore to the list in *char*. "W" adds all printable characters to the list in *char*. "K" means to keep rather than remove the characters in *char*. "WK" means to remove everything except the printable characters.
COMPBL(*string*)	Convert consecutive spaces to a single space.
REVERSE(*string*)	Reverse the order of the characters in *string*.
TRIM(*string*)	Remove trailing spaces from *string*.
STRIP(*string*)	Remove leading and trailing spaces from *string*.
TRANSLATE(*string, to, from*)	Replace each character in *from* with the corresponding character in *to*.
QUOTE(*string*)	Puts *string* in quotes.
DEQUOTE(*string*)	Extracts a string from within quotes.
Functions for concatenating strings	
string_1||*string_2*	Concatenate *string_1* and *string_2*.
CAT(*string1* <,...*string-n*>)	Concatenate character strings.
CATT(*string1* <,...*string-n*>)	Concatenate character strings, removing trailing blanks.
CATS(*string1* <,...*string-n*>)	Concatenate character strings, removing leading and trailing blanks.
CATX(*separator, string1* <,...*string-n*>)	Concatenate character strings, remove leading and trailing blanks, and add *separator* between the strings.
Functions that extract a substring from a string	
SUBSTR(*string, k* <,*n*>)	Substring of *n* consecutive characters of *string*, beginning with the *k*-th. If the argument *n* is missing, all characters starting with the *k*-th are returned.
SCAN(*string, i, char*)	Return the *i*-th element of *string* if *char* is used as a delimiter. If *i* is negative, the *i*-th element from the end is selected.
Functions for converting characters to numbers and numbers to characters	
INPUT(*string, informat*)	Convert *string* to an equivalent number, using *informat*.
PUT(*number, format*)	Convert *number* to an equivalent string, using *format*.

You can find the full list of character functions in the SAS online documentation at

Base SAS → SAS Language Reference: Dictionary
　　→Dictionary of Language Elements→Functions and CALL Routines
　　　　→Functions and CALL Routines by Category

In addition to the character functions, you will find Character String Matching functions and CALL Routines based on Perl regular expressions.

Table 2-5 lists some SAS functions for extracting information from a character variables. The next code sample illustrates the use of these functions to check the validity of the phone numbers in JES.Contacts.

```
DATA Contacts_2; SET JES.Contacts;
  FORMAT Phone_Ck $6.;
  IF LENGTH(Number) NE 12                      THEN Phone_Ck="Length"; ❶
  IF VERIFY(STRIP(Number), '.-0123456789')>0 THEN Phone_Ck="Char"; ❷
RUN;
```

❶ The LENGTH function returns the length of the Number variable, which is always 12 for valid phone numbers in the U.S., and sets the Phone_Ck variable to "Length" if any other number is found.

❷ The VERIFY function returns the location of the first character in Number which is not "-" or "." or a number. The STRIP function is used to remove leading and trailing spaces from Number before applying the VERIFY function.

The Phone_Ck variable shows that both phone numbers are invalid, the first because the letter "O" was entered instead of zero, and the second because the length is only 11 characters.

```
Contacts_2

   Name          City        State      Number        Phone_Ck

John X. Doe    Lodi           nj      201-555-O123    Char
Mary Murphy    San Jose       CA      408.555.678     Length
```

The next example uses the PUT and INPUT functions (see Table 2-4) to convert numeric variables to character variables and vice versa.

```
DATA CharNum;
   Char = "7.654";    Num = 3.1415927;  OUTPUT;
   Char = "9.0";      Num = 100;        OUTPUT;
   Char = "9999";     Num = 0.55;       OUTPUT;
RUN;
DATA CharNum_1; set CharNum;
   Char2Num = INPUT(Char, 8.4);
   Num2Char = PUT(Num, 8.2);
RUN;
```

The first DATA step creates the CharNum data set with one numeric and one character variable. The second DATA step uses the INPUT function to convert Char to a numeric variable, Char2Num, using informat 8.4, and then uses the PUT function to convert Num to a character variable, Num2Char, using format 8.2.

```
CharNum_1
Char        Num         Char2Num     Num2Char

7.654       3.142       7.6540         3.14
9.0       100.000       9.0000       100.00
9999        0.550       0.9999         0.55
```

Note that because "9999" does not have decimal point, the format 8.4 is applied and the result is 0.9999. The presence of a decimal point in the value in the INPUT function overrides the placement specified in the informat. Without a decimal point in the value, the informat specifies where the decimal point should be placed, counting from the right-most digit.

Table 2-5: Functions That Return Information about a Character Variable

Function	Meaning
LENGTH(*string*),	The number of characters in *string*, excluding trailing spaces.
INDEX(*string, substring*)	Location of the first occurrence of *substring* within *string*. Returns zero if *substring* is not found.
INDEXC(*string, char*)	Location of the first occurrence of any character in *char* within *string*. Returns zero if no characters are found in *char*.
VERIFY(*source, excerpt-1 <,...excerpt-n>*)	Location of the first character in *source* that is not in any *excerpt-i*. Returns zero if every character in *source* is in at least one *excerpt*.
ANYDIGIT(*string <,start>*)	Location of the first occurrence of a digit within *string*. With two arguments, the search begins at the absolute value of *start*. If *start* is negative, the search proceeds to the left. If *start* < - LENGTH(*string*), the search proceeds from the end of *string*. Returns zero if no digits are found.
NOTDIGIT(*string <,start>*)	Location of the first character within *string* which is not a digit. The start parameter works the same as with ANYDIGIT.
See also the similarly defined: ANYALNUM, ANYALPHA, ANYCNTRL, ANYFIRST, ANYGRAPH, ANYLOWER ANYNAME, ANYPRINT, ANYPUNCT, ANYSPACE, ANYUPPER, ANYXDIGIT, and the corresponding NOT versions of these functions.	

Your data might include dates that are not in a standard SAS date format; for example, the Time variable in the JES.TimeStamp data set. Such variables are often encountered in extracts from relational databases.

```
JES.TimeStamp
          Time
Tue Apr 03 08:25:00 MST 2004
Tue Mar 26 17:52:31 MST 2004
Tue Jun 03 08:25:00 MDT 2004
```

The final code sample uses character functions to extract a valid SAS date from the Time variable.

```
DATA TestDates; SET JES.TimeStamp;
  M=UPCASE(SCAN(Time,2,' '));  ❶
  D=SCAN(Time,3,' ');
  Y=SCAN(Time,6);
  DMY=CATS(D, M, Y);  ❷
  TestDate = INPUT(DMY, date9.);  ❸
  FORMAT TestDate mmddyy10.;
RUN;
```

❶ The SCAN function is used to extract the month, day, and year from the Time variable.

❷ The CATS function concatenates D, M, and Y into the DMY variable, which looks like a SAS date.

❸ The INPUT function converts the character variable DMY to the numeric variable TestDate, and the FORMAT statement formats the integer TestDate variable as a recognizable date.

```
TestDates
          Time                     M    D    Y      DMY         TestDate
Tue Apr 03 08:25:00 MST 2004      APR   03   2004   03APR2004   04/03/2004
Tue Mar 26 17:52:31 MST 2004      MAR   26   2004   26MAR2004   03/26/2004
Tue Jun 03 08:25:00 MDT 2004      JUN   03   2004   03JUN2004   06/03/2004
```

You would generally want to drop the M, D, Y, and DMY variables after TestDate is computed, but they are included here to make it easier to follow the steps.

2.4.4 The LAG*n*() and DIF*n*() Functions

The functions described in the previous sections all work with variable values in a single row of a data set, and return results to the same row. SAS provides two families of functions that enable you to use values from previous rows when computing the value of a variable in the current row.

LAG*n*(X) = The value of the variable X from *n* rows before the current row.
DIF*n*(X) = The current value of X minus the value of X from *n* rows before the current row.

These functions return a missing value if the referenced previous row does not exist.

The code for this section is in ~\JES\sample_code\ch_2\lag_retain.sas. The example uses the JES.Poisson data set, which contains a table of Poisson probabilities similar to the one created in Section 2.2.3.

```
JES.Poisson

K        F          P
0     0.00674    0.00674
1     0.04043    0.03369
2     0.12465    0.08422
3     0.26503    0.14037
4     0.44049    0.17547
5     0.61596    0.17547
```

```
DATA Lag_Table; SET JES.Poisson;
  Dif_F = DIF(F);  ❶
  Lag2_P = LAG2(P);  ❷
  IF K IN (0,2) THEN Lag_F=LAG(F);  ❸
RUN;
```

❶ The DIF function computes the difference between successive values of F, which, as expected, is the same as the value of P.

❷ The LAG2 function returns the value of P from two rows before the current row.

❸ The second LAG function returns the value of F from *the last time that it was called*, and it is called only when K=0 or 2. So when K=2, the value returned is the value of F for K=0.

```
Lag_Table

Obs    K       F          P         Dif_F      Lag2_P       Lag_F

 1     0    0.00674    0.00674        .           .            .
 2     1    0.04043    0.03369     0.03369        .            .
 3     2    0.12465    0.08422     0.08422     0.00674     .006737947
 4     3    0.26503    0.14037     0.14037     0.03369         .
 5     4    0.44049    0.17547     0.17547     0.08422         .
 6     5    0.61596    0.17547     0.17547     0.14037         .
```

CAUTION: As pointed out in the note (❸), the value of Lag_F is not what you might have expected. It is best to avoid using the LAG or DIF functions in statements that might not be executed.

2.5 The RETAIN Statement

Another way to use variable values from earlier rows is to use a RETAIN statement, which tells SAS to retain the value of a variable from the previous row for use in computing values in the current row. The code below uses a RETAIN statement to calculate the cumulative sum of the P variable in Table.

```
DATA Retain_Table; SET JES.Poisson; RETAIN Sum_P 0;  ❶
   Sum_P = Sum_P + P;  ❷
RUN;
```

❶ The RETAIN statement specifies Sum_P as a RETAIN variable and initializes its value to 0.

❷ The variable Sum_P is set equal to the value of Sum_P retained from the previous row, plus P.

```
Retain_Table

K       F         P         Sum_P
0     0.00674   0.00674   0.00674
1     0.04043   0.03369   0.04043
2     0.12465   0.08422   0.12465
3     0.26503   0.14037   0.26503
4     0.44049   0.17547   0.44049
5     0.61596   0.17547   0.61596
```

As expected, the Sum_P variable contains the cumulative sum of the values of P, which is the same as the value of F. Note that RETAIN might not act as expected if the RETAIN variable is already in the data set. If you rerun the same code using the Retain_Table data set, which already contains the Sum_P variable, the results are not what you might expect.

```
DATA Retain_Table_2; SET Retain_Table;
   RETAIN Sum_P 0;
   Sum_P = Sum_P + P;
RUN;
```

When you run this code, the value of Sum_P is no longer the sum of the values of P; it is the sum of the values of P added to the previous value of Sum_P.

```
Retain_Table_2

K       F         P         Sum_P
0     0.00674   0.00674   0.01348
1     0.04043   0.03369   0.07412
2     0.12465   0.08422   0.20888
3     0.26503   0.14037   0.40540
4     0.44049   0.17547   0.61596
5     0.61596   0.17547   0.79143
```

You can avoid this problem by dropping the RETAIN variable from the data set named in the SET statement, as shown in the final code sample. The resulting data set, Retain_Table_3, is the same as Retain_Table. The DROP data set option is explained in Section 2.6.

```
DATA Retain_Table_3; SET Retain_Table(DROP=Sum_P);
   RETAIN Sum_P 0;
   Sum_P = Sum_P + P;
RUN;
```

2.6 Selecting Subsets of Data Sets

This section shows how to select a subset of the rows and/or columns of a data set.

2.6.1 Selecting a Subset of Columns

You can eliminate unneeded variables from a data set by using a KEEP statement, or data set option, to name the variables that you want to keep, or a DROP statement, or data set option, to name the variables that you don't want. The KEEP or DROP data set *options* can be used at the beginning of the DATA step to keep or drop variables as the data set is read. The KEEP or DROP *statements* can be used at the end of the DATA step, to keep or drop variables as the new data set is written. The code for this section is in ~\JES\sample_code\ch_2\subsets.sas. The example uses the JES.First data set, shown in Section 9.1.1, which contains six variables: Batch, Sample, Resistance, Fail, Result, and Defects. The code sample shows four different ways to create a new data set containing only the Batch, Sample, and Result variables from JES.First. All of these new data sets are identical to Subset_1.

```
DATA Subset_1; SET JES.First; KEEP Batch Sample Result;     ❶
RUN;
DATA Subset_2; SET JES.First; DROP Resistance Fail Defects; ❷
RUN;
DATA Subset_3; SET JES.First(KEEP=Batch Sample Result);     ❸
RUN;
DATA Subset_4; SET JES.First(DROP=Resistance Fail Defects); ❹
RUN;
```

❶ The KEEP statement keeps only the Batch, Sample, and Result variables in Subset_1.

❷ The DROP statement drops the Resistance, Fail, and Defects variables from Subset_2.

❸ The KEEP option reads in only the Batch, Sample, and Result variables from JES.First.

❹ The DROP option drops the Resistance, Fail, and Defects variables as JES.First is read in.

```
Subset_1 (first three rows)
Batch     Sample    Result
  1         1       Pass
  1         2       Pass
  1         3       Pass
```

You can also use the RENAME data set option to change the name of a variable as you read in a data set. Note that if you also use a KEEP data set option in the SET statement, you must use the original variable name as SAS will not yet recognize the new name.

```
DATA Subset_5; SET JES.First(KEEP=Batch Sample Result
                             RENAME=(Result=Outcome));
RUN;
```

```
Subset_5 (first three rows)
Batch     Sample    Outcome
  1         1       Pass
  1         2       Pass
  1         3       Pass
```

2.6.2 Selecting a Subset of Rows

You can select a subset of the rows of a data set either by specifying the row numbers you want to keep or by specifying one or more conditions on the rows you want to keep. The code below uses two different methods to select only rows 8 through 10 of the JES.First data set.

```
DATA Temp; SET JES.First(FIRSTOBS=8 OBS=10);  ❶
RUN;
DATA Temp_2; SET JES.First;
   IF _N_ >= 8 AND _N_ <= 10;  ❷
RUN;
```

❶ The FIRSTOBS=8 and OBS=10 options are used to limit the data read in to rows 8 through 10.

❷ The DATA step iteration counter, _N_, is used to select only rows 8 through 10. The DATA step iteration counter is available during a DATA step, but is not part of the data set.

The Temp and Temp_2 data sets are identical.

```
Temp

Batch      Sample     Resistance      Fail     Result      Defects
  1          8          17.54           0       Pass          0
  1          9          16.88           0       Pass          0
  1         10          15.69           0       Pass          0
```

The final code sample creates two data sets at once by using different conditions to select rows for each.

```
DATA LowRes HighRes; SET JES.First(WHERE=(Batch>=23));  ❶
   IF Result="Low Res"   THEN OUTPUT LowRes;  ❷
   IF Result="High Res"  THEN OUTPUT HighRes;
RUN;
```

❶ The DATA statement names two data sets to be created: LowRes and HighRes. The SET statement includes a WHERE clause which limits the rows read in to those satisfying the condition in the parentheses. In this case, only rows with Batch >= 23 are included.

❷ The first IF ... THEN... OUTPUT statement causes any row with Result="LowRes" to be written to the LowRes data set. The second IF ...THEN... OUTPUT statement causes any row with Result= "High Res" to be written to the HighRes data set.

```
LowRes
Batch      Sample     Resistance      Fail     Result      Defects
  23         5          10.49           1      Low Res        0
```

```
HighRes
Batch      Sample     Resistance      Fail     Result      Defects
  24         2          25.94           1      High Res       0
  24         4          23.45           1      High Res       1
  24         6          22.58           1      High Res       0
  25         8          22.85           1      High Res       0
  25        10          24.80           1      High Res       0
```

These two data sets contain all the failing Resistance values, for Batch >=23, from JES.First.

2.7 Sorting Data Sets

You can use PROC SORT to sort a data set by one or more columns in ascending or descending order. The code for this section is in ~\JES\sample_code\ch_2\sort_merge.sas. The data sets used in the examples in this section and the next are JES.Units and JES.Fails, representing the installation and failure data on a population of manufactured units. Quality and reliability data analysis often requires sorting and merging these kinds of data sets. Sorting is discussed in this section, and merging is discussed in Section 2.8.

The JES.Units data set includes the serial number (SN), install date, and manufacturing location (Loc) for each of a population of 10 units. The JES.Fails data set includes the serial number (SN), date of failure, and location of the failure on the unit (Loc) for all failures that have occurred on the units listed in the JES.Units data set. You can easily spot some problems with these data sets. The units with SN=0007, 0015 and 0035 seem to have been installed twice, the unit with SN=0085 is not found in JES.Units, and the same variable name, Loc, is used with different meaning in the two data sets. These problems are included so that the examples can show you how to detect and overcome them.

```
JES.Units
SN         Install      Loc
0035       06/17/2006   CA
0027       06/05/2006   CA
0007       06/16/2006   CA
0035       06/11/2006   NY
0015       06/09/2006   CA
0007       06/09/2006   NY
0016       06/06/2006   NY
0015       06/08/2006   CA
0061       06/09/2006   NY
0005       06/21/2006   NY
```

```
JES.Fails
SN         Fail         Loc
0027       07/17/2006   Top
0016       07/22/2006   Bottom
0061       08/02/2006   Bottom
0035       08/05/2006   Top
0085       08/05/2006   Bottom
0007       09/05/2006   Top
0035       09/06/2006   Top
```

This code uses PROC SORT to sort the JES.Units data set by SN and Install:

```
PROC SORT DATA=JES.Units  ❶
    OUT=Units_Sorted;  ❷
  BY SN DESCENDING Install;  ❸
RUN;
```

❶ The DATA= statement of PROC SORT specifies the data set to be sorted, JES.Units.

❷ The OUT= statement causes the sorted data set to be stored in a new data set, Units_Sorted. If the OUT= statement is left out, then the sorted data is stored back into the JES.Units data set.

❸ The BY statement lists the variables to sort by and the order of the sort. By default, variables are sorted in ascending numerical or alphabetical order. The DESCENDING option specifies that Install should be sorted in descending order.

```
Units_Sorted
Obs      SN        Install      Loc
 1      0005      06/21/2006    NY
 2      0007      06/16/2006    CA
 3      0007      06/09/2006    NY
 4      0015      06/09/2006    CA
 5      0015      06/08/2006    CA
 6      0016      06/06/2006    NY
 7      0027      06/05/2006    CA
 8      0035      06/17/2006    CA
 9      0035      06/11/2006    NY
10      0061      06/09/2006    NY
```

2.7.1 Automatic Variables: FIRST.var and LAST.var

When a data set is sorted, SAS creates automatic variables, FIRST.var and LAST.var, for each variable in the BY statement, that tell us whether each record contains the first or last occurrence of each of the sorted variables. But these variables are not added to the sorted data set, so you never see them unless you know how to look. The next bit of code makes these automatic variables "visible" by creating variables in the data set that are equal to each of the FIRST. and LAST. values.

```
DATA Units_Sorted_Plus; SET Units_Sorted; BY SN DESCENDING Install;  ❶
   F_SN      = FIRST.SN;  ❷
   L_SN      = LAST.SN;
   F_Install = FIRST.Install;  ❸
   L_Install = LAST.Install;
RUN;
```

❶ The BY SN DESCENDING Install; statement is required to make the automatic variables available to the DATA step. The Units_Sorted data set must first have been sorted with the same BY statement.

❷ The FIRST. SN variable is 1 if the record contains the first occurrence of SN in the data set, and 0 otherwise. The LAST.SN = 1 if the record contains the last occurrence of SN in the data set and 0 otherwise.

❸ FIRST.Install and LAST.Install are defined similarly. Note that FIRST.Install and LAST.Install are both equal to 1 in rows 3 and 4, even though the value of Install is the same (06/09/2006), because the values of the first sort variable, SN, are different.

```
Units_Sorted_Plus
Obs      SN        Install      Loc    F_SN    L_SN    F_Install    L_Install
 1      0005      06/21/2006    NY      1       1         1            1
 2      0007      06/16/2006    CA      1       0         1            1
 3      0007      06/09/2006    NY      0       1         1            1
 4      0015      06/09/2006    CA      1       0         1            1
 5      0015      06/08/2006    CA      0       1         1            1
 6      0016      06/06/2006    NY      1       1         1            1
 7      0027      06/05/2006    CA      1       1         1            1
 8      0035      06/17/2006    CA      1       0         1            1
 9      0035      06/11/2006    NY      0       1         1            1
10      0061      06/09/2006    NY      1       1         1            1
```

2.7.2 Selecting Unique Records and Duplicates

In many cases, logic dictates that there should be only one record associated with each value of a variable. For example, there should be one and only one install date associated with each serial number in the JES.Units data set. The next code sample uses the automatic variables, FIRST and LAST, described in the previous section, to create two data sets: Units_U, which includes only one row for each SN, and Units_Dup, which contains all the rows with duplicate values of SN. The Units_U data set is suitable for further analysis, as it satisfies the uniqueness requirement. The Units_Dup data set can be used to investigate the problem in your data source that led to duplicate SN records.

```
DATA Units_U Units_Dup; SET Units_Sorted; BY SN DESCENDING Install;  ❶
   IF LAST.SN=1 THEN OUTPUT Units_U;  ❷
   IF FIRST.SN=0 OR LAST.SN=0 THEN OUTPUT Units_Dup;  ❸
RUN;
```

❶ The BY SN DESCENDING Install; statement is required to make the automatic variables, FIRST.SN and LAST.SN, available during the DATA step.

❷ The IF LAST.SN=1 condition is satisfied only once for each unique value of SN. Since the Install variable is sorted in descending order, the LAST.SN=1 condition selects the earliest Install date.

❸ This IF condition is true if the value of SN is not the first, or not the last, so that the Units_Dup data set will get every row for which the same value of SN is found in another row.

```
Units_U
Obs      SN        Install      Loc
 1       0005      06/21/2006   NY
 2       0007      06/09/2006   NY
 3       0015      06/08/2006   CA
 4       0016      06/06/2006   NY
 5       0027      06/05/2006   CA
 6       0035      06/11/2006   NY
 7       0061      06/09/2006   NY
```

```
Units_Dup
Obs      SN        Install      Loc
 1       0007      06/16/2006   CA
 2       0007      06/09/2006   NY
 3       0015      06/09/2006   CA
 4       0015      06/08/2006   CA
 5       0035      06/17/2006   CA
 6       0035      06/11/2006   NY
```

The steps described in this example are commonly required as part of a data cleansing process before data can be used for analysis. In practical use, you might use a different selection criterion to extract a data set with unique serial numbers. For example, you might use the last install date rather than the first, or use other variables in the data set to determine the most reliable record to use.

Once you have a clean version of the population data, Units_U, it is often necessary to merge this data with failure data on the corresponding units, represented by the JES.Fails data in the example.

2.8 Merging Data Sets

You can use a MERGE statement in a DATA step to combine records from two or more data sets that have the same value for one or more common variables. The code for this section is in ~\JES\sample_code\ch_2 \sort_merge.sas. The next code sample merges the install records in the Units_U data set with the failure records in JES.Fails by matching on the unit serial number, SN.

```
PROC SORT DATA=Units_U; BY SN; RUN;   ❶
PROC SORT DATA=JES.Fails; BY SN; RUN;
DATA Units_Fails; MERGE Units_U JES.Fails;   ❷
   BY SN;   ❸
RUN;
```

❶ The two PROC SORTs ensure that both data sets are sorted by the SN variable. **Note:** Even if the rows happen to be in the right order, that does not mean they are sorted, so you should always run PROC SORT on each data set before trying to merge them.

❷ The MERGE statement lists the (two or more) data sets to be merged.

❸ The BY statement specifies the variables to match on, and must be the same as the BY statements used in the PROC SORTs. More than one variable can be used in the merge. For example, you might SORT and MERGE using a BY Model SN; statement if there were a Model variable in each data set.

```
UNITS_FAILS

Obs     SN        Install      Loc         Fail

 1     0005     06/21/2006     NY             .
 2     0007     06/09/2006     To         09/05/2006
 3     0015     06/08/2006     CA             .
 4     0016     06/06/2006     Bo         07/22/2006
 5     0027     06/05/2006     To         07/17/2006
 6     0035     06/11/2006     To         08/05/2006
 7     0035     06/11/2006     To         09/06/2006
 8     0061     06/09/2006     Bo         08/02/2006
 9     0085        .           Bo         08/05/2006
```

Note that there are missing values wherever the BY variable, SN, appears in only one of the two data sets. The Fail variable is missing in rows 1 and 3, because the corresponding units, 0005 and 0015, have never failed. This is normal because you don't expect every unit to fail. For real data, you might expect missing values for Fail in most of the rows. However, the missing Install value in row 9 is a problem. You cannot have a failure on a unit that was never installed, so this record should be cleansed before the analysis proceeds. Section 2.8.2 shows how to modify the merge code to detect and eliminate such problems.

Note also that the value of Loc from Fails has overwritten the value of Loc from Units_U, which doesn't make sense. In general you don't want the merged data sets to have any variables in common other than the BY variables to avoid such unpredictable overwrites. This problem can be eliminated by using the RENAME option to rename the Loc variable in JES.Fails to "Position".

```
DATA Units_Fails_2; MERGE Units_U JES.Fails(RENAME=(Loc=Position));
   BY SN;
RUN;
```

See Section 4.2.4 for an alternative way to merge data sets using PROC SQL and Section 5.9 for a comparison of the two methods.

2.8.1 Automatic Variable: IN

When you merge data sets, you can take advantage of the automatic IN variable to identify which rows of the merged data set were found in each of the data sets used in the merge. The next code sample creates the same merge as the previous example and also makes use of the IN variables.

```
DATA Units_Fails_Plus;
MERGE Units_U(IN=in_u)  ❶
    JES.Fails(IN=in_f RENAME=(Loc=Position));  ❷
    BY SN;
    In_Units = in_u;  ❸
    In_Fails = in_f;
RUN;
```

❶ The IN= in_u option defines a temporary variable, in_u, which is equal to 1 if the row of the merged data set is found in the Units_U data set, and 0 otherwise. This variable is available during the DATA step, but is not part of the resulting data set.

❷ The IN=in_f option defines the temporary in_f variable, which is equal to 1 if the row of the merged data set is found in the JES.Fails data set, and 0 otherwise. The RENAME option is used to rename the Loc variable in JES.Fails to avoid overwriting the Loc variable in the Units_U data set.

❸ This statement defines a new variable, In_Units, with the same value as in_u, and the next statement defines In_Fails with the same value as in_f.

```
Units_Fails_Plus
Obs   SN    Install     Loc         Fail    Position    In_Units    In_Fails
 1   0005   06/21/2006  NY            .                    1           0
 2   0007   06/09/2006  NY         09/05/2006  Top         1           1
 3   0015   06/08/2006  CA            .                    1           0
 4   0016   06/06/2006  NY         07/22/2006  Bottom      1           1
 5   0027   06/05/2006  CA         07/17/2006  Top         1           1
 6   0035   06/11/2006  NY         08/05/2006  Top         1           1
 7   0035   06/11/2006  NY         09/06/2006  Top         1           1
 8   0061   06/09/2006  NY         08/02/2006  Bottom      1           1
 9   0085      .                   08/05/2006  Bottom      0           1
```

The In_Units and In_Fails variables were defined just so that you could see what the values of the temporary variables, in_u and in_f, were during the DATA step. You can use the temporary variables during the DATA step without assigning them to permanent variables in the data set, as illustrated in the next example.

2.8.2 Selecting Records by Which Data Sets They Are Found In

The IN variables are useful for selecting records from the merged data set for further analysis. For example, you might want one data set containing the Install and Fail dates for all units installed, another for only the units which failed, and a third data set containing the Fail records for which there are no corresponding Install dates. This code creates all of the required data sets in one pass.

Chapter 2: DATA Step Programming **49**

```
DATA Units_Fails Fails_OK Fails_WO_Install;
  MERGE Units_U(IN=In_U) JES.Fails(IN=In_F RENAME=(Loc=Position));
  BY SN;
  IF In_U=1 THEN OUTPUT Units_Fails; ❶
  IF In_U=1 and In_F=1 THEN OUTPUT Fails_OK; ❷
  IF In_U=0 and In_F=1 THEN OUTPUT Fails_WO_Install; ❸
RUN;
```

❶ This statement causes all rows found in the Units_U data set to be written to the Units_Fails_OK data set.

❷ This statement writes only the rows found in both data sets to the Fails_OK data set.

❸ This statement writes the rows found in JES.Fails, but not found in Units_U, to be written to the Fails_WO_Install data set.

```
Units_Fails_OK
SN        Install       Loc         Fail      Position
0005      06/21/2006    NY            .
0007      06/09/2006    NY          09/05/2006   Top
0015      06/08/2006    CA            .
0016      06/06/2006    NY          07/22/2006   Bottom
0027      06/05/2006    CA          07/17/2006   Top
0035      06/11/2006    NY          08/05/2006   Top
0035      06/11/2006    NY          09/06/2006   Top
0061      06/09/2006    NY          08/02/2006   Bottom
```

```
Fails_OK
SN        Install       Loc         Fail      Position
0007      06/09/2006    NY          09/05/2006   Top
0016      06/06/2006    NY          07/22/2006   Bottom
0027      06/05/2006    CA          07/17/2006   Top
0035      06/11/2006    NY          08/05/2006   Top
0035      06/11/2006    NY          09/06/2006   Top
0061      06/09/2006    NY          08/02/2006   Bottom
```

```
Fails_WO_Install
SN        Install       Loc         Fail      Position
0085          .                     08/05/2006   Bottom
```

This completes the data cleansing process begun in Section 2.7.2. You now have clean versions of the original JES.Units and JES.Fails data sets that can be used for data analysis, for example, using the reliability analysis methods described in Chapter 10. And you also have the Fails_WO_Install and Units_Dup data sets that can be used to investigate and correct the data integrity problems in the original data sets.

2.9 More Than Enough

This section includes very brief introductions to some more advanced topics just to make you aware of other SAS capabilities that you might need.

2.9.1 PROC DATASETS

The DATASETS procedure is useful for managing and extracting metadata about your data sets. The code for this section is in ~\JES\sample_code\ch_2 \proc_datasets.sas. The documentation can be found at

Base SAS→Base SAS Procedures Guide→Procedures→The DATASETS Procedure

The first example uses a CONTENTS statement with PROC DATASETS to extract information about the Units_U data set created in Section 2.7.2.

```
%INCLUDE "&JES.sample_code/ch_2/sort_merge.sas";   ❶
PROC DATASETS LIBRARY=WORK;   ❷
   CONTENTS DATA=Units_U;   ❸
QUIT;   ❹
```

❶ The %INCLUDE statement recreates the temporary data sets used in Sections 2.7 and 2.8.

❷ This statement requests that PROC DATASETS be run for the WORK library.

❸ The CONTENTS statement requests information about the Units_U data set.

❹ The QUIT statement ends the procedure.

When you run the code, this information about the Units_U data set is printed to the Output window.

```
The DATASETS Procedure
Data Set Name         WORK.UNITS_U              Observations           7
Member Type           DATA                      Variables              3
Engine                V9                        Indexes                0
Created               Tuesday, October 28,      Observation Length     16
Last Modified         Tuesday, October 28,      Deleted Observations   0
------(lines omitted)------
Alphabetic List of Variables and Attributes
              #    Variable    Type    Len    Format
              2    Install     Num      8     MMDDYY10.
              3    Loc         Char     2
              1    SN          Char     4     $4.
```

You can also use PROC DATASETS to delete selected data sets, or all data sets, in a specified library.

```
PROC DATASETS LIBRARY=WORK; DELETE Fa: Units_U; QUIT;   ❶
PROC DATASETS LIBRARY=WORK MEMTYPE=DATA GENNUM=ALL KILL; QUIT;   ❷
```

❶ This line deletes the Units_U data set and any data set beginning with "Fa" in the WORK library; for example, Fails_OK and Fails_WO_Install.

❷ This line uses the KILL option to delete all data sets in the WORK library.

2.9.2 PROC COMPARE

You can use PROC COMPARE to quickly and easily compare all rows and columns of two data sets and produce a report listing all differences. The code for this section is in ~\JES\sample_code \ch_2 \compare.sas. The documentation can be found at

Base SAS→Base SAS Procedures Guide→Procedures→The COMPARE Procedure

The code sample compares the Units_U and Units_Fails data sets from Sections 2.7 and 2.8.

```
%INCLUDE "&JES.sample_code/ch_2/sort_merge.sas";  ❶
PROC COMPARE BASE= Units_U COMPARE=Units_Fails; ID SN; RUN;  ❷
```

❶ The INCLUDE statement re-creates the data sets from Sections 2.7 and 2.8.

❷ The PROC COMPARE statement compares the Units_U and Units_Fails data sets, using the SN variable as an ID variable to match the data sets, and sends a report to the Output window.

This is an abridged version of the report that is sent to the Output window when you run the code.

```
Comparison of WORK.UNITS_U with WORK.UNITS_FAILS (several lines omitted)
                       Data Set Summary
     Dataset                  Created           Modified      NVar    NObs
     WORK.UNITS_U        28OCT08:14:16:03   28OCT08:14:16:03    3       7
     WORK.UNITS_FAILS    28OCT08:14:16:03   28OCT08:14:16:03    4       9

                       Variables Summary
     Number of Variables in Common: 3.
     Number of Variables in WORK.UNITS_FAILS but not in WORK.UNITS_U: 1.
     Number of Variables with Differing Attributes: 1.
     Number of ID Variables: 1.

           Listing of Common Variables with Differing Attributes
            Variable    Dataset             Type   Length   Format
            Loc         WORK.UNITS_U        Char      2
                        WORK.UNITS_FAILS    Char      2     $6.
     WARNING: The data set WORK.UNITS_FAILS contains a duplicate observation
              at observation number 7.

     Number of Observations with Some Compared Variables Unequal: 5.
     Number of Observations with All Compared Variables Equal: 2.

                       Values Comparison Summary
     Number of Variables Compared with All Observations Equal: 1.
     Number of Variables Compared with Some Observations Unequal: 1.
     Total Number of Values which Compare Unequal: 5.

                       Variables with Unequal Values
            Variable   Type    Len    Ndif    MaxDif
            Loc        CHAR     2      5
```

The comparison reveals the fact that there is a variable named Loc in each data set, with different formats and values. It also warns that there are two rows in the Units_Fails data set with the same ID value, SN = 0035, which might be a data integrity problem if these are non-repairable units, and not able to fail twice.

2.9.3 PROC TRANSPOSE

The usual meaning of "transpose" is just an interchange of the rows and columns of a matrix; for example:

Matrix				Transpose		
ChiTronix	5000	25		ChiTronix	Duality	Empirical
Duality	1000	7		5000	1000	7000
Empirical	7000	100		25	7	100

This simple kind of transpose would usually not make much sense for a SAS data set, especially because every column (variable) in a data set needs to contain only one kind of information. However, SAS has a TRANSPOSE procedure that allows for much more complex transpose operations. The code for this section is in ~\JES\sample_code\ch_2 \transpose.sas. The documentation can be found at

Base SAS→Base SAS Procedures Guide→Procedures→The TRANSPOSE Procedure

The example uses the JES.Rates data set, which contains test results by quarter for each vendor.

```
JES.Rates
Vendor              GEO       QTR         Rate       Test       Fail
ChiTronix           APAC      Q1          4.98%      5000       249
ChiTronix           APAC      Q2          4.96%      5000       248
ChiTronix           APAC      Q3          4.92%      5000       246
ChiTronix           APAC      Q4          4.92%      5000       246
Duality             EMEA      Q2          2.20%      1000       22
Duality             EMEA      Q3          2.30%      1000       23
Duality             EMEA      Q4          2.00%      1000       20
Empirical           AMER      Q1          0.89%      7000       62
Empirical           AMER      Q2          0.87%      7000       61
Empirical           AMER      Q3          0.97%      7000       68
```

Quality data is commonly found in this format, and there is often a need to transpose such data so that the results from each quarter are shown as separate columns. This is easy to do using PROC TRANSPOSE.

```
PROC TRANSPOSE DATA=JES.Rates OUT = TRates;  ❶
  ID QTR;  ❷
  VAR Fail;
  BY Vendor GEO;  ❸
RUN;
```

❶ This statement requests that the transposed version of JES.Rates be stored in TRates.

❷ The ID QTR; and VAR Fail; statements specify that the QTR and Fail variables be transposed.

❸ The BY Vendor GEO; statement specifies that the transpose be performed for each value of Vendor and GEO.

The resulting data set, TRates, includes four variables, Q1, Q2, Q3, and Q4. These variables contain the values of the Fail variable from JES.Rates for the corresponding values of Vendor, GEO, and QTR.

```
TRates
Vendor              GEO       _NAME_     Q1       Q2       Q3       Q4
ChiTronix           APAC      Fail       249      248      246      246
Duality             EMEA      Fail       .        22       23       20
Empirical           AMER      Fail       62       61       68       .
```

2.9.4 Arrays

If you come to SAS from another programming language, you might miss the ability to directly address the elements of data set with a notation like Temp[I;J], referring to the element in the I-th row and J-th column of Temp. You can get some of this capability with an ARRAY statement. The code for this section is in ~\JES\sample_code\ch_2 \array.sas. The documentation can be found at

> **Base SAS→SAS Language Reference: Dictionary→Dictionary of Language Elements →Statements→ARRAY Statement**

For some data summaries, you might want to compute the cumulative number of Fails by quarter for the values in a data set like TRates. This can be done with simple assignment statements (for example; CF2 = Q1 + Q2), but the coding is sometimes simpler with ARRAY statements, as shown in the code sample.

```
DATA CUM_Fails; set TRates;
   ARRAY Q(4)    Q1-Q4;  ❶
   ARRAY CF(4) CF1-CF4;  ❷
   DO i=1 TO 4;  ❸
       IF Q(i)=. THEN Q(i)=0;
   END;
   CF(1)=Q(1);
   DO i=2 TO 4;  ❹
       CF(i)=CF(i-1)+Q(i);
   END;
   DROP i _NAME_ GEO;
RUN;
```

❶ The first ARRAY statement creates the Q array which takes on the existing values of Q1,…,Q4.

❷ The second ARRAY statement creates the CF array with values CF1,…,CF4 which are not yet defined.

❸ The first DO loop replaces the missing values in Q with zeros. This is necessary because the Q(i) are to be added in the next step, and, if any are missing, the total will be missing.

❹ The second DO loop computes CF(i), i=1,2,3,4 as the cumulative sums of the values of the Q array. These values are stored in the CF1, CF2, CF3, and CF4 variables.

The resulting data set, CUM_Fails, includes four new variables, CF1, CF2, CF3, and CF4, containing the required cumulative sums.

```
CUM_Fails

Vendor           Q1      Q2      Q3      Q4     CF1     CF2     CF3     CF4

ChiTronix       249     248     246     246     249     497     743     989
Duality           0      22      23      20       0      22      45      65
Empirical        62      61      68       0      62     123     191     191
```

An array is a logical grouping of like variables, and the grouping expires when the DATA step ends. You will not be able to refer to Q(i) or CF(i) in subsequent operations on Cum_Fails unless you redefine them with another ARRAY statement.

2.10 Chapter Summary

2.10.1 Recap

After finishing this chapter, you should know how to

- Create and save SAS data sets.
- Use SAS functions to add variables (columns) to a data set.
- Use PROC PRINT to print data sets to the Output window.
- Select a subset of the rows or columns of a data set.
- Sort, merge, and transpose data sets.
- Use the automatic variables _N_, FIRST.var, LAST.var, and IN.

2.10.2 For More Information

Books

Aster, Rick. 2005. *Professional SAS® Programming Shortcuts, Second Edition.* Paoli, PA: Breakfast Communications Corp.

Delwiche, Lora D., and Susan J. Slaughter. 2003. *The Little SAS® Book: A Primer, Third Edition.* Cary, NC: SAS Institute Inc.

- See Chapter 1 for basic information on DATA steps and libraries.
- See Chapter 3 for information on SAS functions, SAS dates, and formats.
- See Chapter 4 for information on PROC SORT and PROC PRINT.
- See Chapter 6 for information on the SET, OUTPUT, and MERGE statements.

Morgan, Derek P. 2006. *The Essential Guide to SAS® Dates and Times.* Cary, NC: SAS Institute Inc.

SAS Institute Inc. 1995. *Combining and Modifying SAS® Data Sets: Examples, Version 6, First Edition.* Cary, NC: SAS Institute Inc.

SAS Conference Papers

Buck, Debbie. 2005. "A Hands-On Introduction to SAS® DATA Step Programming." *Proceedings of the Thirtieth Annual SAS Users Group International Conference.* Cary, NC: SAS Institute Inc. Paper 134-30. **[134-30.pdf]**

Carpenter, Arthur L. 2005. "Looking for a Date? A Tutorial on Using SAS® Dates and Times." *Proceedings of the Thirtieth Annual SAS Users Group International Conference.* Cary, NC: SAS Institute Inc. Paper 255-30. **[255-30.pdf]**

Carr, David W. 2008. "When PROC APPEND May Make More Sense Than the DATA STEP." *Proceedings of the SAS Global Forum 2008 Conference.* Cary, NC: SAS Institute Inc. Paper 085-2008. **[085-2008.pdf]**

Cody, Ronald. 2007. "An Introduction to SAS® Character Functions." *Proceedings of the SAS Global Forum 2007 Conference.* Cary, NC: SAS Institute Inc. Paper 217-2007. **[217-2007.pdf]**

First, Steven. 2008. "The SAS INFILE and FILE Statements." *Proceedings of the SAS Global Forum 2008 Conference*. Cary, NC: SAS Institute Inc. Paper 166-2008.

Heaton, Ed. 2008. "Many-to-Many Merges in the DATA Step." *Proceedings of the SAS Global Forum 2008 Conference*. Cary, NC: SAS Institute Inc. Paper 081-2008. **[081-2008.pdf]**

Malby, Ann, and Sally Williams. 2008. "Save time! Merge SAS® Files to Themselves." *Proceedings of the SAS Global Forum 2008 Conference*. Cary, NC: SAS Institute Inc. Paper 234-2008. **[234-2008.pdf]**

Morgan, Derek. 2008. "The Essentials of SAS® Dates and Times." *Proceedings of the SAS Global Forum 2008 Conference*. Cary, NC: SAS Institute Inc. Paper 168-2008. **[168-2008.pdf]**

Rhodes, Dianne Louise. 2005. "Pretty Dates All in a Row." *Proceedings of the Thirtieth Annual SAS Users Group International Conference*. Cary, NC: SAS Institute Inc. Paper 055-30. **[055-30.pdf]**

SAS Institute Inc. 2003. SAS Technical Support. "SAS® Dates, Times, and Interval Functions." Cary, NC: SAS Institute Inc. **[ts668.pdf]**

Tilanus, Erik W. 2008. "Poor Man's Parallel Processing Using the DATA Step View." *Proceedings of the SAS Global Forum 2008 Conference*. Cary, NC: SAS Institute Inc. Paper 096-2008. **[096-2008.pdf]**

Tilanus, Erik W. 2008. "Sending E-mail from the DATA Step." *Proceedings of the SAS Global Forum 2008 Conference*. Cary, NC: SAS Institute Inc. Paper 038-2008. **[038-2008.pdf]**

Tilanus, Erik W. 2008. "SET, MERGE and Beyond." *Proceedings of the SAS Global Forum 2008 Conference*. Cary, NC: SAS Institute Inc. Paper 167-2008. **[167-2008.pdf]**

2.10.3 Exercises

1. Create a data set, BINOMIAL, containing the CDF values of the Binomial distribution with parameters N=10 and P=.4, for K=0,1,…,10.

2. Create a data set, POISSON , containing the CDF values of the Poisson distribution with mean=4, for K=0,1,…,10.

3. Create a new data set, BIN_POI, by merging BINOMIAL and POISSON by the common parameter K. Make sure that you use different names for the CDF values in each table so that the values from one data set do not overwrite the values from the other.

4. Create two new data sets from the BIN_POI data set:

    - NEAR = the rows for which the difference between the two CDF values is <= .05.
    - FAR = the rows for which the difference between the two CDF values is > .05.

5. Use the LAG() function to compute the PDF for the Binomial and Poisson distributions created above. Use the relationships:

    - CDF(0) = PDF(0)
    - CDF(K) = CDF(K-1) + PDF(K).

Chapter 3

Data Out, Data In—Spreadsheets

3.1 Introduction 58
3.2 Exporting a SAS Data Set to a Spreadsheet 58
3.3 Importing Spreadsheet Data to SAS 62
3.4 More Than Enough 70
3.5 Chapter Summary 74

3.1 Introduction

This chapter explains various methods to export data from SAS data sets to spreadsheets, and to import data from spreadsheets into SAS data sets. Even if you do most of your work in SAS, you will probably need to exchange data with colleagues in spreadsheet form. Section 3.2 shows how to export SAS data sets as spreadsheets, and Section 3.3 shows how to import spreadsheets as SAS data sets.

3.2 Exporting a SAS Data Set to a Spreadsheet

This section describes methods for exporting a SAS data set to a comma-delimited, or CSV, file. A CSV file can easily be opened by any spreadsheet program, such as Excel, Star Office, or Open Office. You can also export data sets to tab-delimited or space-delimited files. If you have SAS/ACCESS Interface for PC Files, you can also export to various other formats, such as Microsoft Excel. See *The Little SAS Book: A Primer, Third Edition* for an explanation of how to export to these other file types. The code for this section is in ~\JES\sample_code\ch_3\ export.sas.

3.2.1 Using PROC EXPORT

The first code sample creates a data set, Week, to be exported as a CSV file.

```
DATA Week;
   FORMAT D 8.0 theDate MMDDYY10. theTime DATETIME20.;
   DO D=0 TO 6;
       theDate = '13APR2007'd + D;
       theTime = theDate*60*60*24 + D*60;
       OUTPUT;
   END;
   LABEL D = "Day Number" theDate="The Date" theTime="Date and Time";
RUN;
```

The Week data set, printed with labels instead of variable names, looks like this.

```
Week
   Day Number        The Date         Date and Time
            0      04/13/2007      13APR2007:00:00:00
            1      04/14/2007      14APR2007:00:01:00
            2      04/15/2007      15APR2007:00:02:00
            3      04/16/2007      16APR2007:00:03:00
            4      04/17/2007      17APR2007:00:04:00
            5      04/18/2007      18APR2007:00:05:00
            6      04/19/2007      19APR2007:00:06:00
```

The next bit of code uses the EXPORT procedure to export the Week data set as a CSV file.

```
PROC EXPORT DATA=Week
   OUTFILE ="&JES.output/Week_Proc_Export.csv" REPLACE;
RUN;
```

The PROC EXPORT DATA=Week statement specifies that the Week data set is to be exported. The OUTFILE statement specifies the path and filename for the exported file. The file extension, .csv, specifies that the exported file should be in CSV (comma separated values) format. The REPLACE option specifies that this CSV file is to overwrite any existing file with the same name. After you run this code, go to your ~\JES\output folder and open the Week_Proc_Export.csv file with your preferred spreadsheet software. Display 3-1 shows Week_Proc_Export.csv opened in Microsoft Excel.

Display 3-1: The Week_Proc_Export.csv File Opened in MS Excel

	A	B	C
1	D	theDate	theTime
2	0	4/13/07	13APR2007:00:00:00
3	1	4/14/07	14APR2007:00:01:00
4	2	4/15/07	15APR2007:00:02:00
5	3	4/16/07	16APR2007:00:03:00
6	4	4/17/07	17APR2007:00:04:00
7	5	4/18/07	18APR2007:00:05:00
8	6	4/19/07	19APR2007:00:06:00

Then if you select **Run→Recall Last Submit** from the menu at the top of your SAS session, the code shown next appears in the Program Editor window. This is the code written by PROC EXPORT to export the Week_Proc_Export.csv file. It does not follow the convention of using uppercase letters for SAS keywords, DATA, FORMAT, etc., which is followed throughout this book. If you need to export similar data sets on a regular basis, you can save the code and rerun it as required. You can also edit the code, for example, to change the variable formats.

```
data _null_;
    %let _EFIERR_ = 0; /* set the ERROR detection macro variable */
    %let _EFIREC_ = 0;  /* clear export record count macro variable */
    file 'c:\JES\output\Week_Proc_Export.csv' delimiter=',' DSD
DROPOVER lrecl=32767;
    if _n_ = 1 then     /* write column names or labels */
      do;
        put
          "D"
        ','
          "theDate"
        ','
          "theTime"
        ;
      end;
   set   WEEK    end=EFIEOD;
      format D 8. ;
      format theDate mmddyy10. ;
      format theTime datetime20. ;
    do;
      EFIOUT + 1;
      put D @;
      put theDate @;
      put theTime ;
      ;
    end;
    if _ERROR_ then call symputx('_EFIERR_',1);  /* set ERROR
detection macro variable */
    if EFIEOD then call symputx('_EFIREC_',EFIOUT);
    run;
```

3.2.2 Using ODS CSVALL

The Output Delivery System (ODS) enables you to package your SAS results for export as CSV, PDF, RTF, HTML, XML, and other types of files. Chapter 7 covers ODS in greater detail, but it makes sense here to include some ODS methods for exporting data sets to spreadsheets.

ODS is incredibly easy to use. You just add some lines before and after your normal SAS code steps, and the output from your SAS code is automatically routed to the destination that you choose.

```
ODS output destination  FILE = "path and filename"
    SAS code that produces output;
ODS output destination  CLOSE;
```

The next code sample uses the CSVALL destination to export the Week data set as a CSV file.

```
ODS CSVALL FILE="&JES.output/ODS_Week.csv";  ❶
  TITLE "The Week Data Set";
  PROC PRINT DATA=Week LABEL NOOBS;  ❷
      VAR D theTime theDate;
  RUN;
ODS CSVALL CLOSE;  ❸
```

❶ The ODS statement specifies that the CSVALL destination be used, and gives the path and filename where the output will be sent.

❷ The PROC PRINT statement prints the rows of the Week data set to the ODS_Week.csv file. The syntax of PROC PRINT is discussed in Section 2.2.5.

❸ The ODS CSVALL CLOSE; statement closes the destination.

When you run this code, you might see a dialog box asking if you want to open or save this file. If so, just select **Cancel**. The new file, ODS_Week.csv, should now be in your ~\JES\output folder.

ODS CSVALL can be used to generate the same CSV file that you created with PROC EXPORT. It also gives you the flexibility to use the LABEL option to output variable labels instead of names, a VAR statement to select and rearrange columns, and a TITLE statement to add descriptive information to the CSV file.

Display 3-2: The ODS_Week.csv File Opened in MS Excel

	A	B	C
1	The Week Data Set		
2			
3	Day Number	Date and Time	The Date
4	0	13APR2007:00:00:00	4/13/07
5	1	14APR2007:00:01:00	4/14/07
6	2	15APR2007:00:02:00	4/15/07
7	3	16APR2007:00:03:00	4/16/07
8	4	17APR2007:00:04:00	4/17/07
9	5	18APR2007:00:05:00	4/18/07
10	6	19APR2007:00:06:00	4/19/07

3.2.3 Using the MSOffice2K Tagset

Another way to get a SAS data set into a spreadsheet is to export it to an XLS file, which can be opened by MS Excel. This approach requires MS Office 2000 or later. The syntax is the same as the previous example except that **CSVALL** is replaced by **TAGSETS.MSOffice2K** and **.csv** is replaced by **.xls**.

```
ODS TAGSETS.MSOffice2K FILE="MSO2K_Week.xls" PATH="&JES.output";
   TITLE "The Week Data Set";
   PROC PRINT DATA=Week LABEL NOOBS;
      VAR D theTime theDate;
   RUN;
ODS TAGSETS.MSOffice2K CLOSE;
```

This code creates MSO2K_Week.xls in your ~\JES\output folder. Display 3-3 shows the file opened in MS Excel. If you change the file extension from .xls to .html it can also be viewed in a Web browser, as shown in Display 3-4. The MSO2K_Week.xls file includes markup code to control the font type, size, and color, background colors, etc. of your spreadsheet. You can control the formatting of the output by using a STYLE statement to choose one of the styles supplied by SAS, or by using PROC TEMPLATE to create your own style. Styles and templates are discussed in Chapter 7.

Display 3-3: The MSO2K_Week.xls File Opened in MS Excel

Display 3-4: The MSO2K_Week.html File Opened in a Web Browser

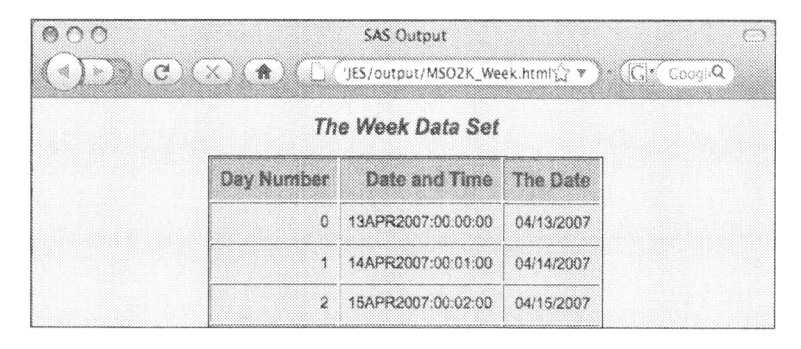

3.3 Importing Spreadsheet Data to SAS

The import process can be more difficult than the export process if the spreadsheet is not correctly formatted for import to a SAS data set. Section 3.3.1 explains the use of PROC IMPORT to import a spreadsheet as a SAS data set, and Section 3.3.2 shows how to handle some of the problems that occur during the import process. The code for this section is in ~\JES\sample_code\ch_3\ import.sas.

3.3.1 Using PROC IMPORT

You can use PROC IMPORT to import comma-delimited, tab-delimited, or space-delimited files as SAS data sets. The example in this section shows how to import the ODS_Week.csv file that was created by ODS CSVALL in Section 3.2.2.

PROC IMPORT expects to find the variable names for the SAS data set in the first row of the CSV file, so first edit the ODS_Week.csv file as follows:

1. Open ODS_Week.csv in a spreadsheet program.
2. Delete the first two rows containing the title and a blank row (see Display 3-2).
3. Save as a CSV file, changing the name to ODS_Week_New.csv.
4. Close the CSV file—you can't import the file while it is in use.

Display 3-5: The ODS_Week_New.csv File Opened in MS Excel

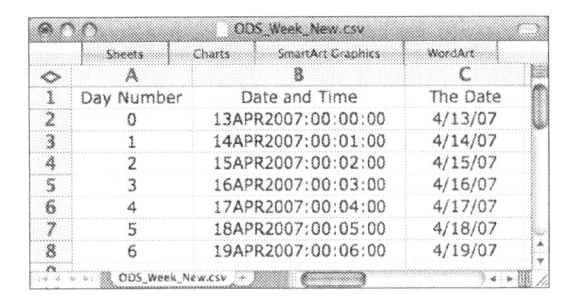

	A	B	C
1	Day Number	Date and Time	The Date
2	0	13APR2007:00:00:00	4/13/07
3	1	14APR2007:00:01:00	4/14/07
4	2	15APR2007:00:02:00	4/15/07
5	3	16APR2007:00:03:00	4/16/07
6	4	17APR2007:00:04:00	4/17/07
7	5	18APR2007:00:05:00	4/18/07
8	6	19APR2007:00:06:00	4/19/07

Now that the CSV file has the variable names in the first row, you are ready to run PROC IMPORT.

```
PROC IMPORT DATAFILE="&JES.output/ODS_Week_New.csv"
  OUT =Week_Import REPLACE;
RUN;
```

The DATAFILE statement specifies the path and filename of the CSV file. The file extension, .csv, specifies that the file to be read is in CSV format. The OUT statement specifies the name of the SAS data set to be created. The REPLACE option specifies that the new version will overwrite any data set with the same name. This code creates a new data set, Week_Import, in the SAS Explorer window.

```
Week_Import (first three rows)
Day_Number    Date_and_Time    The_Date
         0       13APR2007    04/13/2007
         1       14APR2007    04/14/2007
         2       15APR2007    04/15/2007
```

The new data set is not quite the same as the Week data set that was exported in Section 3.2.1.

- The order of the columns was changed by the VAR statement in PROC PRINT (see Section 3.2.2).
- The variable names were changed to the labels by the LABEL option in PROC PRINT. Note that spaces were replaced by underscores to create valid SAS names from the labels.
- The format of the Date_and_Time variable was changed from DATETIME20. to DATE9.

You can verify the format change by right-clicking the data set in the Explorer window and selecting **View Columns**, as shown in Display 3-6.

Display 3-6: The Properties Window for the Week_Import Data Set

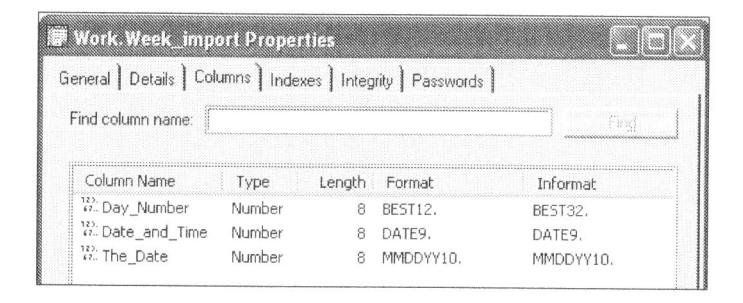

The first two changes were intentional, but the third was not. When you run PROC IMPORT, SAS makes some educated guesses about the appropriate variable names and formats to use in reading the file and creating the new data set. In this example, SAS guessed that DATE9. is the format that you would want for Date_and_Time. You can override these guesses by editing the code created by PROC IMPORT.

If you then select **Run→Recall Last Submit**, the code written by PROC IMPORT will appear in the Program Editor window.

```
data WORK.WEEK_IMPORT                         ;
    %let _EFIERR_ = 0; /* set the ERROR detection macro variable */
    infile 'c:\JES\output\ODS_Week_New.csv' delimiter = ',' MISSOVER DSD
lrecl=32767 firstobs=2 ;
        informat Day_Number best32. ;
        informat Date_and_Time DATE9. ;     ← Change DATE9. to DATETIME20.
        informat The_Date mmddyy10. ;
        format Day_Number best12. ;
        format Date_and_Time DATE9. ;       ← Change DATE9. to DATETIME20.
        format The_Date mmddyy10. ;
    input
                Day_Number
                Date_and_Time
                The_Date
    ;
    if _ERROR_ then call symputx('_EFIERR_',1);  /* set ERROR detection
macro variable */
    run;
```

You can edit this code to change the choices made by PROC IMPORT, and then rerun the code to get exactly what you want. If you change DATE9. to DATETIME20., as shown above in bold, and rerun the code, the new version of the Week_Import data set will have the DATETIME20. format for the Date_and_Time variable. If you need to import similar data sets on a regular basis, for example, a weekly update of a table based on the latest data, you can run PROC IMPORT once, edit the code if necessary, and save the code for subsequent imports. As long as the columns of the new data set are the same as in the original data set, the same code will work.

3.3.2 Fixing Data Import Problems

The import example in Section 3.3.1 worked very smoothly because the CSV file was exported from SAS and therefore was suitably formatted for import. Unfortunately, it is not always so easy. In many cases there will be one or more formatting problems in the file that you want to import, which will cause PROC IMPORT to fail. The example shown here illustrates some of the common problems with import files, and shows you how to fix them. This example uses the sample.csv file which was included in the input_data folder that you downloaded to your ~\JES\ folder. The first four rows of the **sample.csv** file are shown here.

Display 3-7: The First Four Rows of the sample.csv File

	A	B	C	D	E
1	Vendor	SN	Symptom	Install	Fail
2	Duality	2070481	ECC errors	06/09/06	07/29/2006 9:37 EST
3	Duality	2070109	Disk not spin up	06/09/06	07/01/2006 8:52 EST
4	Duality	2070967	No Power	06/09/06	11/25/2006 9:14 EST

Before you get started, select **Edit→Clear All** in the Log window because you will be searching the Log window for clues to fix the problems that you will encounter. Then run the next code sample to import sample.csv.

```
PROC IMPORT DATAFILE="&JES.input_data/sample.csv"
  OUT =Sample_Import REPLACE;
RUN;
```

When you run this code, an error message is written to the Log window.

```
ERROR: Import unsuccessful.  See SAS Log for details.
NOTE: The SAS System stopped processing this step because of errors.
```

In order to find out what went wrong, you should look at

- The "Invalid Data" messages in the Log window.
- The corresponding records in the CSV file that you are trying to import.
- The SAS code which was written by PROC IMPORT.

3.3.2.1 "Invalid data" Messages in the Log Window

Scroll to the top of the Log window, and then scroll down again until you reach the first "Invalid data" note.

```
NOTE: Invalid data for Fail in line 2 37-56. ❶
RULE:     ----+----1----+----2----+----3----+----4----+----5----+----6----+----7----+----8--
--+--
2      Duality,2070481,ECC errors,6/9/2006,07/29/2006  9:37 EST 56 ❷
Vendor=Duality SN=2070481 Symptom=ECC errors Install=06/09/2006 Fail=. _ERROR_=1 _N_=1 ❸
```

    ❶ The NOTE tells you that the problem is the value of Fail in line 2.

    ❷ This line tells you exactly what it found in line 2 of the CSV file (cf row 2 in Display 3-7).

    ❸ This line tells you what values were assigned to each variable. The value assigned to Fail was '.' which means that SAS was expecting a numeric value, but didn't find one, so it substituted with a '.' (period), which represents a missing numeric value.

3.3.2.2 The Corresponding Records in the CSV File

The "Invalid data" message tells you that the first problem is in line 2, so you should look in row 2 of the CSV file. You can see in Display 3-7 that the value of Fail in row 2 is '07/29/2006 9:37 EST'.

3.3.2.3 The SAS Code Written by PROC IMPORT

To see the code written by PROC IMPORT, first clear the Program Editor window, or open a new Program Editor window, and then select **Run→Recall Last Submit**. The code shown below will appear in the Program Editor window, except for the bold text, which contains the recommended fixes discussed below.

```
data WORK.SAMPLE_IMPORT                         ;
    %let _EFIERR_ = 0; /* set the ERROR detection macro variable */
    infile 'c:\JES\input_data\sample.csv' delimiter = ',' MISSOVER DSD
lrecl=32767 firstobs=2 ;
       informat Vendor $7. ;
       informat SN best32. ;
       informat Symptom $22. ;
       informat Install mmddyy10. ;
       informat Fail mmddyy10.     <-- Change to INFORMAT Fail $25.;
       format Vendor $7. ;
       format SN best12. ;
       format Symptom $22. ;
       format Install mmddyy10. ;
       format Fail mmddyy10. ;     <-- Change to FORMAT Fail $25.;
       input
               Vendor $
               SN
               Symptom $
               Install
               Fail                <-- Change to Fail $;
   ;
   if _ERROR_ then call symputx('_EFIERR_',1);  /* set ERROR detection
macro variable */
    run;
```

If you compare the "Invalid data" message, the CSV file, and the import code, the problem should be clear:

- PROC IMPORT is trying to read the Fail variable with the MMDDYY10. informat (see Section 2.4.2).
- The value of Fail in row 2 is '07/29/2006 9:37 EST' which does not match the MMDDYY10. Format because of the '9:37 EST' at the end.

The easiest way to fix this problem is to first read in Fail as a character variable, and then use the character functions described in Section 2.4.3 to extract a numeric date value. To fix the import code, change MMDDYY10. to $25. in the INFORMAT and FORMAT statements for Fail, and add a $ after Fail in the INPUT statement to specify that Fail is a character variable. The $25. format accepts up to 25 characters, which is more than enough to contain values like '07/29/2006 9:37 EST'. These changes are shown above in bold.

Edit the code in your Program Editor window by making the changes shown above, and then proceed to the next step.

Before you run the new code, clear the Log window again (**Edit → Clear All**) so that you can easily find any new error messages. Then run the code and you will find more "Invalid data" messages in the Log window, this time for the SN variable.

```
NOTE: Invalid data for SN in line 202 11-18. ❶
RULE:     ----+----1----+----2----+----3----+----4----+----5----+----6----+----7----+--
202   Empirical,01AB0600,Blank Screen,6/2/2006,10/11/2006  1:21 MT 60 ❷
Vendor=Empiric SN=. Symptom=Blank Screen Install=06/02/2006 Fail=10/11/2006  1:21 MT ❸
```

❶ The NOTE tells you that the problem is the value of SN in line 202.

❷ This line tells you exactly what it found in line 202 of the CSV file. Note that the second comma-separated element is '01AB0600'.

❸ This line tells you what values were assigned to each variable. The value assigned to SN was '.' which means that SAS was expecting a numeric value, but didn't find one, so it substituted with a '.', which represents a missing numeric value.

Now open the sample.csv file and scroll down to line 202.

Display 3-8: Selected Rows of the sample.csv File

	A	B	C	D	E
1	Vendor	SN	Symptom	Install	Fail
2	Duality	2070481	ECC errors	6/9/06	07/29/2006 9:37 EST
3	Duality	2070109	Disk not spin up	6/9/06	07/01/2006 8:52 EST
200	ChiTronix	1000399	ECC errors	6/8/06	08/18/2006 7:50 PST
201	ChiTronix	1000930	No Power	6/16/06	11/06/2006 17:39 PST
202	Empirical	01AB0600	Blank Screen	6/2/06	10/11/2006 1:21 MT
203	Empirical	01AB0036	Disk not spin up	6/1/06	06/30/2006 0:14 MT

By default, PROC IMPORT looks at only the first 20 rows of the file in order to guess the appropriate variable type and format. In this case the first 201 values of SN look like numbers, so SAS decided to use a numeric format for SN. But, starting in row 202, there are some alphameric values of SN. You can fix this by changing the numeric informat and format of SN to a character informat and format as shown below.

```
informat SN best32. ;      ← Change to INFORMAT SN $8.;
format SN best12. ;        ← Change to FORMAT SN $8.;
input
SN                         ← Change to SN $
```

After you have made these changes, clear the Log window and run the code once again. This time the import appears to be successful as there are no error messages in the Log window. However, if you open the new data set, Sample_Import, and scroll down, you will see that starting in line 101, the value of Vendor has been truncated to only seven characters—"ChiTron" instead of "ChiTronix".

```
Duality  2070480 No Power          06/08/2006 11/13/2006  8:22 EST
Duality  2070152 Disk not spin up  06/16/2006 06/20/2006 17:29 EST
ChiTron  1000362 No Power          06/17/2006 11/14/2006 17:53 PST
ChiTron  1000183 Disk not spin up  06/16/2006 06/30/2006 17:29 PST
```

This happened because, after finding Vendor = "Duality" in the first 20 rows, SAS guessed that seven characters were enough to contain the Vendor variable, and decided to use the $7. INFORMAT. You can fix this by changing the INFORMAT and FORMAT of Vendor to $9. , which is enough for all of your vendor names.

The code shown next includes the changes to fix all of these import problems.

```
     data SAMPLE_IMPORT                                         ;
        %let _EFIERR_ = 0; /* set the ERROR detection macro variable */
        infile 'c:\JES\input_data\sample.csv' delimiter = ',' MISSOVER DSD
lrecl=32767 firstobs=2 ;
           informat Vendor $9. ;
           informat SN $8. ;
           informat Symptom $22. ;
           informat Install mmddyy10. ;
           informat Fail $25. ;
           format Vendor $9. ;
           format SN $8. ;
           format Symptom $22. ;
           format Install mmddyy10. ;
           format Fail $25. ;
        input
                   Vendor $
                   SN $
                   Symptom $
                   Install
                   Fail $
        ;
        if _ERROR_ then call symputx('_EFIERR_',1);  /* set ERROR detection
macro variable */
        run;
```

The final step is to use the INPUT and SCAN functions (see Section 2.4.3) to extract a numeric date, Fail_Date, from the character variable Fail.

```
DATA Sample_Import; SET Sample_Import;
  FORMAT Fail_Date MMDDYY10.;
  Fail_Date = INPUT(SCAN(Fail, 1, " "), MMDDYY10.); ❶
RUN;
```

❶ The SCAN function extracts the date part of the Fail variable, for example "07/29/2006" in the first row, and then the INPUT function converts this character variable to a valid SAS date.

Note that two of the problems you had to fix were caused by the fact that, by default, SAS only looks at the first 20 rows before guessing the appropriate variable type and format. You can use a GUESSINGROWS statement to override the default.

```
PROC IMPORT DATAFILE="&JES.input_data\sample.csv"
  OUT =Sample_Import_2 REPLACE;
  GUESSINGROWS=500;
RUN;
```

In this code, the GUESSINGROWS=500 statement directs SAS to look at the first 500 rows before deciding on the variable type and format. If you had used this statement in the first place, SAS would have chosen the correct formats for SN and Vendor, and you would have had to fix only the problem with the Fail format.

3.3.3 Problems with Control Characters

Even when the import is successful, there might be problems lurking if the CSV file contains any control characters, such as carriage return characters or tab characters. These characters might or might not cause problems, depending on how you use the data set. One common problem arises when you try to re-export the data set to a CSV file. The next code sample exports the Sample_Import data set to the sample_export.csv file.

```
PROC EXPORT DATA=Sample_Import
   OUTFILE ="&JES.output/sample_export.csv" REPLACE;
RUN;
```

After you run this code, go to your ~\JES\output\ folder and open the new CSV file. You will see that the second line from your data set was split across two lines of the CSV file, as shown in Display 3-9. If you scroll down, you will see many other examples of the same problem. These line splits are caused by the fact that there is a carriage return character after the word "not" in "Disk not spin up".

Display 3-9: The sample_export.csv File Opened in MS Excel

	A	B	C	D	E	F
1	Vendor	SN	Symptom	Install	Fail	Fail_Date
2	Duality	2070481	ECC errors	6/9/06	07/29/2006 9:37 EST	7/29/06
3	Duality	2070109	Disk not			
4	spin up		6/9/06	07/01/2006 8:52 EST	7/1/06	
5	Duality	2070967	No Power	6/9/06	11/25/2006 9:14 EST	11/25/06
6	Duality	2070216	No Power	6/20/06	12/08/2006 21:09 EST	12/8/06

You can use the COMPRESS function (Section 2.4.3) to remove control characters as shown here:

```
DATA Sample_Import_OK; SET Sample_Import;
   Symptom=COMPRESS(Symptom, , 'WK'); ❶
RUN;
PROC EXPORT DATA=Sample_Import_OK ❷
   OUTFILE ="&JES.output/sample_export_ok.csv" REPLACE;
RUN;
```

❶ The COMPRESS function removes control characters from the Symptom variable. The "WK" modifier means keep (code 'K') all printable characters (code 'W') and remove all other characters, including carriage return, line feed, tab, etc. from the Symptom variable.

❷ PROC EXPORT writes the new Sample_Import_OK data set to the sample_export_ok.csv file.

After you run this code, open the new sample_import_ok.csv file in a spreadsheet to verify that the problem has been fixed.

3.3.4 A Trial-and-Error Process for Importing CSV Files

It is likely that many of the CSV files you need to import will contain problems that are similar to those described in the previous sections, and that causes PROC IMPORT to fail on the first attempt. But these problems are easily fixed by following the process summarized below.

1. Try PROC IMPORT.
   - Clear the Log window with the **Edit → Clear All** command.
   - Run PROC IMPORT.
     - Use a GUESSINGROWS statement to force SAS to look at all rows before guessing the appropriate data type and format.
   - If there are errors in the Log window, proceed to Step 2, otherwise go on to Step 3.
2. Fix problems and rerun the import code.
   - Select **Edit → Clear All** to clear the Program Editor window, or open a new window.
   - **Run→Recall Last Submit** to bring back the code created by PROC IMPORT.
   - Search the Log window for the first "Invalid data" problem.
   - Open the CSV file and scroll to the line with the first problem.
   - Fix the code in the Program Editor window. Simply changing a numeric format to a character format fixes most problems.
   - Close the CSV file, clear the Log window, and submit the new code.
   - If you still have error messages, repeat Step2; otherwise proceed to Step 3.
3. After the import runs without error messages:
   - Open the data set and check for possible truncation of character variables.
     - This should not happen if you use a GUESSINGROWS statement as recommended above.
     - Fix the code and rerun as required.
   - If you used a character format to read in numeric data, create the numeric variables as required. (Sections 2.4.3 and 3.3.2 include examples of using the SCAN and INPUT functions to extract a SAS date from a character string.)
   - Use COMPRESS(*variable*, , 'WK') to remove lurking control characters as required.
   - Save the code in case you need to read an updated version of the same CSV file.

3.4 More Than Enough

The subject of exchanging data between SAS and various MS Office products is a high priority with SAS users and developers alike. You've seen some of the results already with the ODS Tagsets.MSOffice2K destination. But the way of the future is XML—an obvious choice as it allows SAS to write to the published XML Spreadsheet format ("XML Spreadsheet Reference," Microsoft Developer Network (http://msdn.microsoft.com), Microsoft Corp).

3.4.1 Using the ExcelXP Tagset to Export SAS Data Sets

The ExcelXP tagset builds output based on the XML Spreadsheet format, and gives you the ability to control many aspects of the final appearance of the exported spreadsheet. In particular, you can export two or more tables into different worksheets of the same workbook. The code for this section is in ~\JES\sample_code\ch_3 \more.sas. The first code sample creates an XML file, Fail_Data_1.xml, containing the data from two of the SAS data sets, Units_U and Fails, created in the examples in Section 2.7 and 2.8

```
%INCLUDE "&JES.sample_code/ch_2/sort_merge.sas"; ❶
DATA Fails; SET JES.Fails(RENAME=(Loc=Position)); RUN;
ODS TAGSETS.ExcelXP FILE="Fail_Data_1.xml" PATH="&JES.output"; ❷
  TITLE "Units Installed"; FOOTNOTE "data collected 08/01/06"; ❸
  PROC PRINT DATA=Unts_U LABEL NOOBS; RUN;
  TITLE "Units Failed"; FOOTNOTE "data collected 09/12/06";
  PROC PRINT DATA=Fails NOOBS; RUN;
ODS TAGSETS.ExcelXP CLOSE; ❹
```

❶ The %INCLUDE statement re-creates the Units_U and Fails data sets from Sections 2.7 and 2.8, and the DATA step renames the Loc variable to fix the problem noted in Section 2.8.

❷ The ODS TAGSETS.ExcelXP statement specifies that the ExcelXP Tagset will be used to write the Fail_Data_1.xml file to the ~/JES/output folder.

❸ PROC PRINT is used to print the Units_U and Fails data sets.

❹ This line closes the ODS output destination.

When you run this code, the Fail_Data_1.xml file is written to your ~\JES\output folder. Display 3-10 shows the two different worksheets in Fail_Data_1.xml.

Display 3-10: The Fail_Data_1.xml File Opened in MS Excel

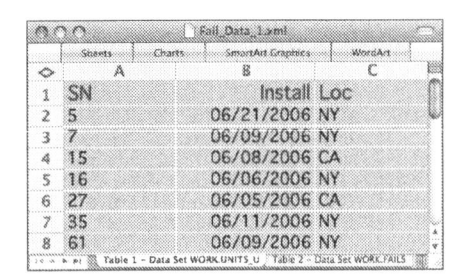

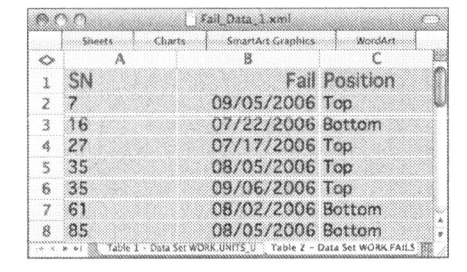

Note that the leading zeros in the SN variable have been lost, e.g. SN= '5' instead of '0005' as in the Units_U data set (see Section 2.7.2). This can be fixed with a STYLE statement as shown in the next example.

3.4.1.1 Titles, Footnotes, Styles, and Formatting

In addition to writing data to multiple worksheets, you can exercise detailed control over the formatting of the XML output. The next code sample shows how to retain the leading zeros in the SN variable and apply a SAS style to the output.

```
ODS TAGSETS.ExcelXP STYLE=Banker ❶ FILE="Fail_Data_2.xml"
PATH="&JES.output"
   OPTIONS(EMBEDDED_TITLES='YES' EMBEDDED_FOOTNOTES='YES'); ❷
   TITLE "Units Installed"; FOOTNOTE "data collected 08/01/06";
   PROC PRINT DATA=Unique LABEL NOOBS;
      VAR SN   / STYLE={tagattr="\0000"}; ❸
      VAR Install Loc;
   RUN;
   TITLE "Units Failed"; FOOTNOTE "data collected 09/12/06";
   PROC PRINT DATA=Fails NOOBS;
      VAR SN   / STYLE={tagattr="\0000"};
      VAR Fail Position;
   RUN;
ODS TAGSETS.ExcelXP CLOSE;
```

❶ The STYLE=Banker option specifies that the SAS predefined "Banker" style will be used to control several attributes of the spreadsheet, including font size and color, background color, etc.

❷ The OPTIONS statement specifies that titles and footnotes be written to the spreadsheet.

❸ The STYLE={tagattr="\0000"} statements control the cell formatting to restore the leading zeros to the SN variable.

When you run the code, the Fail_Data_2.xml file is written to your ~\JES\output folder. Display 3-11 shows the two different worksheets in Fail_Data_2.xml.

Display 3-11: The Fail_Data_2.xml File Opened in MS Excel

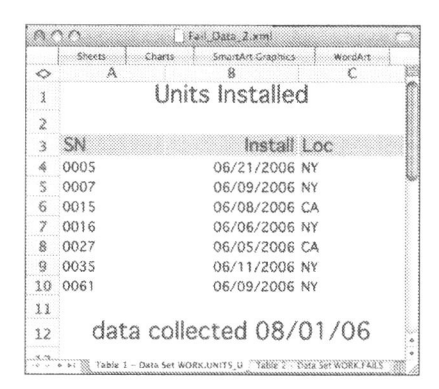

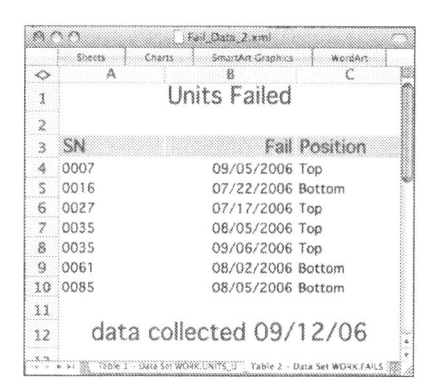

In addition to "Banker", SAS provides more than 40 other styles to choose from. If you run the PROC TEMPLATE code shown here, the list of available styles will be sent to the Output window. You can rerun the above code, replacing "Banker" with various other styles to see how they affect the output. You can also use PROC TEMPLATE, which is discussed in Chapter 7, to edit these styles or create your own.

```
PROC TEMPLATE; LIST STYLES; RUN;
```

3.4.1.2 Customizing the Worksheet Names

Sending SAS data sets to different worksheets in the same spreadsheet is great, but the automatically assigned sheet names, "Table 1—Data Set WORK.UNITS_U" are usually not the names you would have chosen. You can use the SHEET_NAME tagset option to specify the names to be used for each sheet.

```
ODS TAGSETS.ExcelXP FILE="&JES.output/Fail_Data_3.xml";
  ODS TAGSETS.ExcelXP OPTIONS(EMBEDDED_TITLES='NO'
EMBEDDED_FOOTNOTES='NO'); ❶
  ODS TAGSETS.ExcelXP OPTIONS(SHEET_NAME='Install Data'); ❷
  PROC PRINT DATA=Units_U LABEL NOOBS;
      VAR SN   / STYLE={tagattr="\0000"};
      VAR Install Loc;
  RUN;
  ODS TAGSETS.ExcelXP OPTIONS(SHEET_NAME='Fail Data'); ❸
  PROC PRINT DATA=Fails NOOBS;
      VAR SN   / STYLE={tagattr="\0000"};
      VAR Fail Position;
  RUN;
ODS TAGSETS.ExcelXP CLOSE;
```

❶ The EMBEDDED_TITLES and EMBEDDED_FOOTNOTES are set to 'NO'. The tagset options set in the previous example remain in effect until they are either turned off or set to another value.

❷ The SHEET_NAME='Install Data' option specifies the name of the next sheet to be created. The first PROC PRINT creates the first worksheet, with the name specified by SHEET_NAME.

❸ The SHEET_NAME='Fail Data' option specifies the name of the next sheet to be created. The second PROC PRINT creates the second worksheet, with the name specified by SHEET_NAME.

Display 3-12 shows the Fail_Data_3.xml file opened in MS Excel. Note that the sheet names are now 'Install Data' and 'Fail Data' as specified. From within Excel, you can select **File→Save As** to save the file as an ordinary Excel Workbook.

Display 3-12: The Fail_Data_3.xml File Opened in MS Excel

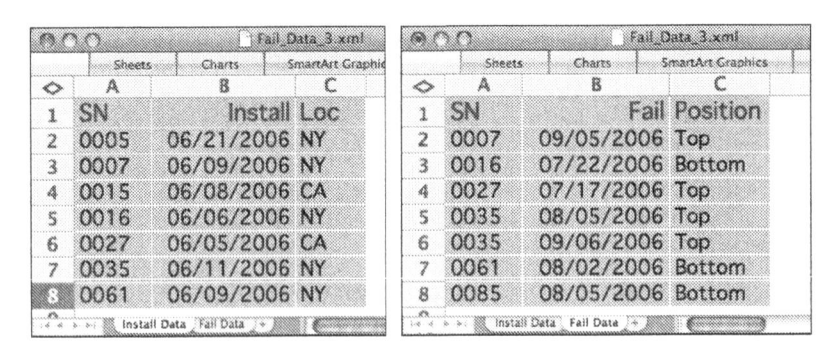

3.4.2 Using the %XLXP2SAS Macro to Import SAS Data

As you might expect, the Output Delivery System does not provide the capability to import data from spreadsheets. However, because SAS is increasingly XML-literate, and Excel spreadsheets can be saved as XML files, it is now possible to take advantage of XML format standards to simplify the import of data from spreadsheets to SAS data sets. One of the papers by Vincent DelGobbo (136-30.pdf) listed in the "Chapter Summary" section includes a description of the %***xlxp2sas*** macro, which can be used to import data from a multisheet Excel spreadsheet, saved as an XML file, into multiple SAS data sets. The paper asks you to download two files, the code for the %***xlxp2sas*** macro, and a special file called **excelxp.map** that simplifies the conversion process. Both of these have already been downloaded to your ~\JES\utility_macros folder, so you are ready to go. However, by the time you read this there will probably be a newer version of this macro. As with anything related to exchanging information between SAS and MS Office, it's a good idea to check the SAS Web site for the latest information and code.

The final code sample of this section uses this macro to re-import the data from the Fail_Data_3.xml spreadsheet created in the previous example. First save the file under a different name as follows:

- Open Fail_Data_3.xml in MS Excel.
- Select **File→Save As**, and save the file as Re_Import.xml to the same folder, ~\JES\output, making sure that the format is .xml, for example "XML Spreadsheet (*.xml)".
- Close the Re_Import.xml file. Otherwise you will get a "file in use" error when you try to import.

Then run the sample code.

```
%INCLUDE "&JES.utility_macros/xlxp2sas.sas"; ❶
%xlxp2sas(excelfile=&JES.output/Re_Import.xml,
          mapfile=&JES.utility_macros/excelxp.map); ❷
```

❶ The %INCLUDE statement defines the %xlxp2sas macro.

❷ This line runs the %xlxp2sas macro to create the Install_Data and Fail_Data data sets. The new data set names are taken from the corresponding worksheet names, with the spaces replaced by underscores so that the result is a valid SAS name

When you run this code, the Install_Data and Fail_Data datasets are created in your WORK library.

```
INSTALL_DATA                             FAIL_DATA
Obs    SN     Install       Loc         Obs    SN       Fail       Position
 1      5    06/21/2006     NY           1      7    09/05/2006       Top
 2      7    06/09/2006     NY           2     16    07/22/2006      Bottom
 3     15    06/08/2006     CA           3     27    07/17/2006       Top
 4     16    06/06/2006     NY           4     35    08/05/2006       Top
 5     27    06/05/2006     CA           5     35    09/06/2006       Top
 6     35    06/11/2006     NY           6     61    08/02/2006      Bottom
 7     61    06/09/2006     NY           7     85    08/05/2006      Bottom
```

Of course this macro will do nothing to resolve the kinds of data import problems discussed in Section 3.2.1. For example, you will note that you have lost the leading zeros in the SN variable. But if you are able to control the formatting of your spreadsheets to avoid such problems, this macro will provide a very useful and powerful way to bring in your data. **Note:** I was not able to get this macro to work in the UNIX environment.

3.5 Chapter Summary

3.5.1 Recap

After finishing this chapter, you should know how to

- Export data from SAS data sets into spreadsheets
  - using PROC EXPORT
  - using ODS CSVALL
  - using the MSOffice2K Tagset
- Import data from a spreadsheet using PROC IMPORT
- Fix the inevitable data import problems

3.5.2 For More Information

Books

Delwiche, Lora D., and Susan J. Slaughter. 2003. *The Little SAS® Book: A Primer, Third Edition.* Cary, NC: SAS Institute Inc.

- See Chapter 2 for the PROC IMPORT and other means for data import not covered here.
- See Chapter 9 for the PROC EXPORT and other means for data export not covered here.

SAS Conference Papers

These papers contain a lot more information about exchanging data between SAS data sets and various MS Office formats, most of which are compatible with the corresponding Star Office and Open Office formats. This is an area of active development, so be sure to check the latest SAS Global Forum papers for more up-to-date information. The filenames are shown in bold below—e.g., **[134-30.pdf]**.

Brown, David. 2005. "%sas2xl: A Flexible SAS® Macro That Uses Tagsets to Produce Complex, Multi-Tab Excel Spreadsheets with Custom Formatting." *Proceedings of the Thirtieth Annual SAS Users Group International Conference.* Cary, NC: SAS Institute Inc. Paper 092-30. **[092-30.pdf]**

DelGobbo, Vincent. 2003. "A Beginner's Guide to Incorporating SAS® Output in Microsoft Office Applications." *Proceedings of the Twenty-Eighth Annual SAS Users Group International Conference.* Cary, NC: SAS Institute Inc. Paper 52-28. **[p52-28.pdf]**

DelGobbo, Vincent. 2004. "From SAS® to Excel via XML." *Proceedings of the Seventeenth Annual NorthEast SAS Users Group Conference.* Baltimore, MD. **[DelGobbo_ExcelXML.pdf]**

DelGobbo, Vincent. 2005. "Moving Data and Analytical Results between SAS® and Microsoft Office." *Proceedings of the Thirtieth Annual SAS Users Group International Conference.* Cary, NC: SAS Institute Inc. Paper 136-30. **[136-30.pdf]**

DelGobbo, Vincent. 2008. "Tips and Tricks for Creating Multi-Sheet Microsoft Excel Workbooks the Easy way with SAS®." *Proceedings of the SAS Global Forum 2008 Conference.* Cary, NC: SAS Institute Inc. Paper 192-2008. **[192-2008.pdf]**

Gebhart, Eric A. 2005. "Tagset Spelunking and Cartography: Debugging and Exploring Tagsets with Battery-Powered Headlamps." *Proceedings of the Thirtieth Annual SAS Users Group International Conference.* Cary, NC: SAS Institute Inc. Paper 14-30. **[14-30.pdf]**

Gebhart, Eric. 2008. "The Devil Is in the Details: Styles, Tips, and Tricks That Make Your Microsoft Excel Output Look Great!" *Proceedings of the SAS Global Forum 2008 Conference*. Cary, NC: SAS Institute Inc. Paper 036-2008. **[036-2008.pdf]**

Parker, Chevell. 2003. "Generating Custom Excel Spreadsheets Using ODS." *Proceedings of the Twenty-Eighth Annual SAS Users Group International Conference*. Cary, NC: SAS Institute Inc. Paper 12-28. **[p12-28.pdf]**

Parker, Chevell. 2004. "SAS 9.1 MS OFFICE Integration." **[office91.pdf]**

Zender, Cynthia. 2004. "Markup 101: Markup Basics." *Proceedings of the Twenty-Ninth Annual SAS Users Group International Conference*. Cary, NC: SAS Institute Inc. Paper 2603-29. **[p2603-29.pdf]**

3.5.3 Exercises

1. Rerun the code from Section 2.2.3, "Creating SAS Data Sets," to create the POISSON_TABLE data set, and then export the data set to a spreadsheet:
   - using PROC EXPORT
   - using ODS CSVALL
   - using the MSOffice2K Tagset
   - using the ExcelXP Tagset

2. Re-import the POISSON_TABLE data from the spreadsheet that you just created to a SAS data set:
   - using PROC IMPORT
   - using the %XLXP2SAS macro

3. Import the import_this.csv file in your ~\JES\input_data folder as a SAS data set, following the process outlined in Section 3.3.4.

Chapter 4

Data Out, Data In—Relational Databases

- 4.1 Introduction 78
- 4.2 Using PROC SQL with SAS Data Sets 79
- 4.3 Using SAS with a Database Management System 84
- 4.4 More Than Enough 94
- 4.5 Chapter Summary 96

4.1 Introduction

Statistical data analysis often requires working with data stored in a relational database management system (DBMS). A DBMS is a lot like a SAS library, but with different terms for the various elements.

SAS	DBMS
Library	Database
Data Set	Table
Observation or Row	Record
Variable or Column	Field

Display 4-1 shows the three data sets that are used for the examples in this chapter. One could diagram a simple DBMS using exactly the same picture.

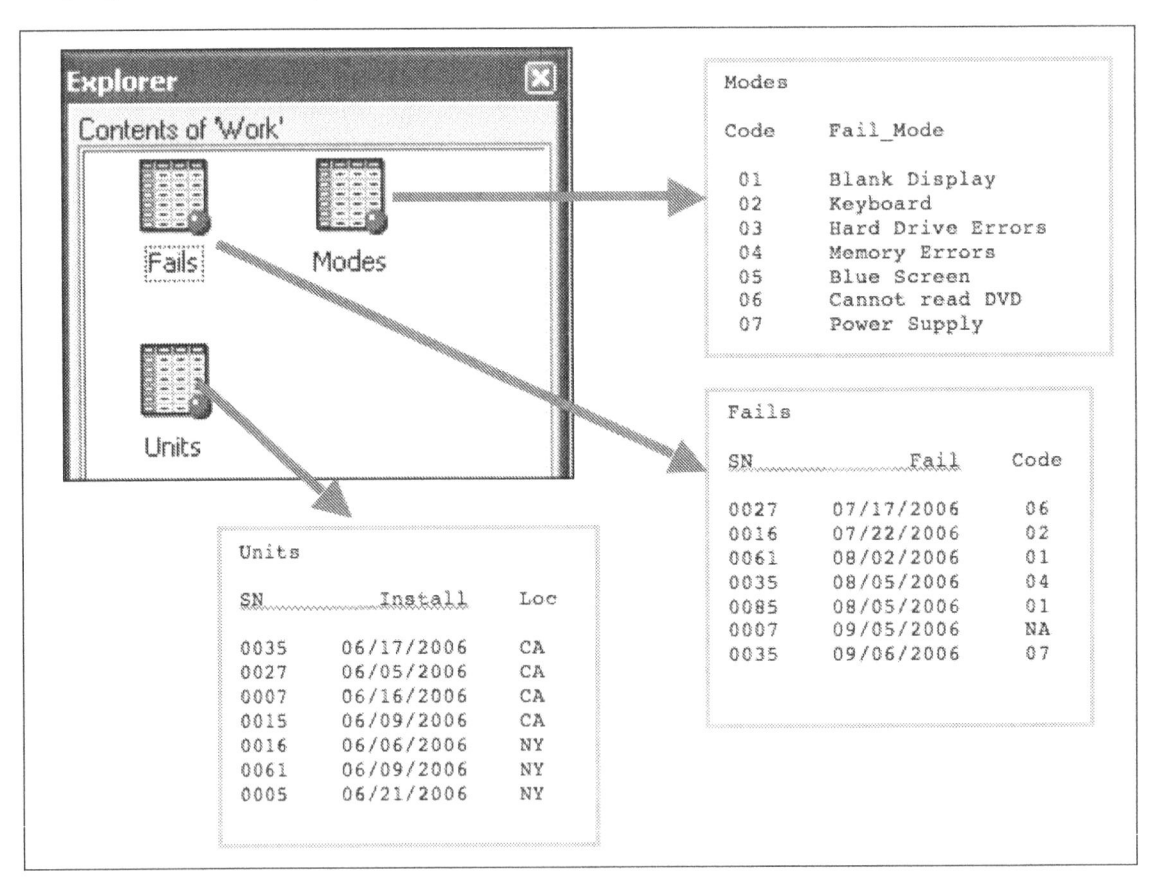

Display 4-1: The Units, Fails, and Modes Data Sets

```
Explorer
Contents of 'Work'

   Fails        Modes


   Units
```

```
Modes

Code    Fail_Mode

01      Blank Display
02      Keyboard
03      Hard Drive Errors
04      Memory Errors
05      Blue Screen
06      Cannot read DVD
07      Power Supply
```

```
Fails

SN          Fail          Code

0027        07/17/2006    06
0016        07/22/2006    02
0061        08/02/2006    01
0035        08/05/2006    04
0085        08/05/2006    01
0007        09/05/2006    NA
0035        09/06/2006    07
```

```
Units

SN          Install       Loc

0035        06/17/2006    CA
0027        06/05/2006    CA
0007        06/16/2006    CA
0015        06/09/2006    CA
0016        06/06/2006    NY
0061        06/09/2006    NY
0005        06/21/2006    NY
```

4.1.1 SQL and PROC SQL

Several different software packages, such as Oracle, DB2, and MySQL, can be used to construct and access database management systems. Fortunately, all of the DBMS software providers have agreed on an ANSI standard set of commands for extracting data. These commands, known as Structured Query Language, or SQL (pronounced like sequel), are based on simple English expressions that are very easy to understand and use, for example

```
select PRICE from PRICE_LIST where NAME = 'iPod'
```

The SAS SQL procedure supports the same set of ANSI standard commands, and makes it easy to extract records from a DBMS and save them to a SAS data set. PROC SQL can also manipulate SAS data sets using the same SQL syntax that is used with a DBMS, and can even be used to join rows of a DBMS to rows of a SAS data set.

Section 4.2 explains the use of PROC SQL to manipulate SAS data sets, and Section 4.3 explains the use of PROC SQL to extract records from a DBMS.

4.2 Using PROC SQL with SAS Data Sets

This section shows how to use PROC SQL to extract rows from a SAS data set and how to join rows from two or more data sets. The code for this section is in ~\JES\sample_code\ch_4 \proc_sql.sas. The first code sample creates the Units, Fails, and Modes data sets, which are used in the examples.

```
PROC DATASETS LIBRARY=WORK MEMTYPE=DATA GENNUM=ALL KILL; QUIT;
%INCLUDE "&JES.sample_code/ch_4/random_data.sas";
```

The first line deletes any data sets in the WORK library using the method described in Section 2.9.1, and the %INCLUDE statement creates the Units, Fails, and Modes data sets. After you run the code, the Explorer window will contain the three data sets shown in Display 4-1.

4.2.1 Extract Rows and Columns from a Data Set (or Table)

The next code sample uses PROC SQL to extract some rows and columns from the Units data set.

```
PROC SQL;  ❶
  CREATE TABLE NY_Recent AS  ❷
  SELECT SN, Install  ❸
  FROM Units  ❹
  WHERE Loc = "NY"
  AND Install >= '09JUN2006'd ;  ❺
QUIT;  ❻
```

❶ The PROC SQL statement begins the SQL procedure.

❷ The CREATE TABLE statement specifies the name of the SAS data set to be created, NY_Recent.

❸ The SELECT clause specifies the variables to be included, SN and Install. Note that the variables are separated by commas.

❹ The FROM clause specifies that the rows are to be selected from the Units data set.

❺ The WHERE clause, including the AND clause, specifies restrictions on the rows to extract.

❻ PROC SQL ends with a QUIT; statement rather than a RUN; statement.

When you run this code, the NY_Recent data set is created.

```
NY RECENT

Obs    SN         Install
1     0061      06/09/2006
2     0005      06/21/2006
```

You can get the same result with a DATA step using the methods explained in Section 2.6, as shown in the next code sample. The advantage of using PROC SQL can be seen in more complicated extracts and, more importantly, in its ability to extract data from a DBMS.

```
DATA NY_Recent; set Units;
  IF Loc="NY";
  IF Install >= '09JUN2006'd;
  KEEP SN Install;
RUN;
```

4.2.2 Joining Rows from Multiple Data Sets

The reason they call them "relational" databases is that the key to understanding, constructing, and using them is in the relationships among the fields in the various tables. For example, the Units and Fails tables are related by the SN variable, or field, in the sense that the records in each table that have the same value of SN are referring to the same physical object. So you can see that the unit with serial number (SN) 0035 was installed on June 17 in California, and later failed on August 5 with failure code "04" and again on September 6 with failure code "07".

```
Units                            Fails
SN        Install     Loc        SN         Fail       Code
0035      06/17/2006  CA         0027       07/17/2006 06
0027      06/05/2006  CA         0016       07/22/2006 02
0007      06/16/2006  CA         0061       08/02/2006 01
0015      06/09/2006  CA         0035       08/05/2006 04
0016      06/06/2006  NY         0085       08/05/2006 01
0061      06/09/2006  NY         0007       09/05/2006 NA
0005      06/21/2006  NY         0035       09/06/2006 07
```

This is a very practical and efficient way to store the data because it allows you, in this example, to store all of the information about the manufacturing and shipment of the units in one table, and all of the information about the subsequent failures and servicing in another table. But of course you need to bring these tables together in order to analyze the unit failure rates, and this is where SQL comes in.

4.2.3 Inner Joins

An inner join merges the rows from two or more data sets that have matching values of one or more variables. This example merges the Units and Fails data sets based on the common variable SN.

```
PROC SQL;
   CREATE TABLE Join AS
      SELECT Units.*, Fails.Fail, Fails.Code  ❶
      FROM Units, Fails  ❷
      WHERE Units.SN = Fails.SN;  ❸
QUIT;
```

❶ The SELECT clause uses a *DataSet.Variable* notation to specify variables to be returned. Units.* means take *all* variables from the Units data set. Fails.Fail means take the Fail variable from the Fails data set.

❷ The FROM clause lists the data sets from which rows will be extracted, separated by commas.

❸ The WHERE clause ensures that only rows with the same value of SN are joined in the result.

```
JOIN
Obs   SN      Install     Loc         Fail       Code
 1    0035    06/17/2006  CA          08/05/2006 04
 2    0035    06/17/2006  CA          09/06/2006 07
 3    0027    06/05/2006  CA          07/17/2006 06
 4    0007    06/16/2006  CA          09/05/2006 NA
 5    0016    06/06/2006  NY          07/22/2006 02
 6    0061    06/09/2006  NY          08/02/2006 01
```

The rows in either table that don't have a matching value of SN in the other table are left out of the result. For example, the rows with SN='0005' and '0015' in Units are left out because these SN do not appear in Fails, and the row with SN='0085' in Fails is left out because this SN does not appear in Units. Section 4.2.4 shows how to use outer joins to include these unmatched rows.

4.2.4 Outer Joins

A *left outer join* includes the matching rows from each data set, plus any unmatched rows from the left-hand data set, i.e., the first data set listed in the FROM statement. A *right outer join* includes the matching rows, plus any unmatched rows from the right-hand data set, i.e. the second data set listed in the FROM statement. A *full outer join* includes all of the matched and unmatched rows from both data sets. The next code samples illustrate each of these types of outer join. The first example is a left outer join.

```
PROC SQL;
   CREATE TABLE Join_Left AS
        SELECT Units.*, Fails.Fail, Fails.Code
        FROM Units LEFT JOIN Fails  ❶
        ON Units.SN = Fails.SN;  ❷
QUIT;
```

❶ The FROM clause of the inner join example in Section 4.2.3 is "FROM Units, Fails". For a left outer join, replace the comma by " LEFT JOIN".

❷ The WHERE clause of the inner join example in Section 4.2.3 is "WHERE Units.SN=Fails.SN". For a left outer join, the "WHERE" is replaced by "ON". Note that you can also have a WHERE clause in addition to the ON clause, for example:

```
ON Units.SN = Fails.SN
WHERE Units.Install > '09JUN2006'd;
```

When you run this code, the Join_Left data set is created. This data set includes all of the rows from the Units data set, with missing values of Fail and Code for the SN that do not appear in the Fails data set.

```
Join_Left
SN        Install      Loc         Fail       Code
0005      06/21/2006   NY                     .
0007      06/16/2006   CA          09/05/2006 NA
0015      06/09/2006   CA                     .
0016      06/06/2006   NY          07/22/2006 02
0027      06/05/2006   CA          07/17/2006 06
0035      06/17/2006   CA          09/06/2006 07
0035      06/17/2006   CA          08/05/2006 04
0061      06/09/2006   NY          08/02/2006 01
```

The next code sample performs a right outer join.

```
PROC SQL;
   CREATE TABLE Join_Right AS
        SELECT Units.*, Fails.Fail, Fails.Code
        FROM Units RIGHT JOIN Fails
        ON Units.SN = Fails.SN;
QUIT;
```

The syntax is the same as for the left outer join, except that "LEFT JOIN" is replaced by "RIGHT JOIN" in the FROM clause. When you run this code, the Join_Right data set is created. This data set includes all of the rows from the Fails data set, with missing values of the Install and Loc variables for the SN (0085) that does not appear in the Units data set.

```
Join_Right
 SN         Install      Loc         Fail     Code
 0007       06/16/2006   CA          09/05/2006   NA
 0016       06/06/2006   NY          07/22/2006   02
 0027       06/05/2006   CA          07/17/2006   06
 0035       06/17/2006   CA          09/06/2006   07
 0035       06/17/2006   CA          08/05/2006   04
 0061       06/09/2006   NY          08/02/2006   01
                .                    08/05/2006   01
```

Note that the SN variable is also missing for the row that does not appear in the Units data set because the SELECT clause requested only the Fail and Code variables from the Fails data set. You can use the COALESCE function of PROC SQL to include such missing values by rewriting the SELECT clause like this:

```
SELECT COALESCE(Units.SN, Fails.SN) AS SN,
   Units.Install, Units.Loc, Fails.Fail, Fails.Code
```

The COALESCE function returns the first nonmissing value in its argument list, as illustrated in the final example, which performs a full outer join.

```
PROC SQL;
  CREATE TABLE Join_Full AS
      SELECT COALESCE(Units.SN,Fails.SN) AS SN,
      Units.Install, Units.Loc, Fails.Fail, Fails.Code
      FROM Units FULL JOIN Fails
      ON Units.SN = Fails.SN;
QUIT;
```

The syntax is the same as for the left and right outer joins, except that "FULL JOIN" is used in the FROM clause, and the COALESCE function is used to ensure that the SN values found in either data set are retained. When you run this code, the Join_Full data set is created.

```
Join_Full
 SN         Install      Loc         Fail     Code
 0005       06/21/2006   NY              .
 0007       06/16/2006   CA          09/05/2006   NA
 0015       06/09/2006   CA              .
 0016       06/06/2006   NY          07/22/2006   02
 0027       06/05/2006   CA          07/17/2006   06
 0035       06/17/2006   CA          09/06/2006   07
 0035       06/17/2006   CA          08/05/2006   04
 0061       06/09/2006   NY          08/02/2006   01
 0085              .                 08/05/2006   01
```

You can get the same result using a MERGE statement in a DATA step, as described in Section 2.8.

```
PROC SORT DATA=Units; BY SN; RUN;
PROC SORT DATA=Fails; BY SN; RUN;
DATA Join_Full_2; MERGE Units Fails; BY SN; RUN;
```

However, the results are NOT the same if there are duplicate values of the join variable on both data sets. For example, if you join the JES.Units and JES.Fails data sets (Section 2.7) which each have two rows with SN='0035', you will get different results using PROC SQL than using a MERGE statement. See Section 5.9 for a comparison of the two methods.

4.3 Using SAS with a Database Management System

4.3.1 SAS/ACCESS Software

The SAS/ACCESS Interface to Relational Databases is actually a family of interfaces, one for each of several different DBMS systems, e.g. Oracle, DB2, MySQL, SYBASE, etc. Each interface is licensed separately, so you will only be able to access the types of DBMS included in your license. As noted in Section 1.8, you can run PROC SETINIT to see which types of DBMS are covered in your site license.

There are two ways to move data between SAS data sets and a DBMS:

- The *PROC SQL Pass-Through Facility* allows you to use PROC SQL to send SQL commands to a DBMS using the syntax expected by that DBMS, with extracted data returned to a SAS data set.
- A *SAS/ACCESS LIBNAME Statement* allows you to treat a DBMS as if it were a SAS library (see Section 2.3 for a discussion of SAS libraries).

The SAS/ACCESS LIBNAME approach is the newer method, and is recommended by SAS because it takes less code and does not require knowledge of SQL. However, they also point out that the Pass-Through Facility can accommodate extensions to ANSI Standard SQL which might be used by your DBMS, and will enable the DBMS to optimize queries and take advantage of indexes to process queries more quickly. I would suggest that you start by using the SAS/ACCESS LIBNAME method, but be prepared to use the Pass-Through Facility if you have queries which are not possible, or take too long, with the LIBNAME method.

The examples shown here will be based on the SAS/ACCESS Interface to Oracle. There are some minor variations in syntax for other types of DBMS—for example, DB2 requires the keyword DSN instead of PATH in the LIBNAME statement. To find the correct syntax for your DBMS, for example, MySQL, follow this path in the SAS online documentation.

> SAS/ACCESS Software → SAS/ACCESS for Relational Databases: Reference
> → DBMS-Specific Reference → SAS/ACCESS for MySQL
> → LIBNAME Statement Specifics for MySQL

4.3.2 Accessing a DBMS

Before you can access a DBMS, you will need to get permission and configure your system for access. The details will vary not only by DBMS type, but also by the practices in your organization. There will typically be a database administrator (DBA) who can help you through this process. The DBA will provide you with a user ID, password, and the name of the database you can access. You will also need to configure your computer to access the database. This might involve setting some environmental variables and adding files in your UNIX home directory, or running a configuration program on your Windows PC. When these steps are complete, you should first try to connect to the DBMS using a simple utility such as SQLPlus, DB2, or MySQL for an Oracle, DB2, or MySQL DBMS, respectively. That way you can iron out any non SAS problems with the connection before trying to access the database with SAS.

4.3.3 Exporting SAS Data Sets to a DBMS Using a LIBNAME Statement

This section shows how to use SAS to create a database in your DBMS with the tables that are used in the examples in the following sections. To do this, you will need to ask your DBA for permission to write to the DBMS. But if you can't get write permission, don't be too concerned. Just skip to Section 4.3.4 on extracting data from a DBMS, and try to follow the examples by doing similar extracts on the actual tables that you plan to work with. The code for this section is in ~\JES\sample_code\ch_4\jesdb.sas. If you do have write authority, and have tested your connection to the DBMS with a utility such as MySQL, then run the code sample to create some tables in your personal database.

```
%INCLUDE "&JES.sample_code/ch_4/random_data.sas"; ❶
%LET USER= my_user; %LET PW = my_password; %LET PATH=my_path; ❷
 LIBNAME Jesdb ORACLE USER=&user PW=&pw PATH=&path; ❸
DATA Jesdb.Units; SET Units; RUN; ❹
DATA Jesdb.Fails; SET Fails; RUN;
DATA Jesdb.Modes; SET Modes; RUN;
```

❶ The INCLUDE statement creates the Units, Fails, and Modes data sets.

❷ This line creates macro variables used in the next line. Before running the code, replace *my_user*, *my_password*, and *my_path* with the values that you got from your DBA.

❸ The LIBNAME statement is different for different DBMS types, so you might need to consult the SAS documentation for your DBMS. For example, the corresponding statement for DB2 is

"LIBNAME Jesdb DB2 UID=&user PWD=&pw DSN=&path;"

After you run the LIBNAME statement, you should see a message in the Log window telling you that "Libref JESDB was successfully assigned".

❹ The DATA steps add three tables to your new database.

When you run this code, the Explorer window will contain what looks like a SAS library with three data sets, as shown in Display 4-2. But in fact you are peeking into an Oracle database and looking at the three tables you just created. The only clue is the small blue dot on the "Jesdb" icon telling you that this is a database rather than a SAS library.

Display 4-2: The Jesdb Database Shown in the Explorer Window

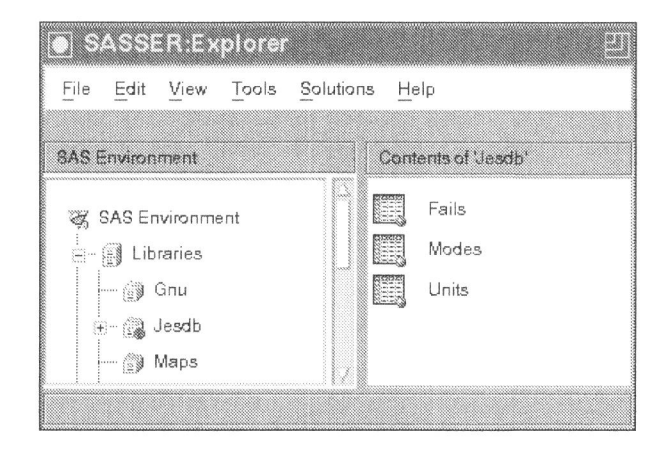

4.3.4 Importing Data from a DBMS Using the SQL Pass-Through Facility

The next example uses the PROC SQL Pass-Through Facility to create a SAS data set using a simple query to a DBMS. If you were not able to create your own DBMS tables in Section 4.3.3, you will need to change this code to match the table and fields for which you have read access. Try to include a date field in your extract. The code for this section is in ~\JES \sample_code\ch_4 \pass_thru.sas.

```
%LET user= my_user; %LET pw = my_password; %LET path=my_path;  ❶
PROC SQL;
  CONNECT TO ORACLE(USER=&user PW=&pw PATH="&path");  ❷
    CREATE TABLE NY_Recent AS SELECT * FROM CONNECTION TO ORACLE  ❸
    (
      SELECT SN, Install  ❹
      FROM Units
      WHERE Loc = 'NY'
      AND Install >= '09-JUN-2006'
    );
  %PUT SQLXMSG;  ❺
  DISCONNECT FROM ORACLE;
QUIT;
```

❶ Replace *my_user*, *my_password* and *my_path* with the values that you got from the DBA.

❷ The CONNECT TO statement establishes the connection to the DBMS. You will need to make changes to connect to a different type of DBMS. For example, for DB2 the syntax is

```
CONNECT TO DB2 (UID=&user PWD=&pw DSN="&path");
```

❸ The CREATE TABLE statement specifies that the results will go to the SAS data set NY_Recent.

❹ The lines within the parentheses are the SQL code which will be sent to the DBMS. Note that this is not SAS code, and does not follow SAS syntax. If you compare this code to the SAS code in Section 4.2.1, you will see that a few changes are required to convert from SAS to SQL syntax.
- Loc="NY" is changed to Loc='NY' because Oracle does not allow the double quotes.
- '09JUN2006'd is changed to '09-JUN-2006' to conform to the Oracle date format.

❺ The %PUT &SQLXMSG; statement is not required, but generates useful error messages if something goes wrong. The final two lines disconnect from the DBMS and end the SQL procedure.

When you run this code, the NY_Recent data set is created in your WORK library. If you right-click NY_Recent in the Explorer window and select **Open**, you will see that this version of NY_Recent is slightly different than the version you created using PROC SQL on SAS data sets in Section 4.2.1. The Install variable was stored and retrieved as a DateTime variable rather than a Date variable.

Display 4-3: The NY_Recent Data Set in a Viewtable Window

	SN	INSTALL
1	0061	09JUN2006:00:00:00
2	0005	21JUN2006:00:00:00

You can right-click NY_Recent and select **View Columns** to verify that Install is a DateTime variable.

Display 4-4: The NY_Recent Data Set in the Properties Window

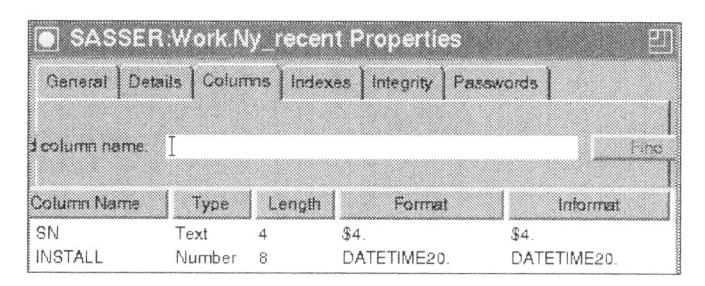

Column Name	Type	Length	Format	Informat
SN	Text	4	$4.	$4.
INSTALL	Number	8	DATETIME20.	DATETIME20.

This DateTime problem is very common in extracts from Oracle, and is easily fixed using the DATEPART function as discussed in Section 2.4.2. The next code sample changes Install to a Date variable.

```
DATA NY_Recent; SET NY_Recent;
   FORMAT Install DATE9.;
   Install=DATEPART(Install);
RUN;
```

The remaining code samples in pass_thru.sas use the Pass-Through Facility to re-create all of the joins shown in Section 4.2. Only a few examples are shown here.

```
PROC SQL;
   CONNECT TO ORACLE(USER=&user PW=&pw PATH="&path");
     CREATE TABLE Join_Left AS SELECT * FROM CONNECTION TO ORACLE
     (
       SELECT Units.*, Fails.Fail
       FROM Units, Fails
       WHERE Units.SN = Fails.SN(+)
     );
   %PUT SQLXMSG;
   DISCONNECT FROM ORACLE;
QUIT;
```

This code uses the Oracle specific (+) notation, which means match rows if possible, but don't exclude a row if you can't find it in this table. If you are joining three or more tables, then LEFT and RIGHT are not enough to cover all the possibilities, while the (+) notation can be used in joins of any number of tables.

```
PROC SQL;
   CONNECT TO ORACLE(USER=&user PW=&pw PATH="&path");
     CREATE TABLE Join_Full AS SELECT * FROM CONNECTION TO ORACLE
     (
       SELECT Units.*, Fails.Fail
       FROM Units FULL OUTER JOIN Fails
       ON  Units.SN = Fails.SN
     );
   %PUT SQLXMSG;
   DISCONNECT FROM ORACLE;
QUIT;
```

This code illustrates the Oracle Full Outer Join syntax, which was new in Oracle 9i.

4.3.5 Using SAS to Explore the Contents of a DBMS

Before you can write SAS code to access a DBMS, you need to know the names of the tables and fields that are available to you. This section provides code for creating SAS data sets that can serve as a guide to the tables and fields in your DBMS. The code is specific to Oracle, and will require syntax changes to work with other DBMS types. The code for this section is in ~\JES \sample_code\ch_4 \explore.sas.

4.3.5.1 Tables and Views

The first code sample creates a data set, DB_Tables, containing information about all the tables in your DBMS.

```
%LET user=my_user; %LET pw =my_password; %LET path=my_path;
 PROC SQL;
    CONNECT TO ORACLE(USER=&user PW=&pw PATH="&path");
    CREATE TABLE DB_Tables AS
    SELECT * FROM CONNECTION TO ORACLE
       (SELECT OWNER, TABLE_NAME, NUM_ROWS, AVG_ROW_LEN  FROM
ALL_TABLES);
    %PUT &SQLXMSG;
    DISCONNECT FROM ORACLE;
 QUIT;
```

The SELECT OWNER,…, FROM ALL_TABLES statement is Oracle specific. Your DBA should be able to provide the equivalent syntax for your DBMS.

```
DB_Tables

OWNER            TABLE_NAME            NUM_ROWS         AVG_ROW_LEN
smith            COST                    185746              26
smith            PRICE                    42564             145
rutledge         FAILS                        7              14
rutledge         UNITS                        7              14
rutledge         MODES                        7              27
```

The resulting data set, DB_Tables, includes the three tables you created, as well as all other tables for which you have read access, in this case two tables owned by 'smith'.

In addition to *tables*, a DBMS will usually have *views* which are like virtual tables. They don't really exist until you run a query, but are constructed on the fly from the tables when you run a query. The next bit of code creates the DB_Views data set, which is similar to DB_Tables, but contains a list of all the available views in your DBMS.

```
PROC SQL;
    CONNECT TO ORACLE(USER=&user PW=&pw PATH="&path");
    CREATE TABLE DB_Views AS
    SELECT * FROM CONNECTION TO ORACLE
       (SELECT OWNER, VIEW_NAME FROM ALL_VIEWS);
    %PUT &SQLXMSG;
    DISCONNECT FROM ORACLE;
 QUIT;
```

4.3.5.2 Fields

When you know the names of the tables and views, you can run another query to extract the fields in each table or view that you are interested in. The next code sample creates the Units_Fields data set, which contains information about each field in the Units table.

```
PROC SQL;
    CONNECT TO ORACLE(USER=&user PW=&pw PATH="&path");
    CREATE TABLE Units_Fields AS
    SELECT * FROM CONNECTION TO ORACLE
    (SELECT OWNER, TABLE_NAME, COLUMN_NAME, DATA_TYPE, DATA_LENGTH
          FROM ALL_TAB_COLUMNS   WHERE TABLE_NAME='UNITS' );
    %PUT &SQLXMSG;
    DISCONNECT FROM ORACLE;
QUIT;
```

The SELECT … FROM …WHERE statement is Oracle syntax that must be changed to be used with a different type of DBMS. Note that the Oracle syntax is case sensitive, so you must specify the TABLE_NAME as "UNITS", not "Units". The same query works for Tables or Views. The resulting data set gives you the name, data type, and length of each field.

```
Units_Fields

OWNER          TABLE_NAME    COLUMN_NAME    DATA_TYPE    DATA_LENGTH

rutledge       UNITS         SN             VARCHAR2     4
rutledge       UNITS         INSTALL        DATE         7
rutledge       UNITS         LOC            VARCHAR2     2
```

4.3.5.3 Records

Finally, it is often useful to get a quick preview of some actual data records, or rows, to give you a better feel for the content of the table or view. The next code sample writes the first three rows of Units to the data set Units_Rows.

```
PROC SQL;
    CONNECT TO ORACLE(USER=&user PW=&pw PATH="&path");
    CREATE TABLE Units_Rows AS
    SELECT * FROM CONNECTION TO ORACLE
    (SELECT * FROM Units    WHERE ROWNUM <= 3);
    %PUT &SQLXMSG;
    DISCONNECT FROM ORACLE;
QUIT;
```

The WHERE ROWNUM <= 3 clause is Oracle syntax for selecting the first three rows of the table.

Again, the same query can be used to extract rows from a Table or a View.

```
Units_Rows

SN                   INSTALL          LOC

0035        17JUN2006:00:00:00        CA
0027        05JUN2006:00:00:00        CA
0007        16JUN2006:00:00:00        CA
```

4.3.6 Importing Data from a DBMS Using a LIBNAME Statement

As noted earlier, the newer and easier way to extract records from a DBMS is to use a SAS/ACCESS LIBNAME statement. This makes a DBMS look like just another SAS library, and allows you to use SAS syntax in your extract code instead of having to learn the syntax required by your DMBS. The code for this section is in ~\JES \sample_code\ch_4 \libname.sas.

```
%LET USER=my_user; %LET pw =my_password; %LET path=my_path;
  LIBNAME Jesdb ORACLE USER=&user PW=&pw PATH=&path;
```

In the first line, replace *my_user*, *my_password* and *my_path* with the values that you got from your DBA. The LIBNAME statement will be different for different DBMS types. For example, the corresponding statement for DB2 is

```
"LIBNAME Jesdb DB2  UID=&user PWD=&pw DSN=&path;"
```

When you run this code, the Jesdb database shows up as a library in the Explorer window.

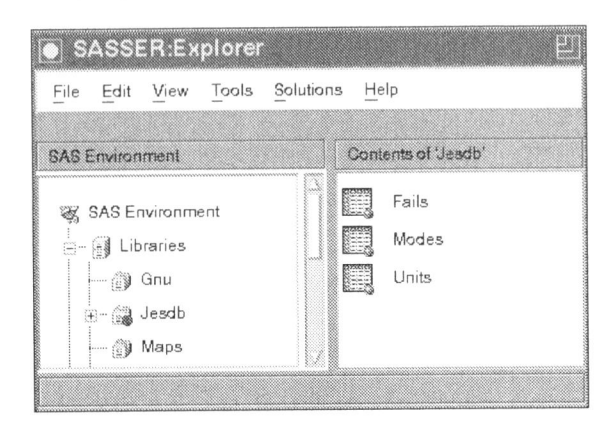

Display 4-5: The Jesdb Database Shown in the Explorer Window

The small blue dot on the Jesdb icon indicates that this is a connection to a DBMS instead of a SAS library. If you compare this to the DB_Tables data set in Section 4.3.5.1, you will notice that the Jesdb library includes only the tables which you created (OWNER='rutledge'). The tables owned by 'smith' do not show up. In order to see the 'smith' tables, you need to add the SCHEMA option to your LIBNAME statement as shown below.

```
LIBNAME Smith ORACLE USER=&user PW=&pw PATH=&path SCHEMA="smith";
```

The SCHEMA statement specifies that the Smith library should include all tables owned by 'smith'.

After running this statement, you will see another SAS library, Smith, in your Explorer window, which contains the tables owned by 'smith'.

You might or might not run into similar problems with your DBMS. Even if you are using Oracle, the SCHEMA='owner' option might not be required, depending on how the Oracle tables are set up. If the SCHEMA option is required, you can use the following steps when you need to access a new DBMS and want to use the SAS/ACCESS LIBNAME method:

1. Create a DB_Tables and/or DB_Views data set using the code shown in Section 4.3.5.

2. Look up the value of OWNER for any tables that you want to access.

3. Include the SCHEMA=*"owner"* option in your LIBNAME statement.

Once you get the LIBNAME statement working correctly, extracting data is very simple. Your sample code shows the LIBNAME version of all the queries shown in Section 4.3.4. Only selected examples are shown here to illustrate some of the differences among the various methods.

This code sample is almost identical to the PROC SQL code shown in Section 4.2.1 for extracting rows from a SAS data set.

```
PROC SQL;
        CREATE TABLE NY_Recent_1 AS
        SELECT SN, Install
        FROM Jesdb.Units
        WHERE Loc = "NY"
        AND DATEPART(Install) >= '09JUN2006'd;
QUIT;
```

The Loc = "NY" condition can use double quotes as this is SAS syntax, not Oracle syntax. The '09JUN2006'd is also SAS syntax. Compare this to Install >='09-JUN-2006' in the Pass-Through version of the query. You still need the DATEPART function because Install was stored as a datetime in the DBMS.

You can get the same result using a DATA step instead of PROC SQL.

```
DATA NY_Recent_2; SET Jesdb.Units;
    IF Loc="NY";
    IF DATEPART(Install) >= '09JUN2006'd;
    KEEP SN Install;
RUN;
```

Either way, you need to apply the DATEPART function to get the Install variable as a SAS date instead of a datetime.

```
DATA NY_Recent_1; SET NY_Recent_1;
    FORMAT Install DATE9.;
    Install=DATEPART(Install);
RUN;
```

The final example shows a simple join using the LIBNAME method, and illustrates the use of aliases for table names.

```
PROC SQL;
        CREATE TABLE Join AS
            SELECT u.*, f.Fail, f.Code
            FROM Jesdb.Units u, Jesdb.Fails f
            WHERE u.SN = f.SN;
QUIT;
```

The 'FROM Jesdb.Units u, Jesdb.Fails f' clause defines 'u' and 'f' as aliases for the Units and Fails tables respectively. These aliases can then be used in the SELECT and WHERE clauses. For example,

- 'u.*' instead of 'Units.*'
- 'f.Fail' instead of 'Fails.Fail'

The use of aliases for the table names is not too important in this small example, but would be very convenient if you had to extract 27 fields from a 'CURRENT_USA_PRICE_LIST' table.

4.3.7 Extracting Records That Match Values in a List

It is often necessary to extract records that match a list of values, for example, all records in the Units table for which the value of SN is 0027, 0016 or 0156. The SQL syntax for this query is shown in the first code sample. The code for this section is in ~\JES\sample_code\ch_4\list_match.sas.

```
%LET USER=my_user; %LET pw =my_password; %LET path=my_path;
 PROC SQL;
    CONNECT TO ORACLE(USER=&user PW=&pw PATH="&path");
      CREATE TABLE SN_Match AS SELECT * FROM CONNECTION TO ORACLE
      (
        SELECT * FROM Units
        WHERE SN IN ('0027','0016','0156')  ❶
      );
    %PUT SQLXMSG;
    DISCONNECT FROM ORACLE;
 QUIT;
```

❶ The SQL syntax uses an IN clause, with the items in the list enclosed in single quotes.

This code creates the SN_Match data set, which contains the requested records.

```
SN_Match
SN                INSTALL          LOC
0027       05JUN2006:00:00:00      CA
0016       06JUN2006:00:00:00      NY
```

This approach works fine for a list of 3 values of SN, but would not be practical for a list of 300 values of SN. Typically, you will have the list in some computer-readable form, for example a spreadsheet or a SAS data set. As long as you can import the list into a SAS data set, you can run the query using one of the following methods. The next bit of sample code (not shown here) creates the SN_Table data set.

```
SN_Table
Obs      SN
 1      0027
 2      0016
 3      0156
```

When the SN list is in a SAS data set, you can perform the same query by using the SAS/ACCESS LIBNAME method to join a DBMS table with a SAS data set.

```
LIBNAME Jesdb ORACLE USER=&user PW=&pw PATH=&path;
PROC SQL;
      CREATE TABLE SN_Match_1 AS
          SELECT Units.*
          FROM Jesdb.Units, SN_Table
          WHERE Units.SN = SN_Table.SN;
QUIT;
```

Caution! This code will first bring *all* the rows from the Units table into your SAS session, and then join them with SN_Table. This is not a problem here, but the query could take a very long time if Units had several million rows. You can avoid this problem by using either the DBKEY option or the MULTI_DATASRC_OPT option in your query, as illustrated in the next two examples.

```
PROC SQL;
    CREATE TABLE SN_Match_2 AS
        SELECT Units.*
        FROM Jesdb.Units (DBKEY=SN), SN_Table
        WHERE Units.SN = SN_Table.SN;
QUIT;
```

The (DBKEY=SN) causes PROC SQL to generate a new result set for each value of SN in SN_Table.

```
LIBNAME Jesdb ORACLE USER=&user PW=&pw PATH=&path
    MULTI_DATASRC_OPT=IN_CLAUSE;
PROC SQL;
    CREATE TABLE SN_Match_3 AS
        SELECT Units.*
        FROM Jesdb.Units, SN_Table
        WHERE Units.SN = SN_Table.SN;
QUIT;
```

The clause MULTI_DATASRC_OPT=IN_CLAUSE in the LIBNAME statement tells PROC SQL to construct an IN clause similar to the IN clause used in the simple join example.

Another alternative is to construct the IN clause yourself. The next code sample creates a macro variable, &snlist, which contains the list of distinct values of SN in SN_Table, in the format required for the DBMS SQL query. (Using PROC SQL to create macro variables is explained in Section 11.2.2.3.) Then you can use &snlist in your query as shown in the final code sample.

```
PROC SQL NOPRINT;
    SELECT DISTINCT TRANSLATE(QUOTE(TRIM(LEFT(SN))), "'", '"')
      INTO :snlist SEPARATED BY ','
      FROM SN_Table;
QUIT;
%PUT snlist=&snlist;
PROC SQL;
   CONNECT TO ORACLE(USER=&user PW=&pw PATH="&path");
      CREATE TABLE SN_Match_4 AS SELECT * FROM CONNECTION TO ORACLE
      (SELECT * FROM Units WHERE SN IN (&snlist));
   DISCONNECT FROM ORACLE;
QUIT;
```

The value of &snlist is written to the Log window by the %PUT statement.

```
11   %put snlist=&snlist;
snlist='0016','0027','0156'
```

All three of these methods are more efficient than the simple join shown in the second example, but it is difficult to predict which will work best for your particular queries. If you will be making similar queries on a regular basis, and run time is an issue, you can experiment with all three and select the method that best suits your needs. In the example shown below, the &SNLIST method is the clear winner in terms of run time, but for different data sets and record extractions, the results might be very different. Also, be sure to read the papers by Garth Helf that are referenced in the "Chapter Summary" section.

Sample Run Times for Joining 50 SN values to a Table with 2.7 Million Rows			
Method	DBKEY=SN	MULTI_DATASRC_OPT	SN in (&snlist)
Run Time: HH:MM:SS	1:13:08	00:08:50	00:03:32

4.4 More Than Enough

You can combine the methods from Sections 3.4 and 4.3.5 to create a spreadsheet listing all the tables and fields that you need to work within your DBMS. This guide can be very useful when you construct your SQL queries. The code for this section is in ~\JES \sample_code\ch_4 \dbms_guide.sas. The sample code creates a guide to the Jesdb database created in Section 4.3.3.

4.4.1 Step 1: Create the SAS Data Sets

Modify the first bit of sample code to extract information on the Tables and Fields of interest in your DBMS, as discussed in Section 4.3.5.

```
%LET USER=my_user; %LET pw =my_password; %LET path=my_path;
 PROC SQL;
    CONNECT TO ORACLE(USER=&user PW=&pw PATH="&path");
    CREATE TABLE DB_Tables as
    SELECT * FROM CONNECTION TO ORACLE
       (SELECT OWNER, TABLE_NAME, NUM_ROWS, AVG_ROW_LEN  FROM
ALL_TABLES);
    %PUT &SQLXMSG;
    DISCONNECT FROM ORACLE;
QUIT;
PROC SQL;
    CONNECT TO ORACLE(USER=&user PW=&pw PATH="&path");
    CREATE TABLE Units_Fields as
    SELECT * FROM CONNECTION TO ORACLE
       (SELECT OWNER, TABLE_NAME, COLUMN_NAME, DATA_TYPE, DATA_Length
           FROM ALL_TAB_COLUMNS   WHERE TABLE_NAME='UNITS' );
    %PUT &SQLXMSG;
    DISCONNECT FROM ORACLE;
QUIT;
PROC SQL;
    CONNECT TO ORACLE(USER=&user PW=&pw PATH="&path");
    CREATE TABLE Units_Rows as
    SELECT * FROM CONNECTION TO ORACLE
    (SELECT * FROM Units  WHERE ROWNUM <= 50  );
    %PUT &SQLXMSG;
    DISCONNECT FROM ORACLE;
QUIT;
----(Similar code to create Fails_Fields, Fail_Rows, Modes_Fields, and
Modes_Rows is omitted)----
```

4.4.2 Step 2: Export the SAS Data Sets to a Multi-Sheet Workbook

The next code sample uses the methods from Section 3.4.1 to export the new SAS data sets to a spreadsheet, myDBMS.xml.

```
ODS LISTING CLOSE;
ODS TAGSETS.ExcelXP FILE="myDBMS.xml" PATH="&JES.output" ;
ODS TAGSETS.ExcelXP OPTIONS(sheet_name='DB_Tables');
    TITLE "DB_Tables";    PROC PRINT DATA=DB_Tables    NOOBS; RUN;
ODS TAGSETS.ExcelXP OPTIONS(sheet_name='Units_Fields');
    TITLE "Units_Fields"; PROC PRINT DATA=Units_Fields NOOBS; RUN;
ODS TAGSETS.ExcelXP OPTIONS(sheet_name='Units_Rows');
    TITLE "Units_Rows";   PROC PRINT DATA=Units_Rows   NOOBS; RUN;
```

```
----(Similar code to export  Fails_Fields, Fail_Rows, Modes_Fields,
Modes_Rows omitted)----
ODS TAGSETS.ExcelXP CLOSE;
ODS LISTING;
```

4.4.3 Step 3: Customize Your Workbook

After you have created your spreadsheet, you can add comments to document, for example

- Which fields you need to extract.
- The meaning of each field.
- How the tables should be joined.

Display 4-6: The myDBMS.xml Database Guide Opened in MS Excel

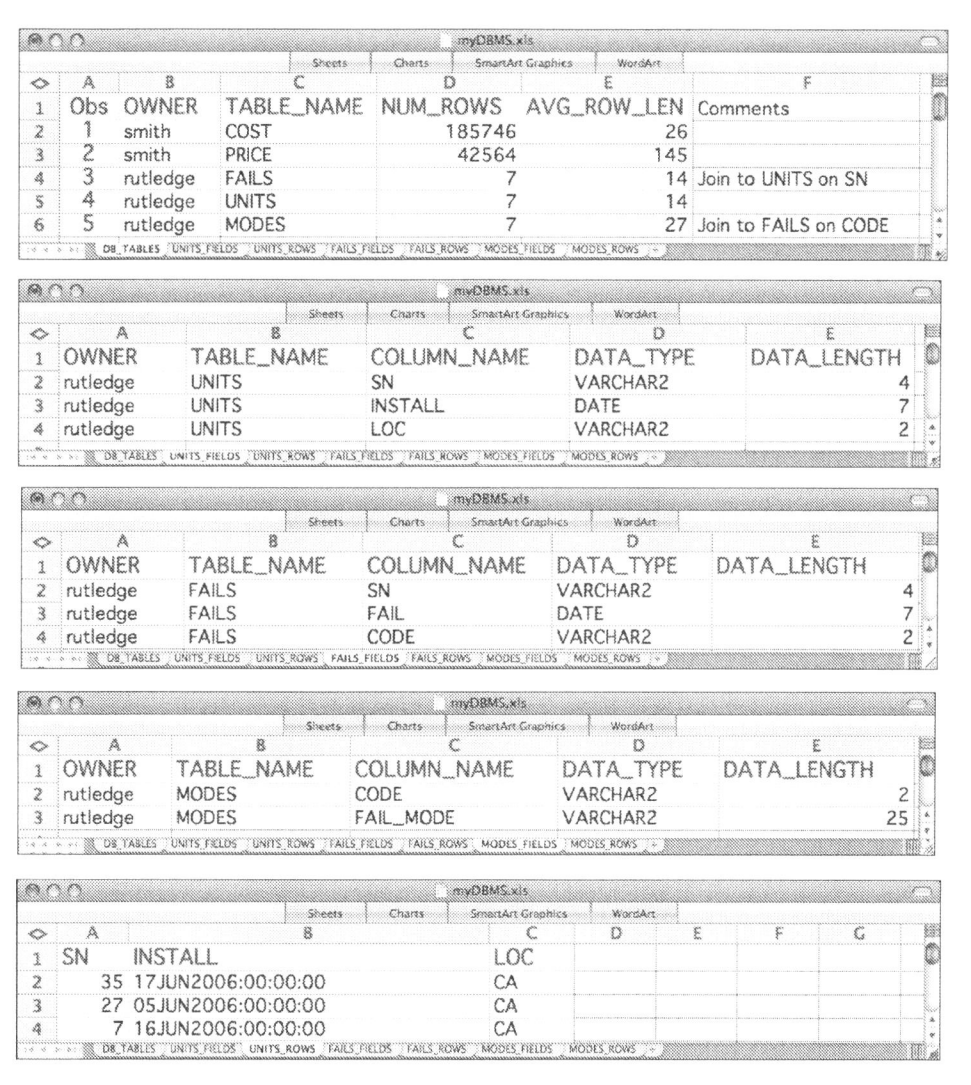

	Obs	OWNER	TABLE_NAME	NUM_ROWS	AVG_ROW_LEN	Comments
2	1	smith	COST	185746	26	
3	2	smith	PRICE	42564	145	
4	3	rutledge	FAILS	7	14	Join to UNITS on SN
5	4	rutledge	UNITS	7	14	
6	5	rutledge	MODES	7	27	Join to FAILS on CODE

OWNER	TABLE_NAME	COLUMN_NAME	DATA_TYPE	DATA_LENGTH
rutledge	UNITS	SN	VARCHAR2	4
rutledge	UNITS	INSTALL	DATE	7
rutledge	UNITS	LOC	VARCHAR2	2

OWNER	TABLE_NAME	COLUMN_NAME	DATA_TYPE	DATA_LENGTH
rutledge	FAILS	SN	VARCHAR2	4
rutledge	FAILS	FAIL	DATE	7
rutledge	FAILS	CODE	VARCHAR2	2

OWNER	TABLE_NAME	COLUMN_NAME	DATA_TYPE	DATA_LENGTH
rutledge	MODES	CODE	VARCHAR2	2
rutledge	MODES	FAIL_MODE	VARCHAR2	25

SN	INSTALL	LOC
35	17JUN2006:00:00:00	CA
27	05JUN2006:00:00:00	CA
7	16JUN2006:00:00:00	CA

The first worksheet shown in Display 4-6, DB_TABLES, shows all of the tables in the database, and includes some manually entered comments. The next three worksheets show the fields in each table. The last worksheet contains the first fifty rows of the Units table, but only the first three rows are shown here.

4.5 Chapter Summary

4.5.1 Recap

After finishing this chapter, you should know how to

- Use PROC SQL with SAS data sets:
  - Extract data from SAS data sets.
  - Perform inner and outer joins on SAS data sets.
- Use SAS/ACCESS software to extract data from a DBMS:
  - Using the PROC SQL Pass-Through Facility.
  - Using a SAS/ACCESS LIBNAME statement.
  - Extract records that match the values of a variable in a SAS data set.

4.5.2 For More Information

Books
Lafler, Kirk Paul. 2004. *PROC SQL: Beyond the Basics Using SAS®*. Cary, NC: SAS Institute Inc.

SAS Conference Papers
Droogendyk, Harry, and Faisal Dosani. 2008. "Joining Data: Data Step Merge or SQL?" *Proceedings of the SAS Global Forum 2008 Conference*. Paper 178-2008. **[178-2008.pdf]**

Feng, Ying. 2006. "The SQL Procedure: When and How to Use It?" *Proceedings of the Thirty-First Annual SAS Users Group International Conference*. Cary, NC: SAS Institute Inc. Paper 044-31. **[044-31.pdf]**

Lafler, Kirk Paul. 2006. "A Hands-On Tour Inside the World of PROC SQL." *Proceedings of the Thirty-First Annual SAS Users Group International Conference*. Cary, NC: SAS Institute Inc. Paper 114-31. **[114-31.pdf]**

The Lafler paper and book will show you how to use PROC SQL with SAS data sets. The following papers discuss the use of PROC SQL to access a DBMS. The first two papers discuss the %DBMSLIST macro that Garth Helf wrote to optimize queries that extract records which match the values in a SAS data set. The final method shown in Section 4.3.7 (SN in (&snlist)) was taken from Helf's paper. The full macro, which does a lot more, is included in your ~\JES\utility_macros folder (see Section 11.4). The paper by Levine also discusses techniques for optimizing SQL queries to relational databases.

Helf, Garth W. 2001. "Joining SAS® and DBMS Tables Efficiently." *Proceedings of the Twenty-Sixth Annual SAS Users Group International Conference*. Cary, NC: SAS Institute Inc. Paper 127-26. **[p127-26.pdf]**

Helf, Garth W. 2002. "Can't Relate? A Primer on Using SAS® with Your Relational Database." *Proceedings of the Twenty-Seventh Annual SAS Users Group International Conference*. Cary, NC: SAS Institute Inc. Paper 155-27. **[p155-27.pdf]**

Levine, Fred. 2001. "Using the SAS/ACCESS® LIBNAME Technology to Get Improvements in Performance and Optimizations in SAS/SQL® Queries." *Proceedings of the Twenty-Sixth Annual SAS Users Group International Conference*. Cary, NC: SAS Institute Inc. Paper 110-26. **[p110-26.pdf]**

4.5.3 Exercises

- Contact your local DBA and get permission to access the DBMS that contains the data you need.
- Use the methods from Section 4.3.5 to locate the Tables and Fields that contain the data you need. You might well need help from local experts to decipher the true meaning of various fields, and how tables are intended to be joined. It is rarely obvious.
- Use the method explained in Section 4.4 to create a personalized guide to the DBMS.
- Try writing some of your own queries.
- If you experience very long run times for your queries, consult your DBA to see if the SQL which SAS passes to the DBMS can be optimized. Also read the papers by Helf and Levine for clues on how to improve performance.

Chapter 5

Summarizing Your Data

- 5.1 Introduction 100
- 5.2 PROC MEANS 102
- 5.3 PROC TABULATE 112
- 5.4 PROC REPORT 114
- 5.5 PROC BOXPLOT 116
- 5.6 PROC ANOM 120
- 5.7 PROC UNIVARIATE 124
- 5.8 PROC FREQ 130
- 5.9 More Than Enough 132
- 5.10 Chapter Summary 134

5.1 Introduction

This chapter provides a brief introduction to the SAS procedures listed in Table 5-1, which can be used to summarize and report on the records in a data set. One of the most important objectives in the analysis of quality and reliability data is the detection of significant differences among subgroups of the total population of units. For example, you might need to know whether the units received from a particular vendor, or manufactured during a particular time period, are significantly worse than other units. The procedures discussed here can often be used as a first step for identifying and reporting on the differences among groups, and the examples in this chapter will focus on these kinds of tasks.

Table 5-1: SAS Procedures for Summarizing Data and Comparing Groups

Procedure	Description
PROC MEANS	Computes various statistics for numeric variables of a data set. Statistics are computed for all rows of the data set, and for any subset of rows defined by the values of one or more class variables.
PROC TABULATE	Computes statistics, as in PROC MEANS, but allows for more complex tabular output.
PROC REPORT	Computes many of the same statistics as PROC MEANS and PROC TABULATE, but provides a different set of options for formatting the results.
PROC BOXPLOT	Creates box plot representations of numerical variables.
PROC ANOM	Uses Analysis of Means to test for significant differences among group means.
PROC UNIVARIATE	Fits a Normal, Lognormal, Weibull, Gamma, Exponential, or Beta distribution to a sample of data and tests the goodness of fit. Creates various plots, including histograms, probability plots, and quantile-quantile plots, of the sample data and fitted distributions.
PROC FREQ	Performs chi-squared and other tests of association for contingency tables.

There is a considerable amount of overlap among the outputs of these procedures, but it is a good idea to become familiar with each so that you can select whichever is best suited to the task at hand. In addition to producing finished reports, PROC MEANS is an invaluable tool for processing raw data into a form that is suitable for use in the next step of your analysis.

5.1.1 Sample Component Data

The examples in this section are based on measurements of components purchased from each of three (fictitious) vendors: ChiTronix Components, Duality Logic, and Empirical Engineering. Each month a sample of the components from each vendor is selected for inspection, measurement, and life testing. First, the number of defects on each component is counted. Then the signal time delay of a test path is measured to characterize the performance of the component. It is known that resistance affects delay, so the resistance between two test points is also measured. Finally, each component is subjected to an accelerated life test consisting of 150 hours of operation at high temperature.

The JES.Results data set contains all of the test results for 2008. Check the Explorer window to make sure that the JES library is defined, and that it contains the Results data set. If not, follow the instructions in Section 1.5 to create the JES library. Several examples in this chapter use only the last three months of data from JES.Results, so, for convenience, a separate data set JES.Results_Q4 data set was created to contain only that data.

```
JES.Results (first five rows)

Vendor          Month       Mon         Unit        Delay       Resistance      Result
ChiTronix       1           2008-01     1           258         22.60           High Res
ChiTronix       1           2008-01     2           211         18.10           Pass
ChiTronix       1           2008-01     3           222         17.81           Pass
ChiTronix       1           2008-01     4           195         12.30           Low Res
ChiTronix       1           2008-01     5           219         18.45           Pass

                                                    Process         Process
    Fail    Lifetime    Survive     Defects         Date            Temp
    1       26          0           0               01JAN2008       25.0
    0       150         1           1               02JAN2008       25.4
    0       86          0           1               03JAN2008       26.5
    1       150         1           1               04JAN2008       27.3
    0       1           0           2               05JAN2008       26.5
```

The test month is recorded as the numeric variable Month and also as the character variable Mon. The Unit variable records the order in which the components were tested for each month and vendor. The specification for these components states that Resistance must be between 12.5 and 22.5. The Result variable is set to 'Low Res' if the Resistance is below 12.5, 'High Res' if Resistance is above 22.5, and 'Pass' otherwise. The Fail variable is then set to 1 for units that failed the specification, and 0 for units that did not fail. The Lifetime variable records the time to failure for those units that failed the life test, and the total test time (150 hours) for those that survived. The Survive variable is 1 for units that survived the life test, and 0 for those that did not. The Defects variable represents the number of defects found on each unit, and is independent of the Fail and Survive variables. The ProcessDate and ProcessTime variables represent the date when the unit was manufactured, and the recorded temperature during a critical processing step.

The first several examples use a smaller data set, JES.Results_1, with only eight rows, containing the test results for the first unit measured for each Vendor and Month, so that it will be easier to see exactly what is going on. Later examples use the JES.Results_Q4 data set, which contains all the data from the last three months of 2008. The full JES.Results data set is used in Chapters 6, 7, and 8.

```
JES.Results_1

Vendor          Month       Mon         Unit        Delay       Resistance      Result
ChiTronix       1           2008-01     1           258         22.60           High Res
ChiTronix       2           2008-02     1           213         18.86           Pass
ChiTronix       3           2008-03     1           182         16.72           Pass
Duality         1           2008-01     1           232         15.39           Pass
Duality         2           2008-02     1           214         14.40           Pass
Empirical       1           2008-01     1           193         16.37           Pass
Empirical       2           2008-02     1           197         17.00           Pass
Empirical       3           2008-03     1           204         16.34           Pass

                                                    Process         Process
    Fail    Lifetime    Survive     Defects         Date            Temp
    1       26          0           0               01JAN2008       25.0
    0       22          0           2               01FEB2008       26.0
    0       150         1           2               01MAR2008       26.3
    0       150         1           0               01JAN2008       27.5
    0       150         1           2               01FEB2008       27.8
    0       138         0           1               01JAN2008       26.2
    0       88          0           1               01FEB2008       26.2
    0       150         1           1               01MAR2008       27.8
```

5.2 PROC MEANS

PROC MEANS can be used to quickly and easily produce useful summaries of your data. In some cases the summary created by PROC MEANS will be enough to answer the question at hand, but it is also very useful in the intermediate steps of analysis to create data sets required for the later steps. Many of the examples in this book use PROC MEANS to prepare data for plotting or further processing. The code for this section is in ~\JES \sample_code\ch_5 \means_examples.sas.

5.2.1 Run PROC MEANS to Summarize Numeric Variables

The first example begins with the simplest possible invocation of PROC MEANS, which computes five statistics for each numerical variable in the JES.Results_1 data set.

```
PROC MEANS DATA=JES.Results_1;  ❶
RUN;  ❷
```

❶ The PROC MEANS statement requests an analysis of the data set JES.Results_1.

❷ The RUN; statement is required before the PROC is actually run.

When you run the code, this table is written to the Output window.

```
The MEANS Procedure
Variable        N        Mean         Std Dev       Minimum       Maximum
-----------------------------------------------------------------------------
Month           8     1.8750000      0.8345230     1.0000000     3.0000000
Unit            8     1.0000000              0     1.0000000     1.0000000
Delay           8   211.7541744     24.0399993   182.0980163   257.8451684
Resistance      8    17.2112644      2.5283688    14.3965808    22.6000000
Fail            8     0.1250000      0.3535534             0     1.0000000
Lifetime        8   109.2500000     56.6486666    22.0000000   150.0000000
Survive         8     0.5000000      0.5345225             0     1.0000000
Defects         8     1.1250000      0.8345230             0     2.0000000
ProcessDate     8       17558.63    25.1051361      17532.00      17592.00
ProcessTemp     8    26.6000000      1.0014276    25.0000000    27.8000000
-----------------------------------------------------------------------------
```

The output includes five statistics computed for each numeric variable in the input data set.

- N = the number of nonmissing values of the variable.
- Mean, Standard Deviation, Minimum, and Maximum, computed for the nonmissing values of each variable.

5.2.2 Use an OUTPUT Statement to Create an Output Data Set

It is usually more useful to send the results of this procedure to another data set than to the output window.

```
PROC MEANS DATA=JES.Results_1 NOPRINT;  ❶
  OUTPUT OUT=Tab_1;  ❷
RUN;
```

❶ The NOPRINT option, which requests that the results not be sent to the Output window, is not required, but it is a good idea to include this option to avoid cluttering up the Output window.

❷ The OUTPUT statement tells SAS to store the results in the Tab_1 data set.

```
Tab_1
_TYPE_    _FREQ_    _STAT_    Month      Unit     Delay    Resistance

  0         8       N         8.00000     8          8        8.00
  0         8       MIN       1.00000     1        182       14.40
  0         8       MAX       3.00000     1        258       22.60
  0         8       MEAN      1.87500     1        212       17.21
  0         8       STD       0.83452     0         24        2.53

                                                  Process    Process
 Fail     Lifetime   Survive     Defects           Date       Temp

   8       8.000     8.00000        8           09JAN1960    8.0000
   0      22.000     0.00000        0           01JAN2008   25.0000
   1     150.000     1.00000        2           01MAR2008   27.8000
   0     109.250     0.50000        1           27JAN2008   26.6000
   0      56.649     0.53452        1           26JAN1960    1.0014
```

The TAB_1 data set includes the same five statistics that were created in the first example, but in a transposed form. Note that two new variables, _TYPE_ and _FREQ_, are also created. These variables will be discussed in later examples, starting in Section 5.2.5.

5.2.3 Use a VAR Statement to Select Variables for Analysis

PROC MEANS computes the same five statistics for every numerical variable in the input data set, but this is usually not what you want. For example, the statistics for the Month variable in JES.Results_1 are not particularly useful. You can use a VAR statement to request statistics for only the variables of interest.

```
PROC MEANS DATA=JES.Results_1 NOPRINT;
   VAR Defects Resistance Fail;  ❶
   OUTPUT OUT=Tab_2;
   FORMAT Defects Fail 6.2;  ❷
RUN;
```

❶ The VAR statement tells SAS to compute statistics only for Defects, Resistance, and Fail.

❷ The FORMAT statement specifies that Defects and Fail should be formatted as 6.2. Otherwise, these variables would inherit the 8.0 format of the Defects and Fail variables in JES.Results_1, which would not give adequate precision for these statistics. Note that the mean values of Defects and Fail are rounded to 1 and 0 in Tab_1.

```
Tab_2
_TYPE_    _FREQ_    _STAT_    Defects    Resistance    Fail
  0         8       N           8.00        8.00       8.00
  0         8       MIN         0.00       14.40       0.00
  0         8       MAX         2.00       22.60       1.00
  0         8       MEAN        1.13       17.21       0.13
  0         8       STD         0.83        2.53       0.35
```

5.2.4 Use Keywords to Select the Required Statistics

By default, PROC MEANS computes five statistics: N, MIN, MAX, MEAN, and STD. But you can choose exactly the statistics you want by using keywords shown in Table 5-2. The code below illustrates how you might select the statistics that you need for each variable.

```
PROC MEANS DATA=JES.Results_1 NOPRINT;
  VAR Defects Resistance Fail;
  OUTPUT OUT=Tab_3
    N(Defects)=N  ❶
    SUM(Defects) = N_Def  ❷
    MEAN(Defects) = M_Def  ❸
    MEAN(Resistance) = M_Res
    SUM(Fail) = N_Fail
    MEAN(Fail) = P_Fail;
    FORMAT M_Def P_Fail 6.2;  ❹
RUN;
```

❶ The N(Defects)=N statement stores the number of nonmissing values of Defects in the variable N. The variable N will be used as the sample size for the Defects, Resistance, and Fail measurements in some of the analysis to follow. In this example there are no missing data for any of these variables so it will not cause a problem, but in general you should check for missing data in the other variables. For example, use N(Resistance) to compute the number of nonmissing values of Resistance.

❷ The SUM(Defects)=N_Def statement computes the sum of the Defects variable and stores the result as a new variable, N_Def.

❸ The MEAN(Defects)=M_Def statement computes the mean of the Defects variable and stores the result as a new variable, M_Def.

❹ The FORMAT statement specifies that M_Def and P_Fail should be formatted as 6.2. Otherwise, these means would inherit the 8.0 format of the variables Defects and Fail, which would not give adequate precision for these means.

This example uses very short names, e.g., N_Def, M_Res, for the computed statistics. This was done to ensure that the example tables would fit on the printed page. It is generally a good idea to use more descriptive names in your code, e.g., Num_Defects, Mean_Resistance. As noted in Section 1.6, variable names can be up to 32 characters.

Tab_3							
TYPE	_FREQ_	N	N_Def	M_Def	M_Res	N_Fail	P_Fail
0	8	8	9	1.13	17.21	1	0.13

Note that when specific statistics are requested, the shape of the output is different, with each statistic in a separate column (cf. Tab_2 on the previous page). This format is more convenient to work with, so you will usually want to use keywords to request statistics, even if the statistics you want are among the defaults.

Table 5-2 gives the default definition of each statistic available in PROC MEANS. The SAS online documentation contains alternate definitions for some of these statistics. The "X" variable in the definitions can be replaced by the name of any numeric variable in the input data set. The default statistics, N, MIN, MAX, MEAN, and STDDEV are highlighted.

Table 5-2: Statistics Available in PROC MEANS

	Descriptive Statistics
N	Number of nonmissing values of X
NMISS	Number of missing values
MIN	Minimum value of X
MAX	Maximum value of X
RANGE	Range = MAX - MIN
SUM	Sum of the values of X
SUMWGT	Sum of the values of a WEIGHT variable, if one has been defined
MEAN	Mean of the values = SUM/N
USS	Uncorrected sum of squares = Sum of the squares of X
CSS	Corrected Sum of Squares = Sum of the squares of [X-MEAN]
VAR	Variance of the values = CSS/[N-1]
STDDEV	Standard Deviation of X = Square root of VAR
STDERR	Standard Error of the Mean = STDDEV/SQRT[N]
CV	Percent Coefficient of Variation = 100*STDDEV/MEAN
SKEWNESS	Skewness = Third central moment of X
KURTOSIS	Kurtosis = (approximately) [Fourth central moment of X] – 3. See the online documentation for the exact formula.
	Confidence Intervals and Hypothesis Tests for the Mean of X
	Confidence Intervals—based on Student's t statistics with N-1 degrees of freedom. The default value of ALPHA is .05.
CLM	Two sided 100(1-ALPHA) percent confidence interval for the mean of X. This keyword can be used only for results sent to the output window, as in Section 5.2.1. It cannot be used to create variables in an output data set.
LCLM	One sided 100(1-ALPHA) percent lower confidence limit for the mean of X.
UCLM	One sided 100(1-ALPHA) percent upper confidence limit for the mean of X.
	Note: If both LCLM and UCLM are requested, then each will result in 100(1-ALPHA/2) confidence limits, so that together they form a 100(1-ALPHA) two-sided confidence interval.
T	Student's t statistic to test the hypotheses that the mean of X is equal to 0.
PROBT	The two-tailed *p*-value for the t statistic, T. PROBT is equal to the probability of a value more extreme than T if the mean of X is equal to 0.
	Percentiles of the Distribution of X
P1	Pi = the *i*-th percentile of the distribution of X, for i =1, 5, 10, 25, 50, 75, 90, 95, 99.
P5	
P10	
P25 | Q1	
P50 | MEDIAN	Q1, MEDIAN, and Q3 can be used in place of P25, P50, and P75, respectively.
P75 | Q3	
P90	
P95	
P99	
QRANGE	Interquartile Range = Q3 – Q1

5.2.5 Use a CLASS Statement to Compute Statistics for Subsets

The real power of PROC MEANS lies in its ability to simultaneously calculate statistics for subsets of a data set based on the values of one or more class variables in the same data set. The next code sample computes separate statistics for each value of the Vendor variable, using the JES.Results_Q4 data set.

```
PROC MEANS DATA=JES.Results_Q4 NOPRINT;
  CLASS Vendor; ❶
  VAR Defects Resistance Fail;
  OUTPUT OUT=Tab_4
    N(Defects)=N
  SUM(Defects) = N_Def
  MEAN(Defects)=M_Def
  MEAN(Resistance)=M_Res
  SUM(Fail) = N_Fail
  MEAN(Fail) = P_Fail ;
  FORMAT M_Def P_Fail 6.2;
RUN;
```

❶ The CLASS statement requests separate statistics for each distinct value of Vendor.

Tab_4								
Vendor	_TYPE_	_FREQ_	N	N_Def	M_Def	M_Res	N_Fail	P_Fail
	0	192	192	413	2.15	17.80	35	0.18
ChiTronix	1	90	90	187	2.08	20.78	16	0.18
Duality	1	57	57	205	3.60	15.06	11	0.19
Empirical	1	45	45	21	0.47	15.31	8	0.18

- The first row of Tab_4 contains the requested statistics computed for all values in the data set.
- The next three rows contain the same statistics computed only for the rows with the corresponding values of Vendor.
- The _TYPE_ variable allows us to distinguish the overall statistics (_TYPE_=0) from the statistics for a particular vendor (_TYPE_=1).
- The _FREQ_ variable shows the total number of observations (rows) in each class.

You can use two or more variables in a CLASS statement to generate statistics for each distinct combination of values of the corresponding variables. For example, the next bit of code computes statistics for each combination of Vendor and Month.

```
PROC MEANS DATA=JES.Results_Q4 NOPRINT;
  CLASS Vendor Month;
  VAR Defects Resistance Fail;
  OUTPUT OUT=Tab_5
  N(Defects)=N
  SUM(Defects) = N_Def
  MEAN(Defects)=M_Def
  MEAN(Resistance)=M_Res
  SUM(Fail) = N_Fail
  MEAN(Fail) = P_Fail;
  FORMAT M_Def P_Fail 6.2;
RUN;
```

Tab_5 Vendor	Month	_TYPE_	_FREQ_	N	N_Def	M_Def	M_Res	N_Fail	P_Fail
	.	0	192	192	413	2.15	17.80	35	0.18
	10	1	72	72	153	2.13	17.38	15	0.21
	11	1	44	44	67	1.52	19.08	8	0.18
	12	1	76	76	193	2.54	17.45	12	0.16
ChiTronix	.	2	90	90	187	2.08	20.78	16	0.18
Duality	.	2	57	57	205	3.60	15.06	11	0.19
Empirical	.	2	45	45	21	0.47	15.31	8	0.18
ChiTronix	10	3	29	29	56	1.93	20.68	5	0.17
ChiTronix	11	3	29	29	59	2.03	20.91	6	0.21
ChiTronix	12	3	32	32	72	2.25	20.76	5	0.16
Duality	10	3	28	28	87	3.11	14.94	7	0.25
Duality	12	3	29	29	118	4.07	15.17	4	0.14
Empirical	10	3	15	15	10	0.67	15.58	3	0.20
Empirical	11	3	15	15	8	0.53	15.55	2	0.13
Empirical	12	3	15	15	3	0.20	14.79	3	0.20

- The output data set contains statistics computed for
  - all rows of the data set (_TYPE_ = 0)
  - all distinct values of Month, the second variable in the CLASS statement (_TYPE_=1)
  - all distinct values of Vendor, the first variable in the CLASS statement (_TYPE_=2)
  - all distinct combinations of Vendor and Month (_TYPE_ = 3)
- To understand the assignment of the _TYPE_ variables, consider the binary representation:
  - _TYPE_ = 0 (binary '00') for all rows, i.e., not classified by either variable
  - _TYPE_ = 1 (binary '01') for classification by the second variable, Month
  - _TYPE_ = 2 (binary '10') for classification by the first variable, Vendor
  - _TYPE_ = 3 (binary '11') for classification by both variables
- The shading will not appear in your Output window, but was added to clarify the logical partitions of the output by _TYPE_.

The CLASS statement can take more than two variables, but of course the number of rows of output tends to increase exponentially with the number of classes. In many cases you are only interested in the rows for which each CLASS variable is specified, for example the rows with _TYPE_ = 3 in the table. You can get this reduced table by including the NWAY keyword in the first line:

```
PROC MEANS DATA=Results NOPRINT NWAY;
```

You can also use a TYPES statement for greater control over which combinations are included in the output. For example, the statement below restricts the output to only the rows with _TYPE_ = 2 or 3. There is an example of the use of the TYPES statement with PROC MEANS in Section 6.4.8.

```
TYPES Vendor Vendor*Month;
```

PROC MEANS provides a very simple way to get a quick look at how various factors affect the variables you are interested in. In this example, you can quickly see (from the _TYPE_=2 rows) that the average resistance (M_Res) is higher for ChiTronix components than for Duality or Empirical components. Looking at the variation by month for each vendor (_TYPE_=3), you can see that these differences are fairly consistent. You can also see that the Empirical components have a low defect rate (0.47%) compared to 2.08% for ChiTronix and 3.6% for Duality components. These differences look interesting, but the table doesn't tell you whether they are statistically significant differences or just due to sampling variation. In the next section you will add confidence limits on the means to answer these kinds of questions.

5.2.6 Add Normal Confidence Limits on the Means

In order to see whether the differences observed in Section 5.2.5 are statistically significant, you can add 90% two-sided confidence intervals for the defect rate, the mean resistance and the probability of failing the resistance specification for each combination of Vendor and Month. For resistance you can use the LCLM and UCLM keywords (see Table 5.2 in Section 5.2.4) to compute confidence intervals based on Student's t statistics. This is appropriate because it is reasonable to assume that the resistance measurements are normally distributed (but you will also check this assumption in Section 5.7). It is less reasonable to assume that the numbers of defects and failures are normally distributed. It is more appropriate to use methods based on the Poisson and binomial distributions to compute confidence intervals for the defect rate and probability of failure, respectively. These Poisson and binomial limits are not available in PROC MEANS, but they are easy to compute in a DATA step using the output from PROC MEANS, as shown Section 5.2.7.

This code uses LCLM and UCLM keywords to compute confidence limits on the mean resistance. The results are saved to the Tab_6 data set.

```
PROC MEANS DATA=JES.Results_Q4 NOPRINT NWAY❶ ALPHA=.10❷;
  CLASS Vendor Month;
  VAR Defects Resistance Fail;
  OUTPUT OUT=Tab_6
  N(Defects)=N
  SUM(Defects) = N_Def
  MEAN(Defects)=M_Def
  MEAN(Resistance)=M_Res
  LCLM(Resistance) = R_L  ❸
  UCLM(Resistance) = R_U
  SUM(Fail) = N_Fail
  MEAN(Fail) = P_Fail;
  FORMAT M_Def P_Fail 6.2;
RUN;
```

❶ The NWAY option ensures that the output contains only the rows with specified Vendor and Month.

❷ The ALPHA=.10 option specifies 90% two-sided confidence limits when requesting both LCLM and UCLM. If only one of LCLM or UCLM is requested, then a one-sided 90% confidence limit is generated.

❸ The LCLM and UCLM keywords are used to request the desired lower and upper confidence limits, R_L and R_U, on the average resistance.

In the next example, only a lower confidence limit is requested, but the confidence is specified as ALPHA=.05, and so the value of R_L is the same as in the previous example. The result is saved to the Tab_7 data set.

```
PROC MEANS DATA=JES.Results_Q4 NOPRINT NWAY ALPHA=.05;
  CLASS Vendor Month;
  VAR Defects Resistance Fail;
  OUTPUT OUT=Tab_7
  N(Defects)=N
  SUM(Defects) = N_Def
  MEAN(Defects)=M_Def
  MEAN(Resistance)=M_Res
  LCLM(Resistance) = R_L
  SUM(Fail) = N_Fail
  MEAN(Fail) = P_Fail;
  FORMAT M_Def P_Fail R_L 6.2;
RUN;
```

Chapter 5: Summarizing Your Data **109**

```
Tab_6

Vendor       Month     _TYPE_    _FREQ_     N       N_Def    M_Def

ChiTronix     10         3         29       29       56      1.93
ChiTronix     11         3         29       29       59      2.03
ChiTronix     12         3         32       32       72      2.25
Duality       10         3         28       28       87      3.11
Duality       12         3         29       29      118      4.07
Empirical     10         3         15       15       10      0.67
Empirical     11         3         15       15        8      0.53
Empirical     12         3         15       15        3      0.20

  M_Res        R_L        R_U      N_Fail    P_Fail

  20.68       20.11      21.24       5        0.17
  20.91       20.21      21.60       6        0.21
  20.76       20.21      21.31       5        0.16
  14.94       14.05      15.82       7        0.25
  15.17       14.35      15.99       4        0.14
  15.58       14.24      16.93       3        0.20
  15.55       14.16      16.94       2        0.13
  14.79       13.56      16.03       3        0.20
```

```
Tab_7

V
e                                                     N       P
n         M  T  F                                     _       _
d         o  Y  R     N      M      M                 F       F
o         n  P  E     _      _      _        R        a       a
r         t  E  Q     D      D      R        _        i       i
          h  _  _  N  e      e      e        L        l       l
                      f      f      s

ChiTronix 10  3 29 29   56   1.93  20.68   20.11      5      0.17
ChiTronix 11  3 29 29   59   2.03  20.91   20.21      6      0.21
ChiTronix 12  3 32 32   72   2.25  20.76   20.21      5      0.16
Duality   10  3 28 28   87   3.11  14.94   14.05      7      0.25
Duality   12  3 29 29  118   4.07  15.17   14.35      4      0.14
Empirical 10  3 15 15   10   0.67  15.58   14.24      3      0.20
Empirical 11  3 15 15    8   0.53  15.55   14.16      2      0.13
Empirical 12  3 15 15    3   0.20  14.79   13.56      3      0.20
```

5.2.7 Add Confidence Limits on Poisson and Binomial Parameters

Confidence limits for the parameters of the Poisson and binomial distributions are not available in PROC MEANS, but they are easily computed from the sums included in the PROC MEANS output.

Poisson Distribution

It is reasonable to assume that the number of defects on each component has the Poisson distribution with mean D, where D is the unknown defect rate for the corresponding sub-population of components. Therefore, the total number of defects, N_Def, in each sub-population has the Poisson distribution with mean N*D. Using the well-known relationship between the Poisson and chi-squared distributions, a $(1-\alpha)\%$ confidence interval for D, ($D_L <= D <= D_U$), can be found by solving for D_L and D_U, which satisfy these equations:

$$\Pr[X \geq N\_Def \mid D = D_L] = \sum_{k=N\_Def}^{\infty} e^{-N \times D_L} (N \times D_L)^k / k! = \Pr[\chi^2_{2N\_Def} > 2N \times D_L] = \alpha/2$$

$$\Pr[X \leq N\_Def \mid D = D_U] = \sum_{k=0}^{N\_Def} e^{-N \times D_U} (N \times D_U)^k / k! = 1 - \Pr[\chi^2_{2(N\_Def-1)} > 2N \times D_U] = \alpha/2$$

where χ^2_{df} represents a chi-squared variable with df degrees of freedom. These equations can be solved using the SAS function CINV(*p, df*) (see Section 2.4.1), which returns the *p*-th quantile of a chi-squared distribution with df degrees of freedom:

$$D_L = CINV(\alpha/2, 2N\_Def)/2N$$

$$D_U = CINV(1-\alpha/2, 2(N\_Def+1))/2N$$

Binomial Distribution

It is reasonable to assume that the number of failures, N_Fail, in each sub-population has the binomial distribution with parameters N and P, where P is the unknown probability of failure. Using the relation between the binomial distribution and the incomplete beta function, a $(1-\alpha)\%$ confidence interval for P, ($P_L <= P <= P_U$), can be found by solving for P_L and P_U, which satisfy these equations:

$$\Pr[X \geq N\_Fail \mid P = P_L] = \sum_{k=N\_Fail}^{N} \binom{N}{k} P_L^k (1-P_L)^{N-k} = I_{P_L}(N\_Fail, N+1-N\_Fail) = \alpha/2$$

$$\Pr[X \leq N\_Fail \mid P = P_U] = \sum_{k=0}^{N\_Fail} \binom{N}{k} P_U^k (1-P_U)^{N-k} = 1 - I_{P_U}(N\_Fail+1, N-N\_Fail) = \alpha/2$$

where IP(a,b) is the incomplete beta function. These equations can be solved using the SAS function BETAINV(p,a,b) (see Section 2.4.1), which returns the p-th quantile of a beta distribution with parameters a and b:

$$P_L = BETAINV(\alpha/2, N\_Fail, N+1-N\_Fail)$$

$$P_U = BETAINV(1-\alpha/2, N\_Fail+1, N-N\_Fail)$$

The code below uses these equations to add confidence intervals for the defect rate and probability of failure to the Tab_6 data set created in the previous example. The IF conditions are required because the lower confidence limits cannot be computed if there are no fails and, for the binomial distribution, the upper limit cannot be computed if all units fail.

```
DATA Tab_8; SET Tab_6;
   FORMAT D_L D_U 8.2 P_L P_U PERCENT7.0;
   IF N_Def>0 THEN D_L=CINV(.05, 2*N_Def)/(2*N);
   D_U=CINV(.95, 2*(N_Def+1))/(2*N);
   IF N_Fail>0 THEN P_L=BETAINV(.05, N_Fail,   N+1-N_Fail);
   IF N_Fail<N THEN P_U=BETAINV(.95, N_Fail+1, N-N_Fail);
RUN;
```

```
Tab_8 (selected columns)
 Vendor      Month        D_L         D_U         R_L         R_U         P_L         P_U
ChiTronix     10         1.53        2.41       20.11       21.24         7%         33%
ChiTronix     11         1.62        2.53       20.21       21.60         9%         37%
ChiTronix     12         1.83        2.74       20.21       21.31         6%         30%
Duality       10         2.58        3.71       14.05       15.82        12%         42%
Duality       12         3.47        4.74       14.35       15.99         5%         29%
Empirical     10         0.36        1.13       14.24       16.93         6%         44%
Empirical     11         0.27        0.96       14.16       16.94         2%         36%
Empirical     12         0.05        0.52       13.56       16.03         6%         44%
```

Comparing the upper confidence limit from one group to the lower confidence limit from another is a simple way to determine whether the observed differences are statistically significant. For example,

- We are 95% confident that the defect rate for Empirical units made in month 10 is less than 1.13. (Note that the upper end of the 90% two-sided confidence interval is a 95% upper confidence limit.) And we are 95% confident that the defect rate for ChiTronix units made in month 10 is greater than 1.53. Thus, we are >95% confident that the defect rate is lower for Empirical than for ChiTronix in Month 10.
- Applying the same logic to the defect confidence limits for the other months, you can see that Empirical is significantly better than ChiTronix, and also better than Duality, for each month of production.

This method of looking for significant differences has two statistical difficulties:

- The overlapping confidence interval method is very conservative and is not the best test for comparing two populations. For example, a *t*-test would be a better way to test whether two normal distributions have the same mean.
- If you apply the method to each pair of populations, you are performing many tests simultaneously (28 distinct tests in the example), and so the likelihood of one or more differences appearing significant by chance is much greater than it would be in any single test.

However, this is still an excellent way to quickly and easily discover significant differences among sub-populations. The first concern above can be addressed by performing the appropriate two-sample test after an interesting difference has been observed. The second concern is addressed by Analysis of Means, as described in Section 5.6. See also the paper by Schenker and Gentleman, "On Judging the Significance of Differences by Examining the Overlap Between Confidence Intervals," referenced in the "Chapter Summary" section. In Section 6.3.3 you will learn how to plot these confidence intervals to simplify the task of detecting significant differences.

5.3 PROC TABULATE

PROC TABULATE offers an alternative way to summarize SAS data sets with categorical variables. It produces tables that are more compact and easier to read, but the coding is more complex, so it generally takes a bit of effort to get things arranged just the way you want them. This section gives a simple example using PROC TABULATE, without much explanation of the syntax, just so that you can see the possibilities. See the books and papers referenced in the "Chapter Summary" section for more details on the syntax of PROC TABULATE. The code for this section is in ~\JES \sample_code\ch_5 \tabulate_examples.sas. The first code sample uses PROC TABULATE to create a two-way table that includes the same average resistance values produced by PROC MEANS in Section 5.2.5.

```
PROC TABULATE DATA=JES.Results_Q4;   ❶
  CLASS Vendor Month;   ❷
  VAR Resistance;   ❸
  TABLES Vendor, MEAN*Resistance*Month;   ❹
RUN;
```

❶ The PROC TABULATE statement requests an analysis of the JES.Results_Q4 data set.

❷ The CLASS statement requests statistics for each distinct value of Vendor and Month.

❸ The VAR statement specifies that statistics be computed for the Resistance variable.

❹ The TABLES statement requests a table of mean Resistance by Vendor and Month.

When you run this code, you will see this table in the Output window. The format of this output table is more compact and easier to read, but it includes only the average values of Resistance, and not the other statistics computed by PROC MEANS.

	Mean Resistance		
	------	------	------
	Month		
	10	11	12
Vendor			
ChiTronix	20.68	20.91	20.76
Duality	14.94	.	15.17
Empirical	15.58	15.55	14.79

With a bit more work you can add the overall average for each Vendor and Month, and get rid of the unnecessary "Mean" and "Resistance" titles.

```
PROC TABULATE DATA=JES.Results_Q4;
  CLASS Vendor Month;
  VAR Resistance;
  TABLES Vendor ALL❶, MEAN=''*Resistance=' '❷*(Month ALL)
    / BOX='Average Resistance';   ❸
RUN;
```

❶ The ALL options create the required overall means.

❷ The MEAN='' and Resistance='' options remove the unnecessary titles.

❸ The BOX= option puts text in the box at the upper left.

The new output table is shown below.

Average Resistance	Month 10	Month 11	Month 12	All
Vendor				
ChiTronix	20.68	20.91	20.76	20.78
Duality	14.94	.	15.17	15.06
Empirical	15.58	15.55	14.79	15.31
All	17.38	19.08	17.45	17.80

Finally, we add the average percent failing the resistance specification to the table.

```
PROC TABULATE DATA=JES.Results_Q4;
   CLASS Vendor Month;
   VAR Resistance Fail; ❶
   TABLES Vendor ALL, ((MEAN=''*F=6.2*Resistance) ❷
   ((MEAN=''*F=PERCENT7.1)*Fail='% Fail'))*(Month ALL)
      / BOX='Test Results' MISSTEXT='N/A'; ❸
RUN;
```

❶ The VAR statement specifies two analysis variables, Resistance and Fail.

❷ The F=6.2 and F=PERCENT7.1 options specify the formats of the averages.

❸ The MISSTEXT='N/A' specifies the text to be used in cells with no data.

Test Results	Resistance				% Fail			
	Month				Month			
	10	11	12	All	10	11	12	All
Vendor								
ChiTronix	20.68	20.91	20.76	20.78	17.2%	20.7%	15.6%	17.8%
Duality	14.94	N/A	15.17	15.06	25.0%	N/A	13.8%	19.3%
Empirical	15.58	15.55	14.79	15.31	20.0%	13.3%	20.0%	17.8%
All	17.38	19.08	17.45	17.80	20.8%	18.2%	15.8%	18.2%

If you compare this table to the PROC MEANS output in Section 5.2.5, you will see that it contains the same mean values as the M_Res and P_Fail variables in Tab_5, but in a form which makes it easier to see the differences among and within vendors and months. You might have also noticed that the syntax is getting more than a bit complex. It generally takes a bit of trial and error to get it right, but the end result is worth the effort when it comes time to understand and communicate the results.

5.4 PROC REPORT

PROC REPORT provides yet another way to create a tabular summary of a data set. In this section you will use PROC REPORT to summarize the same JES.Results_Q4 data set used in the previous sections. The code for this section is in ~\JES \sample_code\ch_5 \report_examples.sas.

The simplest form of PROC REPORT can be used to print selected columns for all rows of a data set.

```
PROC REPORT DATA=JES.Results_Q4 NOWINDOWS;  ❶
   COLUMN Vendor Month Resistance Fail;  ❷
RUN;
```

❶ The PROC REPORT statement requests a report on the JES.Results_Q4 data set. The NOWINDOWS option specifies that the batch version of PROC REPORT will be run. Without this option, PROC REPORT launches an interactive report generating facility.

❷ The COLUMN statement specifies the columns to be printed, and the order in which they appear.

```
                          Resistan
Vendor            Month         ce      Fail
ChiTronix            10      20.35         0
ChiTronix            10      20.60         0
ChiTronix            10      19.52         0
ChiTronix            10      19.61         0
ChiTronix            10      20.95         0
ChiTronix            10      17.05         0
ChiTronix            10      21.67         0
ChiTronix            10      20.81         0
(partial listing)
```

The output looks identical to what you would get from PROC PRINT using a VAR statement similar to the COLUMN statement. However, there are advantages to using PROC REPORT in that certain options can be invoked to enhance the output. For example, you can add computed columns to the output, or embed HTML links and apply "traffic lighting" to cells when the output is directed to HTML (see the examples in Sections 7.4.2 and 7.4.3).

The next sample of code uses PROC REPORT to compute the mean of Resistance and Fail by Vendor and Month.

```
PROC REPORT DATA=JES.Results_Q4 NOWINDOWS HEADLINE;  ❶
   COLUMN Vendor Month Resistance Fail;
   DEFINE Vendor/GROUP;  ❷ ❸
   DEFINE Month/ GROUP;
   DEFINE Resistance / ANALYSIS ❹'Avg./Resistance' WIDTH=12 MEAN;  ❺
   DEFINE Fail / ANALYSIS 'Percent/Failure' FORMAT = 8.3 WIDTH=8 MEAN;
   BREAK AFTER Vendor / SUMMARIZE OL SKIP;  ❻
   RBREAK AFTER / SUMMARIZE DOL;  ❼
RUN;
```

❶ The HEADLINE option causes a line to be drawn under the column headers.

❷ The DEFINE statements specify options to be applied to each column.

❸ The GROUP option specifies that statistics be computed for each unique combination of values of the GROUP variables.

❹ The ANALYSIS option specifies that statistics will be computed for this variable.

❺ The MEAN keyword specifies that the statistic to be computed is the mean. The WIDTH option specifies column width. If WIDTH is not specified, the width of numeric columns defaults to the numeric format, which is what caused the unnatural break of "Resistance" in the first table.

❻ The BREAK statement creates and formats a summary line for each value of Vendor. The AFTER option specifies that the break will come *after* each value of Vendor. The SUMMARIZE option specifies that summary statistics will be included. The OL option specifies an overline above the summary statistics. The SKIP option specifies a blank line after the summary statistics.

❼ The RBREAK statement creates and formats a summary line for the whole data set. The SUMMARIZE option specifies that summary statistics will be computed. The AFTER option specifies that the summary will come *after* the entire report. The DOL option specifies a double overline above the summary statistics.

```
                         Avg.        Percent
Vendor         Month     Resistance  Failure

ChiTronix      10        20.68       0.172
               11        20.91       0.207
               12        20.76       0.156
               ─────────────────────────────
ChiTronix                20.78       0.178

Duality        10        14.94       0.250
               12        15.17       0.138
               ─────────────────────────────
Duality                  15.06       0.193

Empirical      10        15.58       0.200
               11        15.55       0.133
               12        14.79       0.200
               ─────────────────────────────
Empirical                15.31       0.178

                         ============ ========
                         17.80        0.182
```

This table contains the same mean values as in the output from PROC MEANS and PROC TABULATE in Sections 5.2.5 and 5.3, except that the mean by Month for all Vendors is not included. See the books and papers referenced in the "Chapter Summary" section for more details on the syntax of PROC REPORT.

You now have three different ways to compute and display the same results. This might seem like overkill, but in practice it provides a great deal of flexibility in producing exactly the right report required in various circumstances. And, as noted in Section 5.1, PROC MEANS is also invaluable for producing various data sets needed in intermediate steps of a more complex analysis.

5.5 PROC BOXPLOT

Box-and-whisker plots (or simply box plots) were introduced by John Tukey as a means to visualize differences in the distribution of a numeric variable among several groups. A box plot consists of

- The Box: A rectangle which runs from the 25th percentile (P25) to the 75th percentile (P75) of the distribution, with a horizontal line drawn at the median and a plotted symbol at the mean. The length of the box, P75-P25, is called the interquartile range (IQR).
- The Whiskers: Lines extending from the lower and upper edges of the box to the minimum and maximum values of the variable.

This type of box plot is called a *skeletal* box plot. Tukey also recommended a variation of the box plot, called a *schematic* box plot, which displays and identifies "outliers", defined as points that are more than [1.5*(P75-P25)] beyond the upper or lower end of the box. The whiskers for a schematic box plot end at the minimum and maximum values that are not outliers.

PROC BOXPLOT, which is a SAS/STAT procedure, can be used to create both types of box plots. The code for this section is in ~\JES\sample_code\ch_5\boxplot_examples.sas.

5.5.1 Skeletal Box Plots

The code below creates a skeletal box plot (Figure 5-1) of Resistance for each value of Vendor.

```
GOPTIONS RESET=ALL GUNIT=PCT HTEXT=3 FTEXT='Arial';   ❶
TITLE1 HEIGHT=5 "Skeletal Box Plot";
PROC BOXPLOT DATA=JES.Results_Q4;   ❷
   PLOT Resistance*Vendor /NAME="F5_1_" BOXSTYLE=SKELETAL;   ❸
RUN; QUIT;   ❹
```

❶ The GOPTIONS and TITLE statements set some graphic options for the plots. These options will be explained in Chapter 6.

❷ The PROC BOXPLOT statement requests box plots of the JES.Results_Q4 data set.

❸ The PLOT statement requests a box plot of Resistance (the measurement variable) for each value of Vendor (the group variable). The PLOT options are listed after the "/". The NAME option specifies a name for the plot to be created. Plot names are explained in Section 6.2. The BOXSTYLE option specifies that this should be a skeletal box plot.

❹ The QUIT; statement specifies that the interactive procedure be ended.

The next example illustrates some of the many options that you can use to enhance your box plots. The resulting graph is shown in Figure 5-2.

```
PROC BOXPLOT DATA=JES.Results_Q4;
   PLOT Resistance*Vendor / BOXSTYLE=SKELETAL
   BOXCONNECT=MEAN   ❶
   BOXWIDTH=7   ❷
   NOTCHES;   ❸
RUN; QUIT;
```

❶ The BOXCONNECT option requests a line joining the means of each group.

❷ The BOXWIDTH option controls the width of the boxes.

❸ The NOTCHES option requests a notched box plot. The notches are drawn at the median +/- 1.58*IQR/SQRT(N). The medians of two groups are significantly different at approximately the 5% significance level if the corresponding notches do not overlap. See the referenced paper by McGill et al. in the "Chapter Summary" section for the derivation of this approximation.

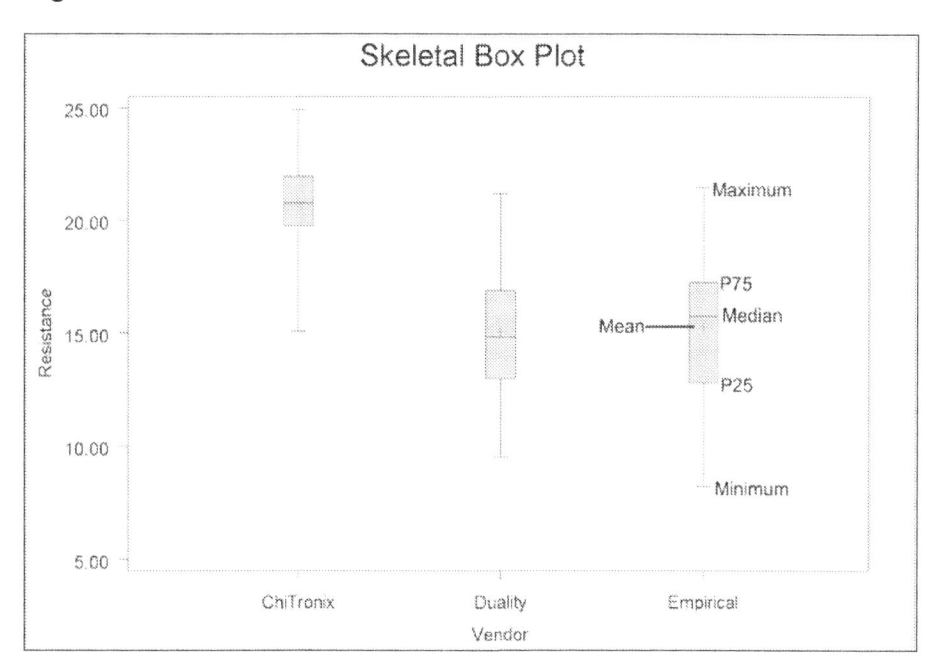

Figure 5-1: Skeletal Box Plot

The annotations, "Mean", "P25", etc., were added in Adobe Photoshop and are not in the Graph window.

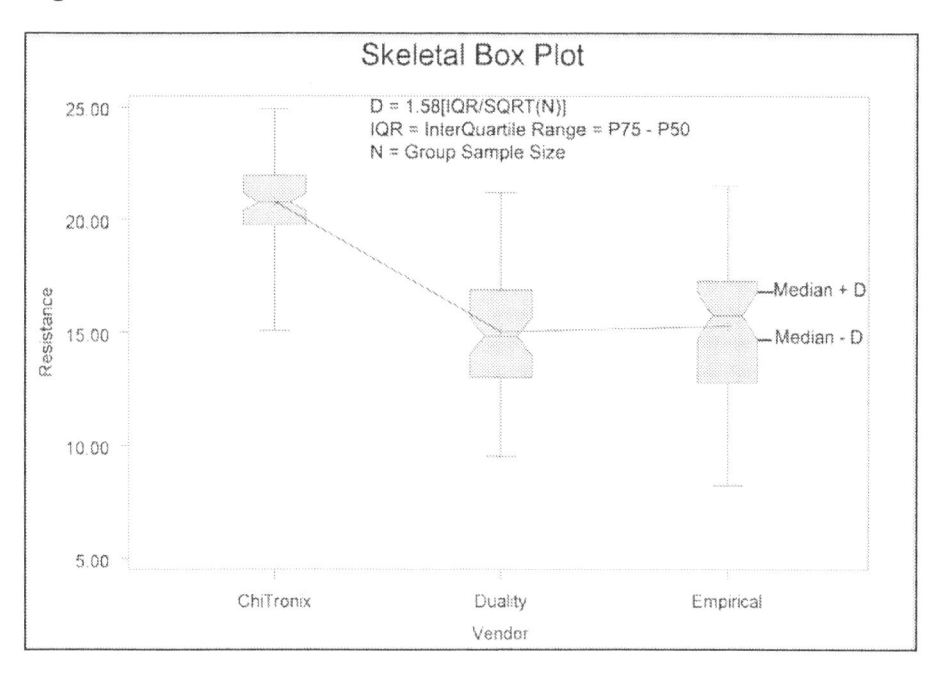

Figure 5-2: Skeletal Notched Box Plot

Because the notches do not overlap, you can see that the median resistance is significantly higher for ChiTronix units than for Duality or Empirical units.

5.5.2 Schematic Box Plots

To create a schematic box plot, use the BOXSTYLE=SCHEMATIC option as shown in the next example.

```
TITLE1 HEIGHT=5 "Schematic Box Plot";
PROC BOXPLOT DATA=Results;
   PLOT Resistance*Vendor /BOXSTYLE=SCHEMATIC;
RUN; QUIT;
```

When you run this code, the plot shown in Figure 5-3 is displayed in the Graph window. The annotations, "P25", "Outside", etc., were added in Adobe Photoshop and do not appear in the Graph window. The plot is the same as the first skeletal box plot on the previous page except that

- The whiskers stop at the extreme points within P25-1.5*IQR and P75+1.5*IQR. These values are what Tukey calls "fences", which he proposed as a rule of thumb for identifying "outliers". For a normal distribution, the fences are approximately two standard deviations above and below the mean.
- The outlier values beyond the fences are plotted individually.

A schematic box plot can be enhanced using the same options as for skeletal box plots, some of which were shown on the previous page. The code below illustrates a few more options, including the ability to attach labels to the outliers.

```
PROC BOXPLOT DATA=JES.Results_Q4;
   PLOT Resistance*Vendor / BOXSTYLE=SCHEMATICID  ❶
   CBOXFILL=GRAYD0  ❷
   CBOXES=GREEN
   IDSYMBOL=dot  ❸
   IDCOLOR=RED
   BOXWIDTH=10  ❹
   BOXWIDTHSCALE=.5  ❺
   BWSLEGEND;
   ID Resistance;  ❻
RUN; QUIT;
```

❶ The BOXSTYLE=SCHEMATICID option requests a schematic box plot in which the outliers are labeled with the value of an ID variable.

❷ The CBOXFILL and CBOXES options specify the fill and outline colors for the boxes. SAS color and gray-scale definitions are discussed in Section 6.3.2.6.

❸ The IDSYMBOL and IDCOLOR options specify the type and color of symbols to be used for outliers. SAS symbol definitions are also discussed in Section 6.3.2.6.

❹ The BOXWIDTH option controls the average width of the boxes.

❺ The BOXWIDTHSCALE=.5 option specifies that each box width should be proportional to the square root of the group size, n. The BWSLEGEND option displays a legend identifying the rule for relative box width. The legend is shown in the lower left of the plot area.

❻ The ID statement specifies that the value of Resistance be used to label the outliers. Any variable in the Results data set could have been chosen. For example, if the data set included Test_Date or Operator, those might have been more useful labels.

Figure 5-3: Schematic Box Plot

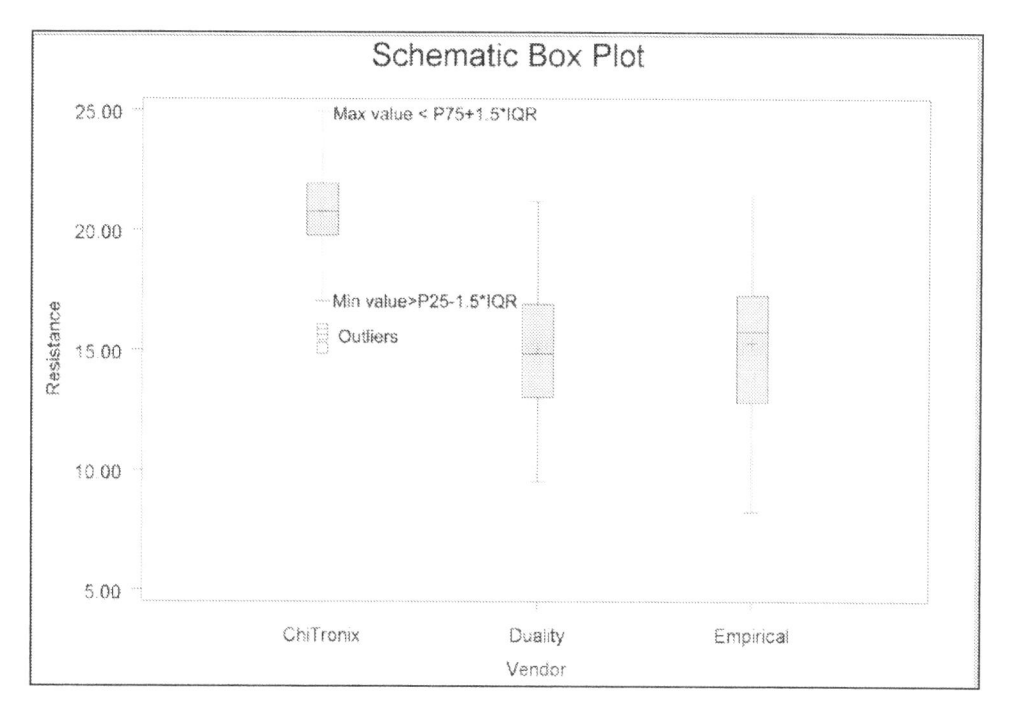

Figure 5-4: Schematic Box Plot with Outliers Labeled

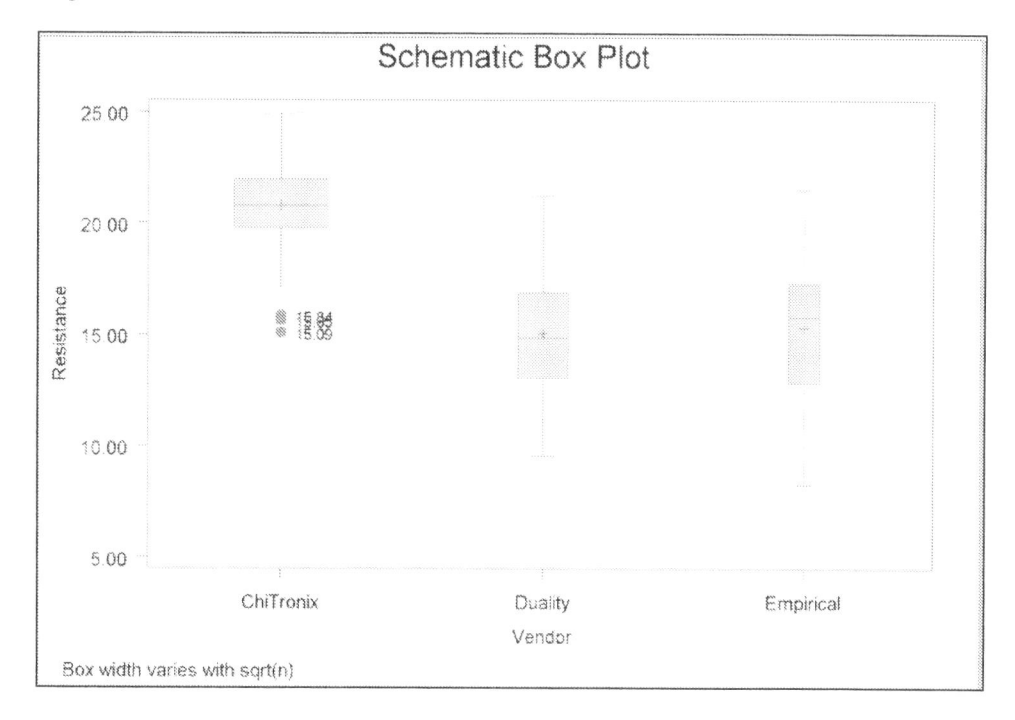

5.6 PROC ANOM

The example in Section 5.2.7 used non-overlapping confidence limits on group means to detect significant differences among groups. As noted in that section, this method is conservative when applied to any two groups and will tend to produce false positives when applied to several groups simultaneously. We can overcome these difficulties by using PROC ANOM for simultaneous comparison of several group means.

PROC ANOM, which is a SAS/QC procedure, uses the Analysis of Means methods developed by Ellis Ott to provide a graphical and statistical method for simultaneously comparing several group (or "treatment") means to the overall population mean at a specified significance level alpha. This procedure tests the hypothesis that all group means are equal and, if the hypothesis is rejected, identifies which groups are significantly different. The significance level, alpha, is the probability that, under the null hypothesis of no group differences, at least one of the group means will exceed the decision limits. The code for this section is in ~\JES \sample_code\ch_5 \anom_examples.sas.

PROC ANOM uses three different statements to analyze three different variable types:

- XCHART for continuous variables, assumed to follow the normal distribution. This is an appropriate assumption for the resistance measurements in the JES.Results_Q4 data set.
- UCHART for rates defined as counts per unit, where the counts are assumed to follow the Poisson distribution. This is an appropriate assumption for the defect counts in the JES.Results_Q4 data set.
- PCHART for proportions X/N, where X is assumed to follow the binomial distribution. This is an appropriate assumption for the failure rates, N_Fail/N, computed from the JES.Results_Q4 data set.

5.6.1 XCHART for Continuous Variables

Use PROC ANOM with an XCHART statement to look for significant differences in the mean resistance among the three vendors.

```
PROC ANOM DATA=JES.Results_Q4;  ❶
   XCHART Resistance*Vendor / ALPHA=.05;  ❷
   BOXCHART Resistance*Vendor / VREF=12.5 22.5;  ❸
RUN;
```

❶ The PROC ANOM statement requests an analysis of means on the JES.Results_Q4 data set.

❷ The XCHART statement requests the plot shown in Figure 5-5, with Resistance as the response variable and Vendor as the group variable. The ALPHA = .05 option specifies the significance level of the test. The plot includes a horizontal line plotted at the overall mean resistance for all vendors, labeled $\bar{\bar{x}}$, and a vertical line drawn from the overall mean to the mean for each vendor. There are also upper-decision limit (UDL) and lower-decision limit (LDL) lines for each vendor. Any group mean that is outside the decision limits is significantly different from the overall mean.

❸ The BOXCHART statement requests the plot shown in Figure 5-6. The VREF= option adds reference lines along the vertical axis at the specification limits of 12.5 and 22.5. The plot includes a skeletal box-and-whiskers plot (see Section 5.5.1) along with the same decision limits as in the XCHART plot. The plotted mean points (+) can be compared to the UDL and LDL limits to identify significant differences as in the XCHART plot. The reference lines at 12.5 and 22.5 can be used to compare each distribution to the specification limits.

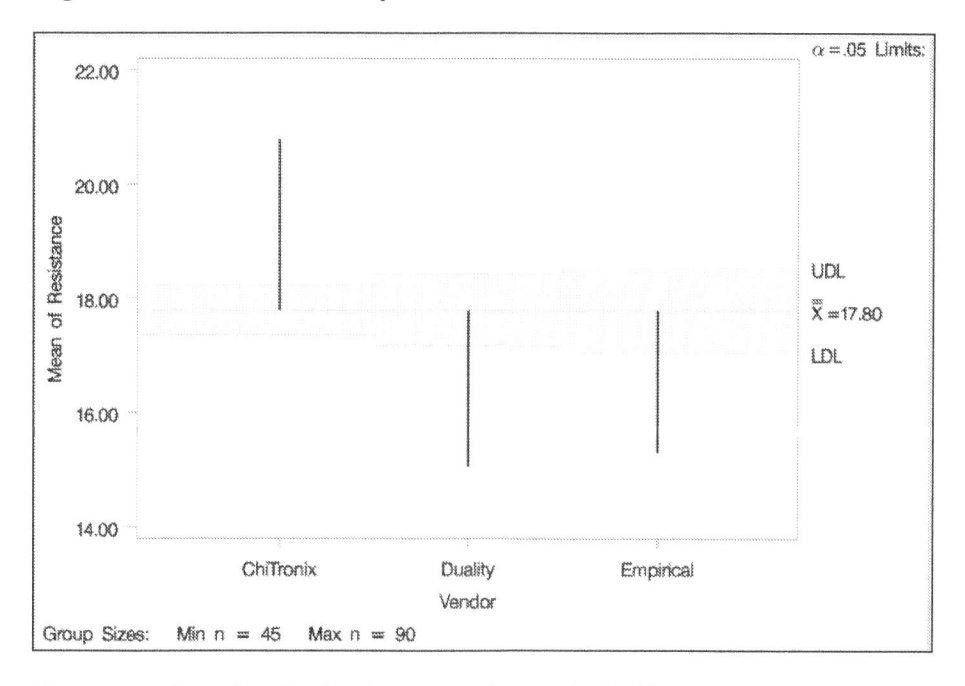

Figure 5-5: Plot Created by the XCHART Statement in PROC ANOM

You can see from this plot that the mean resistance is significantly higher for ChiTronix units, and significantly lower for Duality and Empirical units.

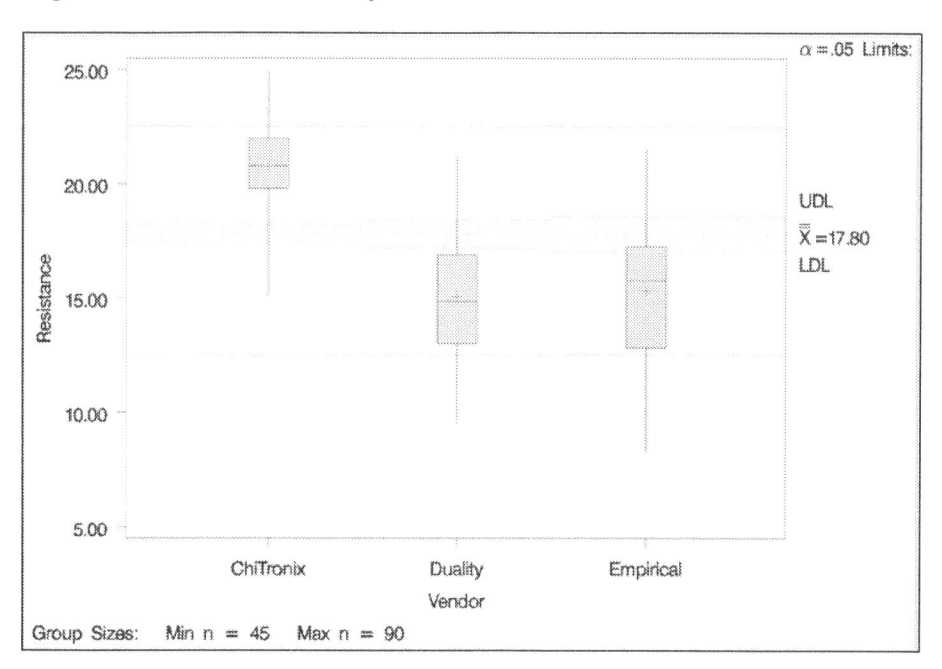

Figure 5-6: Plot Created by the BOXCHART Statement in PROC ANOM

This plot includes the same analysis of means as in Figure 5-5. In addition, the box plots clearly show the differences in the distribution of resistance among the vendors, and the relationship of each distribution to the upper and lower specification limits of 12.5 and 22.5.

5.6.2 UCHART Statement for Rates from Group Counts

Use PROC ANOM with a UCHART statement to detect differences in defect rate among the vendors. First create a table of group counts that are required by PROC ANOM for analysis of rates or proportions.

```
PROC MEANS DATA=JES.Results_Q4 NOPRINT NWAY;
   CLASS Vendor;
   VAR Defects Fail;
   OUTPUT OUT=Tab_9
   N(Defects)=N
   SUM(Defects) = N_Def
   SUM(Fail) = N_Fail;
RUN;
```

```
Tab_9
 Vendor      _TYPE_     _FREQ_        N        N_Def       N_Fail
ChiTronix       1         90         90         187          16
Duality         1         57         57         205          11
Empirical       1         45         45          21           8
```

The new data set, Tab_9, contains the defect counts, N_Def, and units per group, N, which are required when using the UCHART statement.

```
PROC ANOM DATA=Tab_9;
   UCHART N_Def*Vendor/GROUPN=N;  ❶
RUN;
```

❶ N_Def, the number of defects, is the response variable, Vendor is the group variable, and the GROUPN=N option specifies that the variable N contains the number of units in the group.

Figure 5-7: Plot Created by the UCHART Statement in PROC ANOM

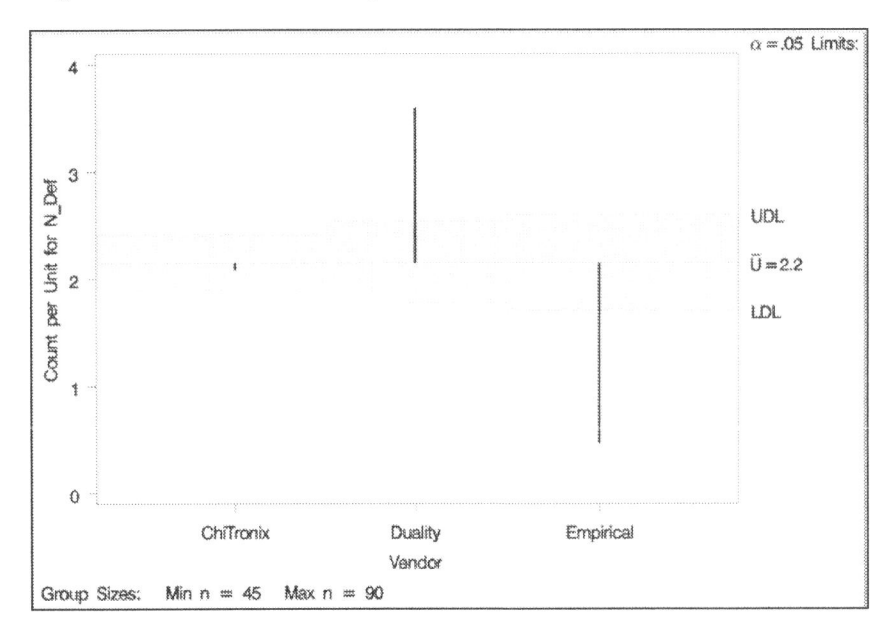

The interpretation of the UCHART plot is the same as for the XCHART plot in Figure 5-5. A group mean, shown as the end of the corresponding vertical line, is significantly different from the overall mean if it lies outside the decision limits. In this example, all differences are significant.

5.6.3 PCHART for Proportions from Group Counts

Use PROC ANOM with a PCHART statement to look for significant differences in the probability of failure among the three vendors. The required table of counts, Tab_9, was created in the previous example.

```
PROC ANOM DATA=Tab_9;
   PCHART N_Fail*Vendor/GROUPN=N;
RUN;
```

As in the previous example, N_Fail, the number of failing units, is the response variable, Vendor is the group variable, and the GROUPN=N option specifies that the variable N contains the number of units in the group.

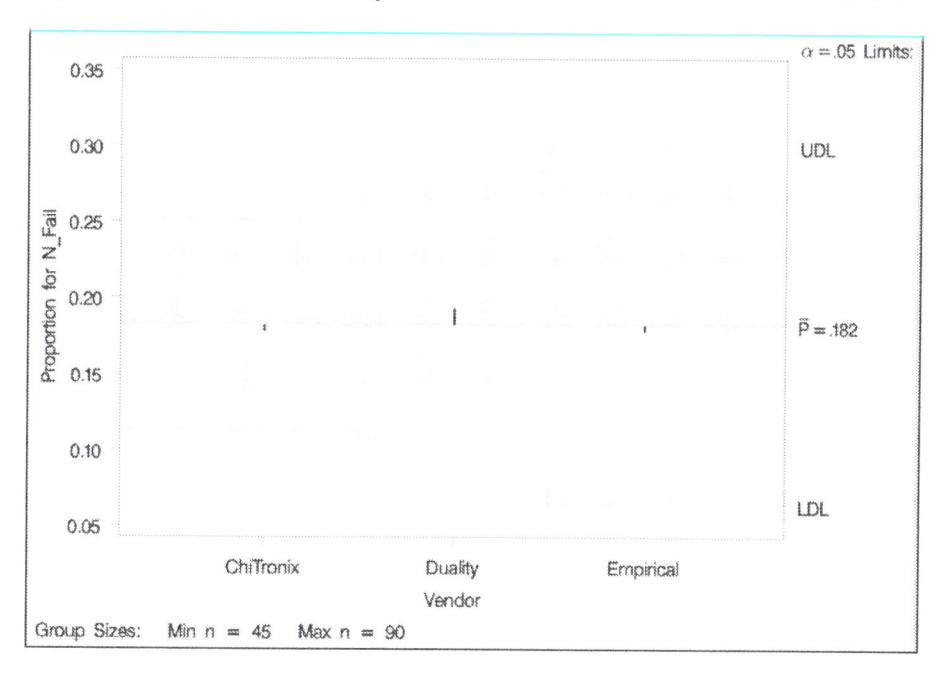

Figure 5-8: Plot Created by the UCHART Statement in PROC ANOM

The interpretation of the PCHART plot is the same as for the XCHART and UCHART plots. At the default significance level of 5%, the ANOM test does not reject the hypotheses that the failure rate is the same for all vendors.

For details on the theory behind the decision limits, and the many options available for customizing the ANOM plots and creating tabular output, consult the SAS online documentation:

SAS/QC → SAS/QC User's Guide → The ANOM Procedure

5.7 PROC UNIVARIATE

PROC UNIVARIATE provides a comprehensive set of tools for analyzing and plotting samples from continuous univariate distributions. The examples in this section use PROC UNIVARIATE to test whether the resistance measurements from the previous example follow a normal distribution, and to compare the distribution of resistance among the different vendors. The code for this section is in ~\JES \sample_code\ch_5 \univariate_examples.sas. The first example uses PROC UNIVARIATE to create several tables of statistics for the distribution of resistance.

```
PROC UNIVARIATE DATA=JES.Results_Q4;
   VAR Resistance; ❶
RUN;
```

❶ The VAR statement specifies that the variable to be analyzed is Resistance. To analyze more than one variable, add variable names to the VAR statement. For example:

```
VAR Resistance Defects;
```

When you run this code, five tables are created in the Output window. The first two tables include the basic descriptive statistics for the sample data. The definitions of these statistics are the same as for the statistics created by PROC MEANS as shown in Table 5-2.

```
The UNIVARIATE Procedure
Variable:  Resistance
                            Moments
N                             192    Sum Weights                192
Mean                    17.7986361    Sum Observations    3417.33812
Std Deviation           3.68851639    Variance            13.6051531
Skewness                -0.2804873    Kurtosis            -0.8639843
Uncorrected SS          63422.5417    Corrected SS        2598.58425
Coeff Variation         20.7235901    Std Error Mean       0.26619574

             Basic Statistical Measures
     Location                    Variability
Mean     17.79864     Std Deviation            3.68852
Median   18.43911     Variance                13.60515
Mode            .     Range                   16.67093
                      Interquartile Range      6.00199
```

The next table gives the results of Student's t test of the hypothesis that the mean is equal to zero, and two tests, the Sign test and the Wilcoxon Signed Rank test, of the hypothesis that the median is equal to zero.

```
Tests for Location: Mu0=0
Test           -Statistic-     -----p Value------
Student's t    t  66.86296     Pr > |t|    <.0001
Sign           M        96     Pr >= |M|   <.0001
Signed Rank    S      9264     Pr >= |S|   <.0001
```

Each test has a *p*-value <.0001 which means that the hypothesis is rejected at the 0.01% significance level. Of course MU0=0 is not a very interesting hypothesis for the resistance measurements, but you can test for any other mean or median value by using the MU0 option. For example, to test the hypotheses that the mean or median is equal to 16, use

```
PROC UNIVARIATE DATA=JES.Results_Q4 MU0=16;
```

The next two tables provide the estimated quantiles and the extreme observations of the sample.

```
Quantiles (Definition 5)

Quantile         Estimate
100% Max         24.92810
99%              24.40082
95%              23.22808
90%              22.05705
75% Q3           20.77805
50% Median       18.43911
25% Q1           14.77605
10%              12.53766
5%               11.94936
1%                9.52923
0% Min            8.25717


              Extreme Observations
------Lowest------        -----Highest-----

    Value       Obs         Value       Obs

   8.25717      192       23.7719       11
   9.52923      102       24.0266       25
  10.09115      157       24.3261       23
  10.46886      132       24.4008       30
  10.47894      123       24.9281       52
```

You can test the hypothesis that the sample comes from a normal distribution by adding the NORMALTEST keyword as shown below.

```
PROC UNIVARIATE DATA=JES.Results_Q4 NORMALTEST;
   VAR Resistance;
RUN;
```

This code generates all of the same tables shown above, plus the four tests of normality shown below.

```
                    Tests for Normality

Test                  --Statistic---     -----p Value------

Shapiro-Wilk          W     0.970347     Pr < W      0.0004
Kolmogorov-Smirnov    D     0.08455      Pr > D     <0.0100
Cramer-von Mises      W-Sq  0.346872     Pr > W-Sq  <0.0050
Anderson-Darling      A-Sq  1.979493     Pr > A-Sq  <0.0050
```

The *p*-values tell us that the hypothesis of normality can be rejected at the 1% significance level. The next two examples create a histogram and a probability plot that provide some insight into the apparent non-normality of the distribution.

PROC UNIVARIATE can also create distribution plots which can help you to better understand the distribution of your variables.

```
GOPTIONS RESET=ALL GUNIT=PCT HTEXT=4 FTEXT='Arial' BORDER;
TITLE HEIGHT=5 "Distribution of Resistance - All Vendors";
PROC UNIVARIATE DATA=JES.Results_Q4;
  VAR Resistance;
  HISTOGRAM Resistance / NORMAL(MU=EST SIGMA=EST);  ❶
  INSET N MEAN STD SKEWNESS KURTOSIS NORMAL(AD ADPVAL)  ❷
  / HEIGHT=2.5 FORMAT = 5.3 POSITION=NW;  ❸
RUN;
```

❶ The HISTOGRAM statement requests a histogram of the values of Resistance. The NORMAL(MU=est SIGMA=est) option requests a fitted normal distribution, with the mean MU and standard deviation SIGMA estimated from the sample data.

❷ The INSET statement specifies statistics to be displayed in the inset. The NORMAL(AD ADPVAL) option requests the statistic and corresponding *p*-value of the Anderson-Darling test for normality.

❸ The HEIGHT, FORMAT, and POSITION options control the appearance and position of the inset box. Position is specified as a compass direction, for example NW for Northwest as shown here.

Figure 5-9: Plot Created by the HISTOGRAM Statement in PROC UNIVARIATE

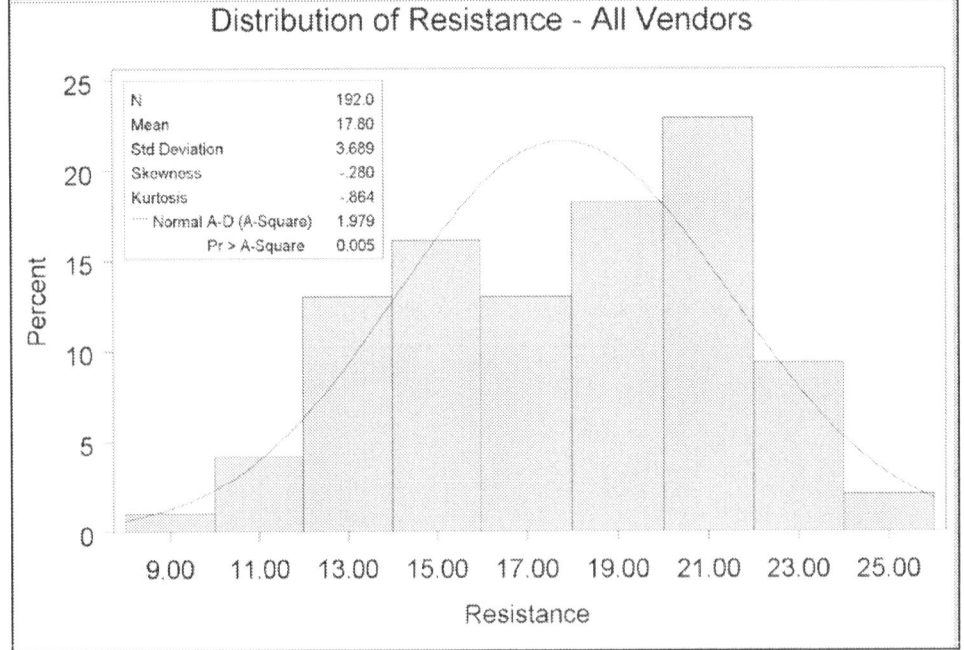

The normal density curve is not a very good fit because the histogram is relatively flat in the center and does not have a peak near the mean. This type of distribution often occurs when the sample contains units from different sources or different time periods. You have already seen in the previous sections (e.g., Figure 5-1) that the resistance values are much higher for ChiTronix units, and that is causing the non-normality of the aggregate sample.

You can also use a PROBPLOT statement to create a probability plot that provides another visual assessment of normality.

```
TITLE HEIGHT=5 "Probability Plot for Resistance - All Vendors";
PROC UNIVARIATE DATA=JES.Results_Q4;
  VAR Resistance;
  PROBPLOT Resistance / NORMAL(MU=EST SIGMA=EST) PCTLMINOR;  ❶
  INSET N MEAN STD SKEWNESS KURTOSIS
  / HEIGHT=2.5 FORMAT = 5.3 POSITION=NW;
RUN;
```

❶ The PROBPLOT statement requests a probability plot of the resistance variable. The NORMAL(MU=EST SIGMA=EST) option requests a fitted normal distribution, with the mean MU and standard deviation SIGMA estimated from the sample data. The PCTLMINOR option is used to add the minor tick marks to the normal percentile axis.

The Anderson-Darling statistics are not included in the inset. The goodness-of-fit tests are only valid after the HISTOGRAM statement.

Figure 5-10: Plot Created by the PROBPLOT Statement in PROC UNIVARIATE

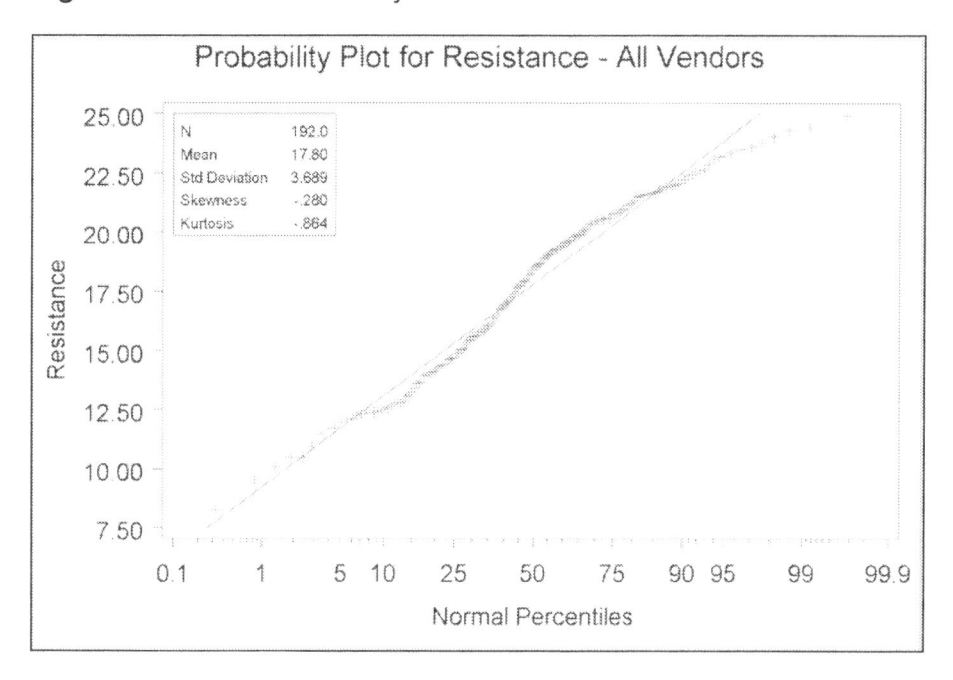

The straight line represents the best fit normal distribution, with mean=17.8 and standard deviation=3.689, while the individual plot points (+) represent the empirical distribution function (EDF) computed from the sample data. If the sample came from a normal distribution, we would expect that the plot of the EDF would be very close to the best fit line. The fact that the EDF differs significantly from a straight line indicates that the normal distribution is not a good fit.

You can also estimate parameters for several populations simultaneously, and plot the distributions of each on the same scale.

```
TITLE HEIGHT=5 "Distribution of Resistance by Vendor";
PROC UNIVARIATE DATA=JES.Results_Q4;
   CLASS Vendor; ❶
   VAR Resistance;
   HISTOGRAM Resistance / NORMAL(MU=est SIGMA=est) NROW=3; ❷
   INSET N MEAN STD
   / HEIGHT=3 FORMAT = 5.2 POSITION=NW;
RUN;
```

❶ The CLASS statement is used to create a separate analysis and plot for each value of Vendor.

❷ The NROW=3 option specifies that three plots will appear together in three rows on the same page.

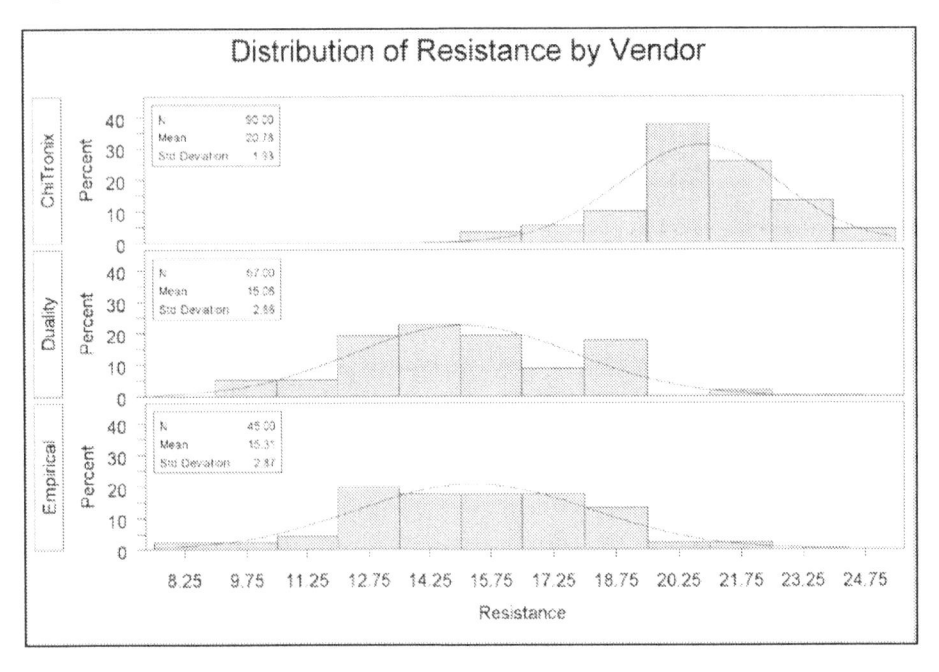

Figure 5-11: Using a CLASS Statement with PROC UNIVARIATE

This plot includes the same mean values of resistance for each vendor, which were computed with PROC MEANS, TABULATE, and REPORT in Sections 5.2–5.4, but the histograms provide a much more compelling visual display of the differences. These plots also make it clear that the variability (standard deviation) of the resistance measurements is less for the ChiTronix units.

Use two class variables to create a two-way layout of histograms of resistance by Vendor and Month.

```
TITLE HEIGHT=5 "Distribution of Resistance by Vendor and Month";
PROC UNIVARIATE DATA=JES.Results_Q4;
   CLASS Vendor Month; ❶
   VAR Resistance;
   HISTOGRAM Resistance / NORMAL(MU=est SIGMA=est) NROW=3 NCOL=3; ❷
   INSET N MEAN
   / HEIGHT=3 FORMAT = 5.3 POSITION=NE;
RUN;
```

❶ The CLASS statement calls for a separate plot for each combination of Vendor and Month.

❷ The NROW=3 NCOL=3 options specify the layout of the plots in three rows and three columns.

Figure 5-12: Using Two CLASS Variables with PROC UNIVARIATE

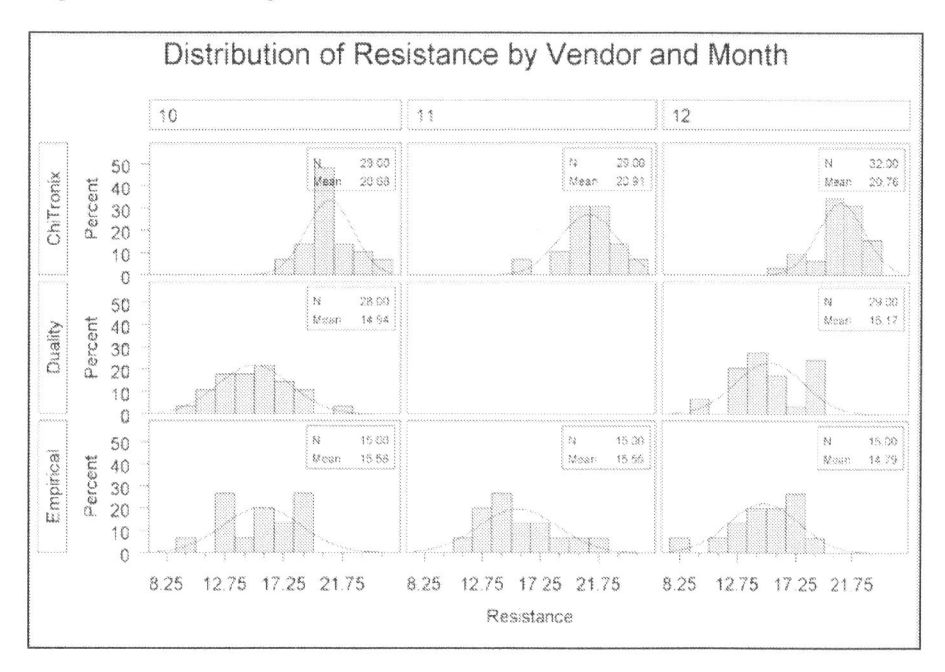

This plot enables you to simultaneously compare the distributions of resistance for every combination of Vendor and Month. For example, you can see that the resistance values for the ChiTronix units have a higher mean and lower standard deviation, for each month, compared to units from the other two vendors.

Some of the other capabilities of PROC UNIVARIATE include

- Plots, fits and tests for Beta, Exponential, Gamma, Lognormal, and Weibull distributions
- Goodness-of-fit tests, including Shapiro-Wilk, Kolmogorov-Smirnov, and Cramer-von Mises
- Quantile-Quantile Plots, Kernel density estimates, and plots

For more information consult the SAS online documentation:

Base SAS → Base SAS Procedures Guide: Statistical Procedures → The UNIVARIATE Procedure

5.8 PROC FREQ

PROC FREQ is used to analyze contingency tables by performing all of the standard tests of association, including chi-square, Fisher's exact test, and many others. The example in this section uses PROC FREQ to test whether the likelihood of failing the resistance specification is independent of Vendor. Each record in the JES.Results_Q4 data set (see Section 5.1) can be classified by the values of Vendor (ChiTronix, Duality or Empirical) and Fail (0 = did not fail or 1 = failed). The code for this section is in ~\JES \sample_code\ch_5 \freq_examples.sas.

```
%INCLUDE "&JES.sample_code\ch_5\random_data_5.sas";
PROC FREQ DATA=JES.Results_Q4;
   TABLES Vendor*Fail /CHISQ FISHER; ❶
RUN;
```

❶ The TABLES statement requests a contingency table with rows containing the values of Vendor, and columns containing the values of Fail. The CHISQ option requests a chi-square test, and the FISHER option requests Fisher's exact test.

When this code is run, the tabular results are displayed in the Output window. The first part of the output is the 3 X 2 contingency table of Vendor by Fail. There are four numbers in each cell, corresponding to the four items listed in the upper-left area of the output. For example, in the top left cell we have:

- Frequency = 74
- Percent = 74/192 = 38.54%
- Row Pct = 74/90 = 84.22%
- Col Pct = 74/157 = 47.13%

Following the contingency table there are the results of whatever tests have been requested. First, because the CHISQ option was specified, there are three different versions of the chi-square test, followed by three measures of association. The "Prob" values are the likelihood of the observed chi-square statistic (or a higher value) under the null hypothesis. These values are all around 96%, so the null hypothesis of the failure rate is the same for all vendors and cannot be rejected at any significance level. In Section 5.6.3 PROC ANOM was used with a PCHART statement to test essentially the same hypothesis, with similar results.

The next output table gives the results from Fisher's exact test. More precisely, as the Fisher test is defined for 2 X 2 tables only, these are the results from the Freeman-Halton extension of Fisher's exact test to the case of R X C tables. The "Table Probability(P)" is the likelihood of getting exactly the observed values under the null hypothesis, which is not too interesting. The more important number is "Pr <= P", which is the likelihood of the observed values, or any more extreme set of values. As this number is 97%, this test also does not reject the hypothesis of independence.

Note that Fisher's exact test is usually recommended as an appropriate alternative to the chi-square test for contingency tables in which one or more of the cell counts are small, e.g., 5 or less. So it is not really needed here, but is included to illustrate the syntax.

```
The FREQ Procedure

Table of Vendor by Fail
Vendor     Fail
Frequency |
Percent   |
Row Pct   |
Col Pct   |        0|        1|  Total
-----------+---------+---------+
ChiTronix  |     74 |     16 |     90
           |  38.54 |   8.33 |  46.88
           |  82.22 |  17.78 |
           |  47.13 |  45.71 |
-----------+---------+---------+
Duality    |     46 |     11 |     57
           |  23.96 |   5.73 |  29.69
           |  80.70 |  19.30 |
           |  29.30 |  31.43 |
-----------+---------+---------+
Empirical  |     37 |      8 |     45
           |  19.27 |   4.17 |  23.44
           |  82.22 |  17.78 |
           |  23.57 |  22.86 |
-----------+---------+---------+
Total            157       35      192
              81.77    18.23   100.00

Statistics for Table of Vendor by Fail

Statistic                      DF       Value       Prob
-----------------------------------------------------------
Chi-Square                      2      0.0622     0.9694
Likelihood Ratio Chi-Square     2      0.0616     0.9697
Mantel-Haenszel Chi-Square      1      0.0022     0.9625
Phi Coefficient                        0.0180
Contingency Coefficient                0.0180
Cramer's V                             0.0180

        Fisher's Exact Test
-----------------------------------
Table Probability (P)     0.0291
Pr <= P                   0.9703
Sample Size = 192
```

For information on the many other options available in PROC FREQ consult the SAS online documentation:

Base SAS → Base SAS Procedures Guide: Statistical Procedures → The FREQ Procedure

5.9 More Than Enough

PROC SQL can be used for some of the data summarization tasks described in this chapter as well as some of the concatenation, sorting, merging, and selection tasks described in Chapter 2. Tables 5-3 and 5-4 show the syntax for some of the tasks which can be done either with PROC SQL or by some other method, such as a DATA step or PROC MEANS. The code for this section is in ~\JES \sample_code\ch_5 \sql_examples.sas. These examples use the JES.First data set shown in Section 9.1.1, the JES.Units and JES.Fails data sets shown in Section 2.7, and the JES.Units_U data set, which is the same as the Units_U data set created in Section 2.7.2. In each example, the Dset data set created with PROC SQL is identical to the Dset2 data set created with other methods, except as noted.

Table 5-3: Performing the Same Operation with PROC SQL or Other Methods

PROC SQL	Other Methods
Create New Variables	

`PROC SQL;`
`    CREATE TABLE Dset AS`
`    SELECT A.*, Fail+Defects AS Tot`
`    FROM JES.First A;`
`QUIT;`
	`DATA Dset2; SET JES.First;`
	`    Tot=Fail+Defects;`
	`RUN;`

Select Columns	

`PROC SQL;`
`    CREATE TABLE Dset AS`
`    SELECT Batch, Sample, Result`
`    FROM JES.First;`
`QUIT;`
	`DATA Dset2;`
	`    SET JES.First;`
	`    KEEP Batch Sample Result;`
	`RUN;`

Select Rows	

`PROC SQL;`
`    CREATE TABLE Dset AS`
`    SELECT * FROM JES.First`
`    WHERE Result="Low Res";`
`QUIT;`
	`DATA Dset2;`
	`    SET JES.First;`
	`    IF Result="Low Res";`
	`RUN;`

Select Distinct Values of a Variable	

`PROC SQL;`
`    CREATE TABLE Dset AS`
`    SELECT distinct SN from`
`JES.Units;`
`QUIT;`
	`PROC SORT DATA=JES.Units OUT=Dset2;`
	`    BY SN;`
	`RUN;`
	`DATA Dset2(keep=SN); SET Dset; BY SN;`
	`    IF LAST.SN;`
	`RUN;`

Select Rows with the Earliest Install Date	

`PROC SQL;`
`    CREATE TABLE Dset AS`
`    SELECT * FROM JES.Units`
`    GROUP BY SN`
`    HAVING Install=MIN(Install);`
`QUIT;`
	`PROC SORT DATA=JES.Units OUT=Dset2;`
	`    BY SN Install;`
	`RUN;`
	`DATA Dset2; SET Dset2; BY SN INSTALL;`
	`    IF First.SN;`
	`RUN;`

Compute Summary Statistics ❶	

`PROC SQL;`
`    CREATE TABLE Dset AS`
`    SELECT Vendor, Month,`
`        SUM(Defects) AS Sum,`
`        MEAN(Defects) AS Avg`
`    FROM JES.Results`
`    GROUP BY Vendor, Month;`
`QUIT;`
	`PROC MEANS DATA=JES.Results NWAY;`
	`    CLASS Vendor Month;`
	`    OUTPUT OUT=Dset2`
	`    SUM(Defects)=Sum`
	`    MEAN(Defects)=Avg;`
	`RUN;`

❶ Dset2 has two additional variables, _TYPE_ and _FREQ_.

Table 5-4: Performing the Same Operation with PROC SQL or Other Methods

PROC SQL	Other Methods
Concatenate Data Sets ❶	
```PROC SQL; CREATE TABLE Dset AS SELECT * from JES.First UNION SELECT * from JES.Second;QUIT;```	```DATA Dset2; SET JES.First JES.Second;RUN;```
Merge Data Sets – Inner Join ❶ ❷	
```PROC SQL; CREATE TABLE Dset AS SELECT a.SN, a.Install, a.Loc, b.Fail FROM JES.Units_U a, JES.Fails b WHERE a.SN = b.SN;QUIT;```	```PROC SORT DATA=JES.Units_U (KEEP=SN Install Loc) OUT=Units; BY SN;RUN;PROC SORT DATA=JES.Fails(KEEP=SN Fail) OUT=Fails; BY SN;RUN;DATA Dset2; MERGE Units(IN=u) Fails(IN=f); BY SN; IF u AND f;RUN;```
Merge Data Sets – Outer Join ❶ ❷	
```PROC SQL NOPRINT; CREATE TABLE Dset AS SELECT COALESCE(a.SN, b.SN) AS SN, a.Install, a.Loc, b.Fail FROM JES.Units_U AS a FULL JOIN JES.Fails AS b ON a.SN=b.SN;QUIT;```	```PROC SORT DATA=JES.Units_U (KEEP=SN Install Loc) OUT=Units; BY SN;RUN;PROC SORT DATA=JES.Fails(KEEP=SN Fail) OUT=Fails; BY SN;RUN;DATA Dset2; MERGE Units Fails; BY SN;RUN;```

❶ The Dset2 data set has the same rows as Dset, but in a different order.

❷ The Dset and Dset2 data sets will NOT be the same if the variable you are matching on, SN in the example, has duplicate values in both data sets. For example, the JES.Units and JES.Fails data sets (Section 2.7) both have two rows with SN='0035'. If you run the same code with JES.Units instead of JES.Units_U, the Dset data set will have four rows with SN='0035', while Dset2 will have only two rows with SN='0035', as shown here for the inner join example. PROC SQL includes a row for each possible combination of rows with SN='0035' from the two data sets, while the MERGE statement includes only one match for each row with SN='0035'.

```
Dset
SN        Install         Fail
0007      06/16/2006      09/05/2006
0007      06/09/2006      09/05/2006
0016      06/06/2006      07/22/2006
0027      06/05/2006      07/17/2006
0035      06/17/2006      08/05/2006
0035      06/11/2006      08/05/2006
0035      06/17/2006      09/06/2006
0035      06/11/2006      09/06/2006
0061      06/09/2006      08/02/2006
```

```
Dset2
SN        Install         Fail
0007      06/16/2006      09/05/2006
0007      06/09/2006      09/05/2006
0016      06/06/2006      07/22/2006
0027      06/05/2006      07/17/2006
0035      06/17/2006      08/05/2006
0035      06/11/2006      09/06/2006
0061      06/09/2006      08/02/2006
```

5.10 Chapter Summary

5.10.1 Recap

After finishing this chapter you should know how to

- Compute the sum, mean, standard deviation, and other statistics for numeric variables in a data set:
    - for all records in the data set
    - separately for each group defined by a unique combination of one or more classification variables
- Prepare a report on these results using PROC MEANS, PROC TABULATE or PROC REPORT.
- Test for the significance of differences in the means among such groups:
    - by comparing confidence limits for each pair of group
    - using Analysis of Means for simultaneous testing of all groups
    - using PROC FREQ for Count variables, e.g., number failing in each group
- Use PROC UNIVARIATE to test whether a sample comes from a normal distribution.

5.10.2 For More Information

Books

Carpenter, Art, 2007. *Carpenter's Complete Guide to the SAS® Report Procedure*, Cary NC: SAS Institute Inc.

Cody, Ronald P., and Jeffrey K. Smith. 2006. *Applied Statistics and the SAS® Programming Language*. Upper Saddle River, NJ: Pearson Prentice Hall.

Delwiche, Lora D., and Susan J. Slaughter. 2003. *The Little SAS® Book: A Primer, Third Edition*. Cary, NC: SAS Institute Inc. Chapter 4, Sections 4.9- 4.21 give an excellent introduction to PROC MEANS, FREQ, TABULATE, and REPORT.

Haworth, Lauren E. 1999. *PROC TABULATE by Example*. Cary, NC: SAS Institute Inc.

Lafler, Kirk Paul. 2004. *PROC SQL: Beyond the Basics Using SAS®*. Cary, NC: SAS Institute Inc.

SAS Institute Inc. 1995. *Logistic Regression Examples Using the SAS® System, Version 6, First Edition*. Cary, NC: SAS Institute Inc.

Stokes, Maura E., Charles S. Davis, and Gary G. Koch. 2000. *Categorical Data Analysis Using the SAS® System, Second Edition*. Cary, NC: SAS Institute Inc.

SAS Conference Papers

These papers can be found in ~\JES\docs. The filenames are shown in bold below—e.g. **[134-30.pdf]**

Carpenter, Arthur L. 2005. "PROC REPORT Basics: Getting Started with the Primary Statements." *Proceedings of the Western Users of SAS Software Thirteenth Annual Conference*. San Jose, CA. **[how_proc_report_basics.pdf]**

Carpenter, Arthur L. 2008. "PROC REPORT: Compute Block Basics—Part I Tutorial."
Proceedings of the SAS Global Forum 2008 Conference. Cary, NC: SAS Institute Inc.
Paper 031-2008. **[031-2008.pdf]**

Droogendyk, Harry, and Faisal Dosani. 2008. "Joining Data: Data Step Merge or SQL?"
Proceedings of the SAS Global Forum 2008 Conference. Cary, NC: SAS Institute Inc.
Paper 178-2008. **[178-2008.pdf]**

Haworth, Lauren. 2003. "SAS Reporting 101: REPORT, TABULATE, ODS and Microsoft
Office." *Proceedings of the Twenty-Eighth Annual SAS Users Group International
Conference*. Cary, NC: SAS Institute Inc. Paper 71-28. **[p71-28.pdf]**

Lafler, Kirk Paul. 2006. "A Hands-On Tour Inside the World of PROC SQL." *Proceedings of the
Thirty-First Annual SAS Users Group International Conference*. Cary, NC: SAS Institute
Inc. Paper 114-31. **[114-31.pdf]**

Pass, Ray, and Daphne Ewing. 2006. "So You're Still Not Using PROC REPORT. Why Not?"
Proceedings of the Thirty-First Annual SAS Users Group International Conference. Cary,
NC: SAS Institute Inc. Paper 235-31. **[235-31.pdf]**

Williams, Christianna S. 2008. "PROC SQL for DATA Step Die-hards." *Proceedings of the SAS
Global Forum 2008 Conference*. Cary, NC: SAS Institute Inc. Paper 185-2008. **[185-2008.pdf]**

Other Papers

McGill, Robert, John W. Tukey, and Wayne A. Larsen. 1978. "Variations of Box Plots." *The
American Statistician,* February 1978, Vol. 32, No. 1. pp. 12-16.

Schenker, Nathaniel, and Jane F. Gentleman. 2001. "On Judging the Significance of Differences
by Examining the Overlap Between Confidence Intervals." *The American Statistician,*
August 2001, Vol. 55, No. 3. pp. 182-186.

5.10.3 Exercises

The JES.Results_Ex data set contains vendor test data in the same form as the JES.Results data set (see Section 5.1.1). Use the methods described in this chapter to answer the following questions for the JES.Results_Ex data:

- Estimate the mean value of resistance, defect rate, and probability of failure for each vendor, for each month, and for each combination of vendor and month.
- Compute confidence limits on the means computed above.
- Test whether the differences among vendors are statistically significant:
    - by comparing the confidence limits on the means for each vendor
    - using notched box plots
    - by using Analysis of Means
- Test the hypothesis that the values of resistance are normally distributed:
    - for the complete data set
    - for the values from each vendor separately
- Use PROC TABULATE and PROC REPORT to prepare summary tables of the means by Vendor and Month.

Chapter 6

Plotting Your Data with SAS/GRAPH Software

- 6.1 Introduction 138
- 6.2 Viewing, Saving, and Naming Your Graphic Ouput 142
- 6.3 PROC GPLOT 146
- 6.4 PROC GCHART 180
- 6.5 More Than Enough 198
- 6.6 Chapter Summary 208

6.1 Introduction

The best way to understand your data is to plot it in such a way that the important relationships are clearly visible. For example, Figure 6-1 shows the distribution of Resistance and Delay for each vendor, the relationship of Delay to Resistance, and how the relationship varies among the vendors. Figure 6-2 shows the increasing defect rate for ChiTronix components, the statistical significance of the trend, and its relation to the specification limit. The BOXPLOT, ANOM, and UNIVARIATE procedures discussed in Chapter 5 produce graphical output that is helpful for understanding your data, but these procedures create only a limited number of graph types. In this chapter we focus on the more general-purpose plotting capabilities of the procedures available in SAS/GRAPH software.

Table 6-1 lists the types of graphs available in SAS/GRAPH. Section 6.2 shows how to view and save the graphs created with any of these procedures. The next two sections cover the most commonly used graph types. Section 6.3 explains how to use the PLOT and PLOT2 statements with PROC GPLOT to create a wide variety of scatter plots and line plots, including those shown in Figures 6-1 and 6-2. Section 6.4 explains how to use the HBAR and VBAR statements with PROC GCHART to create horizontal and vertical bar charts. Section 6.5 gives a brief introduction to many of the other graph types available in SAS/GRAPH, including choropleth, contour and surface plots that can be used, for example, to plot wafer maps.

This chapter shows you how to save your graphic output in various image formats, including GIF, JPEG, PNG, and EPS, which can be imported or pasted into your documents. In Chapter 7, "The Output Delivery System," you will see how to export your graphical output directly into HTML, PDF, and RTF documents.

Four new statistical graphics, or "SG", procedures were introduced with SAS 9.2. One of these procedures, SGPLOT, provides most of the same capabilities as PROC GPLOT and PROC GCHART, but with major improvements. SGPLOT takes advantage of ODS graphic templates to create consistent and good looking graphical output by default, with less coding than would be required using GPLOT or GCHART. The SG procedures are discussed in Chapter 8, "Plotting Your Data with ODS Graphics."

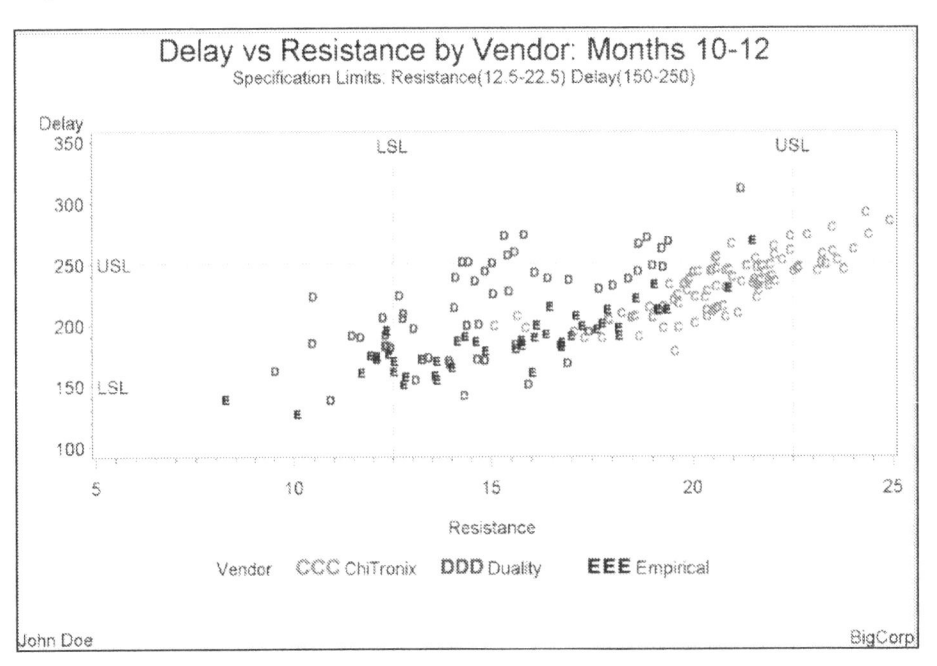

Figure 6-1: Scatter Plot of Delay vs. Resistance

Table 6-1: Graph Types Available in SAS/GRAPH Software

Procedure	Statement	Graph Type
GAREABAR	HBAR	Horizontal bar chart with bar length and width representing two variables.
	HBAR3D	Same as HBAR, but with a 3-D effect on the bars.
	VBAR	Same as HBAR, but with vertical bars.
	VBAR3D	Same as VBAR, but with a 3-D effect on the bars.
GBARLINE	BAR	Vertical bar chart.
	PLOT	Line plot overlaid on the bar chart created with the BAR statement.
GCHART	HBAR	Horizontal bar chart.
	HBAR3D	Horizontal bar chart with 3-D effect on the bars.
	VBAR	Vertical bar chart.
	VBAR3D	Vertical bar chart with 3-D effect on the bars.
	BLOCK	Blocks set in a 2-D grid defined by two variables, with height proportional to a third variable.
	PIE	Pie chart.
	PIE3D	Pie chart with 3-D effect.
	DONUT	Pie chart with a hole in the middle for explanatory text.
	STAR	Straight lines radiating from the center of a circle, with each line length proportional to the value of another variable.
GCONTOUR	PLOT	Contour plot representing the levels of one variable for positions on the plane defined by the values of two other variables.
GMAP	CHORO	Two-dimensional maps in which values of a variable are represented by areas of varying patterns and colors.
	BLOCK	Three-dimensional block maps on which values of a variable are represented by blocks of varying height, pattern and color.
	PRISM	Three-dimensional prism maps on which values of a variable are represented by polyhedrons of varying height, pattern and color.
	SURFACE	Three-dimensional surface maps on which values of a variable are represented by spikes of varying height.
GPLOT	PLOT	Line or scatter plot of one variable, Y, on the vertical axis vs. another variable, X, on the horizontal axis.
	PLOT2	Overlays a plot of a second Y variable, with a different vertical axis, vs. the same X variable, on a graph created with a PLOT statement.
	BUBBLE	An array of circles with the location of the centers determined by the values of two variables, and the radii proportional to the value of a third variable.
	BUBBLE2	Adds a second bubble plot to a graph created with a BUBBLE statement.
GRADAR	CHART	Straight lines radiating from the center of a circle, with each line representing a category based on one variable, and another line connecting points on each radial line representing the relative frequency of a second variable.
G3D	PLOT	Surface described by the values of two horizontal variables and a third vertical variable.
	SCATTER	Similar to a surface plot, but with the data represented by points instead of a surface.

6.1.1 Data for the Examples

The examples in this chapter use the same JES.Results data set used in Chapter 5 (see Section 5.1.1), and also the JES.Results_Tab data set, which contains summary statistics created by running PROC MEANS on the JES.Results data set.

```
PROC MEANS DATA=JES.Results NOPRINT NWAY ALPHA=.10;
  CLASS Vendor Month Mon;
  VAR Defects Resistance Fail;
  OUTPUT OUT=JES.Results_Tab
    N(Defects)=N
  SUM(Defects) = N_Def
  MEAN(Defects)=M_Def
  MEAN(Delay)=M_Del
  LCLM(Delay) = D_L
  UCLM(Delay) = D_U
  MEAN(Resistance)=M_Res
  LCLM(Resistance) = R_L
  UCLM(Resistance) = R_U
  SUM(Fail) = N_Fail
  MEAN(Fail) = P_Fail;
  FORMAT M_Def P_Fail 6.2;
RUN;
DATA JES.Results_Tab; SET JES.Results_Tab;
  LABEL
      Mon="Month"
      M_Res="Mean Resistance"
      R_L="Lower Confidence Limit"
      R_U="Upper Confidence Limit"
      N_Fail="Number of Fails"
      M_Del="Mean Delay"
      D_L="Lower Confidence Limit"
      D_U="Upper Confidence Limit";
  DROP _TYPE_ _FREQ_;
RUN;
```

This code computes the mean value, together with 90% confidence limits, for the Delay and Resistance variables, and the sum and mean for the Fail and Defects variables. This code is explained in Section 5.2.6. The second DATA step adds labels to several variables. These will be used by default to label the axes of the plots created in the examples. If labels are not defined, the variable names are used to label the axes.

```
JES.Results_Tab (first five rows)

Vendor      Month     Mon        N       N_Def     M_Def     M_Del        D_L

ChiTronix      1    2008-01     20        19       0.95       215         206
ChiTronix      2    2008-02     20        24       1.20       205         197
ChiTronix      3    2008-03     23        32       1.39       215         206
ChiTronix      4    2008-04     23        31       1.35       217         210
ChiTronix      5    2008-05     23        24       1.04       223         216

    D_U        M_Res         R_L         R_U       N_Fail     P_Fail

    223        18.59        17.71       19.48        2         0.10
    214        17.95        17.23       18.67        0         0.00
    223        18.52        17.70       19.35        1         0.04
    223        18.84        18.23       19.46        0         0.00
    230        19.78        19.11       20.45        2         0.09
```

The plot in Figure 6-1 uses the full JES.Results data set. The plot in Figure 6-2 uses the summary data set, JES.Results_Tab, to create a trend plot of defect rate by month, together with confidence intervals. This kind of plot enables you to see significant differences much more easily than by looking at the numbers in the table. The code used to create this plot is shown in Section 6.3.3.4.

Figure 6-2: Trend Plot of Defect Rate for ChiTronix Units

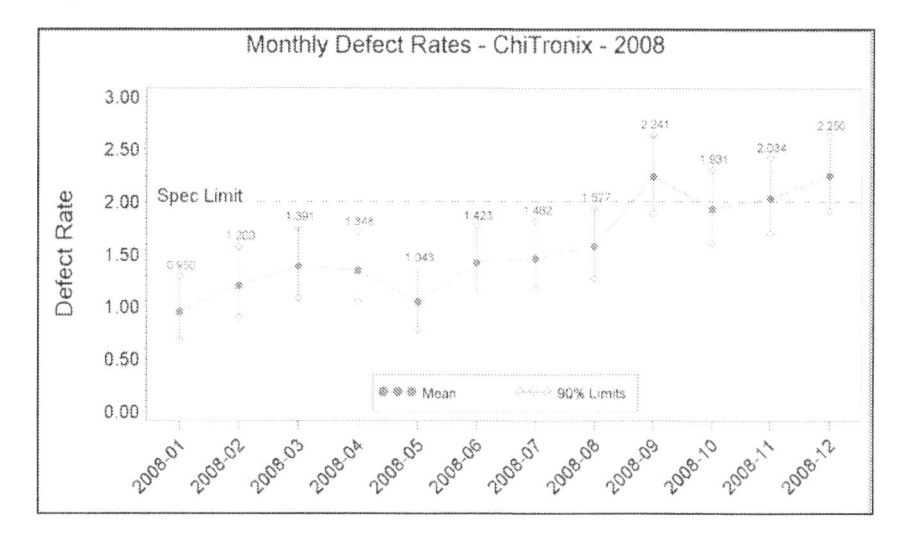

142 *Just Enough SAS: A Quick-Start Guide to SAS for Engineers*

6.2 Viewing, Saving, and Naming Your Graphic Output

The graphic output from SAS/GRAPH procedures can be viewed in the Graph window, and can also be saved in any of several image formats, including GIF, PNG, JPEG and Encapsulated PostScript (EPS). The code for this section is in ~\JES \sample_code\ch_6 \view_save.sas.

The first few lines of code initialize the environment to a known state. The same lines are included at the beginning of each of the code samples in this chapter.

```
PROC DATASETS MEMTYPE=CAT NOLIST; DELETE GSEG; RUN; QUIT;  ❶
GOPTIONS RESET=ALL BORDER;  ❷
```

❶ The PROC DATASETS statement deletes the WORK.GSEG catalog, which contains any previously created graphic output. A detailed discussion of the GSEG catalog is beyond the scope of this book, but you should know how to use this line to delete it. The catalog is temporary and will in any case be deleted when you close the SAS session. The reason why it should be deleted before running the examples is because the names assigned to graphs can be affected by any previously assigned names, as explained in the discussion following Display 6-2.

❷ The GOPTIONS statement resets all graphics options to their default values, and then sets the BORDER option, which specifies that a rectangular border be drawn around graphic output. Graphic options are discussed in Section 6.3.2.

The next section of code creates a trend plot of average resistance (M_Res) vs. Month by Vendor. The statements used to create the graph will be explained later. The point of this example is just to explain how to view your graphic output.

```
GOPTIONS GUNIT=PCT HTEXT=3 FTEXT='Arial' HTITLE=3;
SYMBOL1 VALUE=dot    HEIGHT=2 COLOR=green   WIDTH=2 INTERPOL=JOIN;
SYMBOL2 VALUE=square HEIGHT=2 COLOR=red     WIDTH=2 INTERPOL=JOIN;
SYMBOL3 VALUE=plus   HEIGHT=2 COLOR=blue    WIDTH=2 INTERPOL=JOIN;
PROC GPLOT DATA =JES.Results_Tab;
   PLOT M_Res*Month=Vendor;
RUN; QUIT;
```

When this code is run, the graph shown in Display 6-2 is displayed in the Graph window. If the Graph window does not open automatically, you can open it either by selecting **View → Graph** from the menu at the top of the screen, or by double-clicking the link **Plot of M_Res...** in the Results window, as shown in Display 6-1. If the Explorer window is on top, you will have to click on **Results** at the bottom to bring the Results window to the front.

Display 6-1: You Can Use the Results Window to Open Your Graphs in the Graph Window

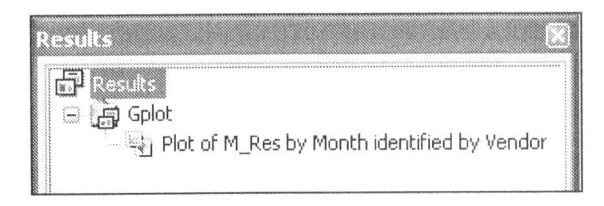

The Results window contains a separate entry for each item that you create in the Output or Graph windows. This provides a convenient way to locate and view your results. The items in the Output, Graph, and Results windows are temporary and are deleted when you close the SAS session.

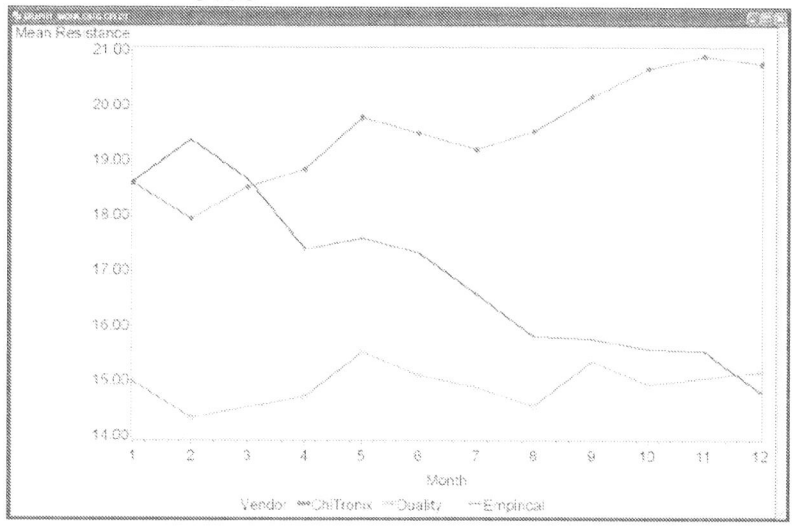

Display 6-2: The Graph Window Displaying the Trend Plot Created by the Sample Code

SAS assigns a name to each graph that you create. The default name is the same as the name of the procedure that created it. In the bar at the top of the Graph window, you can see "WORK.GSEG.GPLOT", which means that this is the graph named GPLOT in the GSEG catalog in the WORK library. (This name is barely legible in Display 6-2, but easy to read in your Graph window.) If you create more than one graph with the same procedure, SAS stores the graphs in the same catalog, and adds a number to the end of the name of each subsequent graph. For example, if you run the same sample code two more times, the new graphs will be named GPLOT1 and GPLOT2. Deleting WORK.GSEG resets the counter so that the next plot name will be GPLOT again. That is why the sample code begins by deleting GSEG—to ensure that the graph names you see in your output are the same as those shown in the book.

The next example shows how to use the NAME= option to assign your own names to your graphs.

This example shows how to save your graph as an image file with a name of your own choosing rather than the default GPLOT*n*.

```
FILENAME Fig "&JES.figures/Chapter_6/";   ❶
GOPTIONS DEVICE=GIF   ❷
   XMAX=6IN YMAX=3.375IN   ❸
   GSFNAME=Fig   ❹
   GSFMODE=REPLACE;   ❺
PROC GPLOT DATA =JES.Results_Tab;
   PLOT M_Res*Month=Vendor / NAME="F6_3_";   ❻
RUN; QUIT;
```

❶ The FILENAME statement defines a file reference (*fileref*), Fig, for the directory where you want to send the output.

❷ The GOPTIONS statement specifies the characteristics of the image file to be saved. The DEVICE=GIF option specifies that the GIF device driver be used. A few of the more useful device drivers are listed in Table 6-2. Consult the SAS online documentation for a complete list of available device drivers.

❸ The XMAX and YMAX options specify the dimensions of the graph to be 6 inches by 3.375 inches.

❹ The GSFNAME=Fig option specifies that the image file be saved in the folder referred to by the Fig fileref.

❺ The GSFMODE=REPLACE option specifies that the new image file will replace any existing file with the same name.

❻ The NAME option specifies that the graph be named "F6_3_". If you rerun the same code several times, the new graphs are named F6_3_1, F6_3_2, etc.

When you run this code, the f6_3_.gif image shown in Figure 6-3 is saved in the **~JES\figures\Chapter_6 folder**. The next bit of code uses the DEVICE=PSLEPSFC option to save the same graph as an EPS file, named f6_4_.ps, in the same folder. This graph is shown in Figure 6-4.

```
GOPTIONS DEVICE=PSLEPSFC XMAX=6IN YMAX=3.375IN
   GSFNAME=Fig GSFMODE=REPLACE;
PROC GPLOT DATA = JES.Results_Tab;
   PLOT M_Res*Month=Vendor / NAME="F6_4_";
RUN; QUIT;
```

Table 6-2: Selected Device Drivers for Graphic Image Output

Driver	Image Type
Vector Formats	
PSLEPS	Encapsulated PostScript - Grayscale
PSLEPSFC	Encapsulated PostScript - Color
Bitmap Formats	
GIF	Graphics Interchange Format (GIF)
JPEG	Joint Photographers Export Group (JPEG)
PNG	Portable Network Graphics (PNG)

Vector formats store graphics as a set of geometric shapes or lines, and can produce very high resolution output, depending on the capability of the display or printer used to render the image. Bitmap formats store graphics as the lightness and color values of each pixel, and the output resolution is limited by the number of pixels stored. In general, vector formats are recommended for data charts and plots such as those created in this book, while bitmap formats are recommended for more complex images, such as photographs, which are not easily represented by simple shapes. Note, however, that bitmap formats are very convenient to work with, and are more than adequate for charts used in most presentations and Web pages. The plot in Figure 6-3 looks much better on a Web page than it does on this page. Therefore, we recommend the use of GIF, JPEG, or PNG images for most of your graphical output, but have used EPS images to create most of the figures in this book. Chapter 7, "The Output Delivery system," will show you how to easily export your bitmap graphics to HTML, RTF, or PDF documents.

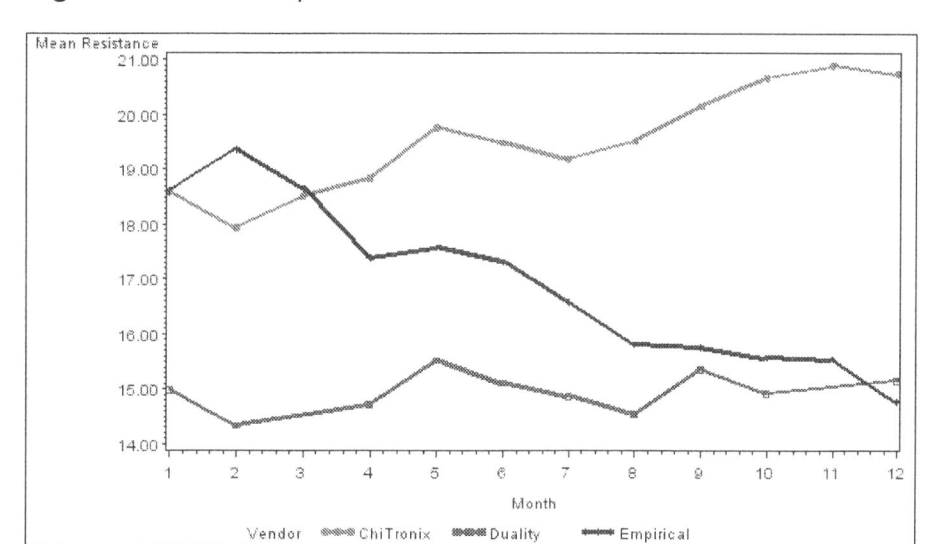

Figure 6-3: This Graph Was Saved in GIF Format

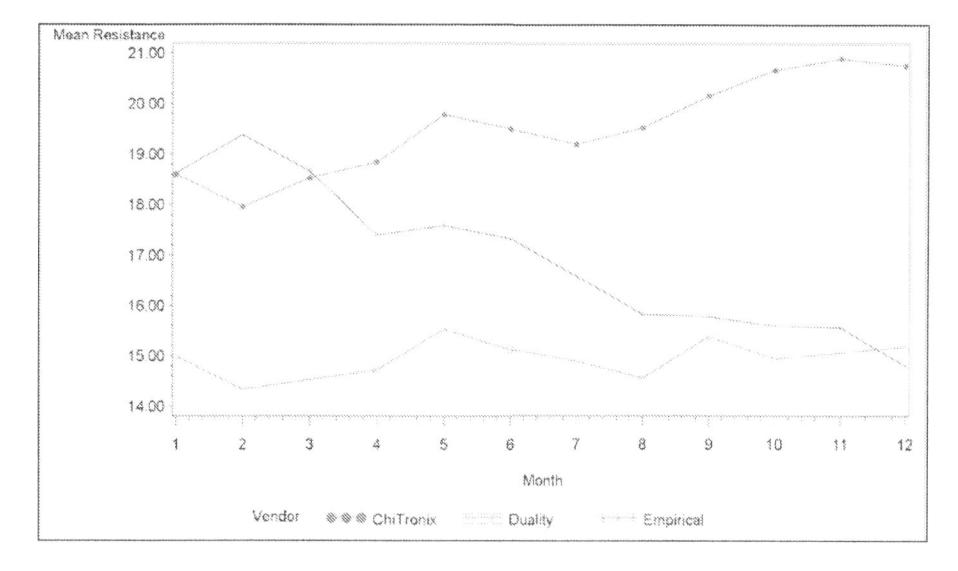

Figure 6-4: This Graph Was Saved in EPS Format

6.3 PROC GPLOT

As shown in Table 6-1, PROC GPLOT can be used with PLOT and PLOT2 statements to create scatter plots or line plots, or with BUBBLE and BUBBLE2 statements to create bubble plots. This section covers only the PLOT and PLOT2 statements. Section 6.5.3 contains a brief example using BUBBLE and BUBBLE2 statements.

Section 6.3.1 explains the basic forms of plots that can be created using the PLOT and PLOT2 statements in PROC GPLOT. Section 6.3.2 shows how to apply various options to control the appearance of your plots, including titles, plot symbols, axes, legends, etc. Finally, Section 6.3.3 presents an extended example of plotting group confidence intervals of the type shown in Figure 6-2. These plots provide a visual comparison of groups that is a very useful complement to the group comparison plots in Section 5.5 (box plots) and Section 5.6 (ANOM plots).

6.3.1 The Basic Forms of the PLOT and PLOT2 Statements

PROC GPLOT can be used to create scatter plots or line plots of the values in a data set. Each plot point is determined by the values of two variables in the same row of the data set. A variable which is plotted on the horizontal axis is referred to as an X variable, and a variable which is plotted on the vertical axis is referred to as a Y variable. Using this notation, the PLOT and PLOT2 statements can be used to create five basic forms of plot:

1. A simple scatter plot or line plot of Y vs. X.
2. Separate plots of Y vs. X, on separate graphs, for each value of a third variable Z.
3. Separate plots of Y vs. X, on the same graph, for each value of a third variable Z.
4. Plots of two or more Y variables vs. X, using the same Y axis.
   - Or two or more X variables vs. the same Y variable, using the same X axis.
5. Plots of two or more Y variables vs. X, using two different Y axes.
   - Or two or more X variables vs. the same Y variable, using two different X axes.

This section will cover the syntax for each of these plot types. The code for this section is in ~\JES\sample_code\ch_6\gplot_forms.sas.

The examples shown here are very basic, and will illustrate how to get all the points plotted correctly, but with very little attention to the appearance of the resulting graph. Section 6.3.2 will show how to make these plots much more presentable through the use of customized titles, footnotes, symbols, legends, etc. Some SYMBOL statements are used to make the plots easier to understand, but these statements are not explained until Section 6.3.2.6.

The examples in this section use numerical X and Y variables for each plot. You can also plot character variables using the same syntax. Sections 6.3.2.7 and 6.3.3 contain examples where the X or Y variable is a character variable.

6.3.1.1 Simple Plot of Y vs. X

The first form simply plots each pair of values in two specified columns of a data set. The first two lines set some graphic options to make sure that the points and text in the output will be large enough to read when reproduced here. Graphic options will be discussed in Section 6.3.2.

```
GOPTIONS GUNIT=PCT HTEXT=3 FTEXT='Arial' HTITLE=3;
SYMBOL1 VALUE=dot HEIGHT=2 COLOR=Green WIDTH=2 INTERPOL=NONE;
PROC GPLOT DATA = JES.Results_Tab;  ❶
   PLOT M_Res*Month  ❷
/ NAME="F6_5_";  ❸
RUN;  ❹
QUIT;  ❺
```

❶ The PROC GPLOT statement specifies that a plot be created using the JES.Results_Tab data set.

❷ The PLOT statement requests a plot of M_Res, the Y variable, vs. Month, the X variable.

❸ The NAME= option sets the name of the graph to "F6_5_".

❹ The RUN; statement causes the preceding lines to run, but does not actually end the procedure.

❺ The QUIT; statement ends the procedure.

Figure 6-5: Plot of Average Resistance (M_Res) vs. Month

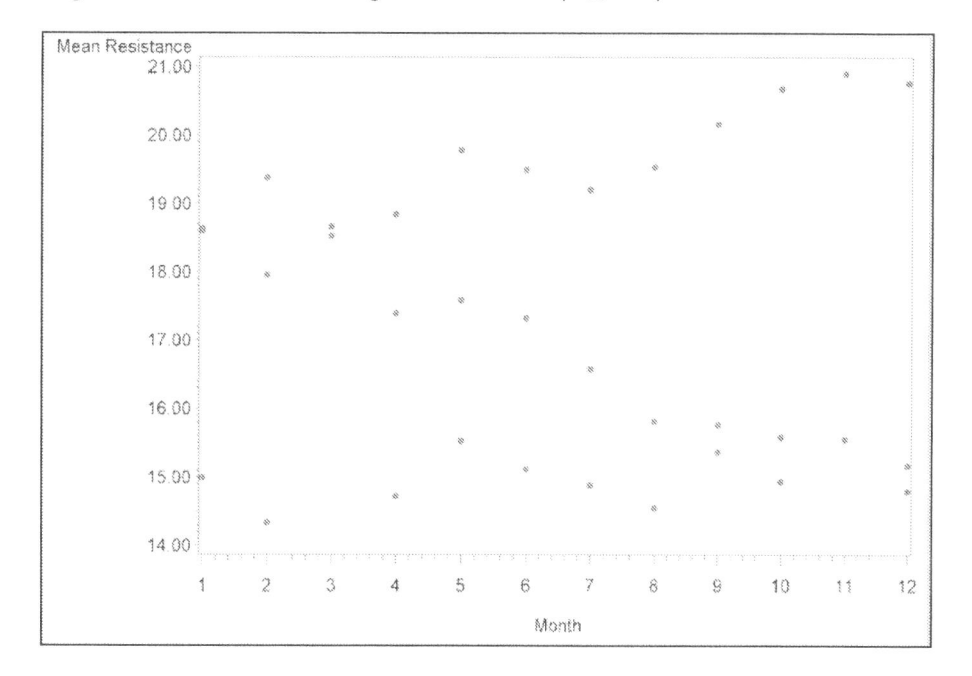

This plot shows the mean value of Resistance by Month for each Vendor. It is correct, but not particularly informative as we cannot distinguish which points are associated with each Vendor. In the next section you will see how to create separate plots for each Vendor.

6.3.1.2 Plots on Separate Graphs for Each Value of a Third Variable

You can use a BY statement to create plots of Y vs. X on separate graphs for each value of a third variable in the data set.

```
PROC GPLOT DATA = JES.Results_Tab; BY Vendor;  ❶
  PLOT M_Res*Month / NAME="F6_6_";
RUN; QUIT;
```

❶ The BY Vendor statement requests separate plots of M_Res vs Month for each distinct value of the variable Vendor. This kind of BY GROUP processing is available for many other SAS procedures, so anytime you need to run the same procedure for each unique value of a variable in the data set, there is a good chance that a BY statement will simplify the process. Consult the SAS documentation for use of BY statements with other procedures.

When this code is run, three separate graphs are created, with the names F6_6_, F6_6_1, and F6_6_2, as shown in Figure 6-6.

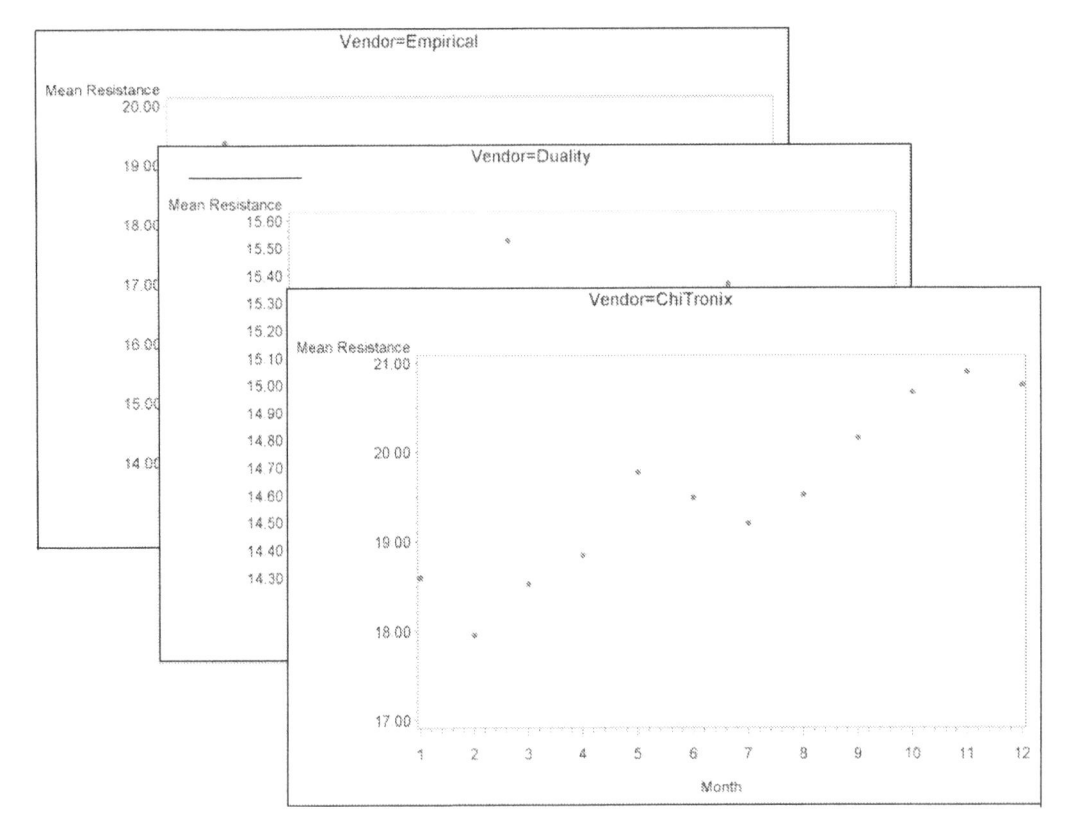

Figure 6-6: Plots of M_Res vs. Month for Each Vendor on Separate Graphs

Note that a default title has been added to each plot to show which value of Vendor was used for that plot.

6.3.1.3 Plots on the Same Graph for Each Value of a Third Variable

It is easier to compare these plots if they are overlaid on the same graph, instead of being plotted on separate graphs.

```
SYMBOL1 VALUE=dot     HEIGHT=2 COLOR=green  WIDTH=2 INTERPOL=JOIN;  ❶
SYMBOL2 VALUE=square  HEIGHT=2 COLOR=red    WIDTH=2 INTERPOL=JOIN;
SYMBOL3 VALUE=plus    HEIGHT=2 COLOR=blue   WIDTH=2 INTERPOL=JOIN;
PROC GPLOT DATA = JES.Results_Tab;
   PLOT M_Res*Month=Vendor  ❷
/ NAME="F6_7_";
RUN; QUIT;
```

❶ The SYMBOL statements specify the type and color of symbols and lines to be used in the plot. These statements are not necessary, but have been included to make the resulting output easier to see. The syntax of these SYMBOL statements is discussed in Section 6.3.2.6.

❷ The '= Vendor' specifies that there should be a separate plot of M_Res vs Month for each distinct value of Vendor.

Figure 6-7: Plots of M_Res vs. Month for Each Vendor on the Same Graph

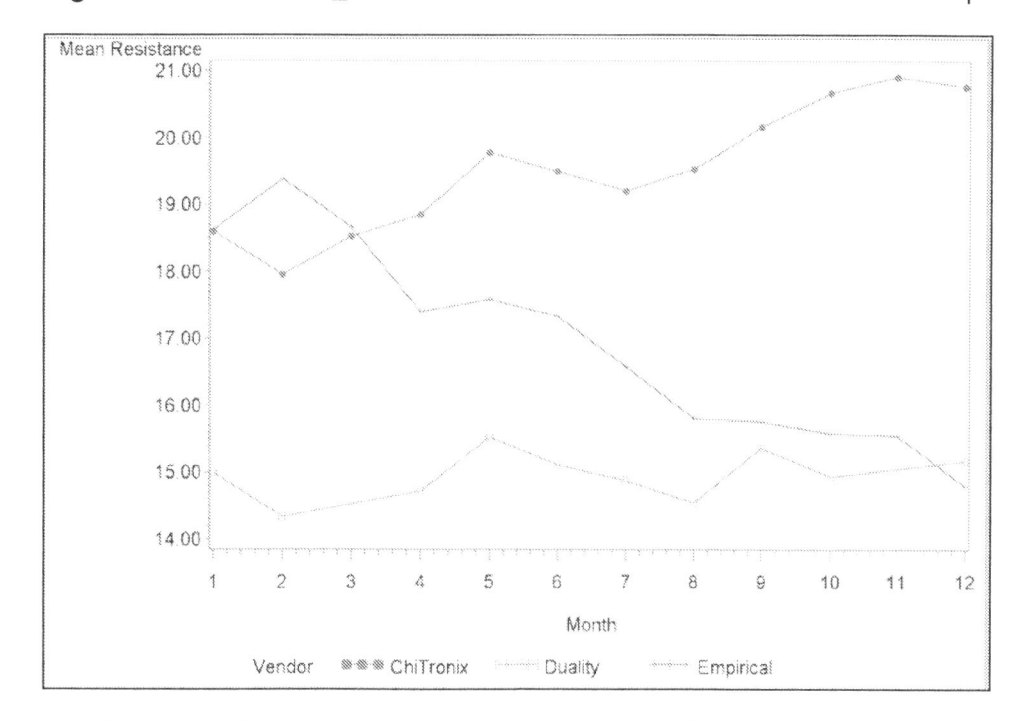

Note that a default legend has been added to indicate which value of Vendor is represented by each curve.

6.3.1.4 Plot Multiple Y Variables vs. X Using the Same Y Axis

The previous examples plotted only one Y variable, M_Res vs. one X variable, Vendor. In many cases you need to plot two or more Y variables vs. the same X variable. In the example, you might want to compare the Delay measurements and Resistance measurements over the same time period.

```
PROC GPLOT DATA = JES.Results_Tab(WHERE=(Vendor="ChiTronix"));  ❶
   PLOT (M_Res M_Del)*Month  ❷
/ NAME="F6_8_" OVERLAY;  ❸
RUN; QUIT;
```

❶ The WHERE clause is used to select only the data for ChiTronix, just to simplify the plot.

❷ (M_Res and M_Del)*Month requests that both M_Res and M_Del be plotted vs. Month.

❸ The OVERLAY option causes the plot lines to be overlaid on the same graph. Without this option, two separate graphs would be created.

Figure 6-8: Plot of M_Res and M_Del vs. Month on the Same Y Axis

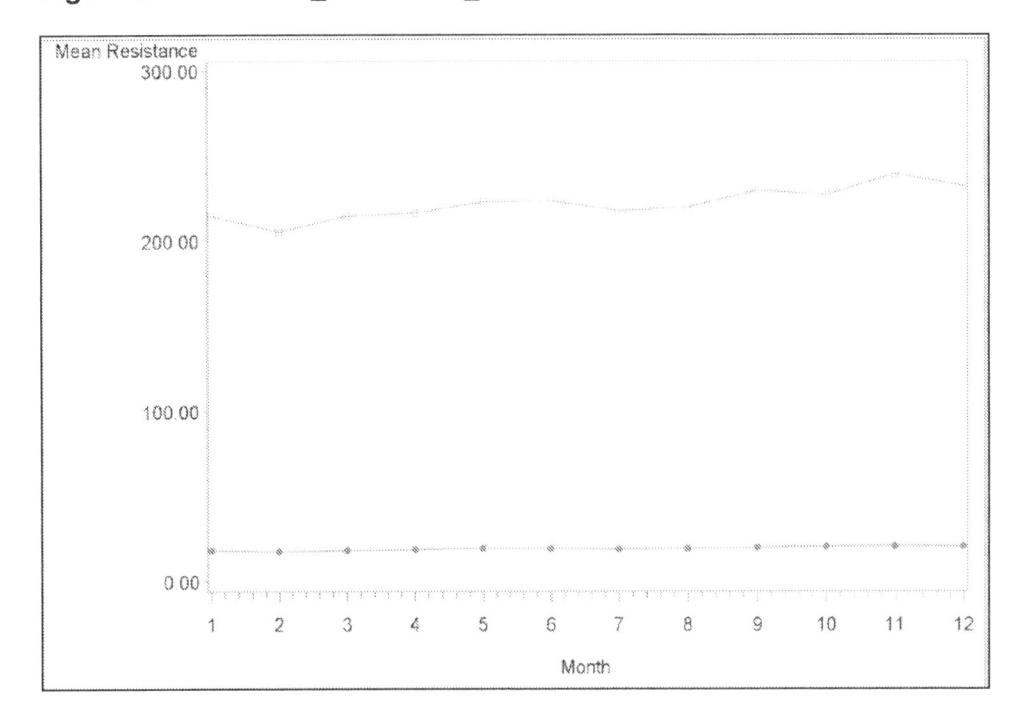

In Section 6.3.2.4 you will see how to add a legend to identify which lines represent Resistance and Delay.

You can also plot a single Y variable vs. multiple X variables as shown in the code below. The output from this code is not shown here, but you can run the example in your sample code to see the result.

```
PROC GPLOT DATA = JES.Results_Tab(WHERE=(Vendor="ChiTronix"));
   PLOT Month*(M_Res M_Del) / NAME="F6_A_" OVERLAY;
RUN; QUIT;
```

6.3.1.5 Plot Multiple Y Variables Using Two Different Y Axes

In Figure 6-8, Delay and Resistance are both plotted against the same Y axis. This makes it difficult to see the variation in either variable because the range of values is so different. You can use a PLOT2 statement to plot the two variables on different Y axes.

```
PROC GPLOT DATA = JES.Results_Tab(WHERE=(Vendor="ChiTronix"));
  PLOT M_Res*Month / NAME="F6_9_";
  PLOT2 M_Del*Month;   ❶
RUN; QUIT;
```

❶ The PLOT2 statement requests a plot of M_Del vs. Month using a different Y axis than that used for the PLOT statement. The Y axis for the PLOT2 statement is shown on the right side of the graph. The X variable in the PLOT2 statement, Month in the example, must be the same as the X variable in the PLOT statement.

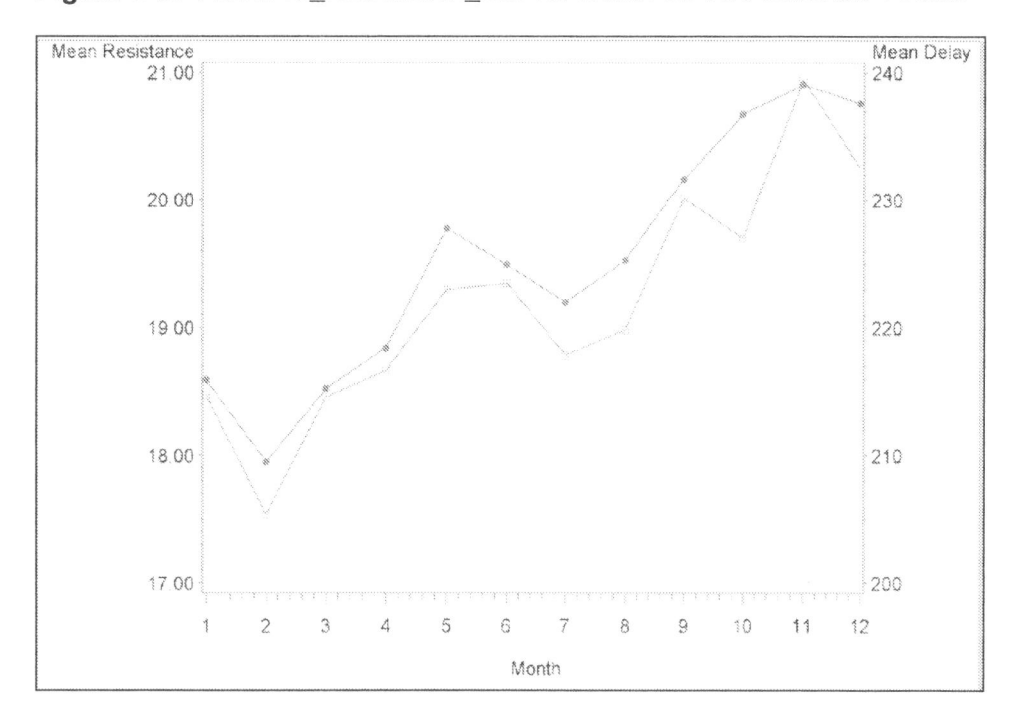

Figure 6-9: Plot of M_Res and M_Del vs. Month on Two Different Y Axes

As you can see, PROC GPLOT has created two different Y axes: one axis on the left for plotting the values of M_Res, and an additional axis on the right for plotting the values of M_Del. This makes it much easier to see the variation in each variable, and the relationship between the variables. It is clear that Delay is strongly correlated with Resistance.

Note that you cannot use a PLOT2 statement to create a second X axis. The code shown below will generate an error message.

```
PROC GPLOT DATA = JES.Results_Tab(WHERE=(Vendor="ChiTronix"));
  PLOT  Month*M_Res / NAME="F6_X_";
  PLOT2 Month*M_Del;
RUN; QUIT;
```

6.3.1.6 Combinations of the Basic Forms
The previous sections showed three ways to plot two or more lines on the same graph.

Section 6.3.1.3: Separate plots of Y vs. X for each value of a third variable Z:

```
PLOT Y*X=Z;
```

Section 6.3.1.4: Plot multiple Y variables vs. the same X variable, on the same Y axis:

```
PLOT(Y1 Y2)*X / OVERLAY;
```

Section 6.3.1.5: Plot multiple Y variables vs. the same X variable, on two different Y axes:

```
PLOT  Y1*X;
PLOT2 Y2*X;
```

Either of the first two forms can be combined with a PLOT2 statement to create more complex plots, as shown in the next two examples. The first example uses a PLOT2 statement with PLOT Y*X=Z.

```
SYMBOL4 VALUE=dot    HEIGHT=2 COLOR=green  WIDTH=2 INTERPOL=JOIN
LINE=2;
SYMBOL5 VALUE=square HEIGHT=2 COLOR=red    WIDTH=2 INTERPOL=JOIN
LINE=2;
SYMBOL6 VALUE=plus   HEIGHT=2 COLOR=blue   WIDTH=2 INTERPOL=JOIN
LINE=2;
PROC GPLOT DATA = JES.Results_Tab;
  PLOT  M_Res*Month = Vendor / NAME="F6_10_";
  PLOT2 M_Del*Month = Vendor;
RUN; QUIT;
```

Figure 6-10: M_Res and M_Del vs. Month, by Vendor, on Two Different Y Axes

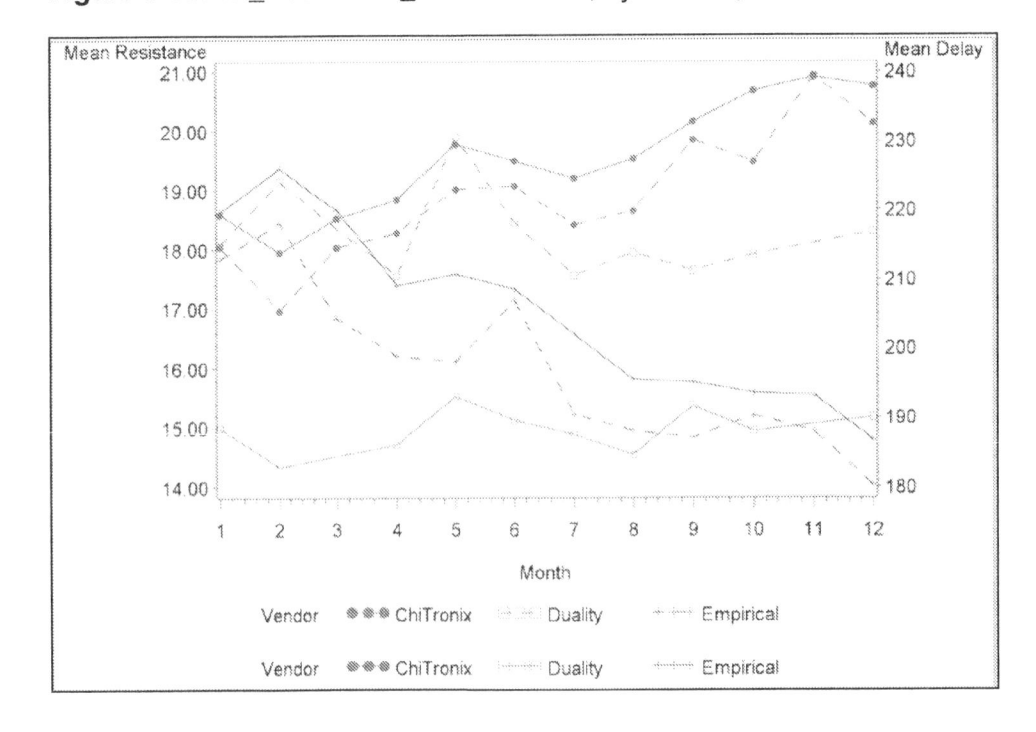

The second example uses a PLOT2 statement with PLOT (Y1 Y2)*X.

```
SYMBOL1 VALUE=dot    HEIGHT=2 COLOR=green WIDTH=2 INTERPOL=JOIN;
SYMBOL2 VALUE=dot    HEIGHT=2 COLOR=green WIDTH=2 INTERPOL=JOIN LINE=2;
SYMBOL3 VALUE=dot    HEIGHT=2 COLOR=green WIDTH=2 INTERPOL=JOIN LINE=2;
SYMBOL4 VALUE=square HEIGHT=2 COLOR=red   WIDTH=2 INTERPOL=JOIN;
SYMBOL5 VALUE=square HEIGHT=2 COLOR=red   WIDTH=2 INTERPOL=JOIN LINE=2;
SYMBOL6 VALUE=square HEIGHT=2 COLOR=red   WIDTH=2 INTERPOL=JOIN LINE=2;

PROC GPLOT DATA = JES.Results_Tab(WHERE=(Vendor="ChiTronix"));
   PLOT  (M_Res R_L R_U)*Month / NAME="F6_11_" OVERLAY;
   PLOT2 (M_Del D_L D_U)*Month / OVERLAY;
RUN; QUIT;
```

Figure 6-11: M_Res and M_Del vs. Month, with 90% Confidence Limits, on Two Different Y Axes

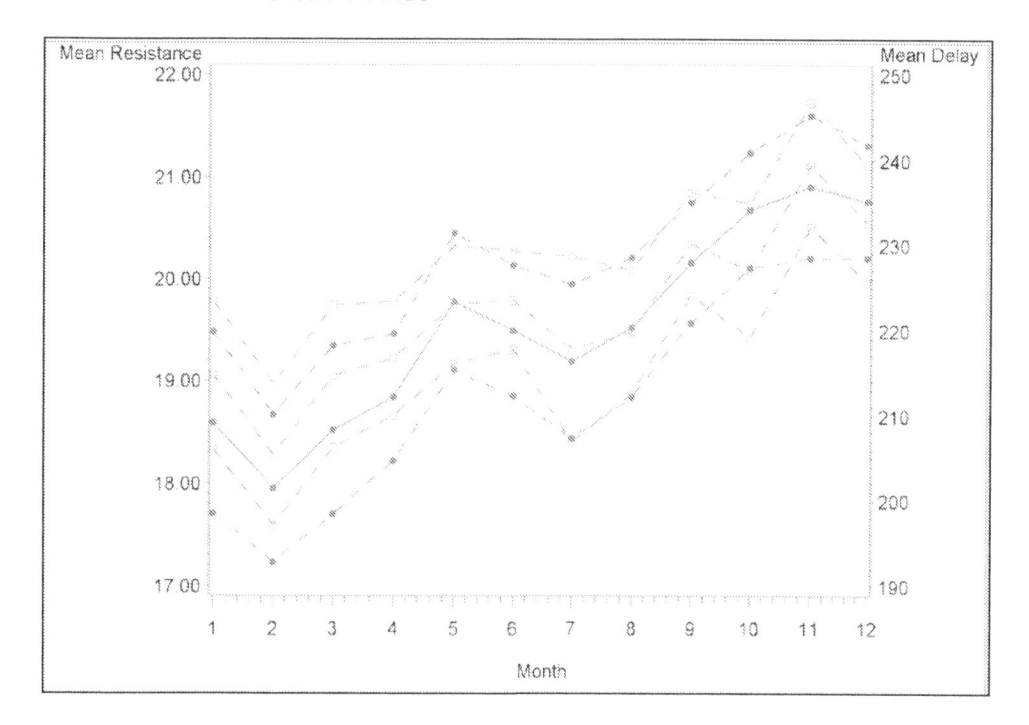

The first two plot forms, however, cannot be used together. When you run the next code sample, a WARNING message, "OVERLAY option specified conflicts with the Y*X=Z type plot request" appears in the Log window, and separate graphs (not shown here) are created for M_Res vs. Month and M_Del vs. Month.

```
PROC GPLOT DATA = JES.Results_Tab;
   PLOT (M_Res M_Del)*Month=Vendor / NAME="F6_B_" OVERLAY;
RUN; QUIT;
```

6.3.2 Options to Customize GPLOT Output

This section shows how to enhance the appearance of your GPLOT output by adding custom titles, footnotes, legends, axes and reference lines, and by selecting the symbols and line types to be used for the points and lines in the plot. The code for this section is in ~\JES \sample_code\ch_6 \gplot_options.sas.

6.3.2.1 Global Graphic Options

Use a GOPTIONS statement to set options which will be used each time you call PROC GPLOT. These options remain in effect until they are changed by another GOPTIONS statement, but they can be temporarily overridden, for example by a TITLE statement which controls the appearance of a particular title. Table 6.3 lists some of the more useful options available with GOPTIONS.

Table 6-3: Selected Graphic Options

Option	Meaning	Some Possible Values
RESET	Resets graphic options to the default values.	ALL
BORDER	Causes a border to be drawn about the plot area.	No value required
GUNIT	Sets the default units used to specify text height.	PCT, CM, IN, CELLS, PT
HTEXT	Sets the height for all text in the units set by GUNIT.	Any number
FTEXT	Sets the font for all text. SAS/GRAPH fonts will be discussed in Section 6.3.2.6.	SIMPLEX, DUPLEX, SWISSB, 'Arial'
CTEXT	Sets the default color for all text. A list of valid SAS color names can be found in Table 6-9.	Red, Green, Black, Blue
INTERPOL	Sets the default interpolation value for the SYMBOL statement. See Section 6.3.2.6.	NONE, JOIN, STEPJ, SPLINE
DEVICE	Sets the type of graphic to be created. See Table 6-2.	JPEG, GIF, PNG, PSLEPS, PSLEPSFC
GSFNAME	Specifies the fileref of the location to which graphics stream file (GSF) records are written.	Any fileref
GSFMODE	Specifies the disposition of records written to a graphics stream file.	APPEND, REPLACE, PORT

The first two lines in the sample code, which initialize the environment to a known state, are explained in Section 6.2.

```
PROC DATASETS MEMTYPE=CAT NOLIST; DELETE GSEG; RUN; QUIT;
FILENAME Fig "&JES.figures/Chapter_6/";
```

The next section of code illustrates the syntax of the GOPTIONS statement.

```
GOPTIONS RESET=ALL  ❶
  BORDER  ❷
  GUNIT=PCT  ❸
  HTEXT=3  ❹
  FTEXT='Arial'  ❺
  CTEXT=Black  ❻
  INTERPOL=JOIN  ❼
  DEVICE=GIF XMAX=6IN YMAX=4.5IN  ❽
  GSFNAME=Fig GSFMODE=REPLACE;  ❾
PROC GPLOT DATA = JES.Results_Tab;
  PLOT M_Res*Month=Vendor / NAME="F6_12_";
RUN; QUIT;
```

❶ The RESET=ALL option resets all graphic options to the default values.

❷ The BORDER option causes a border to be drawn around the graphics area.

❸ The GUNIT=PCT option specifies that the default unit of measurement for text height be PCT, the percentage of the graphics output area.

❹ The HTEXT=3 option sets the default text height to be 3% of the graphics output area.

❺ The FTEXT='Arial' sets the default text font to be Arial.

❻ The CTEXT=Black option sets the default color for text and for the border to be black.

❼ The INTERPOL=JOIN option specifies that plot points be joined with straight line segments.

❽ The DEVICE=GIF sets the graphic output file to be a GIF. The XMAX and YMAX options specify the size of the GIF as 6 inches by 4.5 inches.

❾ The GSFNAME=Fig option specifies that the GIF be saved in the folder referred to by the Fig fileref. The GSFMODE=REPLACE option specifies that the new image file will replace any existing file with the same name.

Figure 6-12: Plot Created Using the DEVICE=GIF Option

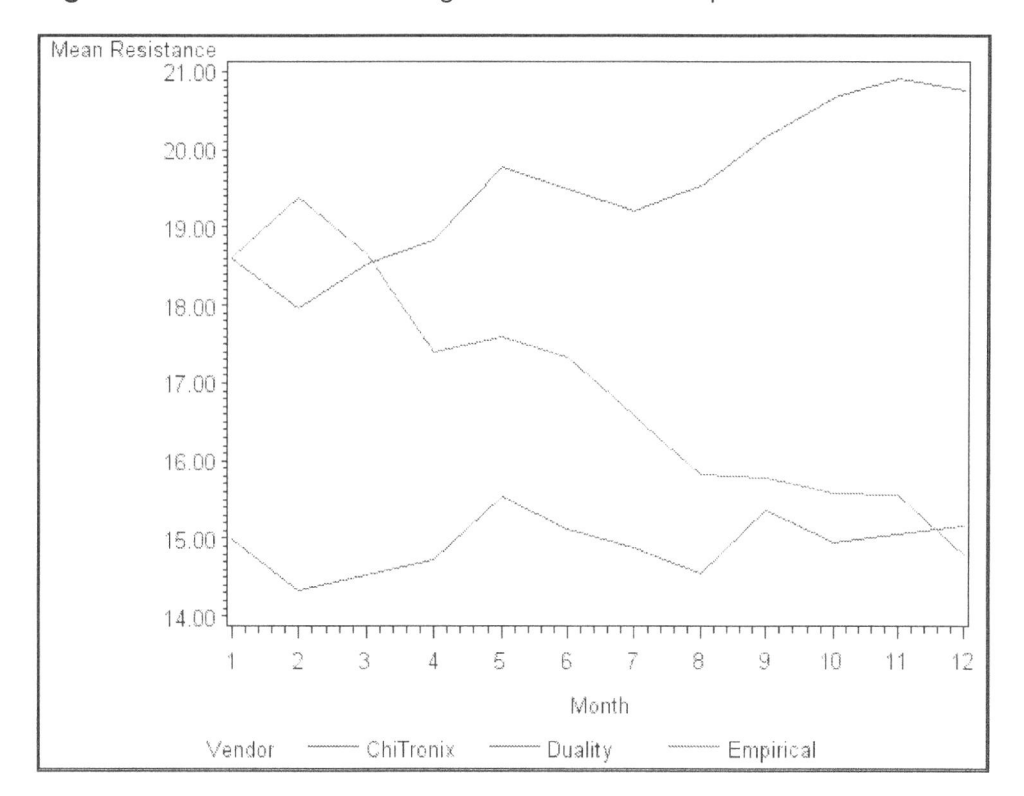

As noted in Section 6.2, images created with DEVICE=GIF work very well in Web pages and presentations, but the resolution is not high enough to reproduce well on this page. Most of the figures in this book were created using the DEVICE=PSLEPSFC option to create Encapsulated PostScript (EPS) images.

6.3.2.2 Titles and Footnotes

Use TITLE and FOOTNOTE statements to add as many as 10 title lines and 10 footnote lines to your plot. The simplest TITLE statement is just TITLE<n>, where n=1, 2, ..., 10, followed by the text of the title.

```
TITLE1 "This is my Title";
```

Table 6-4 shows some of the options that you can use to control the characteristics of the title or footnote text.

Table 6-4: Selected Options for TITLE and FOOTNOTE Statements

Option	Meaning	Some Possible Values
FONT	Sets the font of the text.	SIMPLEX, DUPLEX, SWISSB, 'Arial'
HEIGHT	Sets the size of the text, in the units set by GUNIT.	Any number
COLOR	Sets the color of the text.	Black, Blue, Green
JUSTIFY	Sets the justification of the text.	LEFT, CENTER, RIGHT
ANGLE	Sets the angle of the baseline of the text string with respect to the horizontal.	0, 45, 90, 270
ROTATE	Sets the angle at which each character is rotated with respect to the baseline of the text string.	0, 45, 90, 270
UNDERLIN	Underlines subsequent text, with a relative line thickness of 0, 1, 2, or 3. The value 0 halts underlining for subsequent text.	0, 1, 2, 3
LINK	Specifies a URL that the title or footnote links to. See Section 7.4.2 for an example using the LINK option.	"../detail.html" "datafile.csv"

The HEIGHT, FONT, and COLOR option temporarily override any HTEXT, FTEXT, or CTEXT options that might have been set with a GOPTIONS statement. This code illustrates the syntax of TITLE and FOOTNOTE statements.

```
GOPTIONS RESET=ALL BORDER GUNIT=PCT HTEXT=3 FTEXT='Arial';  ❶
SYMBOL1 VALUE=dot    HEIGHT=2 COLOR=green  WIDTH=2 INTERPOL=JOIN;
SYMBOL2 VALUE=square HEIGHT=2 COLOR=red    WIDTH=2 INTERPOL=JOIN;
SYMBOL3 VALUE=plus   HEIGHT=2 COLOR=blue   WIDTH=2 INTERPOL=JOIN;
TITLE1 HEIGHT=5 "Mean Resistance by Vendor";  ❷
TITLE2 FONT=SWISSB COLOR= Blue "12 Month Period Beginning Jan. 2008"; ❸
FOOTNOTE1 JUSTIFY=LEFT "John Doe" JUSTIFY=RIGHT "BigCorp";  ❹
FOOTNOTE2   "Do"  ❺
   COLOR=Red UNDERLIN=2 "NOT"  ❻
   COLOR=Black UNDERLIN=0 "  Distribute This Report";  ❼
PROC GPLOT DATA = JES.Results_Tab;
   PLOT M_Res*Month=Vendor / NAME="F6_13_";
RUN; QUIT;
```

❶ The GOPTIONS statement resets all graphic options to the default values, and then specifies the font height to be 3% of the graphic output area and the font type to be Arial. (See Section 6.3.2.1.)

❷ The TITLE1 statement specifies the first title line. The HEIGHT=5 option sets the text height to 5% of the graphics output area. The font type is not specified, so Arial is used because that was specified in the GOPTIONS statement.

❸ The TITLE2 statement sets the font type to SWISSB, and the font color to blue, for the second title line. The font height is not specified, so the value set by the GOPTIONS statement is used.

❹ The FOOTNOTE1 statement creates the first footnote line, using the JUSTIFY option to left justify the person's name and to right justify the company name.

❺ The FOOTNOTE2 statement defines the text of the second footnote. The footnote begins with "Do", using the default text height and color.

❻ The COLOR=Red and UNDERLIN=2 cause the subsequent text to be red and underlined.

❼ The COLOR=Black and UNDERLIN=0 options cause the subsequent text to be black and not underlined.

Figure 6-13: Using TITLE and FOOTNOTE Statements

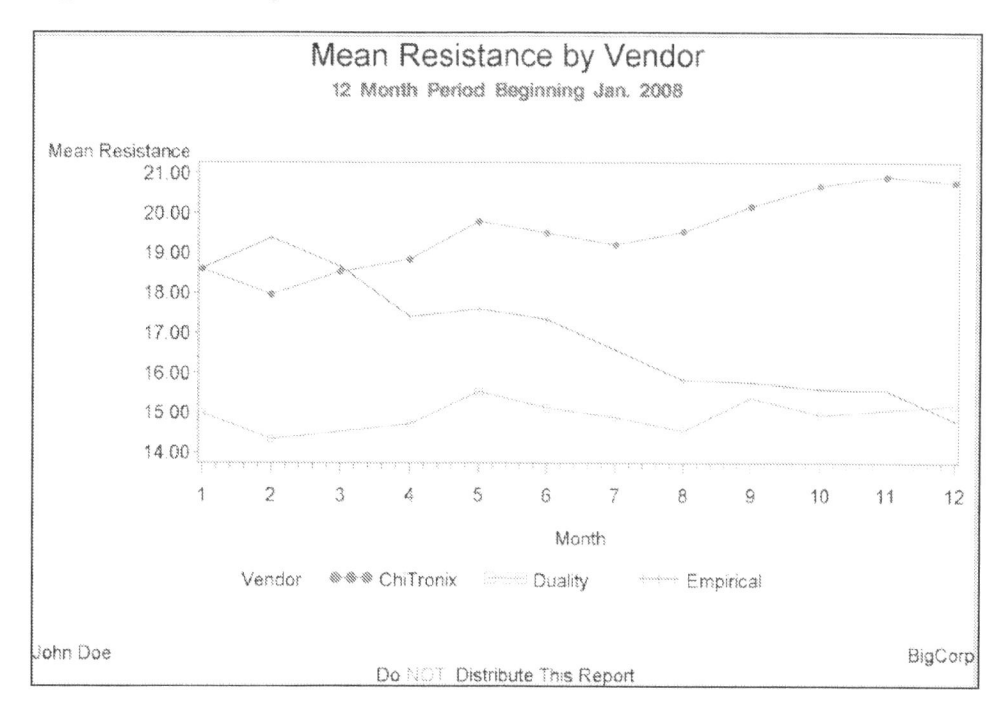

TITLE and FOOTNOTE statements are global and remain in effect until overridden. This is very convenient if you want to use the same titles or footnotes repeatedly, for example to put your name and the date in a footnote on every graph you create. On the other hand, you will sometimes find "old" titles or footnotes showing up where they are not wanted. To avoid this problem, use a TITLE<n> or FOOTNOTE<n> statement to cancel all previous TITLE or FOOTNOTE statements with the same or a higher number. For example, this statement will cancel the FOOTNOTE2 statement, as well as any higher numbered FOOTNOTE statements (FOOTNOTE3, FOOTNOTE4, etc.) that might have been specified, while leaving the FOOTNOTE1 statement in effect.

```
FOOTNOTE2;
```

These statements will cancel all previous TITLE and FOOTNOTE statements.

```
TITLE1;
FOOTNOTE1;
```

6.3.2.3 Axes

Use AXIS statements to specify the range and appearance of the horizontal and vertical axes in your plots. This is a two-step process: first use one or more AXIS statements, AXIS1, AXIS2, etc., to set the desired characteristics, and then use the HAXIS and VAXIS options within GPLOT to specify which AXIS definition should be applied to the horizontal and vertical axes. Some of the more useful AXIS statement options are shown in Table 6-5.

Table 6-5: Selected AXIS Statement Options

Option	Meaning	Some Possible Values
LABEL	Sets the text used to label the axis. *	(HEIGHT=3 FONT='Arial' COLOR=Blue "Month"), NONE
VALUE	Sets the text used to label the major tick marks on the axis. *	(HEIGHT=3 FONT='Arial' ANGLE=90), NONE
ORDER	Sets the range and major tick marks for the axis. **	(5 TO 25), (0 TO 20 BY 2)
MINOR	Sets the number of minor tick marks between each major tick mark.	(N=4), NONE
OFFSET	Sets the distance, in GUNIT units, from the first and last major tick marks to the end of the axis.	(1, 5)
* The FONT, HEIGHT, COLOR, JUSTIFY, ANGLE, and ROTATE options of the TITLE and FOOTNOTE statements, described in Table 6-4, are also suboptions of the LABEL and VALUE options.		
** You can use the HZERO or VZERO option in the PLOT statement to force the horizontal or vertical axis to include zero without using the ORDER option.		

You do not need to specify all of these options in an AXIS statement. Any options not specified will revert to the default values. You can suppress the default labels, values or minor tick marks by using the NONE option, for example LABEL=NONE. The next bit of sample code illustrates the syntax and usage of AXIS statements.

```
AXIS1 LABEL=(JUSTIFY=Center HEIGHT=4 FONT='Arial' COLOR=Black "Month
of Manufacture") ❶
    VALUE = (ANGLE=45 ROTATE=0 HEIGHT=3 FONT='Arial') ❷
    OFFSET= (5, 0); ❸
AXIS2 LABEL =(H=4  A=90  FONT='Arial' 'Average Resistance') ❹
    VALUE = (H=3 FONT='Arial')
    ORDER = (10 TO 25 BY 5) ❺
    MINOR = (N=4); ❻
PROC GPLOT DATA = JES.Results_Tab;
  PLOT M_Res*Mon=Vendor / NAME="F6_14_"
  HAXIS=AXIS1 VAXIS=AXIS2; ❼
  Format M_Res 3.0; ❽
RUN; QUIT;
```

❶ The LABEL option sets the font, height, color, justification, and text for the axis label.

❷ The VALUE option sets the font and height for the axis values. The ANGLE suboption causes the value text to be written at an angle of 45 degrees. Without the ANGLE option, the text would not fit horizontally and would be written vertically instead.

❸ The OFFSET =(5,0) option adds extra space at the left side of the horizontal axis, but not at the right side. This graph would look better with space added at both sides using an OFFSET=(5,5) option.

❹ The LABEL option includes an ANGLE=90 option to write the label text vertically along the axis. Note that in the AXIS2 statement, the HEIGHT and ANGLE options are abbreviated as H and A.

❺ The ORDER option sets the range and major tick marks for the vertical axis.

❻ The MINOR=(N=4) option sets the number of minor tick marks between major tick marks.

❼ The HAXIS and VAXIS options of the PLOT statement specify that AXIS1 and AXIS2 be used for the horizontal and vertical axes, respectively.

❽ The FORMAT M_Res 3.0 statement formats the values on the vertical axis. Without this statement, the values would have the same 8.2 format as M_Res. See Figure 6-13.

Figure 6-14: Using AXIS Statements and the HAXIS and VAXIS Options

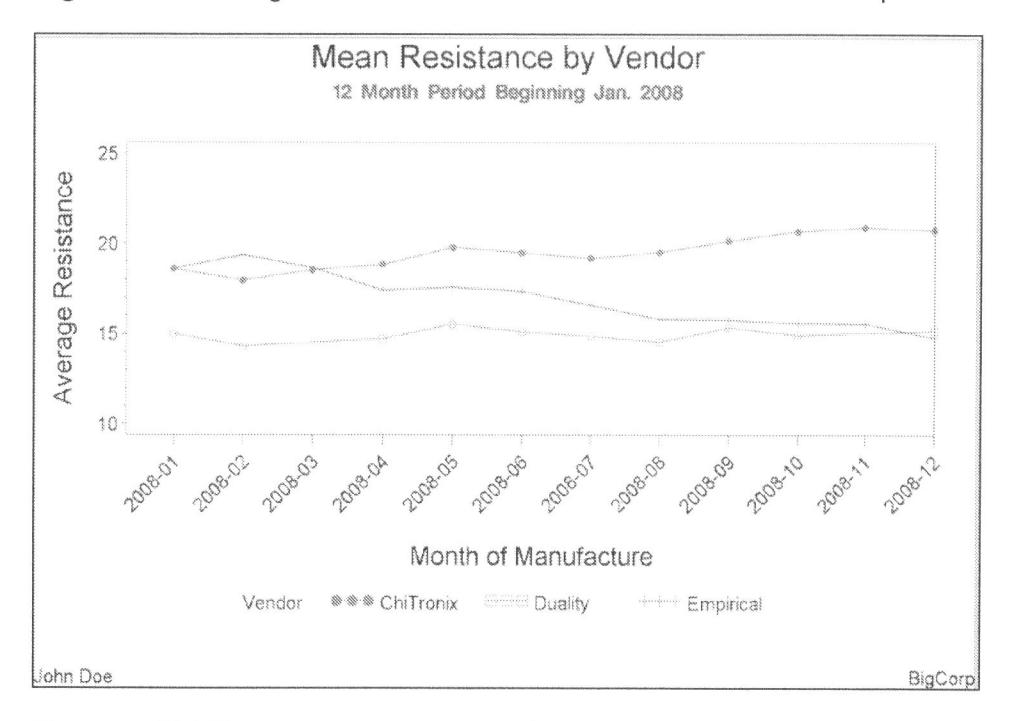

If you use a PLOT2 statement, you can use another AXIS statement to modify the second Y axis.

```
AXIS3 LABEL =(ANGLE=270 H=3 F='Arial' 'Average Delay')
ORDER=(100 TO 250 BY 50) MINOR=(N=4);
PROC GPLOT DATA = JES.Results_Tab;
  PLOT M_Res*Mon=Vendor / NAME="F6_C_" HAXIS=AXIS1 VAXIS=AXIS2;
  Format M_Res 3.0;
  PLOT2 M_Del*Mon=Vendor / VAXIS=AXIS3;
RUN; QUIT;
```

The resulting plot is not shown here, but is displayed in the Graph window when you run the code sample.

6.3.2.4 Legends

Use LEGEND statements to control the legends that identify the meaning of the lines and symbols used in your plots. First, use one or more LEGEND statements, LEGEND1, LEGEND2, etc., to set the desired characteristics, and then use the LEGEND option within GPLOT to specify which LEGEND definition should be used for each PLOT or PLOT2 statement. Some of the more useful LEGEND statement options are shown in Table 6-6.

Table 6-6: Selected LEGEND Statement Options

Option	Meaning	Some Possible Values
LABEL	Sets the text used to label the legend.* Use LABEL=NONE to suppress the legend label.	(H=3 F='Arial' C=blue A=90 "Vendor")
VALUE	Sets the text used to label each symbol in the legend. *	(J=Left C=green H=3 "A" "B" "C")
POSITION	Specifies where the legend will be drawn relative to the plot area: • TOP, MIDDLE, or BOTTOM • LEFT, CENTER, or RIGHT • INSIDE or OUTSIDE the plot area	(TOP RIGHT INSIDE)
MODE	If POSITION=OUTSIDE, then MODE=RESERVE takes space from the graphics area for the legend. If POSITION =INSIDE, then the legend suppresses any graphic elements in the same space if MODE=PROTECT, or coexists with the graphic elements if MODE=SHARE.	PROTECT, SHARE, RESERVE
ACROSS	Sets the number of columns to use for legend entries.	1, 2, 3…
FRAME	Draws a frame around the legend.	No value required

* The FONT, HEIGHT, COLOR, JUSTIFY, ANGLE, and ROTATE options of the TITLE and FOOTNOTE statements, described in Table 6-4, are also suboptions of the LABEL and VALUE options.

This code sample illustrates the use of a LEGEND statement.

```
LEGEND1
  LABEL = (F='Arial' H=3 C=Black J=CENTER POSITION=TOP "Component
Vendor") ❶
    VALUE = (C=Black H=3 "A" "B" "C") ❷
    POSITION = (BOTTOM CENTER INSIDE) ❸
    FRAME ❹
    ACROSS = 3 ❺
    MODE = PROTECT; ❻
PROC GPLOT DATA = JES.Results_Tab;
  PLOT M_Res*Mon=Vendor / NAME="F6_15_"
    HAXIS=AXIS1 VAXIS=AXIS2 LEGEND=LEGEND1; ❼
  Format M_Res 3.0;
RUN; QUIT;
```

❶ The LABEL option sets the text used to label the legend. The POSITION=TOP suboption causes the text to be written above the legend entries.

❷ The VALUE option sets the text used for each legend entry. The text entries, "A", B", "C" are used to label the symbols in alphabetical order. If the text values are left out, the actual values, "ChiTronix", Duality", "Empirical", are used.

❸ The POSITION option causes the legend to be drawn at the bottom center position inside the plot area.

❹ The FRAME option causes a frame to be drawn around the legend.

❺ The ACROSS=3 option specifies that the legend entries should be arranged in three columns.

❻ The MODE=PROTECT option suppresses any plot points which might overlap the legend area.

❼ The LEGEND=LEGEND1 option in the PLOT statement specifies that LEGEND1 be used for the plot.

If no legend is required, use a NOLEGEND option in the GPLOT statement.

Figure 6-15: Using LEGEND Statements and Options

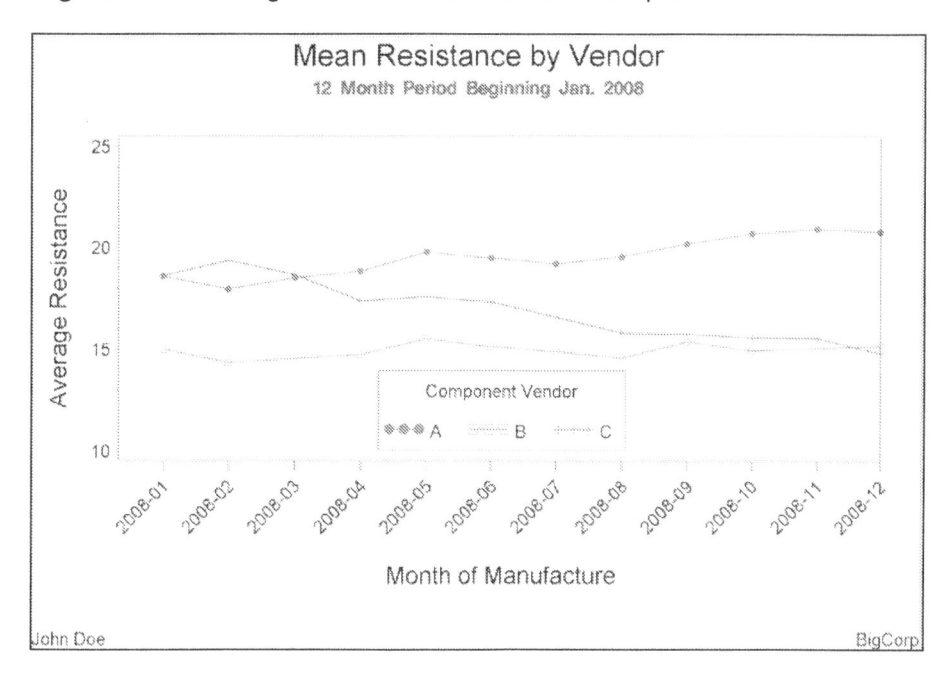

If you use a PLOT2 statement, you can use another LEGEND statement to create a legend for the PLOT2 statement.

```
LEGEND2 POSITION=(TOP CENTER INSIDE) FRAME;
PROC GPLOT DATA = JES.Results_Tab;
  PLOT M_Res*Mon=Vendor / NAME="F6_D_"
  HAXIS=AXIS1 VAXIS=AXIS2 LEGEND=LEGEND1;
  Format M_Res 3.0;
  PLOT2 M_Del*Mon=Vendor / VAXIS=AXIS3 LEGEND=LEGEND2;
RUN; QUIT;
```

The resulting plot is not shown here, but is displayed in the Graph window when you run the code sample.

6.3.2.5 Reference Lines

Use the AUTOHREF and AUTOVREF options of GPLOT to add reference lines to the horizontal or vertical axes of your plot. With these options, SAS decides where to put the reference lines. To control the location of reference lines, use HREF or VREF instead. Your sample code includes an example using AUTOHREF and AUTOREF which is not shown here. The next example illustrates the use of HREF and VREF.

```
TITLE1 H=5 "Delay vs Resistance by Vendor: Months 1-3";
TITLE2 H=3 "Specification Limits: Resistance(12.5-22.5) Delay(150-
250)";
SYMBOL1 FONT=SWISSB VALUE='C' HEIGHT=2 COLOR=green INTERPOL=NONE;
SYMBOL2 FONT=SWISSB VALUE='D' HEIGHT=2 COLOR=navy  INTERPOL=NONE;
SYMBOL3 FONT=SWISSB VALUE='E' HEIGHT=2 COLOR=black INTERPOL=NONE;
AXIS1 ORDER=(5 TO 25 BY 5);
AXIS2 ORDER=(100 TO 350 BY 50);
PROC GPLOT DATA = JES.Results_Q4;
   PLOT Delay*Resistance=Vendor / NAME="F6_16_"
   HAXIS=AXIS1 VAXIS=AXIS2
   HREF=12.5 22.5 VREF=150 250 ❶
   LHREF=2 LVREF=2 CHREF=GRAY80 CVREF=GRAY80; ❷
   FORMAT Resistance 8.0;
RUN; QUIT;
```

❶ The HREF option puts reference lines at 12.5 and 22.5 on the horizontal axis. The VREF option puts reference lines at 150 and 250 on the vertical axis.

❷ The LHREF, LVREF, CHREF, and CVREF, options control the line type and color of the reference lines. Line types and colors are discussed in Sections 6.3.2.6.

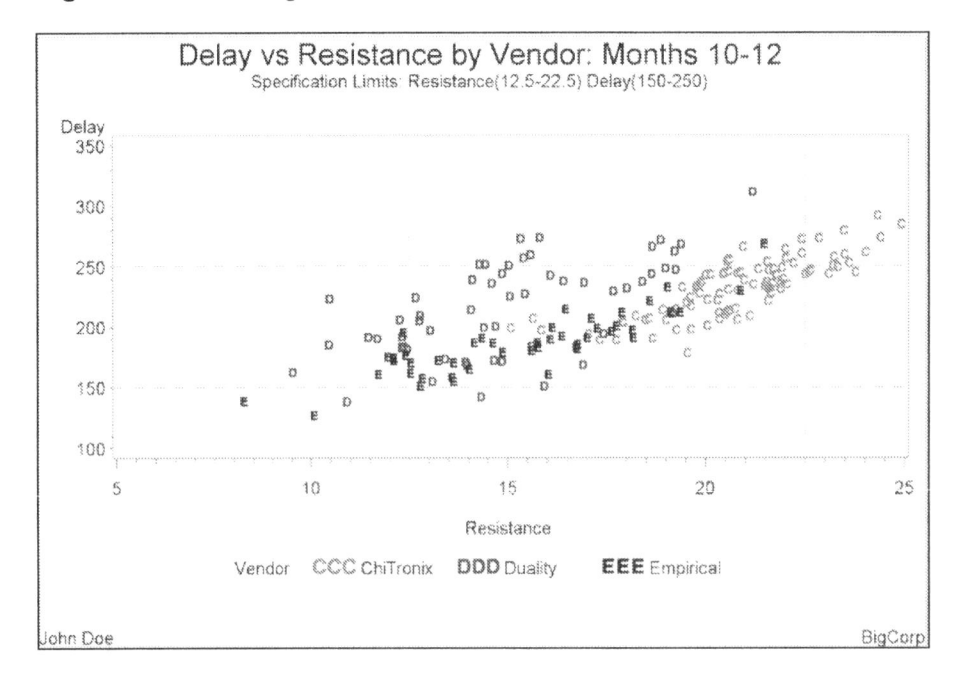

Figure 6-16: Adding Reference Lines

Use the REFLABEL option to specify text labels for your reference lines. The FONT, HEIGHT, COLOR, JUSTIFY, ANGLE, and ROTATE options, described in Table 6-4, can be used to modify the text.

```
AXIS1 ORDER=(5 TO 25 BY 5)
   REFLABEL=(H=3 F='Arial' POSITION=TOP J=CENTER 'LSL' 'USL'); ❶
AXIS2 ORDER=(100 TO 350 BY 50)
   REFLABEL=(H=3 F='Arial' J=LEFT POSITION=MIDDLE ' LSL' ' USL'); ❷
PROC GPLOT DATA = JES.Results_Q4
   PLOT Delay*Resistance=Vendor / NAME="F6_17_"
   HAXIS=AXIS1 VAXIS=AXIS2
   HREF=12.5 22.5 VREF=150 250
   LHREF=2 LVREF=2 CHREF=GRAY80 CVREF=GRAY80;
   FORMAT Resistance 8.0;
RUN; QUIT;
```

❶ The first REFLABEL option specifies 'LSL' and 'USL' as labels for the reference lines on the horizontal axis. The POSITION and JUSTIFY(J) suboptions cause the labels to be centered at the top of the reference lines.

❷ The second REFLABEL option specifies 'LSL' and 'USL' as labels for the reference lines on the vertical axis. The POSITION and JUSTIFY(J) suboptions cause the labels to be in the middle of the lines, on the left side of the plot area.

Figure 6-17: Adding Reference Line Labels

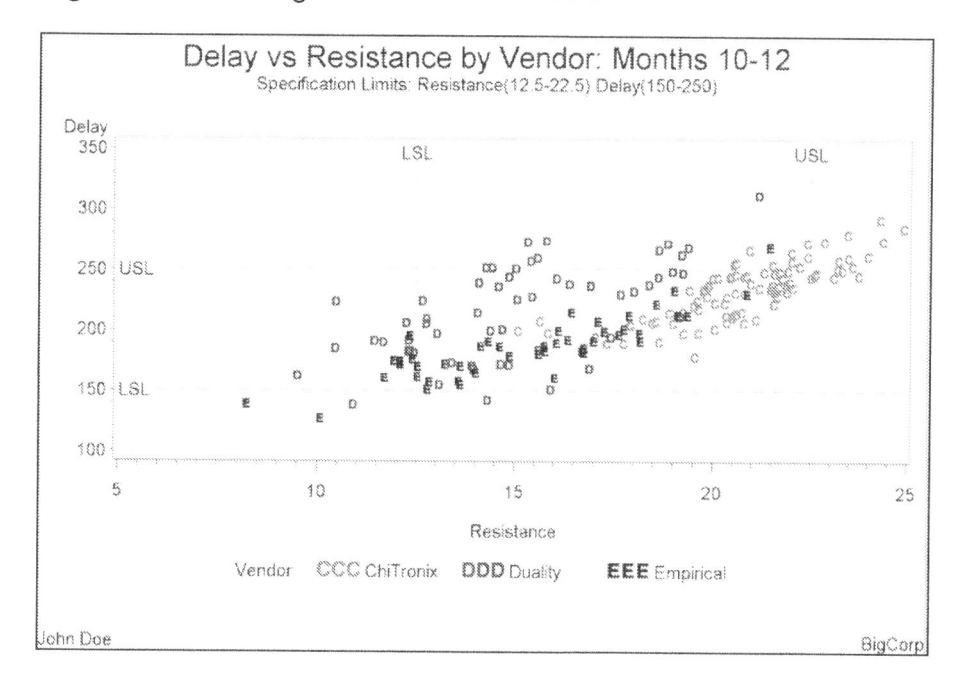

6.3.2.6 Symbols and Lines

Use SYMBOL statements to control the appearance of the symbols and lines in your plots. Table 6-7 includes some of the more useful SYMBOL statement options.

Table 6-7: Selected SYMBOL Statement Options

Option	Meaning	Some Possible Values
FONT	Sets the font used to draw the symbols. Selected SAS fonts are shown in Table 6-8. If no font is specified, the symbols in Display 6-3 are used.	MARKER, SWISS
VALUE	Sets the value from the selected font, or from the symbols in Display 6-3, to be plotted. For symbol or alphabet fonts, enter the corresponding Roman character(s). For the default special symbols, enter the corresponding word or character.	'A', 'B', 'C', 'DOA' PLUS, X, STAR
HEIGHT	Sets the height of the symbols in the units set by GUNIT.	.5, 1, 2
WIDTH	Sets the width of lines (if any) which connect the plot points.	1.5, 2, 3
INTERPOL	Sets the type of line used to join the plot points, for example: • NONE: No line is drawn joining the points. • JOIN: Successive pairs of points are joined by straight lines. • STEPJ: Successive pairs of points are joined by a Step function. • SPLINE: Fits a cubic spline which passes through each point. • BOX<*options*><*nn*>: Create a box-and-whisker plot. (See examples in Section 6.3.2.7.)	NONE, JOIN, STEPJ, SPLINE
LINE	Sets the line type. The first 10 line types are shown in Figure 6-19.	1, 2, ..., 46
COLOR	Sets the color of the symbols and lines. See Table 6-9 for a list of SAS color names.	Red, Blue, Gray,...

This code illustrates the use of SYMBOL statements. In this example, there are separate plots for each value of Vendor. The SYMBOL<*n*> statements are applied to each plot in alphabetical order - SYMBOL1 for "ChiTronix", SYMBOL2 for "Duality", and SYMBOL3 for "Empirical".

```
SYMBOL1 FONT=MARKER VALUE='V'  HEIGHT=2 COLOR=Blue
   INTERPOL=STEPJ WIDTH=2 LINE=2; ❶
SYMBOL2 FONT=SWISS  VALUE='D'  HEIGHT=3 COLOR=Red
   INTERPOL=JOIN WIDTH=1 LINE=1; ❷
SYMBOL3 FONT=       VALUE=DOT  HEIGHT=4 COLOR=Black
   INTERPOL=SPLINE WIDTH=3 LINE=3; ❸
TITLE1 H=5 "Symbols and Line Types";
PROC GPLOT DATA = JES.Results_Tab;
   PLOT M_Res*Month=Vendor / NAME="F6_18_";
   Format M_Res 3.0;
RUN; QUIT;
```

❶ The FONT=MARKER VALUE='V' options specify the symbol corresponding to the letter 'V' in the MARKER font. (See Display 6-4.) The HEIGHT and COLOR options set the size and color of the symbols. The INTERPOL=STEPJ specifies that the points be joined by a Step function. The WIDTH and LINE options specify the width and type of the line that joins the points.

❷ The FONT=SWISSB VALUE='D' options specify that the symbol be the letter 'D' in the SWISSB font. The INTERPOL=JOIN option specifies that the points be joined by straight line segments.

❸ The FONT= Value=DOT options specify the symbol named DOT in the table of Special Symbols, shown in Display 6-3. The INTERPOL=SPLINE option specifies that the points be joined by a smoothing spline. The FONT= option is not needed here because the default symbols are being used. But if a previous SYMBOL3 statement had specified a different font, then that font is used until the FONT= option is used to explicitly reset the option to the default symbols.

Figure 6-18: Using SYMBOL Statements

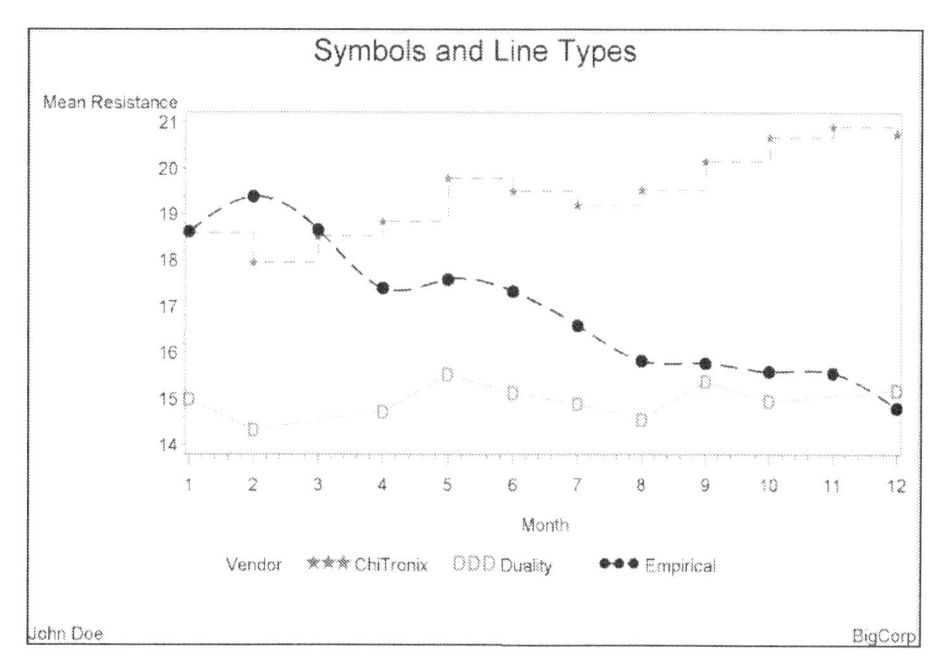

In this example it would be more appropriate to use INTERPOL=JOIN for all three plots, but STEPJ and SPLINE were used just to illustrate the different options.

Figure 6-19: The First 10 Line Types

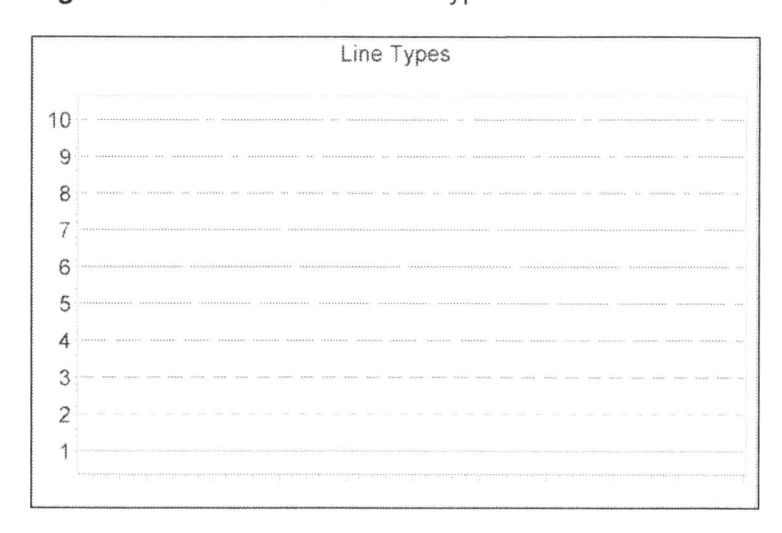

SAS/GRAPH Symbols and Fonts

The default symbols used by PROC GPLOT are the special symbols shown in Display 6-3. The words in parentheses, e.g. (percent), are not part of the syntax and are included to clarify the keystroke required for each symbol. Use the VALUE option to select the desired symbol. For example,

- "VALUE=PLUS" sets the plot symbol to +.
- "VALUE = =" sets the plot symbol to a star shape.

Display 6-3: Special Symbols for Plotting Data Points

VALUE=	Plot Symbol	VALUE=	Plot Symbol
PLUS	+	% (percent)	♣
X	×	& (ampersand)	♣
STAR	✻	' (single quote)	⚘
SQUARE	□	= (equals)	☆
DIAMOND	◇	- (hyphen)	⊙
TRIANGLE	△	@ (at)	☿
HASH	⊞	* (asterisk)	♀
Y	Y	+ (plus)	⊕
Z	Z	> (greater than)	♂
PAW	⋰	. (period)	♃
POINT	·	< (less than)	♄
DOT	●	, (comma)	⚴
CIRCLE	○	/ (slash)	♆
_ (underscore)	⊓	? (question mark)	♇
" (double quote)	♤	((left parenthesis)	☾
# (pound sign)	♡	) (right parenthesis)	⚵
$ (dollar sign)	◇	: (colon)	✺

If a FONT option is used in a SYMBOL statement to select a font, then the characters from that font are used as the plot symbols. You can choose from fonts supplied by SAS or hardware or software fonts available in your operating environment. Some of the fonts supplied by SAS are shown in Table 6-8. For a complete list of fonts supplied by SAS, as well as information on how to specify and use other fonts available in your environment, consult the SAS online documentation.

Table 6-8: Selected SAS Font Names

Roman Alphabet Fonts				
Normal	**Bold**	**Empty**	**Italic**	**Bold Italic**
SWISS	SWISSB	SWISSE	SWISSI	SWISSBI
ZAPF	ZAPFB	ZAPFE	ZAPFI	ZAPFBI
SIMPLEX	DUPLEX			
COMPLEX	TRIPLEX		ITALIC	TITALIC
Symbol Fonts				
MARKER	MARKERE	ELECTRON	MATH	SPECIAL

You can use PROC GFONT to display the characters in any font.

```
GOPTIONS RESET=ALL BORDER GUNIT=PCT HTEXT=3 FTEXT='Arial';
TITLE H=5 "Marker Font";
PROC GFONT NAME=MARKER  HEIGHT=6 CTEXT=blue  ❶
   SHOWROMAN ROMHT=3  ROMCOL=Black ROMFONT=SWISSB NOBUILD;  ❷
RUN;
```

❶ The NAME=MARKER option specifies that the MARKER font is to be displayed. The HEIGHT and CTEXT options control the size and color used to display the MARKER font.

❷ The SHOWROMAN option specifies that the Roman characters should also be displayed. The ROMHT and ROMCOL options control the size and color of the Roman text. The NOBUILD specifies that a new font is not being generated.

When the code is run, the Marker font is displayed in the Graph window. This provides a translation table that tells you which Roman letter to use to produce the desired symbol. For example, FONT=MARKER VALUE='N' specifies a heart-shaped symbol.

Display 6-4: The Marker Font

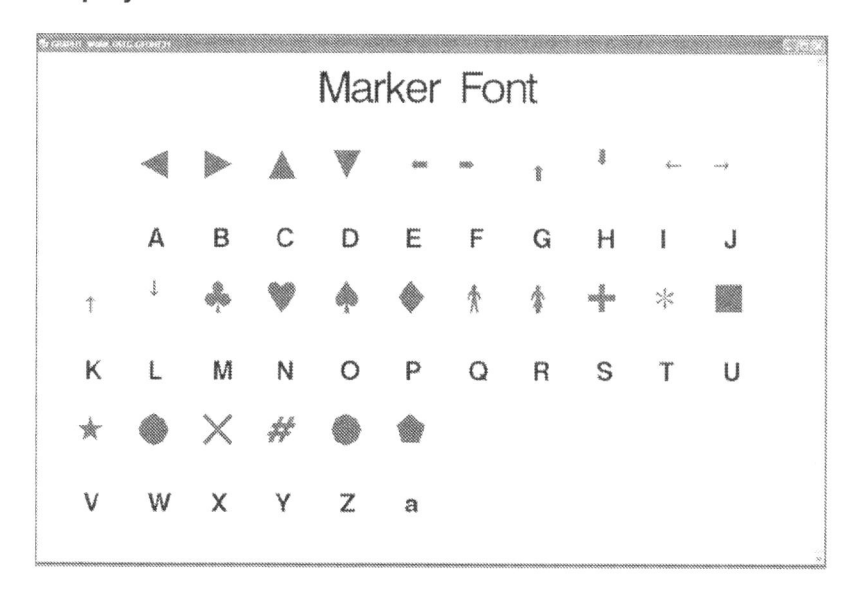

SAS/GRAPH Colors

SAS supports the RGB and CMYK color-naming schemes, as well as several other schemes based on hue, lightness, saturation, and brightness. But the simplest way to specify colors is by using one of the 144 color names provided in the SAS Registry. These color names are a good choice for HTML output because they are supported by most browsers. However, in order to use these names, you first need to know what they are, and some are not so obvious. BurntUmber and CeruleanBlue do not work, but Chocolate, PapayaWhip, and PeachPuff are valid SAS color names.

You can use PROC REGISTRY to list all of the valid SAS color names.

```
PROC REGISTRY LIST
   STARTAT='COLORNAMES';
RUN;
```

When you run this code, the SAS color names are listed in the Log window.

```
PROC REGISTRY LIST
STARTAT='COLORNAMES';
RUN;

NOTE: Contents of SASHELP REGISTRY starting at subkey [COLORNAMES]
[ COLORNAMES]
  Active="HTML"
[   HTML]
   AliceBlue=hex: F0,F8,FF
   AntiqueWhite=hex: FA,EB,D7
   Aqua=hex: 00,FF,FF
   Aquamarine=hex: 7F,FD,D4
   Azure=hex: F0,FF,FF
   Beige=hex: F5,F5,DC
```

The list includes the color name as well as the RGB encoding for each of 144 colors. The complete list of color names is shown in Table 6-9. You can use either the color name or the RGB equivalent, preceded by CX to tell SAS this is an RGB code, to specify a color. For example COLOR=Aqua or COLOR=CX00FFFF are equivalent. In addition to these color names, you can use any of 256 gray-scale color codes, GRAY*hh*, where *hh* is any hexadecimal number in the range 00 through FF.

Run this code to see what these colors, and a selection of gray-scale levels, will look like in your PROC GPLOT output.

```
%INCLUDE "&JES.sample_code/ch_6/colors.sas"; ❶
%Show_Colors ❷
```

❶ The %INCLUDE statement includes the code in colors.sas to define a macro function, %Show_Colors. This code is beyond the scope of this section and will not be explained here.

❷ The %Show_Colors macro creates GPLOT output which is displayed in the Graph window.

The first six pages of GPLOT output each contain 24 colors from one of the columns in Table 6-9. The last page of output is the plot of gray-scale lines shown in Figure 6-20.

Table 6-9: SAS Color Names

\multicolumn{6}{c	}{SAS Color Names}
1-24	25-48	49-72	73-96	97-120	121-144
AliceBlue	DarkGoldenrod	Gainsboro	LightSalmon	NavajoWhite	SaddleBrown
AntiqueWhite	DarkGray	GhostWhite	LightSeaGreen	Navy	Salmon
Aqua	DarkGreen	Gold	LightSkyBlue	O	SandyBrown
Aquamarine	DarkKhaki	Goldenrod	LightSlateGray	Oldlace	SeaGreen
Azure	DarkMagenta	Gray	LightSteelBlue	Olive	Seashell
Beige	DarkOliveGreen	Green	LightYellow	OliveDrab	Sienna
Bisque	DarkOrange	GreenYellow	Lime	Orange	Silver
Black	DarkOrchid	Honeydew	LimeGreen	OrangeRed	SkyBlue
BlanchedAlmond	DarkRed	HotPink	Linen	Orchid	SlateBlue
Blue	DarkSalmon	IndianRed	Magenta	P	SlateGray
BlueViolet	DarkSeaGreen	Indigo	Maroon	PaleGoldenrod	Snow
BR	DarkSlateBlue	Ivory	MediumAquamarine	PaleGreen	SpringGreen
Brown	DarkSlateGray	Khaki	MediumBlue	PaleTurquiose	SteelBlue
Burlywood	DarkTurquoise	Lavender	MediumOrchid	PaleVioletRed	Tan
CadetBlue	DarkViolet	LavenderBlush	MediumPurple	PapayaWhip	Teal
Chartreuse	DeepPink	LawnGreen	MediumSeaGreen	Peachpuff	Thistle
Chocolate	DeepSkyBlue	LemonChiffon	MediumSlateBlue	Peru	Tomato
Coral	DimGray	LightBlue	MediumSpringGreen	Pink	Turquoise
CornFlowerBlue	DodgerBlue	LightCoral	MediumTurquoise	Plum	Violet
Cornsilk	FireBrick	LightCyan	MediumVioletRed	PowerBlue	Wheat
Crimson	FloralWhite	LightGoldenrodYellow	MidnightBlue	Purple	White
Cyan	ForestGreen	LightGreen	MintCream	Red	WhiteSmoke
DarkBlue	Fuchsia	LightGrey	MistyRose	RosyBrown	Yellow
DarkCyan	G	LightPink	Moccasin	RoyalBlue	YellowGreen

Figure 6-20: Selected SAS Gray-Scale Colors

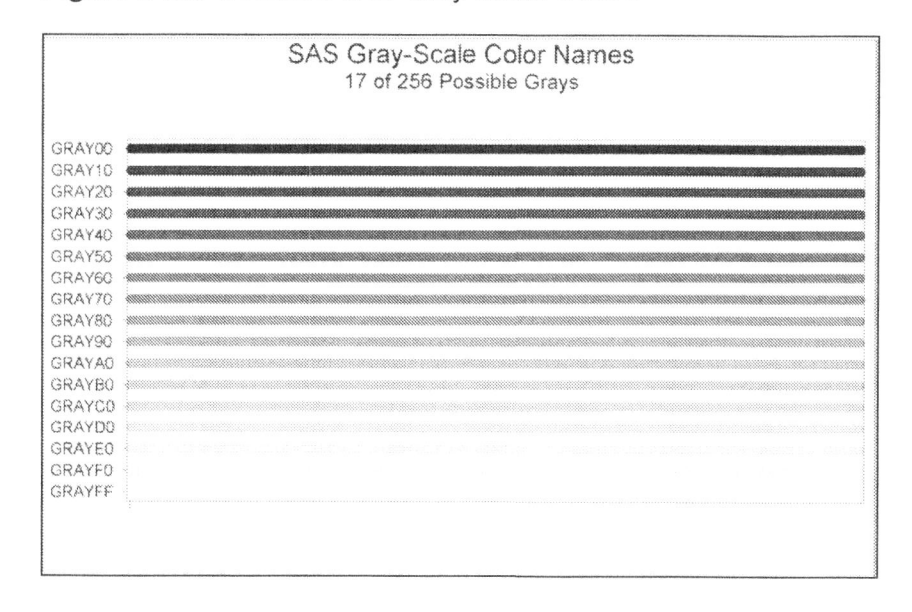

6.3.2.7 Box-and-Whisker Plots

Section 5.5 showed how to use PROC BOXPLOT to create skeletal and schematic box plots. You can also create box plots with PROC GPLOT by using the INTERPOL=BOX<*options*><*nn*> option in a SYMBOL statement.

Table 6-10: Syntax of the INTERPOL=BOX<*options*><*nn*> Option

	Values of <*options*>
F	Fill the box with the color set by the CV option.
J	Join the median points of the boxes with a line.
T	Draw tops and bottoms on the whiskers.
	Values of <*nn*> determine the length of the whiskers
00	**Skeletal box plot**: Whiskers extend to the maximum and minimum data values.
01	Whiskers extend to the 1st percentile on the low side and the 99th percentile on the high side.
05	Whiskers extend to the 5th percentile on the low side and the 95th percentile on the high side.
10	Whiskers extend to the 10th percentile on the low side and the 90th percentile on the high side.
25	No whiskers are produced.
nn omitted	**Schematic Box Plot**: Whiskers extend to the maximum (minimum) data point which is at most 1.5 times the Interquartile Range above (below) the 75th (25th) percentile.
	Other Options
CO	Sets the color of the box outline.
CV	Sets the color of the outside points and the box fill (if any).
CI	Sets the color of the line joining the medians.
WIDTH	Sets the width of the box outline and the line joining the medians.
BWIDTH	Sets the width of the boxes.

```
TITLE1 H=5 "Resistance by Vendor";
TITLE2 H=4 "Schematic Box Plot";
AXIS1 LABEL=NONE VALUE=(F=Arial H=4) OFFSET=(10, 10);
AXIS2 LABEL=(H=5 F='Arial' ANGLE=90 "Resistance")
      VALUE=(F='Arial' H=4)
      ORDER=(5 TO 30 BY 5);   ❶
SYMBOL1 INTERPOL=BOXT BWIDTH=5   ❷
  CO=green WIDTH=2   ❸
  FONT=Marker VALUE='V'  HEIGHT=2 CV=red;   ❹
PROC GPLOT DATA= JES.Results;
  PLOT Resistance*Vendor / NAME="F6_21_"
  HAXIS=AXIS1 VAXIS=AXIS2 VREF=12.5 22.5;
  FORMAT Resistance 8.0;
RUN; QUIT;
```

❶ The ORDER option eliminates any points less than 5 or greater than 30 from the plot and from the calculation of the percentiles used to draw the box and whiskers. In this case no harm is done because all data points are between 5 and 30.

❷ The INTERPOL=BOXT specifies a schematic box plot with cross bars at the end of each whisker. The BWIDTH option specifies the width of the boxes.

❸ The CO and WIDTH options specify the color and width of the box outline.

❹ The FONT, VALUE, HEIGHT, and CV options specify the characteristics of the outside plot points.

The resulting plot is shown in Figure 6-21.

Figure 6-21: Box Plot Created with the INTERPOL=BOXT Option

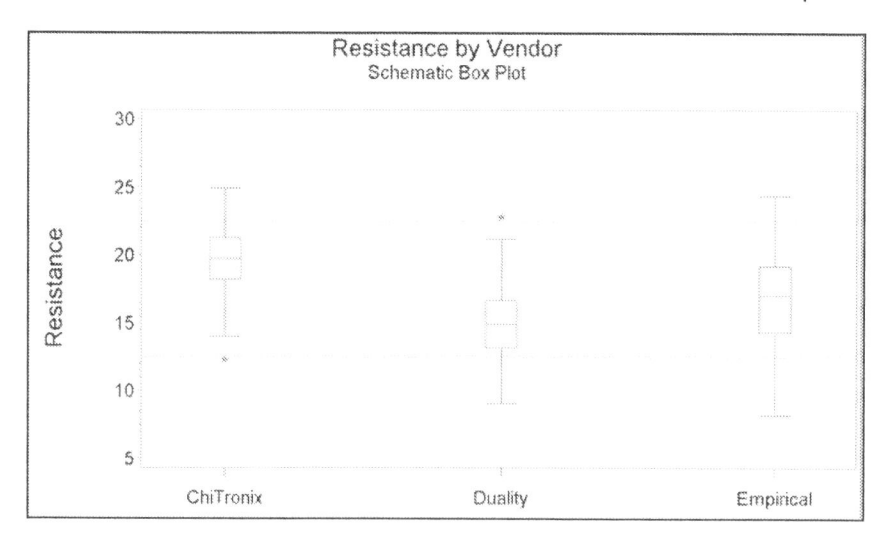

The next code sample uses this SYMBOL1 statement to create the plot shown in Figure 6-22. The INTERPOL=BOX05TJ option controls the appearance of the box plots. The '05' causes the whiskers to be drawn to the 5th and 95th percentiles, and the 'J' causes the medians of the boxes to be joined. The CI and LINE options set the color and type of the line joining the means.

```
SYMBOL1 INTERPOL=BOX05TJ BWIDTH=10
   CO=Blue WIDTH=3 CI=Green LINE=2
   FONT= VALUE='X' H=3 CV=Red;
```

Figure 6-22: Box Plot Created with the INTERPOL=BOX05TJ Option

6.3.3 GPLOT Example: Plotting Group Confidence Limits

One of the best ways to understand group differences is by plotting confidence limits for the mean of each group. By comparing the confidence limits of the different groups, you can easily see which differences are statistically and practically significant. This section presents an extended example illustrating one method for creating group confidence limit plots with PROC GPLOT, and also introduces two more very useful GPLOT options, SKIPMISS and ANNOTATE. The code for this section is in ~\JES \sample_code\ch_6 \conlim.sas.

The first two lines in the sample code, which initialize the environment to a known state, are explained in Section 6.2.

```
PROC DATASETS MEMTYPE=CAT NOLIST; DELETE GSEG; RUN; QUIT;
FILENAME Fig "&JES.figures/Chapter_6/";
```

The next bit of code creates the Tab data set, which contains the mean Resistance, M_Res, and the 90% confidence limits, R_L and R_U, for Month=1, from the JES.Results_Tab data set.

```
DATA Tab; SET JES.Results_Tab;
   IF Month=1;
   KEEP Vendor M_Res R_L R_U;
RUN;
```

```
Tab

Vendor         M_Res          R_L            R_U

ChiTronix      18.59          17.71          19.48
Duality        14.99          14.23          15.75
Empirical      18.62          17.25          19.99
```

By examining the Tab data set, you can see that the upper confidence limit on mean Resistance for the Duality units (R_U=15.75) is less than the lower confidence limits for each of the other two vendors. Therefore, you can conclude that the mean resistance of Duality units is significantly lower than the mean resistance of units from the other two vendors. This difference is much easier to see in a plot than in a table, so the next code sample creates a plot of the mean and confidence limits for each vendor.

```
GOPTIONS RESET=ALL BORDER GUNIT=PCT HTEXT=3 FTEXT='Arial';
TITLE H=5 "Resistance by Vendor - Month 1";
SYMBOL1 VALUE=dot       HEIGHT=3 COLOR=green;
SYMBOL2 VALUE=diamond   HEIGHT=3 COLOR=blue;
SYMBOL3 VALUE=diamond   HEIGHT=3 COLOR=blue;
AXIS1 label=none VALUE=(F=Arial H=4) OFFSET=(10,10);
AXIS2 label=(H=5 F='Arial' "Resistance")
   VALUE=(F='Arial' H=4)
   ORDER=10 TO 26 by 2
   REFLABEL=(H=3 J=Center POSITION=Top 'Lower Limit' 'Upper Limit');
PROC GPLOT DATA=Tab;
PLOT Vendor*(M_Res R_L R_U) / NAME="F6_23_"
   OVERLAY VAXIS=AXIS1 HAXIS=AXIS2 HREF=12.5 22.5;
RUN; QUIT;
```

The plot created by this code is shown in Figure 6-23.

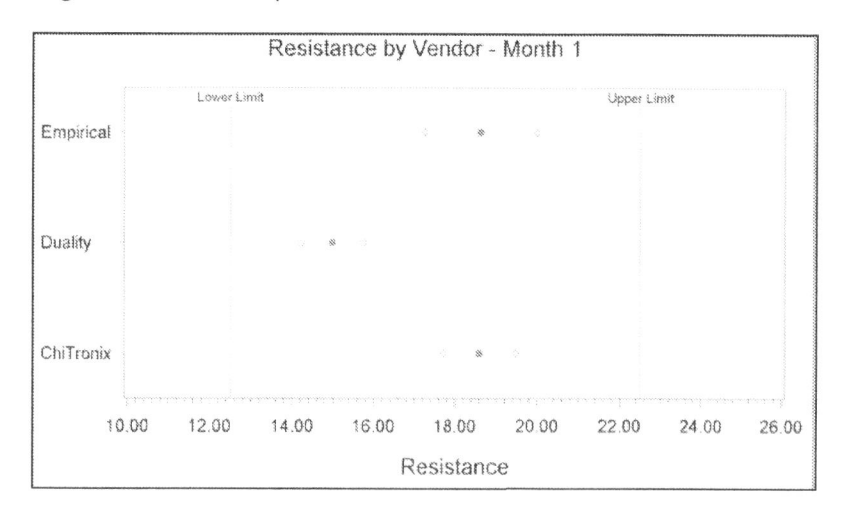

Figure 6-23: Group Confidence Limit Plot

```
                    Resistance by Vendor - Month 1
            Lower Limit                    Upper Limit
  Empirical                        ◊  *  ◊


  Duality           ◊ *  ◊


  ChiTronix                  ◊  *  ◊

           10.00  12.00  14.00  16.00  18.00  20.00  22.00  24.00  26.00
                              Resistance
```

6.3.3.1 Draw a Line Joining the Confidence Limits

Figure 6-23 would be easier to understand if there were lines connecting the upper and lower confidence limits, R_L and R_U, for each vendor. However, there is no way for GPLOT to connect plot points for different variables (columns) in the same data set. The obvious solution is to copy the R_L and R_U variables into the same column.

```
DATA Tab_1;
   SET Tab(RENAME=(R_L=CL)) Tab(RENAME=(R_U=CL)); ❶
RUN;
PROC SORT DATA=Tab_1; BY Vendor CL; RUN; ❷
```

❶ The SET statement reads in the Tab data set twice, storing first the value of R_L, and then the value of R_U, into a new variable, CL.

❷ The SORT statement brings together the records for each vendor so that the connecting lines are drawn in the correct order.

The resulting data set, Tab_1, is shown below.

```
Tab_1

 Vendor        M_Res        CL         R_U        R_L

 ChiTronix     18.59        17.71      19.48      .
 ChiTronix     18.59        19.48      .          17.71
 Duality       14.99        14.23      15.75      .
 Duality       14.99        15.75      .          14.23
 Empirical     18.62        17.25      19.99      .
 Empirical     18.62        19.99      .          17.25
```

With the upper and lower confidence limits in the same column, you can easily draw a line to join them.

```
SYMBOL2 INTERPOL=JOIN;
PROC GPLOT DATA=Tab_1;
   PLOT Vendor*(M_Res CL) / NAME="F6_24_" OVERLAY
   VAXIS=AXIS1 HAXIS=AXIS2 HREF=12.5 22.5;
RUN; QUIT;
```

The INTERPOL=JOIN option causes a line to be drawn joining the CL plot points. The resulting plot is shown in Figure 6-24.

Figure 6-24: Group Confidence Limit Plot with Line Joining the Lower and Upper Limits

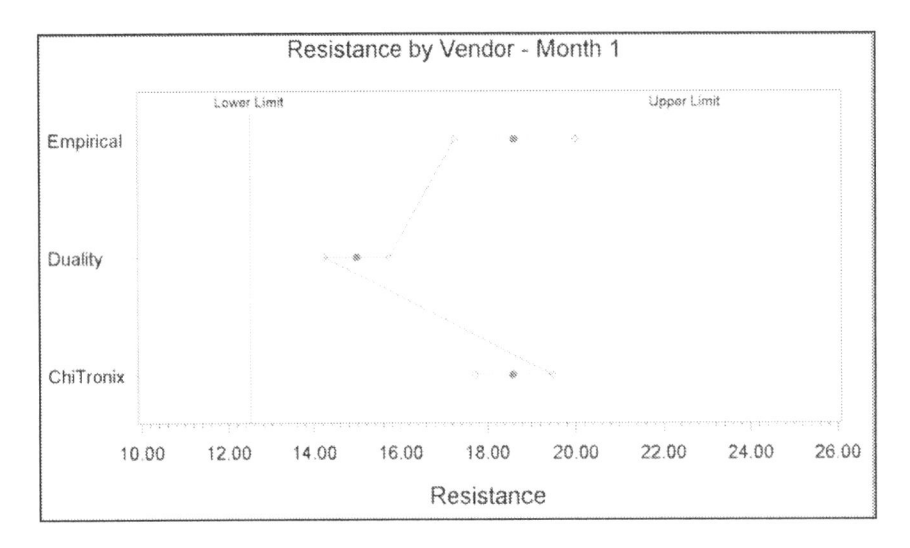

This plot joins the confidence limits as desired, but also adds unwanted lines joining the CL values for different vendors.

6.3.3.2 Use the SKIPMISS Option to Force Breaks in Plotted Lines

When the INTERPOL=JOIN option is used in a SYMBOL statement, all the points to which the SYMBOL statement applies (the Vendor*CL points in this example) will be joined by a line. However, if the SKIPMISS option is used in the PLOT statement, then the line will be broken at any point where there is missing data for either of the plot variables. So you can use SKIPMISS to break the line joining the confidence limits for different vendors, but first you need to insert some records with missing values to force the breaks to occur where you need them.

```
PROC SQL NOPRINT;  ❶
   CREATE TABLE Missing AS SELECT DISTINCT Vendor FROM Tab_1;
QUIT;
DATA Tab_2; set Tab_1 Missing;  RUN;  ❷
PROC SORT DATA=Tab_2; BY Vendor CL; RUN;  ❸
```

❶ The PROC SQL statement creates a data set, Missing, containing only the three vendor names.

❷ The DATA statement appends the Missing data set to Tab_1, creating a row for each vendor with missing values for CL and the other numeric variables.

❸ The PROC SORT statement interleaves the missing values among the CL values for each vendor.

The resulting data set, Tab_2, includes the missing values required to break the lines joining the confidence limits in the appropriate places.

```
Tab_2

  Vendor         M_Res          CL           R_U          R_L

  ChiTronix        .             .            .            .
  ChiTronix      18.59         17.71        19.48          .
  ChiTronix      18.59         19.48          .          17.71
  Duality          .             .            .            .
  Duality        14.99         14.23        15.75          .
  Duality        14.99         15.75          .          14.23
  Empirical        .             .            .            .
  Empirical      18.62         17.25        19.99          .
  Empirical      18.62         19.99          .          17.25
```

You now have a missing value of CL between the plot points for each vendor. The code below takes advantage of these missing values by using the SKIPMISS option to force the required line breaks.

```
PROC GPLOT DATA=Tab_2;
   PLOT Vendor*(M_Res CL) / NAME="F6_25_" OVERLAY
   VAXIS=AXIS1 HAXIS=AXIS2 HREF=12.5 22.5 SKIPMISS;
RUN; QUIT;
```

The plot created by this code is shown in Figure 6-25.

Figure 6-25: Group Confidence Limit Plot Using the SKIPMISS Option

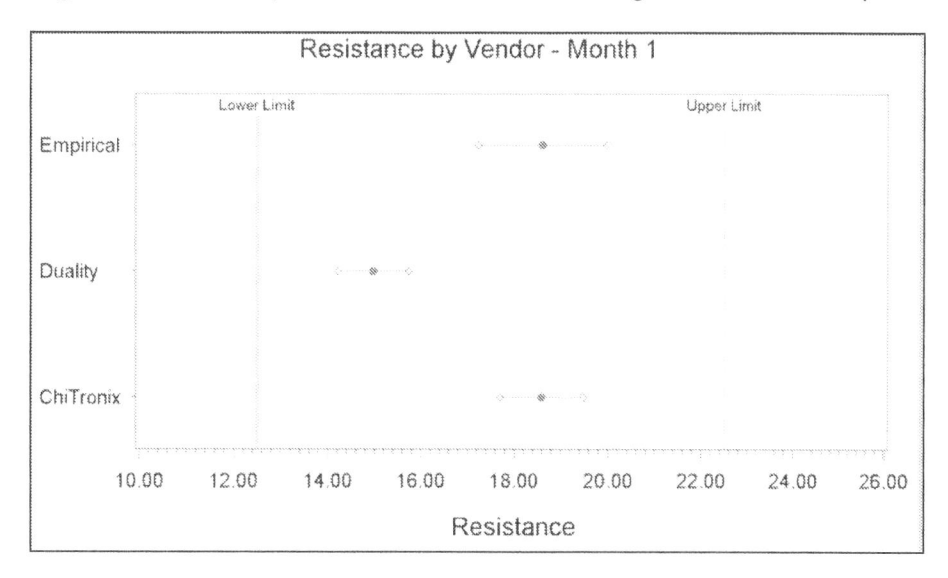

This plot provides a very effective presentation of group differences. Section 6.3.3.3 shows how to enhance the plot by using the ANNOTATE option to add labels to some of the plot points.

6.3.3.3 Use the ANNOTATE Option to Label Selected Plot Points

You can use the ANNOTATE option to add text, symbols, or drawings anywhere on your graphical output. The example in this section shows how to add labels to some of the points in the group confidence limit plot in Figure 6-25. A detailed discussion of the ANNOTATE option is beyond the scope of this book, but the example should be enough to get you started, and there are references in the chapter summary.

To use the ANNOTATE option, you must first create an *Annotate data set* containing variables with predefined names, which contains instructions for drawing objects in the graphics output area. An annotate data set contains three types of variables:

- The Action variable, FUNCTION, tells SAS *what* to do.
- Positioning variables tell SAS *where* to do it.
- Attribute variables tell SAS *how* to do it.

Table 6-11 shows some of the variables which can be used in an Annotate data set.

Table 6-11: Selected Annotate Variables

Variable	Meaning	Some Possible Values
Action Variable		
FUNCTION	Specifies what kind of object to draw, or what action to take.	LABEL, MOVE, SYMBOL
Positioning Variables		
XSYS YSYS	Specifies the coordinate system to be used for the horizontal (X) and vertical (Y) axes. *	'1', '2', ... '9', 'A', 'B', 'C'
HSYS	Specifies the coordinate system used by the SIZE variable. *	
X XC	The position of the object on the X axis. Use X for numerical variables or XC for character variables.	X=Mean
Y YC	The position of the object on the Y axis. Use Y for numerical variables or YC for character variables.	YC=Vendor
Attribute Variables		
POSITION	Sets the placement and alignment of text strings. **	'0'-'9' and 'A'-'F'
STYLE	Specifies the font used to draw text.	SWISS, 'Arial'
SIZE	Sets the size of the object to be drawn, e.g. height of text.	1, 1.25, 2
TEXT	Sets the text string for a LABEL, SYMBOL or COMMENT	'Some Text'

* XSYS='2' and YSYS='2' sets the coordinate system to be the same as for the plot points.
  HSYS='3' sets the units for SIZE to be the percentage of the graphics output area.
  See the SAS OnLine documentation of the meaning of the other codes.
** The POSITION variable causes the object to be drawn to the right, left or center and above or below the plot point. The effect of the POSITION variable is illustrated in an exercise in Section 6.6.3.

This code uses the Tab_2 data set from the previous example to create a new data set, My_Anno, which you can use to annotate the group confidence limit plot.

```
DATA My_Anno; SET Tab_2(WHERE=(R_U>0));
  FUNCTION='LABEL';
  XSYS='2'; YSYS='2'; HSYS='3';
  X= R_U; YC=Vendor;
  POSITION='6';
  STYLE='"Arial"';
  SIZE=3;
  TEXT=" Mean = "||TRIM(LEFT(PUT(M_Res, 6.2)));
  KEEP XSYS YSYS HSYS STYLE FUNCTION POSITION SIZE X YC TEXT;
RUN;
```

```
My_Anno

FUNCTION XSYS YSYS HSYS     X        YC      POSITION  STYLE  SIZE    TEXT

 LABEL     2    2    3   19.4821 ChiTronix      6     "Arial"  3   Mean = 18.59
 LABEL     2    2    3   15.7497 Duality        6     "Arial"  3   Mean = 14.99
 LABEL     2    2    3   19.9927 Empirical      6     "Arial"  3   Mean = 18.62
```

The My_Anno data set contains one line for each label to be written on the graph.

The action variable, FUNCTION, specifies that labels are to be written. The positioning variables, XSYS, YSYS, HSYS, X, and YC specify that the labels are to be placed at (X, YC) on the graph, which is the location of the upper confidence limit for each vendor. The attribute variables STYLE, SIZE and TEXT specify the text to be written and the font and size to be used. Finally, the POSITION variable sets the placement and alignment of the text string relative to the point (X, YC). The value POSITION='6' puts the text string to the right of and slightly below the plot point.

The next bit of sample code uses the My_Anno data set to annotate the confidence limit plot.

```
PROC GPLOT DATA=Tab_2;
  PLOT Vendor*(M_Res CL)/ NAME="F6_28_"
      OVERLAY VAXIS=AXIS1 HAXIS=AXIS2 HREF=12.5 22.5 SKIPMISS
ANNOTATE=My_Anno;  ❶
RUN; QUIT;
```

❶ The ANNOTATE=My_Anno option causes the objects to be drawn on the graphics output area according to the instructions encoded in the My_Anno data set.

Figure 6-26: Group Confidence Limit Plot with Points Annotated

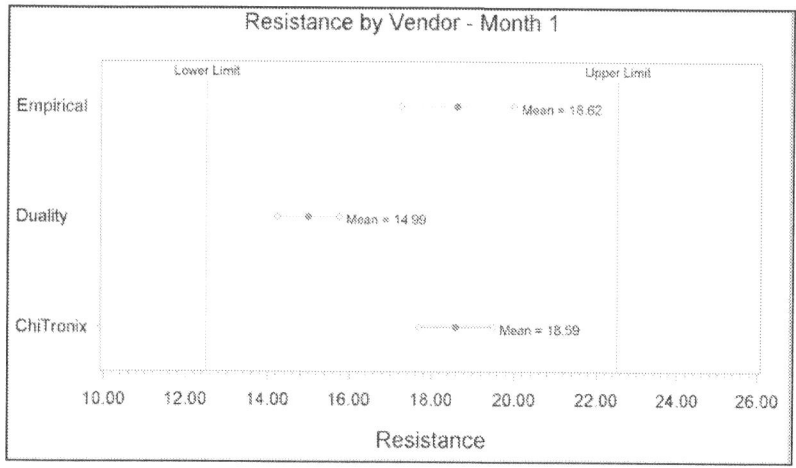

The labels here are somewhat redundant because the same information is contained in the location of the plot points. But of course the labels could include other information, such as the sample size, which is not evident from the plot.

6.3.3.4 Trend Plot of Defect Rate with Confidence Limits

This section brings together the techniques illustrated in the previous sections to provide a more concise example showing how to create a trend plot including confidence limits. The first bit of code creates the data set to be plotted.

```
DATA Tab_3; SET JES.Results_Tab;
  IF Vendor="ChiTronix";
  FORMAT LCL UCL 8.3;
  LCL = CINV(.10, 2*N_Def)/(2*N);
  UCL = CINV(.90, 2*(N_Def+1))/(2*N);
  KEEP Mon N N_Def M_Def LCL UCL;
RUN;
DATA Tab_4;
  SET       Tab_3(RENAME=(LCL=CL) )
        Tab_3(RENAME=(UCL=CL))
        Tab_3(KEEP=Mon M_Def); RUN;
PROC SORT DATA=Tab_4; BY Mon CL; RUN;
```

The first DATA step creates the Tab_3 data set from JES.Results_Tab, adding confidence limits on the mean number of defects using the method shown in Section 5.2.7.

```
Tab_3 (partial listing)

Mon        N        N_Def      M_Def        LCL          UCL

2008-01    20       19         0.95         0.684        1.295
2008-02    20       24         1.20         0.899        1.579
2008-03    23       32         1.39         1.087        1.763
```

The second DATA step creates Tab_4 from Tab_3 adding the CL variable to contain both the upper and lower confidence limits, with some missing values so that you can use SKIPMISS to join the confidence limits as shown in Section 6.3.3.2. Note that there are no missing values for M_Def so that you can draw a line joining the monthly means.

```
Tab_4

Mon        N        N_Def      M_Def        CL           UCL          LCL

2008-01    .        .          0.95         .            .            .
2008-01    20       19         0.95         0.684        1.295        .
2008-01    20       19         0.95         1.295        .            0.684
2008-02    .        .          1.20         .            .            .
2008-02    20       24         1.20         0.899        1.579        .
2008-02    20       24         1.20         1.579        .            0.899
```

The next DATA step creates the Annotate data set My_Anno_2 which contains instructions for labeling each upper confidence limit point with the mean defect rate for that month.

```
DATA My_Anno_2; SET Tab_4(WHERE=(UCL>0));
  FUNCTION='LABEL';
  XSYS='2'; YSYS='2'; HSYS='3'; XC=Mon; Y = UCL; POSITION='2';
  STYLE='"Arial"'; SIZE=2.5; TEXT=PUT(M_Def, 6.3);
  KEEP FUNCTION XSYS YSYS HSYS XC Y POSITION STYLE SIZE TEXT;
RUN;
```

This Annotate data set is similar to the one in Section 6.3.3.3, except that the POSITION variable is set to '2', which causes the labels to be centered over the (XC, Y) plot point.

```
My_Anno_2 (first three rows)

FUNCTION  XSYS  YSYS  HSYS    XC          Y     POSITION   STYLE    SIZE  TEXT

  LABEL    2     2     3   2008-01    1.29513      2      "Arial"   2.5   0.950
  LABEL    2     2     3   2008-02    1.57918      2      "Arial"   2.5   1.200
  LABEL    2     2     3   2008-03    1.76273      2      "Arial"   2.5   1.391
```

The code below uses many of the options described in the previous sections to control the appearance of the plot.

```
TITLE H=5 "Monthly Defect Rates - ChiTronix - 2008";
SYMBOL1 VALUE=dot      HEIGHT=3 COLOR=green INTERPOL=JOIN;
SYMBOL2 VALUE=diamond  HEIGHT=3 COLOR=blue  INTERPOL=JOIN;
LEGEND1 LABEL=NONE POSITION=(BOTTOM CENTER INSIDE) FRAME
        VALUE=('Mean' '90% Limits') MODE=PROTECT;
AXIS1 LABEL=NONE
        VALUE=(ANGLE=45 ROTATE=0 F=Arial H=4) OFFSET=(4,4);
AXIS2 LABEL=(ANGLE=90 H=5 F='Arial' "Defect Rate")
        VALUE=(F='Arial' H=4) ORDER=0 TO 3 BY .5
        REFLABEL=(H=4 J=Left POSITION=Top ' Spec Limit ');
PROC GPLOT DATA=Tab_4;
   PLOT (M_Def CL)*Mon / NAME="F6_27_" OVERLAY LEGEND=LEGEND1
   HAXIS=AXIS1 VAXIS=AXIS2 SKIPMISS
   LVREF=2 CVREF=Black VREF=2 ANNOTATE= my_Anno_2;
RUN; QUIT;
```

The final plot, Figure 6-27, clearly shows the significance of the increasing defect rate trend in relation to the specification limit of 2%.

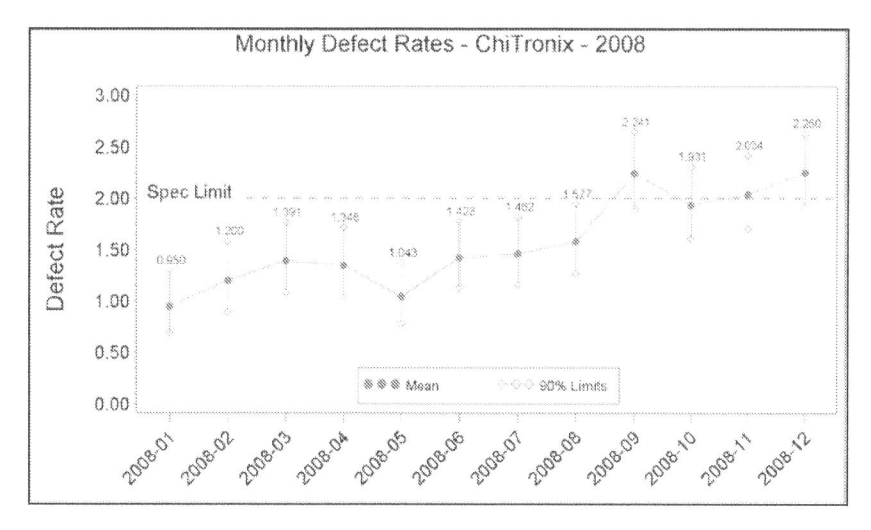

Figure 6-27: Defect Rates by Month with Confidence Limits

6.4 PROC GCHART

PROC GCHART produces nine types of charts, each representing one of six statistics computed separately for different subsets, or categories, of a data set. The variable whose values determine the categories to be plotted is referred to as the *chart variable*. The values of the chart variables that identify categories of data are referred to as the *midpoint values*. Table 6-12 summarizes the possible chart types, statistics and midpoint values. This section will cover only the HBAR and VBAR statements, and only the FREQ, PCT, SUM, and MEAN statistics. There are brief examples using the BLOCK and PIE statements in Section 6.5. Consult the SAS online documentation for information on the other chart types. The code for this section is in ~\JES \sample_code\ch_6 \gchart_examples.sas.

Table 6-12: PROC GCHART Chart Types

GCHART Statements	
Statement	**Chart Type**
HBAR	Horizontal bar chart.
HBAR3D	Horizontal bar chart with 3-D effect.
VBAR	Vertical bar chart.
VBAR3D	Vertical bar chart with 3-D effect.
BLOCK	Blocks with height proportional to the statistic, each set in a square that represents a group (midpoint value) based on one or two variables in the data set.
PIE	Pie chart.
PIE3D	Pie chart with 3-D effect.
DONUT	Pie chart with a hole in the middle where you can place text.
STAR	Star charts represent statistics as the length of lines, one for each midpoint, radiating from the center of a circle towards the perimeter.
MIDPOINT Values	
Variable Type	**Midpoint Values**
Character	One midpoint for each unique value of the variable.
Discrete Numeric	One midpoint for each unique value of the variable.
Continuous Numeric	One midpoint for each range of the variable. For example, if the ranges are 0-10, 10-20 and 20-30, the midpoint values will be 5, 15, and 25. Section 6.4.2 shows how to control the midpoint values for numeric variables.
Chart Statistics	
Option	**Statistic**
TYPE=FREQ	The number of rows in the data set for each midpoint.
TYPE=CFREQ	The cumulative number of rows up to and including the current midpoint.
TYPE=PCT	The percent of rows in the data set for each midpoint.
TYPE=CPCT	The cumulative percent of rows up to and including the current midpoint.
TYPE=SUM	The sum of the values of the variable specified in the SUMVAR= statement for each midpoint.
TYPE=MEAN	The mean of the values of the variable specified in the SUMVAR= statement for each midpoint.
The CFREQ and CPCT statistics are not available for PIE, PIE3D, DONUT, or STAR charts. A SUMVAR= statement is required when TYPE=SUM or MEAN, and invalid otherwise.	

The sample code first initializes the environment to a known state (see Section 6.2).

```
PROC DATASETS MEMTYPE=CAT NOLIST; DELETE GSEG; RUN; QUIT;
FILENAME Fig "&JES.figures/Chapter_6/";
```

The next bit of code uses PROC MEANS to create the Defect_Tab data set, which contains the number of rows (_FREQ_), the number of defects, and the average number of defects per unit, for each Month and Vendor. Defect_Tab is not used in the examples but it may be helpful to refer to the values in Defect_Tab to understand what is being plotted in some of the examples.

```
PROC MEANS DATA=JES.Results NOPRINT;
  CLASS Vendor Month;
  TYPES Vendor Month;
  OUTPUT OUT=Defect_Tab
  SUM(Defects)=Num_Defects
  MEAN(Defects)=Defect_Rate;
  FORMAT Defect_Rate 8.2;
RUN;
```

```
Defect_Tab

                                        Num_      Defect_
Vendor        Month     _TYPE_   _FREQ_  Defects   Rate

              1         1        60      69        1.15
              2         1        60      68        1.13
              3         1        38      41        1.08
              4         1        64      80        1.25
              5         1        64      87        1.36
              6         1        68      115       1.69
              7         1        68      121       1.78
              8         1        68      112       1.65
              9         1        72      164       2.28
              10        1        72      153       2.13
              11        1        44      67        1.52
              12        1        76      193       2.54
ChiTronix     .         2        306     498       1.63
Duality       .         2        268     685       2.56
Empirical     .         2        180     87        0.48
```

6.4.1 Selecting Statistics with the TYPE and SUMVAR Options

The statistics plotted by PROC GCHART are specified by the TYPE option, which can take the values FREQ, CFREQ, PCT, CPCT, SUM, and MEAN. If you use the SUM or MEAN option, then you must also use the SUMVAR option to specify the variable whose sum or mean will be computed. The first code sample illustrates the use of the FREQ, PCT, SUM, and MEAN options. For these examples, the chart variable is Vendor, and the midpoint values are 'ChiTronix', 'Duality', and 'Empirical'.

```
GOPTIONS RESET=ALL BORDER GUNIT=PCT HTEXT=3 FTEXT='Arial';
AXIS1 LABEL=(HEIGHT=3 "Vendor"); ❶
AXIS2 LABEL=(HEIGHT=3 "Number of Units Tested");
TITLE1 H=5 "Number of Units Tested by Vendor";
PROC GCHART DATA=JES.Results;
   HBAR Vendor / NAME="F6_28a" TYPE=FREQ ❷
           MAXIS=AXIS1 RAXIS=AXIS2; ❸
RUN; QUIT;
TITLE1 H=5 "Percent of Units Tested by Vendor";
AXIS2 LABEL=(HEIGHT=3 "Percent of Units Tested");
PROC GCHART DATA=JES.Results;
   HBAR Vendor / NAME="F6_28b" TYPE=PCT MAXIS=AXIS1 RAXIS=AXIS2; ❹
RUN; QUIT;
TITLE1 H=5 "Number of Defects by Vendor";
AXIS2 LABEL=(HEIGHT=3 "Number of Defects");
PROC GCHART DATA=JES.Results;
   HBAR Vendor / NAME="F6_28c" TYPE=SUM SUMVAR=Defects ❺
           MAXIS=AXIS1 RAXIS=AXIS2 ;
RUN; QUIT;
TITLE1 H=5 "Defects per Unit by Vendor";
AXIS2 LABEL=(HEIGHT=3 "Defects per Unit");
PROC GCHART DATA=JES.Results;
   HBAR Vendor /NAME="F6_28d" TYPE=MEAN SUMVAR=Defects ❻
           MAXIS=AXIS1 RAXIS=AXIS2;
   Format Defects 8.1;
RUN; QUIT;
```

❶ The AXIS and TITLE statements follow the syntax described in Sections 6.3.2.2 and 6.3.2.3.

❷ The 'HBAR Vendor' statement requests a horizontal bar chart, with one bar for each value of Vendor. The TYPE=FREQ option specifies that bar length be equal to the number of rows for each Vendor. By default, a table of statistics is placed to the right of the chart. Section 6.4.6 shows how to control the display of statistics. The resulting plot is the first plot in Figure 6-28.

❸ The MAXIS and RAXIS options specify which AXIS definition is applied to the Midpoint and Response axes, respectively.

❹ The TYPE=PCT option specifies that the bar length be equal to the percent of rows for each Vendor. This looks almost identical to the first chart, except for the horizontal axis which shows that these are percents rather than counts. The resulting plot is the second plot in Figure 6-28.

❺ The TYPE=SUM SUMVAR=Defects options specify that the bar lengths be equal to the total number of defects for each Vendor. The resulting plot is the third plot in Figure 6-28.

❻ The TYPE=MEAN SUMVAR=Defects options specify that the bar lengths be equal to the mean number of defects per unit. The resulting plot is the fourth plot in Figure 6-28.

Figure 6-28: Using the HBAR Statement with the FREQ, PCT, SUM, and MEAN Options

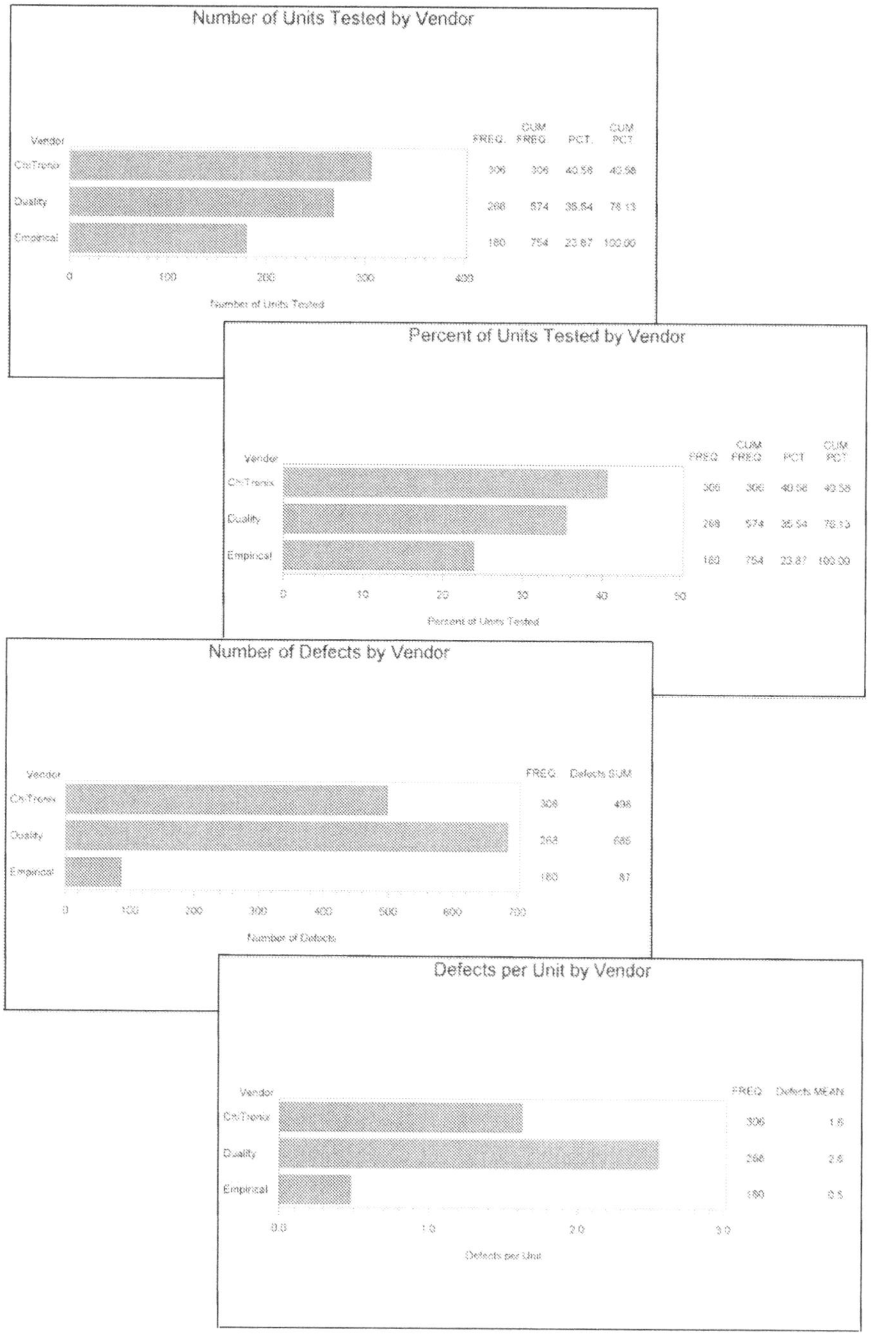

6.4.2 Using the MIDPOINTS and DISCRETE Options

If the chart variable is a character variable, HBAR creates one bar for each distinct value of the variable. The bars are evenly spaced on the vertical axis, in alphabetical order starting from the top, and labeled with the value of the character value. If the chart variable is numeric, HBAR:

- Guesses an appropriate set of equal width ranges for the variable, e.g., 0-10, 10-20, 20-30.
- Computes the appropriate statistics for the values which fall in each of these ranges.
- Labels the bars with the midpoints for each range, e.g., 5, 15, 25.

In either case, if you don't like the default choice of midpoints, you can use the MIDPOINTS option to change them. For example, the next example uses the MIDPOINTS option to create the chart shown in Figure 6-29, which is similar to the first chart in Figure 6-28, but with only the bars for Empirical and ChiTronix, in the order listed in the MIDPOINTS option.

```
TITLE1 H=5 "Number of Units Tested by Vendor";
PROC GCHART DATA=JES.Results;
   HBAR Vendor / NAME= "F6_29_"  TYPE=FREQ MIDPOINTS='Empirical'
'ChiTronix';
RUN; QUIT;
```

The next example uses a numerical variable, Month, as the chart variable.

```
TITLE1 H=5 "Number of Units Tested by Month";
PROC GCHART DATA=JES.Results(WHERE=(Month <=6));
   HBAR Month / NAME= "F6_30_" TYPE=FREQ;
RUN; QUIT;
```

This code creates the plot shown in Figure 6-30. In this case the default choice of midpoints is inappropriate. The midpoint values 1.5, 2.5, etc. don't make sense. You can use the MIDPOINTS option to force integer values for the midpoints.

```
AXIS1 LABEL=(HEIGHT=4 "Month");
PROC GCHART DATA=JES.Results(WHERE=(Month <=6));
   HBAR Month / NAME= "F6_31_" TYPE=FREQ
   MAXIS=AXIS1 MIDPOINTS = 1 TO 6 BY 1;
RUN; QUIT;
```

The resulting plot is shown in Figure 6-31. Note that an AXIS1 statement and the MAXIS option are used to specify the axis label as 'Month', which is more appropriate than the default 'Month MIDPOINT' label used in Figure 6-30.

You can get exactly the same plot by using the DISCRETE option, which creates a bar for each distinct value of the numeric chart variable. The plot created by this code is identical to Figure 6-31.

```
PROC GCHART DATA=JES.Results(WHERE=(Month<=6));
   HBAR Month / NAME="F6_F_" TYPE=FREQ
   MAXIS=AXIS1 DISCRETE;
RUN; QUIT;
```

Note that with the DISCRETE option the bars will be equally spaced, and not necessarily consistent with a numerical scale. For example, if there were no data for Month=2, the DISCRETE option would plot equally spaced bars for Months 1, 3, 4, 5, and 6 without any space left for Month 2.

Figure 6-29: Using the MIDPOINT Option with a Character Chart Variable

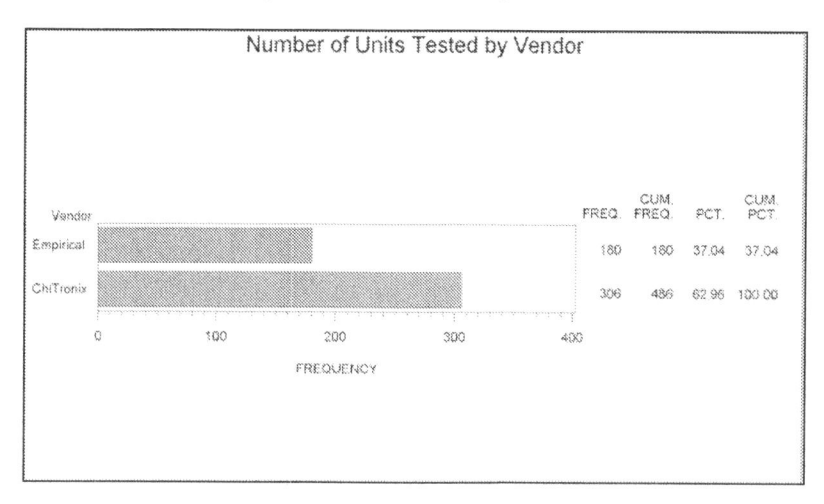

Figure 6-30: Plotting a Numeric Chart Variable with HBAR

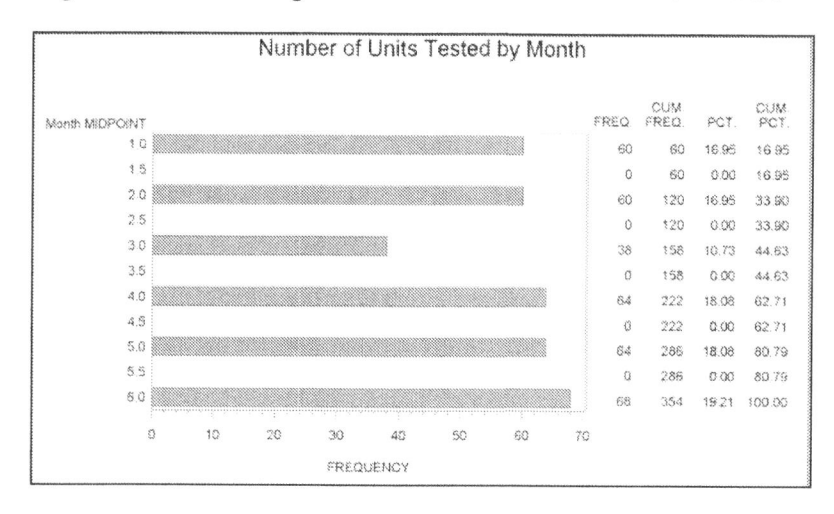

Figure 6-31: Using the MIDPOINT Option with a Numeric Chart Variable

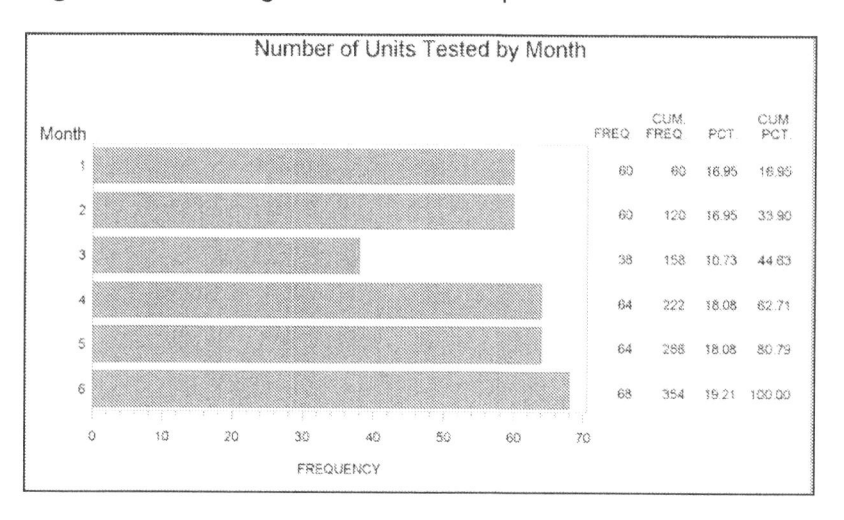

6.4.3 Using PATTERN Statements to Control Bar Fill Patterns

Just as you use SYMBOL statements to control the appearance of the plotted points in GPLOT, you can use PATTERN statements to control the appearance of the bars in GCHART. In the previous examples, the default fill pattern was good enough. But when you use the GROUP and SUBGROUP options (see Sections 6.4.4. and 6.4.7) to create more complex bar charts, several different bar fill patterns are required. In these cases the default patterns might not work well, especially if the chart is reproduced in black and white.

You can use PATTERN<n> statements, $n=1,2,...,255$, to define up to 255 different patterns. Each time you run PROC GCHART, these patterns are used in order, PATTERN1, then PATTERN2, etc. up to as many patterns are required for the chart. You can use the PATTERNID option within PROC GCHART to control how these patterns are applied to each bar. Table 6-13 lists some of the more useful PATTERN and PATTERNID options.

Table 6-13: Selected PATTERN Statement Options and the PATTERNID Option

Selected PATTERN Statement Options		
Option	Meaning	Some Possible values
COLOR	Sets the color of the bar fill.	red, blue, gray (see Section 6.3.2.6)
VALUE	Sets the pattern for the bar fill.	(see below)
Possible values of the VALUE option		
SOLID	Solid color.	
R1, R2, R3, R4, R5	Right slanting lines. R1 produces the lightest shading, R5 the heaviest.	
L1, L2, L3, L4, L5	Left slanting lines. L1 produces the lightest shading, L5 the heaviest.	
X1, X2, X3, X4, X5	Crosshatched lines. X1 produces the lightest shading, X5 the heaviest.	
Selected PATTERNID Option values		
MIDPOINT	Change pattern every time the midpoint value changes.	
GROUP	Change pattern every time the value of the group variable changes **	
SUBGROUP	Change pattern every time the value of the subgroup variable changes**	
** Group and subgroup variables are discussed in Section 6.4.4 and Section 6.4.7		

The next bit of sample code uses the PATTERNID option to force a different fill pattern for each Month.

```
AXIS1 LABEL=(HEIGHT=3 "Month");
AXIS2 LABEL=(HEIGHT=3 "Number of Units");
TITLE1 H=5 "Number of Units Tested by Month";
PROC GCHART DATA=JES.Results;
  HBAR Month / NAME= "F6_32_" TYPE=FREQ
    MAXIS=AXIS1 RAXIS=AXIS2 MIDPOINTS = 1 TO 12 BY 1
    PATTERNID=MIDPOINT;
RUN; QUIT;
```

The resulting plot, shown in Figure 6-32, uses the first 12 default fill patterns. The plot looks good in color, for example in the Graph window, but some of the colors are indistinguishable shades of gray when reproduced in black and white, as in this book. You can use PATTERN statements to control the fill pattern for each bar.

```
PATTERN1 COLOR=SlateBlue VALUE=S;
PATTERN2 COLOR=Tomato    VALUE=S;
PATTERN3 COLOR=GRAY99    VALUE=S;
PATTERN4 COLOR=Blue VALUE=R1;
PATTERN5 COLOR=Red VALUE=R3;
PATTERN6 COLOR=Green VALUE=R5;
PATTERN7 COLOR=Blue VALUE=L1;
```

```
PATTERN8  COLOR=Red   VALUE=L3;
PATTERN9  COLOR=Green VALUE=L5;
PATTERN10 COLOR=Blue  VALUE=X1;
PATTERN11 COLOR=Red   VALUE=X3;
PATTERN12 COLOR=Green VALUE=X5;
```

Figure 6-33 shows the result of rerunning the same PROC GCHART code after the PATTERN statements.

Figure 6-32: Using the PATTERNID Option with the Default Bar Fill Patterns

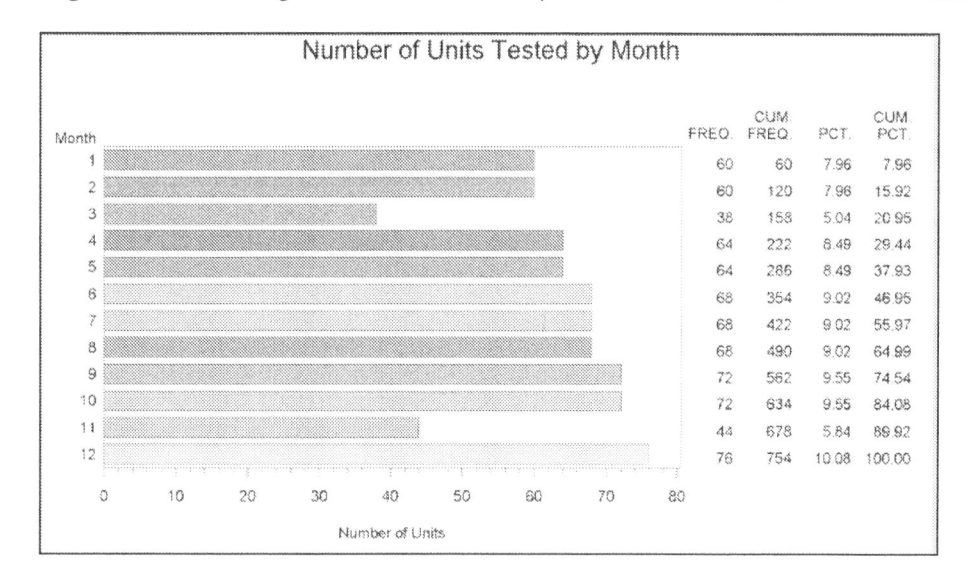

Figure 6-33: Using PATTERN Statements to Control the Bar Fill Patterns

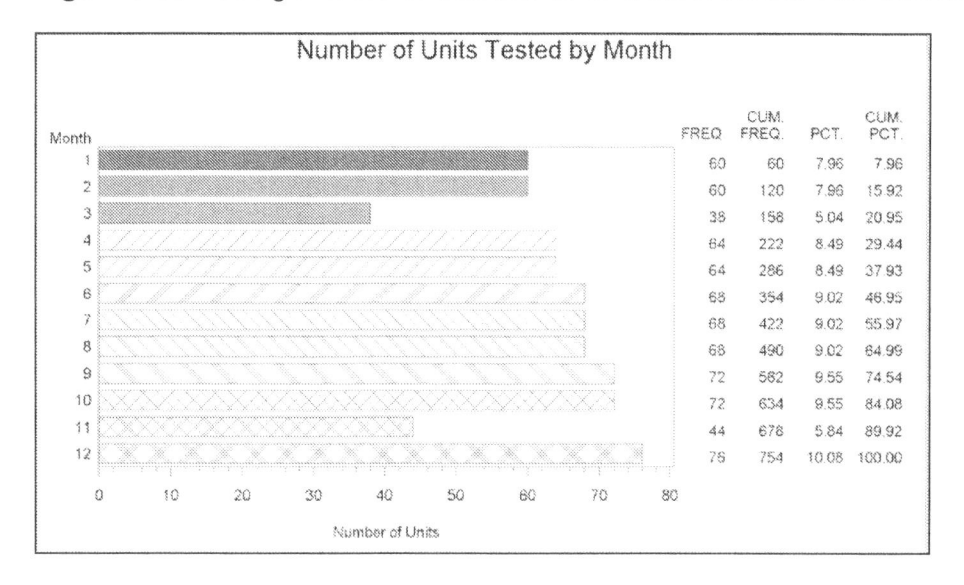

There is really not much point in having different bar fills for each month, but these examples are intended to illustrate the look of the various patterns. The following sections include more practical examples of the use of varying bar fills to distinguish groups and subgroups.

6.4.4 Using the GROUP Option

You can use the GROUP option to create separate groups of bars for each distinct value of a *group variable*, which can be any other variable in the data set. The code below plots the number of records (TYPE=FREQ) for each Vendor separately by Month.

```
PATTERN1 COLOR=Red    VALUE=SOLID;
PATTERN2 COLOR=Blue   VALUE=R3;
PATTERN3 COLOR=Green  VALUE=L3;
AXIS2 LABEL=(HEIGHT=3 "Number of Units");
TITLE1 H=5 "Number of Units Tested by Month and Vendor";
PROC GCHART DATA=JES.Results(WHERE=(Month<=3));
   HBAR Vendor /NAME="F6_34_" TYPE=FREQ RAXIS=AXIS2
   GROUP=Month  ❶
   PATTERNID=GROUP;  ❷
RUN; QUIT;
```

❶ The GROUP=Month option requests a separate set of bars for each value of the variable Month.

❷ The PATTERNID=GROUP option requests a different fill pattern for each value of the group variable, Month.

The resulting plot is shown in Figure 6-34. It is not too useful to have the same pattern for each vendor when the bars are grouped by vendor, but this example is included just to illustrate the meaning of the PATTERN option.

The next bit of code is similar, but interchanges Vendor and Month so that the bars will be grouped by month within Vendor instead of by Vendor within Month. It also uses the PATTERNID=MIDPOINT option so that the fill pattern changes within the groups instead of from group to group.

```
TITLE1 H=5 "Number of Units Tested by Vendor and Month";
PROC GCHART DATA=JES.Results(WHERE=(Month<=3));
   HBAR Month / NAME="F6_35_" TYPE=FREQ RAXIS=AXIS2
   GROUP=Vendor  ❶
   DISCRETE  ❷
   PATTERNID=MIDPOINT;  ❸
RUN; QUIT;
```

❶ The GROUP=Vendor option requests a separate set of bars for each value of the variable Vendor.

❷ The DISCRETE option is required to avoid the inappropriate choice of default midpoint values for Month, like those in Figure 6-30.

❸ The PATTERNID=MIDPOINT option requests a different fill pattern for each value of the chart variable, Month.

The resulting plot is shown in Figure 6-35. In this case the patterns are more useful because they enable you to easily identify the bars corresponding to each month.

Figure 6-34: HBAR with Chart Variable=Vendor and GROUP=Month

Figure 6-35: HBAR with Chart Variable = Month and GROUP=Vendor

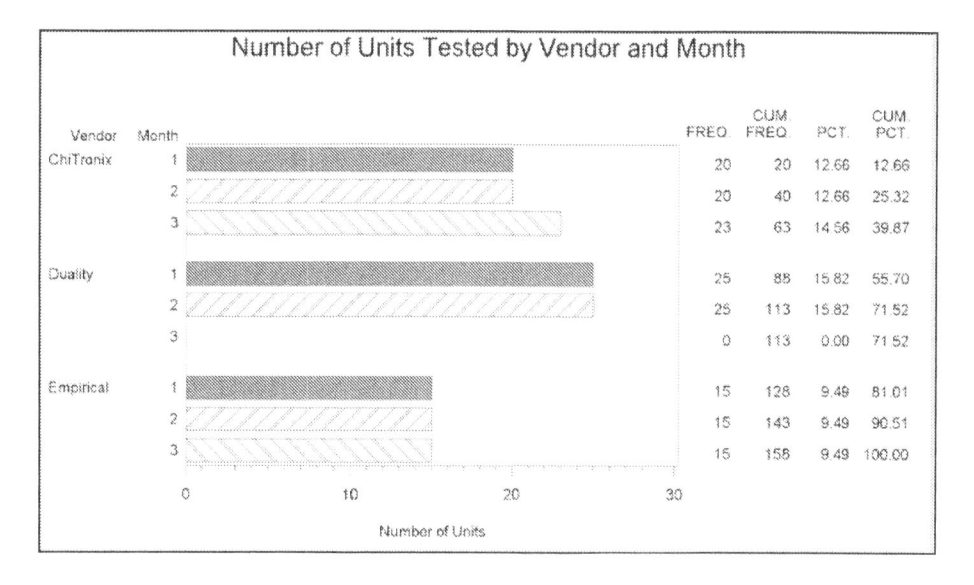

6.4.5 Using VBAR to Create Vertical Bar Charts

If you use a VBAR statement instead of an HBAR statement with PROC GCHART, the output will be similar, except that the bars will be vertical instead of horizontal. The code below creates the VBAR version of the first HBAR chart shown in Figure 6-28. The resulting plot is shown in Figure 6-36.

```
AXIS1 LABEL =(H=4  A=90  FONT='Arial' 'Number of Units')
      VALUE = (H=3 FONT='Arial');
TITLE1 H=5 "Number of Units Tested by Vendor";
PROC GCHART DATA=JES.Results;
   VBAR Vendor / NAME="F6_36_" TYPE=FREQ RAXIS=AXIS1;
RUN; QUIT;
```

Figure 6-36: Using a VBAR Statement with PROC GCHART

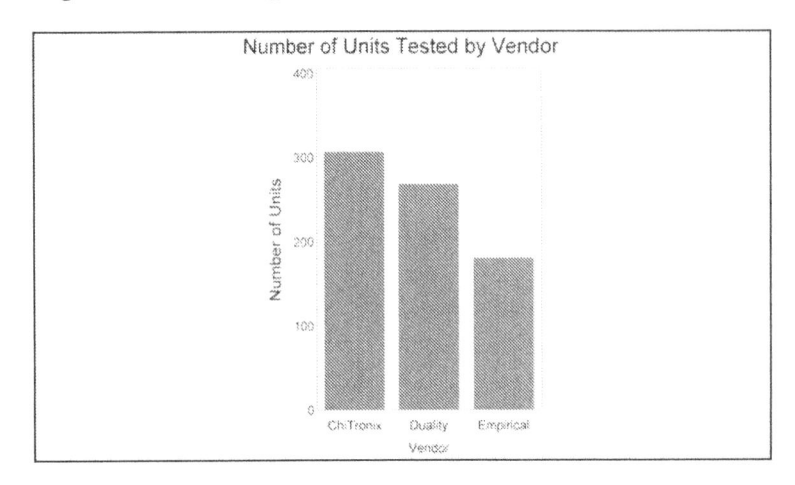

You can use the WIDTH and SPACE options to control the width of the bars and the space between bars.

```
PROC GCHART DATA=JES.Results;
   VBAR Vendor /NAME="F6_37_" TYPE=FREQ RAXIS=AXIS1
     WIDTH=50 SPACE=20;  ❶
RUN; QUIT;
```

❶ The WIDTH=50 option creates bars of width 50, and the SPACE=20 option sets the space between bars to 2, in units of *character cells*. If you use the WIDTH and SPACE options, it is easy to specify more space than is available in the graphics output area. When this happens, you will see a warning in the Log window. If you get this warning, reduce WIDTH and/or SPACE and try again.

> WARNING: There was not enough room to use the specified width of 60 and space of 25 for the bars.

When your plot includes the GROUP option, you can also use the GSPACE option to control the space between groups as shown in the next code sample. The plot is shown in Figure 6-38.

```
TITLE1 H=5 "Number of Units Tested by Vendor and Month";
PROC GCHART DATA=JES.Results(WHERE=(Month<=3));
   VBAR Month /NAME="F6_38_" TYPE=FREQ RAXIS=AXIS1 GROUP=Vendor
     DISCRETE
     PATTERNID=MIDPOINT
     WIDTH=17 SPACE=3 GSPACE=6;
RUN; QUIT;
```

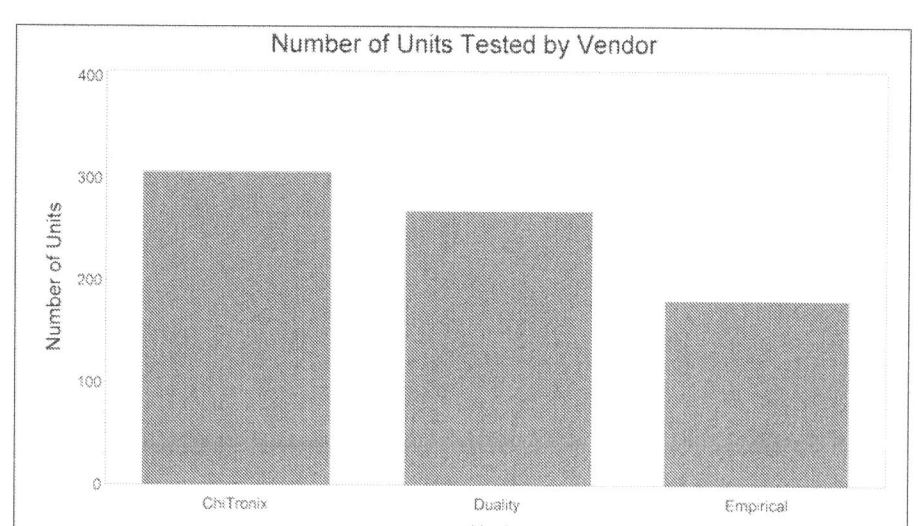

Figure 6-37: Using the WIDTH and SPACE Options

Figure 6-38: Using the GSPACE Option

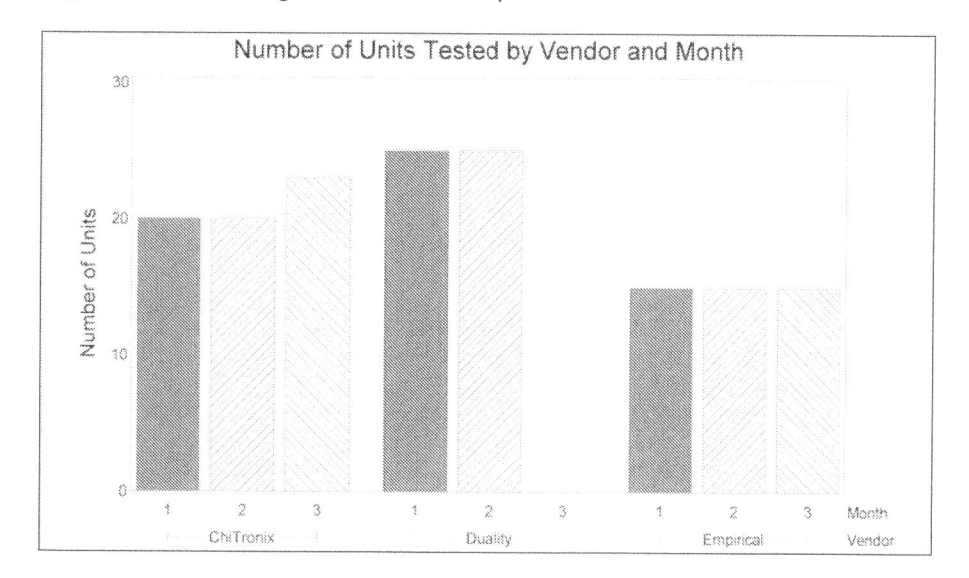

6.4.6 Controlling the Display of Statistics with Bar Charts

By default, HBAR displays a table of statistics to the right of the plot area. The default statistics for each value of TYPE can be seen in the examples in Figure 6-28. You can control which statistics are displayed by including one or more of the statistics options, FREQ, CFREQ, PCT, CPCT, SUM, MEAN, on your PROC GCHART statement. You can also use corresponding label options, e.g. FREQLABEL, to specify the column headers for the table. Use the NOSTATS option to prevent the display of any statistics.

The next code sample illustrates the use of these options.

```
AXIS1 LABEL=(HEIGHT=4 "Vendor");
AXIS2 LABEL=(HEIGHT=4 "Defects per Unit");
TITLE1 H=5 "Defects per Unit by Vendor";
PROC GCHART DATA=JES.Results;
   HBAR Vendor / NAME="F6_39_" TYPE=MEAN SUMVAR=Defects
   MAXIS=AXIS1 RAXIS=AXIS2 WIDTH=10
   FREQ SUM MEAN ❶
   FREQLABEL="Number of Units" ❷
   SUMLABEL="Number of Defects"
   MEANLABEL="Defects per Unit";
   FORMAT Defects 8.1;
RUN; QUIT;
```

❶ The FREQ, SUM, and MEAN options request that those statistics, and no others, be displayed.

❷ The FREQLABEL, SUMLABEL, and MEANLABEL options specify the headers for the displayed table.

By default, VBAR does not display a table of statistics. However, you can request that statistics be displayed either above or within the bars.

```
GOPTIONS HTEXT=3; ❶
TITLE1 H=5 "Defects per Unit by Vendor and Month";
AXIS2 LABEL=(H=4 A=90 "Defects per Unit");
PATTERN1 COLOR=GRAYCC VALUE=SOLID; ❷
PROC GCHART DATA=JES.Results(WHERE=(Month<=3));
   VBAR Month / NAME="F6_40_" TYPE=MEAN SUMVAR=Defects GROUP=Vendor
   DISCRETE WIDTH=18 SPACE=2 GSPACE=6 RAXIS=AXIS2
   OUTSIDE=MEAN INSIDE=SUM; ❸
   FORMAT Defects 8.1;
RUN; QUIT;
```

❶ The GOPTIONS HTEXT=3 statement is used to ensure that the text is small enough to fit above and within the bars. If the text does not fit, a "labels are wider than the bars" warning is written to the Log window, and the statistics are not displayed. If you get this warning, increase the bar width or decrease the text height and try again.

❷ The PATTERN statement sets the bar fill pattern to light gray so that the statistics within the bars will be visible.

❸ The OUTSIDE=MEAN option requests that the mean of the chart variable be printed above each bar. The INSIDE=SUM option requests that the sum of the chart variable be printed inside each bar.

Figure 6-39: Using the Statistics and Label Options to Control the Display of Statistics

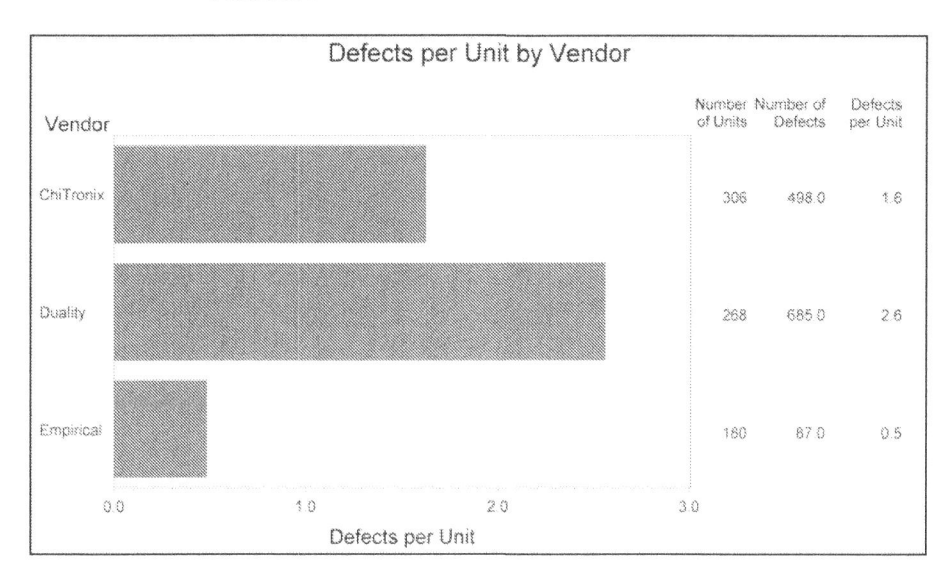

Figure 6-40: Using the INSIDE and OUTSIDE Options to Control the Display of Statistics

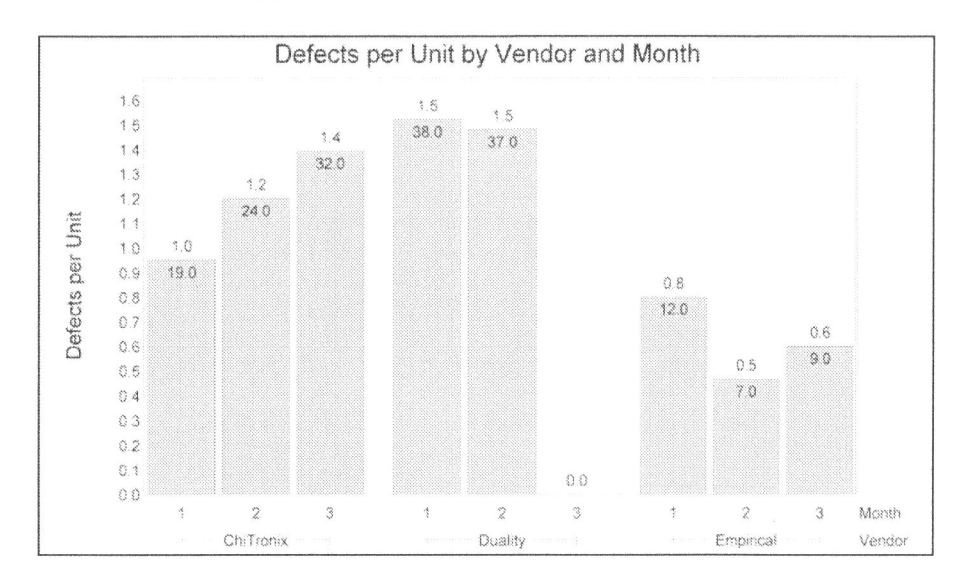

6.4.7 Using the SUBGROUP Option

You can use the SUBGROUP option to divide each bar of a bar chart into segments. The length of each segment is proportional to the contribution of each value of the *subgroup variable* to the total statistic for the bar. The examples in this section use the Fail and Result variables in the JES.Results data set (see Section 5.1.1). The Result variable is equal to "Pass" if the corresponding Resistance measurement is within the specification limits of (12.5, 22.5), and equal to "Low Res" or "High Res" if Resistance is below or above the specification limits. The Fail variable is equal to zero if the Resistance measurement is within the specification limits, and one otherwise.

The first example creates a bar chart (Figure 6-41) of the number of fails by Vendor, with the bars subdivided to show the relative contribution of the two failure modes, "High Res" and "Low Res".

```
GOPTIONS HTEXT=4;
PATTERN1 COLOR=Red      VALUE=L1;
PATTERN2 COLOR=GRAYD0   VALUE=SOLID;
LEGEND1 LABEL=(J=CENTER POSITION=LEFT "Failure Mode") FRAME;
TITLE1 H=5 "Total Number of Resistance Failures by Vendor and Cause";

PROC GCHART DATA= JES.Results(WHERE=(Result NE 'Pass'));
   HBAR Vendor /NAME="F6_41_" TYPE=SUM SUMVAR=Fail  ❶
SUM SUMLABEL="Total # Fails"
   SUBGROUP=Result  ❷
   LEGEND=LEGEND1 WIDTH=10;
RUN; QUIT;
```

❶ The TYPE=SUM and SUMVAR=Fail options request a plot of the number of Fails for each Vendor.

❷ The SUBGROUP=Result option specifies that the bars should be divided according to the value of Result.

This plot provides a useful aid to understanding the contribution of various failure modes to the total number of failures. But it would be far more useful to see the contribution to the failure rate, i.e., the number of fails divided by the number tested. The next example attempts to do this by using the TYPE=MEAN option.

```
TITLE1 H=5 "Average Number of Resistance Failures by Vendor and
Cause";
PROC GCHART DATA= JES.Results(WHERE=(Result NE 'Pass'));
   HBAR Vendor / NAME="F6_42_" TYPE=MEAN SUMVAR=Fail
MEAN MEANLABEL="Avg # Fails"
   SUBGROUP=Result
   LEGEND=LEGEND1 WIDTH=10;
RUN; QUIT;
```

Unfortunately, the result (Figure 6-42) does not provide the required information. The length of each bar segment is the failure rate within the corresponding subgroup. And the failure rate is 1 (100%) in the subgroups defined by Result= "Low Res" and Result="High Res" and zero in the subgroup defined by Result="Pass'. The total length of each bar is equal to the sum of these mean values, which is mathematically correct but not particularly useful. This example makes it clear that it is dangerous to use the SUBGROUP option together with TYPE=MEAN.

Section 6.4.8 shows how to achieve the desired result by first using PROC MEANS to compute the required rates, and then using PROC GCHART to create the plot.

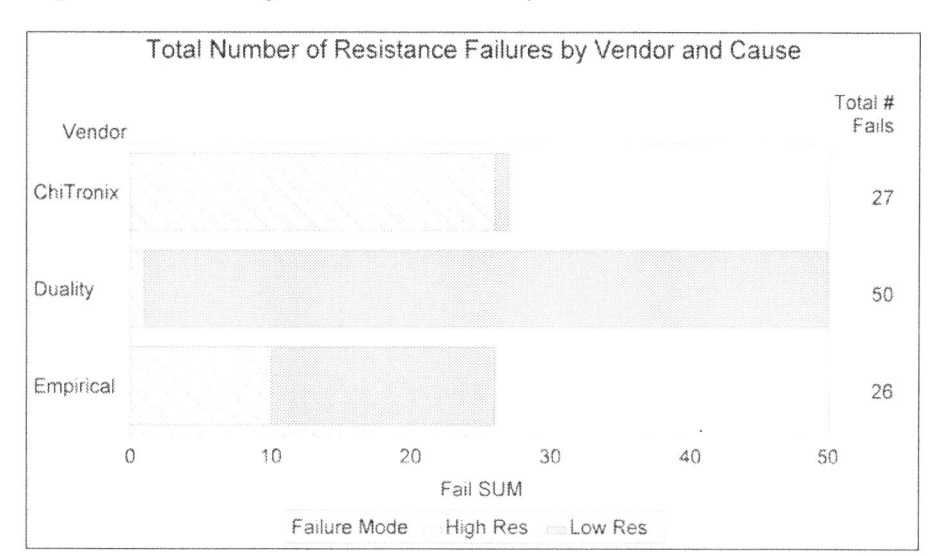

Figure 6-41: Using the SUBGROUP Option with TYPE=SUM

Figure 6-42: Using the SUBGROUP Option with TYPE=MEAN

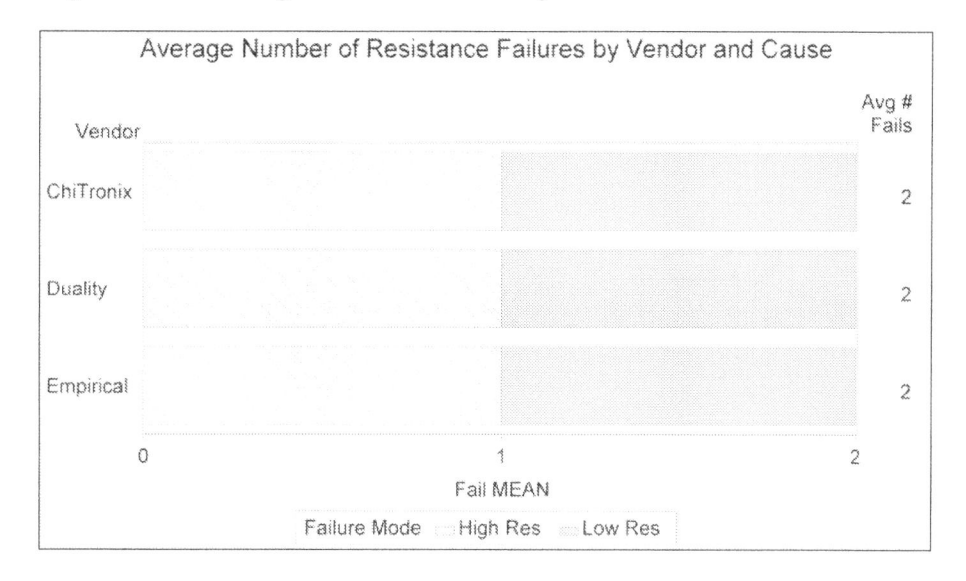

6.4.8 Creating a Stacked Bar Chart for Failure Rates

Stacked bar charts provide a very useful and easily understood way to visualize the relative contributions of different failure modes to the total failure rate. As noted in Section 6.4.7, you can't create a stacked bar chart of rates from the JES.Results data set simply by using the SUBGROUP option. But you can create the desired chart by first using PROC MEANS and a DATA step to compute the required rates.

The next bit of code uses PROC MEANS to simultaneously compute the number tested by Vendor, and the number of fails by Vendor and Result.

```
PROC MEANS DATA= JES.Results NOPRINT;
  CLASS Vendor Result;
  TYPES Vendor Vendor*Result; ❶
  OUTPUT OUT=Tab
    N(Fail)=N_Test SUM(Fail)=N_Fail;
RUN;
```

❶ The TYPES statement requests only a subset of the statistics computed by PROC MEANS, specified by the list following TYPE:

- 'Vendor' requests statistics for each value of Vendor.
- 'Vendor*Result' requests statistics for each combination of Vendor and Result.

```
Tab
 Vendor      Result      _TYPE_    _FREQ_    N_Test    N_Fail
 ChiTronix                  2        306       306       27
 Duality                    2        268       268       50
 Empirical                  2        180       180       26
 ChiTronix   High Res       3         26        26       26
 ChiTronix   Low Res        3          1         1        1
 ChiTronix   Pass           3        279       279        0
 Duality     High Res       3          1         1        1
 Duality     Low Res        3         49        49       49
 Duality     Pass           3        218       218        0
 Empirical   High Res       3         10        10       10
 Empirical   Low Res        3         16        16       16
 Empirical   Pass           3        154       154        0
```

The highlighted values in Tab are required to compute the failure rates. The next DATA step creates the Summary data set by merging the failure counts (N_Fail) for each failure mode (Result) with the number tested (N_Test) and then computing the required failure rates by Vendor and mode.

```
DATA Summary;
  MERGE  Tab(WHERE=(_TYPE_=2) KEEP=Vendor N_Test _TYPE_)
         Tab(WHERE=(_TYPE_=3) KEEP=Vendor Result N_Fail _TYPE_);
  BY Vendor;
  FORMAT P_Fail percent8.2;
  P_Fail=N_Fail/N_Test;
RUN;
```

The MERGE statement brings together the values you need to compute the required rates:

- The value of N_Test from the rows where _TYPE_ = 2.
- The values of Result and N_Fail from the rows where _TYPE_=3.

The required rates are then computed as P_Fail=N_Fail/N_Test.

```
Summary
 Vendor      _TYPE_    N_Test    Result       N_Fail      P_Fail
 ChiTronix      3        306     High Res       26         8.50%
 ChiTronix      3        306     Low Res         1         0.33%
 ChiTronix      3        306     Pass            0         0.00%
 Duality        3        268     High Res        1         0.37%
 Duality        3        268     Low Res        49        18.28%
 Duality        3        268     Pass            0         0.00%
 Empirical      3        180     High Res       10         5.56%
 Empirical      3        180     Low Res        16         8.89%
 Empirical      3        180     Pass            0         0.00%
```

PROC GCHART can be used with the Summary data set to create the desired stacked bar chart.

```
GOPTIONS HTEXT=3;
TITLE1 H=5 "Resistance Test Failure Rate by Vendor and Cause";
AXIS1 LABEL=(H=4 "% Fail") ;
AXIS2 LABEL=(HEIGHT=4 "Vendor") VALUE=(H=4);
PATTERN1 COLOR=Red     VALUE=L1;
PATTERN2 COLOR=GRAYD0  VALUE=SOLID;
PROC GCHART DATA=Summary(WHERE=(Result ne 'Pass'));
   VBAR Vendor / NAME="F6_43_" TYPE=SUM SUMVAR=P_Fail SUBGROUP=Result❶
   WIDTH=30 SPACE=25 RAXIS=AXIS1 MAXIS=AXIS2 COUTLINE=Blue
   OUTSIDE=SUM INSIDE=SUM; ❷
RUN; QUIT;
```

❶ The TYPE=SUM SUMVAR=P_Fail SUBGROUP=Result options request the sum of the failure rate (P_Fail) for each value of Result.

❷ The OUTSIDE=SUM option writes the total failure rate above each bar. The INSIDE=SUM option writes the failure rate by cause within each bar segment, if there is enough room.

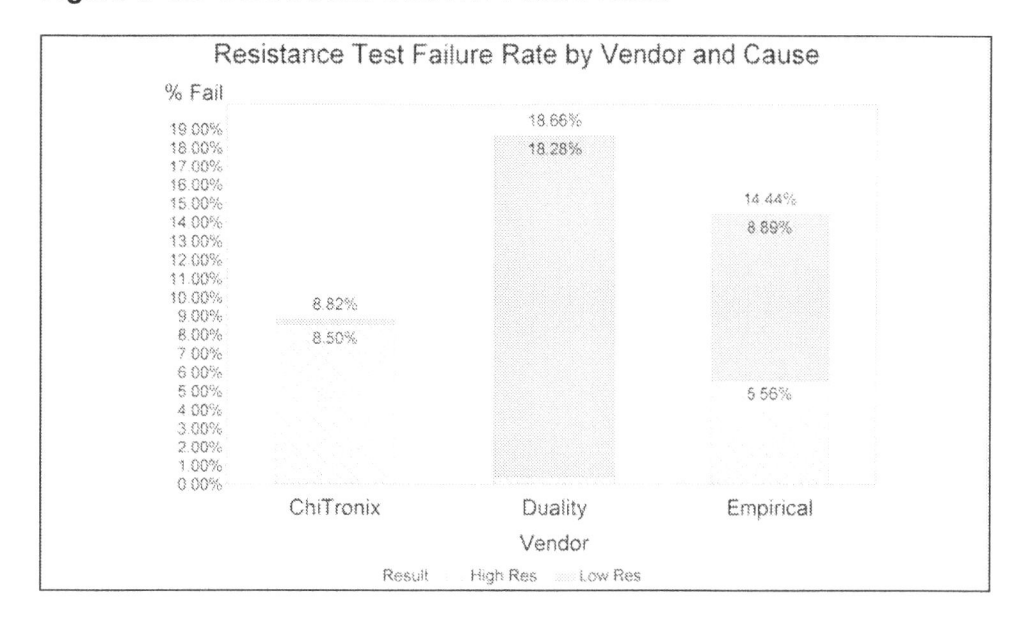

Figure 6-43: Stacked Bar Chart for Failure Rates

6.5 More Than Enough

This section includes examples of several SAS/GRAPH plot types which were not discussed in the previous sections. The examples are given with very little explanation of the syntax, just to give you an idea of what is possible. If you want to use one of these plot types, consult the SAS online documentation for more information. Table 6-14 shows which plot types are included in each section. The code for this section is in ~\JES \sample_code\ch_6 \other_examples.sas.

Table 6-14: Selected Graph Types Available in SAS/GRAPH Software

Section	Procedure	Statement	Graph Type
6.5.1	GCHART	BLOCK	Blocks set in a 2-D grid defined by two variables, with height proportional to a third variable.
6.5.2		PIE	Pie chart.
6.5.3	GPLOT	BUBBLE	An array of circles with the location of the centers determined by the values of two variables, and the radii proportional to the value of a third variable.
		BUBBLE2	Overlays a second bubble plot on a graph created with a BUBBLE statement.
6.5.4	GMAP	CHORO	Two-dimensional maps in which values of a variable are represented by areas of varying patterns and colors.
		BLOCK	Three-dimensional block maps on which values of a variable are represented by blocks of varying height, pattern and color.
		PRISM	Three-dimensional prism maps on which values of a variable are represented by polyhedrons of varying height, pattern and color.
		SURFACE	Three-dimensional surface maps on which values of a variable are represented by spikes of varying height.
6.5.5	GCONTOUR	PLOT	Contour plot representing the levels of one variable for positions on the plane defined by the values of two other variables.
6.5.6	G3D	PLOT	Surface described by the values of two horizontal variables and a third vertical variable.
		SCATTER	Similar to a surface plot, but with the data represented by points instead of a surface.

6.5.1 PROC GCHART: Block Charts

In PROC GCHART, a BLOCK statement creates a set of 3-D blocks set in a 2-D grid defined by two variables, with height proportional to a third variable.

```
GOPTIONS COLORS=(Gray60 GRAYD0) CTEXT=Black HPOS=40 VPOS=40;
TITLE1 H=4 "Resistance Test Results by Vendor and Month";
PROC GCHART DATA= JES.Results_Q4;
   BLOCK Month / NAME="F6_44_" ❶
   GROUP=Vendor ❷
   SUBGROUP=Result ❸
   TYPE=FREQ MIDPOINTS=10 TO 12 COUTLINE=Black CAXIS=Green
BLOCKMAX=50;
RUN; QUIT;
```

❶ The BLOCK statement requests a block plot of the number of records (TYPE=FREQ) for each value of Month.

❷ The GROUP=Vendor option requests a 2-D array of blocks for each combination of Month and Vendor.

❸ The SUBGROUP=Result option specifies that the blocks be subdivided by the number of records corresponding to each value of the Result variable.

Figure 6-44: Using a BLOCK Statement with PROC GCHART

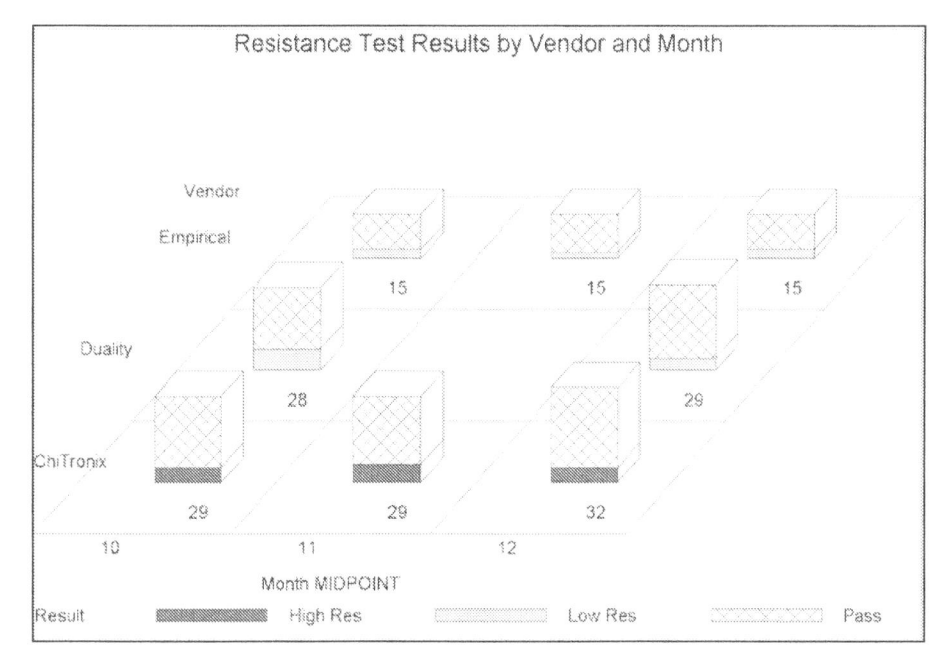

6.5.2 PROC GCHART: Pie Charts

You can use a PIE, PIE3D, or DONUT statement to produce various kinds of pie charts. The PIE statement is illustrated here. A PIE3D statement would give the plot a 3-D appearance, while a DONUT statement would leave a hole in the center of the plot where you can place descriptive text.

```
TITLE1 H=4 "Resistance Test Results by Vendor";
PATTERN1 COLOR=Green  VALUE=P3N135;
PATTERN2 COLOR=Red    VALUE=P3N45;
PATTERN3 COLOR=GRAYD0 VALUE=PSOLID;
PROC GCHART DATA=JES.Results;
   PIE Result / NAME="F6_45_"  ❶
   SUBGROUP=Vendor  ❷
   COUTLINE=Blue NOHEADING;
RUN; QUIT;
```

❶ The PIE statement requests a pie chart with slice size proportional to the number of records for each distinct value of the Result variable.

❷ The SUBGROUP=Vendor option requests separate pies, arranged as concentric circles of increasing size, for each value of Vendor.

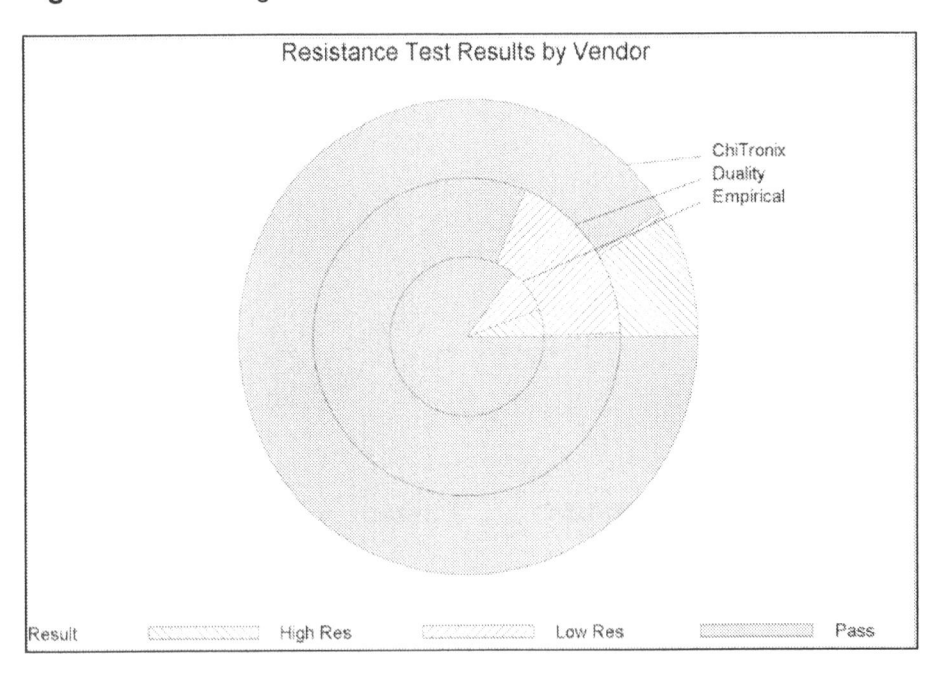

Figure 6-45: Using a PIE Statement with PROC GCHART

6.5.3 PROC GPLOT: Bubble Plots

In PROC GPLOT, a BUBBLE statement creates an array of circles with the location of the centers determined by the values of two variables, and the radii proportional to the value of a third variable. An optional BUBBLE2 statement creates an overlay of circles on the same grid with radii proportional to a different variable in the data set.

```
GOPTIONS YMAX=3.375IN;
TITLE1 H=4 "M_Def and P_Fail by Vendor and Month";
AXIS1 LABEL=('Month') OFFSET=(10, 10);
AXIS2 LABEL=NONE OFFSET=(10, 10);
PROC GPLOT DATA= JES.Results_Tab(WHERE=(Month<=6));
   BUBBLE Vendor*Month=M_Def /NAME="F6_46_"  ❶
      BSIZE=15 HAXIS=AXIS1 VAXIS=AXIS2 BLABEL;
   BUBBLE2 Vendor*Month=P_Fail /  ❷
      BSIZE=10 VAXIS=AXIS2 BCOLOR=Red NOAXIS;
RUN; QUIT;
```

❶ The BUBBLE Vendor*Month=M_Def statement requests an array of circles with centers determined by Vendor and Month, and radii proportional to the mean number of defects, M_Def.

❷ The BUBBLE2 Vendor*Month=P_Fail statement requests an overlaid array of circles with centers determined by Vendor and Month, and radii proportional to P_Fail.

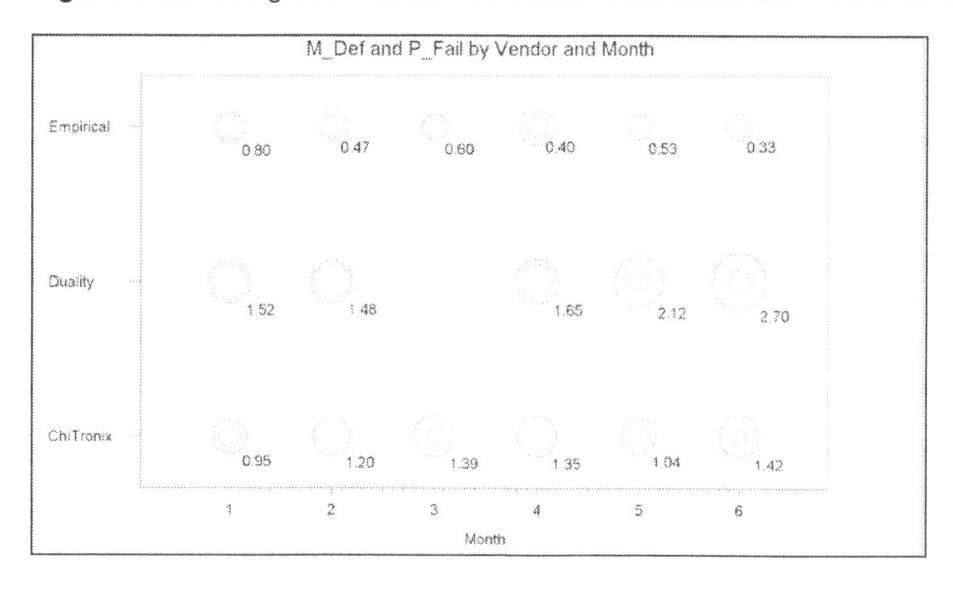

Figure 6-46: Using BUBBLE and BUBBLE2 Statements with PROC GPLOT

6.5.4 PROC GMAP: Choro, Block, Prism, and Surface Plots

PROC GMAP can be used to plot a response variable on a map, for example the median household income for each state in the US, plotted on a map of the US with outlines of each state. In the example used here, the map is a wafer map, which shows the location of each chip on a wafer, and the response variable is the value of some parametric measurement for each chip on the wafer map. PROC GMAP requires two data sets, one to define the outlines of each area on the map, and another to specify the value of the response variable. The required data sets, JES.Wafer_Map and JES.Wafer_Data, were created using this code.

```
DATA JES.Wafer_Map;  ❶
  DO I=1 TO 8;
     DO J= 1 TO 8;
           Loc = J + 8*(I-1);
           X= I;     Y = J;        OUTPUT;
           X= (I+1); Y = J;        OUTPUT;
           X= (I+1); Y = (J+1);    OUTPUT;
           X= I;     Y = (J+1);    OUTPUT;
     END;
  END;
RUN;
DATA JES.Wafer_Data; SET JES.Wafer_Map;  ❷
  IF MOD(_N_,4)=1;
  Parm = SQRT((X-5)*(X-5)+(Y-5)*(Y-5));
RUN;

DATA Anno; LENGTH FUNCTION COLOR STYLE $8.;    SET JES.Wafer_Data;  ❸
  RETAIN FUNCTION 'LABEL' COLOR 'White' WHEN 'A'
      STYLE 'SWISSB' XSYS YSYS '2' SIZE 1;
  X=X+.5; Y=Y+.5;TEXT = PUT(Parm, 5.2);
RUN;
```

❶ The first DATA step creates JES.Wafer_Map, which includes the outline of each chip location on the wafer. The Loc variable will be used to identify each chip on the wafer. For each value of Loc, the X and Y variables define the outline of the corresponding chip.

❷ The second DATA step creates JES.Wafer_Data, which includes the value of Parm for each value of Loc. The SQRT function is used to create a hypothetical parameter that increases with the distance from the center of the wafer.

❸ The third DATA step creates the Anno data set which you can use to annotate the plot.

The first several rows of WafMap, WafData, and Anno are shown below.

```
WafMap
I    J    Loc    X    Y
1    1    1      1    1
1    1    1      2    1
1    1    1      2    2
1    1    1      1    2
1    2    2      1    2
1    2    2      2    2
1    2    2      2    3
1    2    2      1    3
       (partial listing)
```

```
WafData
I    J    Loc    X    Y      Parm
1    1    1      1    1    5.65685
1    2    2      1    2    5.00000
1    3    3      1    3    4.47214
1    4    4      1    4    4.12311
       (partial listing)
```

```
Anno
I    J    Loc    X    Y      Parm
1    1    1      1    1    5.65685
1    2    2      1    2    5.00000
1    3    3      1    3    4.47214
1    4    4      1    4    4.12311
       (partial listing)
```

You can use four different statements: CHORO, BLOCK, PRISM, and SURFACE with PROC GMAP to create four different kinds of plot of the response variable using the same map coordinates. The first example uses the CHORO statement to create a plot where the values of the response variable are represented by the color or pattern in each area of the map.

```
GOPTIONS YMAX=4IN
COLORS=(GRAYF3 GRAYD0 GRAYB0 GRAY90 GRAY70 GRAY50 GRAY30) CTEXT=Black; ❶
PATTERN1; PATTERN2;
TITLE1 H=4 "Parm Measurement by Wafer Location";
LEGEND1 LABEL=("Parameter");
PROC GMAP MAP= JES.Wafer_Map DATA= JES.Wafer_Data ANNOTATE=Anno; ❷
   CHORO Parm / NAME="F6_47_" ❸
MIDPOINTS=0 TO 6 BY 1 LEVELS=7 ❹
COUTLINE=Gray WOUTLINE=3 LEGEND=LEGEND1;
   ID Loc; ❺
RUN; QUIT;
```

❶ The COLORS=(GRAYF0 ...) option specifies the colors to be used in the plot. Colors are assigned in the order specified, with the first color assigned to the lower values of the response variable.

❷ The MAP= DATA= and ANNOTATE= options identify the data sets to be used in the plot.

❸ The CHORO Parm statement specifies that the value of Parm will be represented by the color of each area (chip location) in the wafer map.

❹ The LEVELS option specifies the number of colors used in the plot.

❺ The ID Loc; statement specifies that the Loc variable be used to link the map and the data.

Figure 6-47: Using a CHORO Statement with PROC GMAP

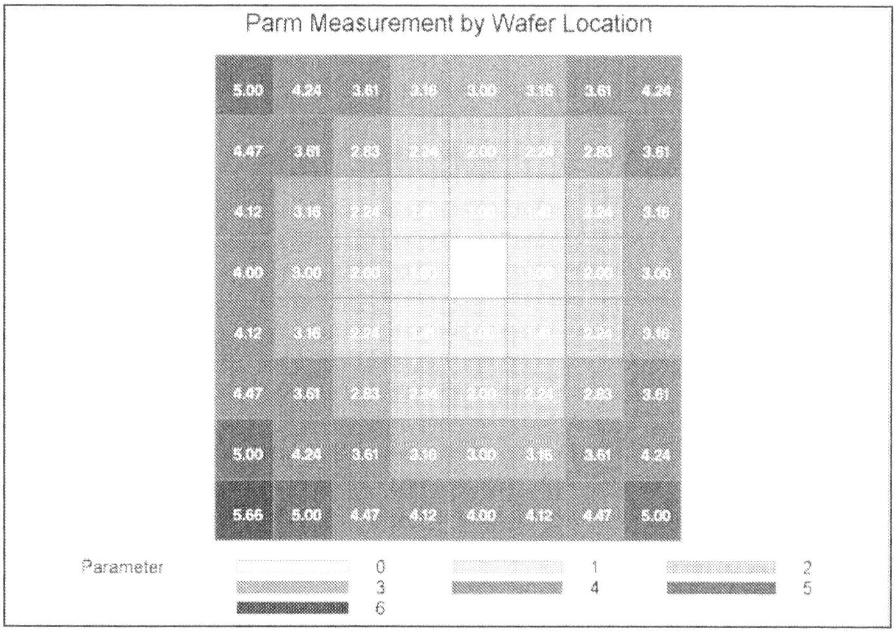

The next three examples plot the same data using the BLOCK, PRISM, and SURFACE statements, respectively.

```
GOPTIONS YMAX=3.375IN;
/*===== PROC GMAP: BLOCK ===============*/
PROC GMAP MAP= JES.Wafer_Map DATA= JES.Wafer_Data;
   BLOCK Parm / NAME="F6_48_" MIDPOINTS=0 TO 6 BY 1 LEVELS=7
COUTLINE=Gray
   LEGEND=LEGEND1 XVIEW=2 YVIEW=-2 ZVIEW=2 BLOCKSIZE=2 SHAPE=CYLINDER;
   ID Loc;
RUN; QUIT;

/*===== PROC GMAP: PRISM ===============*/
PROC GMAP MAP= JES.Wafer_Map DATA= JES.Wafer_Data;
   PRISM Parm / NAME="F6_49_" MIDPOINTS=0 TO 6 BY 1 LEVELS=7
COUTLINE=Gray WOUTLINE=3
   LEGEND=LEGEND1 XVIEW=2 YVIEW=-2 ZVIEW=2 XLIGHT=8 YLIGHT=8;
   ID Loc;
RUN; QUIT;

/*===== PROC GMAP: SURFACE ===============*/
GOPTIONS COLORS=(Black);
PROC GMAP MAP= JES.Wafer_Map DATA= JES.Wafer_Data;
   SURFACE Parm / NAME="F6_50_" LEVELS=7 ROTATE=40 TILT=60 NLINES=75;
   ID Loc;
RUN; QUIT;
```

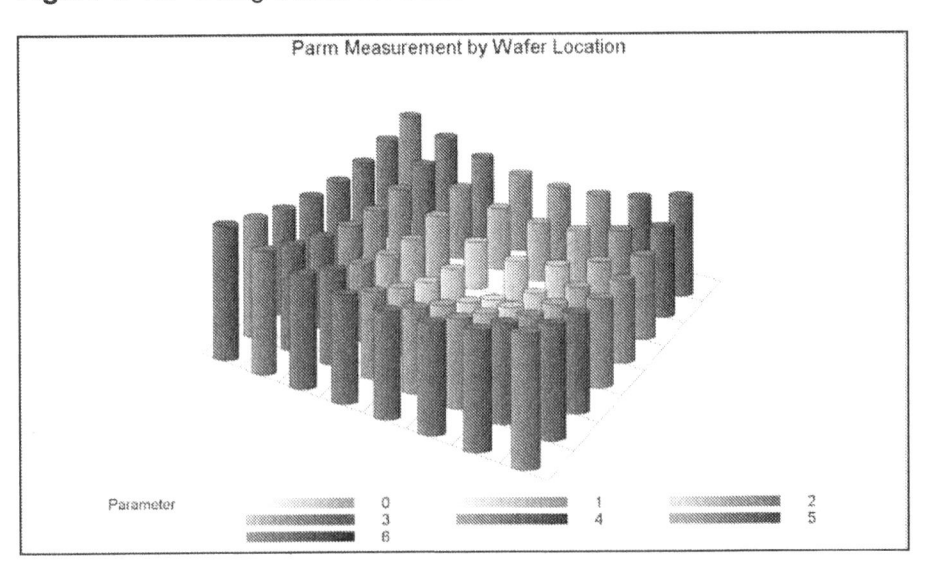

Figure 6-48: Using a BLOCK Statement with PROC GMAP

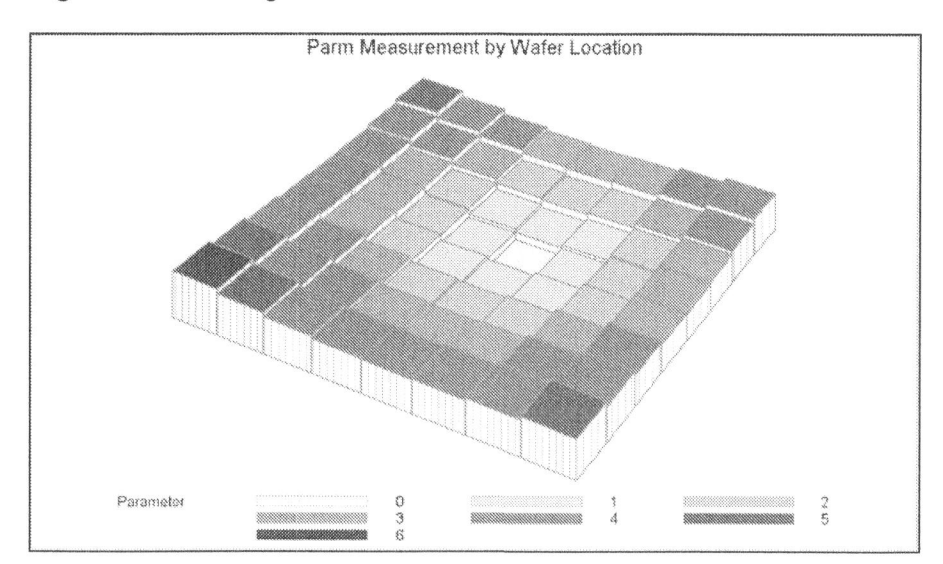

Figure 6-49: Using a PRISM Statement with PROC GMAP

Figure 6-50: Using a SURFACE Statement with PROC GMAP

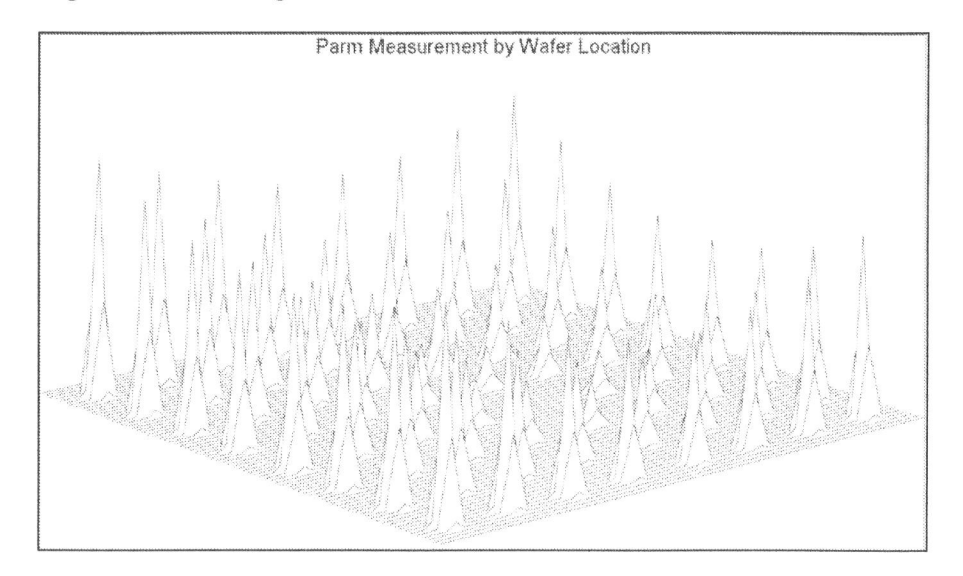

6.5.5 PROC GCONTOUR: Contour Plots

PROC GCONTOUR creates a contour plot with lines indicating the locus of equal values of the response variable on a grid representing the values of two other variables. A common example of a contour plot would be a topographic map with lines representing equal height plotted vs. longitude and latitude. This example uses the same JES.Wafer_Data data set as in the GMAP examples, but not the JES.Wafer_Map data set. The response variable is plotted against the X,Y values, not against areas defined in a map.

```
GOPTIONS COLORS=(GRAYB0 GRAY90 GRAY70 GRAY50 GRAY30);
PROC GCONTOUR DATA= JES.Wafer_Data;
   PLOT X*Y=Parm ❶
/NAME="F6_51_" LEVELS=0 TO 6 AUTOLABEL CAXIS=Blue CTEXT=Black;
RUN; QUIT;
```

❶ The PLOT X*Y=Parm statement requests contour lines in the X,Y grid which represent equal values of the variable Parm.

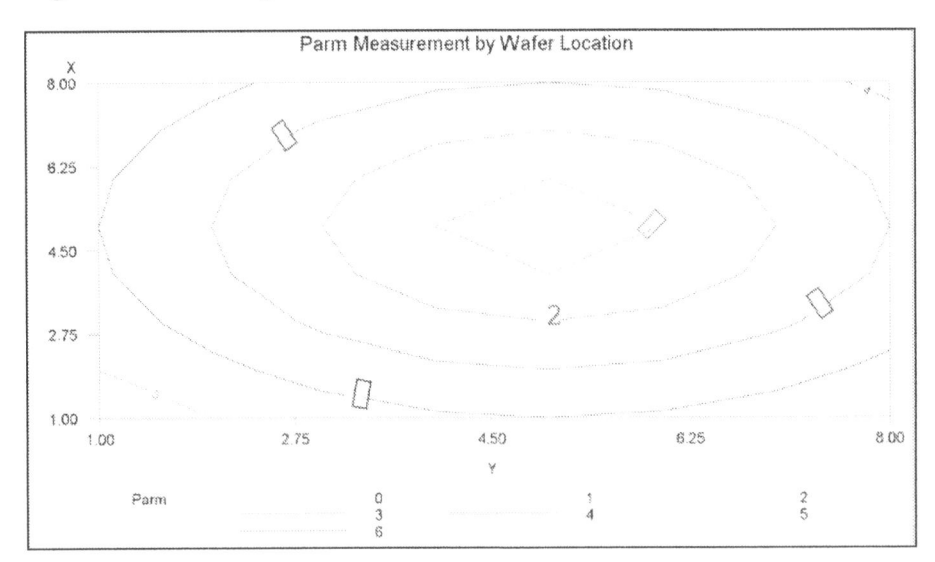

Figure 6-51: Using a PLOT Statement with PROC GCONTOUR

6.5.6 PROC G3D: Surface and Scatter Plots

You can use a PLOT or SCATTER statement with PROC G3D to create two types of 3-D plots with a response variable plotted as the height above a plane defined by two other variables.

```
PROC G3D DATA= JES.Wafer_Data;
   PLOT X*Y=Parm /NAME="F6_52_" CAXIS=Blue CBOTTOM=Red CTOP=Green;
RUN; QUIT;
```

Figure 6-52: Using a PLOT Statement with PROC G3D

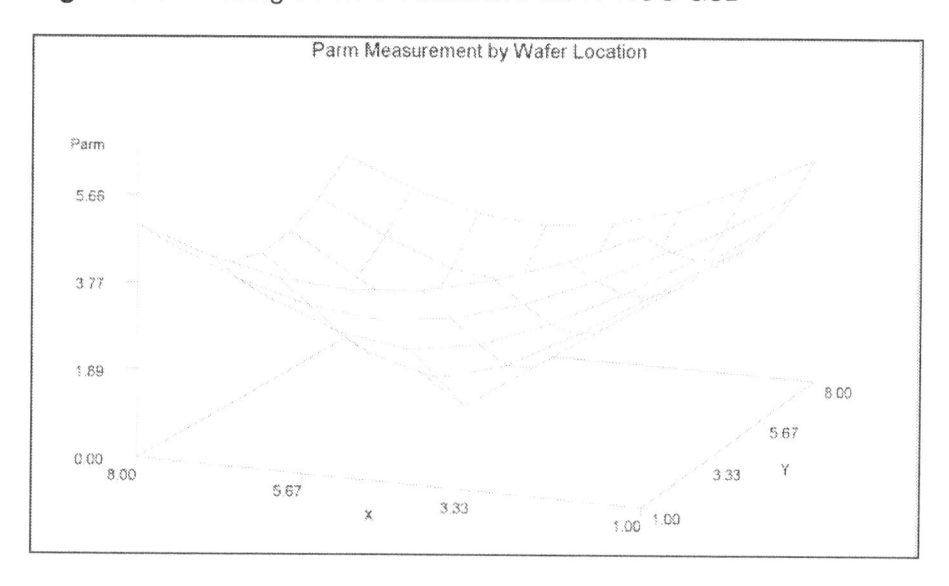

```
PROC G3D DATA= JES.Wafer_Data;
   SCATTER X*Y=Parm /NAME="F6_53_" CAXIS=Blue COLOR='Red' ROTATE=30
SHAPE='HEART';
RUN; QUIT;
```

The SCATTER X*Y=Parm statement requests a 3-D scatter plot of Parm vs. X and Y.

Figure 6-53: Using a SCATTER Statement with PROC G3D

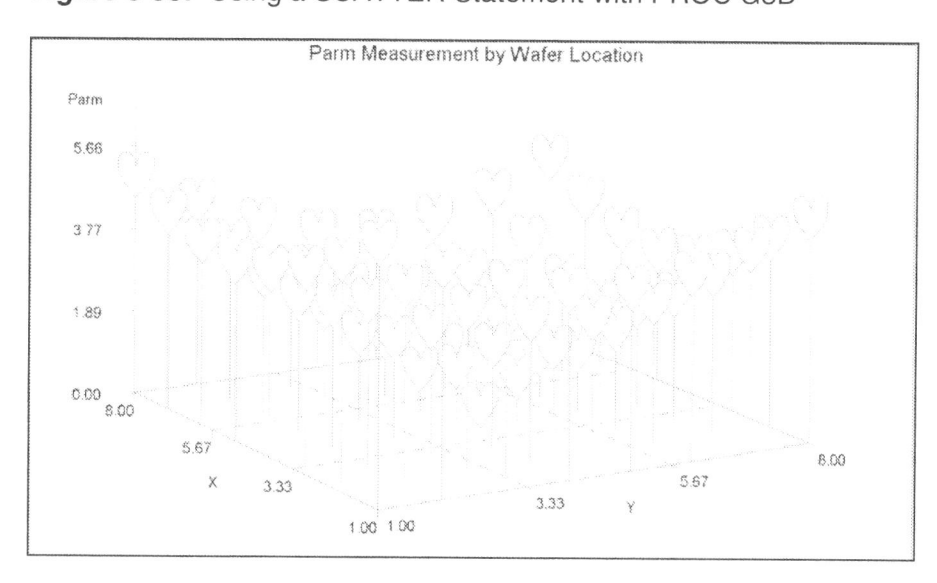

6.6 Chapter Summary

6.6.1 Recap

After finishing this chapter you should know how to

- Use PROC GPLOT to create line and scatter plots and customize:
    - Titles and footnotes
    - Horizontal and vertical axes
    - Legends
    - Reference lines
    - Symbols and line types
    - Fonts and colors for all of the above

    Use the INTERPOL=BOX<*nn*><*options*> option for simple box plots.

    Use the SKIPMISS option to break lines where required.

    Use the ANNOTATE option to place text in your graphic output.

- Use PROC GCHART to create horizontal and vertical bar charts and
    - Select the charting statistic: FREQ, CFREQ, PCT, CPCT, SUM, or MEAN
    - Control the midpoint values of the chart variable
    - Control the bar fill patterns
    - Use the GROUP option to create groups of bars based on a group variable
    - Use the SUBGROUP variable to create stacked bar charts
    - Preprocess your data with PROC MEANS to create stacked bars of defect or failure rates

6.6.2 For More Information

Books

Carpenter, Art. 1999. *Annotate: Simply the Basics.* Cary, NC: SAS Institute Inc.

Carpenter, Arthur L., Charles E. Shipp. 1995. *Quick Results with SAS/GRAPH® Software.* Cary NC: SAS Institute Inc.

Miron, Thomas. 1995. *The How-To Book for SAS/GRAPH® Software*, Cary, NC: SAS Institute Inc.

Zdeb, Mike. 2002. *Maps Made Easy Using* SAS®. Cary, NC: SAS Institute Inc.

SAS Conference Papers

Carpenter, Arthur L. 2006. "Data Driven Annotations: An Introduction to SAS/GRAPH's® Annotate Facility." *Proceedings of the Thirty-First Annual SAS Users Group International Conference.* Cary, NC: SAS Institute Inc. Paper 108-31. **[108-31.pdf]**

Cisternas, Miriam, and Art Carpenter. 2005. "Extreme Graphics Make Over: Using SAS/GRAPH® to Get the Graphical Output You Need." *Proceedings of the Thirtieth Annual SAS Users Group International Conference.* Cary, NC: SAS Institute Inc. Paper 14-30. **[133-30.pdf]**

Massengill, A. Darrell. 2005. "Tips and Tricks: Using SAS/GRAPH® Effectively." *Proceedings of the Thirtieth Annual SAS Users Group International Conference*. Cary, NC: SAS Institute Inc. Paper 14-30. **[090-30.pdf]**

Mink, David, and David J. Pasta. 2006. "Improving Your Graphics Using SAS/GRAPH® Annotate Facility." *Proceedings of the Thirty-First Annual SAS Users Group International Conference*. Cary, NC: SAS Institute Inc. Paper 108-31. **[085-31.pdf]**

Repole, Warren. 2007. "Exporting SAS/GRAPH Output for Inclusion in Web Pages and Other Software Applications." *Proceedings of the SAS Global Forum 2007 Conference*. Cary, NC: SAS Institute Inc. Paper 296-2007. **[sgf2007-gsf.pdf]**

Zdeb, Mike, and Robert Allison. 2006. "SAS/GRAPH® 101." *Proceedings of the Thirty-First Annual SAS Users Group International Conference*. Cary, NC: SAS Institute Inc. Paper 108-31. **[239-31.pdf]**

6.6.3 Exercises

The JES.Results_Ex data set contains vendor test data in the same form as the JES.Results data set (see Section 5.1.1). Use the methods described in this chapter with the JES.Results_Ex data set to explore the differences in the distribution of resistance among the three vendors:

- Create box plots of resistance by vendor (Section 6.3.2.7).
- Plot and annotate the mean resistance with confidence limits for the three vendors (Section 6.3.3).
- Use PROC GCHART to create a histogram of resistance values, grouped by vendor.
- Use ANNOTATE with a PLOT statement in PROC GPLOT to create a plot like Figure 6-54, which shows the effect of the POS variable on the location of LABEL text.

Figure 6-54: Plot Illustrating the Effect of the POS Variable in an Annotate Data Set

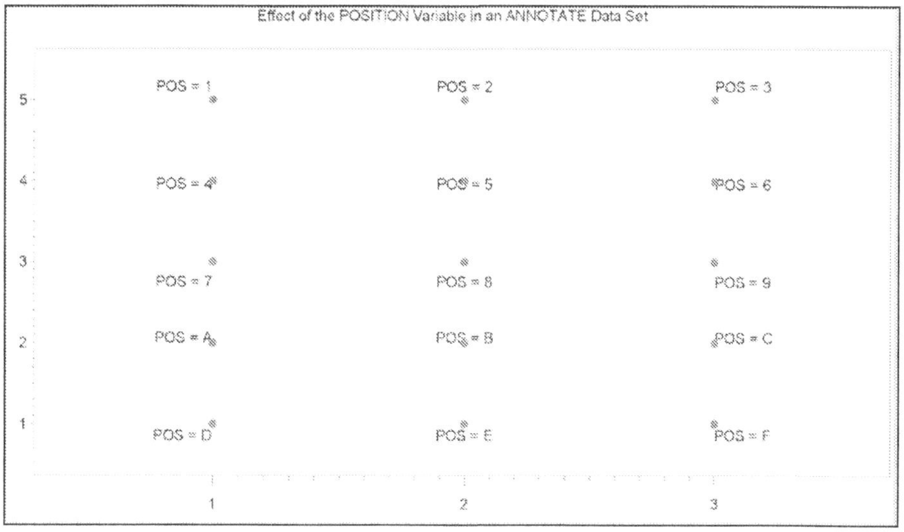

Chapter 7

The Output Delivery System

 7.1 Introduction 212
 7.2 Publishing Your Report in RTF 216
 7.3 Publishing Your Report in PDF 218
 7.4 Publishing Your Report to the Web 220
 7.5 Using ODS to Save and Select Procedure Output 232
 7.6 More Than Enough 238
 7.7 Chapter Summary 244

7.1 Introduction

In Chapters 5 and 6 you learned how to create a variety of tables and graphs to summarize, analyze, and display your data. This chapter will show you how to use the Output Delivery System (ODS) to publish your results as an integrated report in any of several commonly used formats, some of which are shown in Table 7-1. The examples in this chapter show you how to publish your reports using the RTF, PDF, and HTML formats. The data sets used in this chapter are

- JES.Results, the data set containing the 2008 test results for each vendor. (See Section 5.1.1.)
- JES.Results_Q4, a subset of JES.Results containing the data for only the last three months of 2008.
- JES.Results_Tab, a PROC MEANS summary of the JES.Results data set. (See Section 6.1.1.)
- JES.Results_Tab_2, a PROC MEANS summary of JES.Results_Q4. The JES.Results_Tab_2 data set is the same as the Tab_4 data set created in Section 5.2.5, except that appropriate labels are included. The labels are used in later examples.

ODS is remarkably easy to use. Once you have written some SAS code to create the tables and graphs you want to publish, for example any of the code samples in Chapters 5 and 6, all you need to do is add a few lines to tell SAS how you want to publish your results. The first example creates a very simple report in RTF format, just to show you how it works. The code for this section is in ~\JES \sample_code\ch_7 \report_1.sas.

Before you run the sample code, check the setting of your Results options by selecting **Tools → Options → Preferences → Results** from the Tools menu at the top of the screen. If the "View results as they are generated" box is checked, then SAS will attempt to open your report with the appropriate application, e.g. Microsoft Word for RTF documents, as soon as it is created. You might find this helpful or annoying, so try it both ways to see which you prefer.

```
GOPTIONS RESET=ALL BORDER;
OPTIONS NODATE;  ❶
ODS LISTING CLOSE;  ❷
ODS RTF FILE="&JES.ods_output/report_1.rtf";  ❸
  TITLE1 "JES.Resistance Data by Vendor";
  PROC PRINT DATA=JES.Results_Tab_2(WHERE=(_TYPE_=1)) NOOBS;
  VAR Vendor N M_Res N_Fail P_Fail;
  RUN;
ODS RTF CLOSE;  ❹
ODS LISTING;
```

❶ The OPTIONS NODATE statement is not required, but it is used in all the examples in this chapter to prevent the current date from being written at the top of each page of output.

❷ The ODS LISTING CLOSE statement tells SAS to stop sending output to the LISTING destination, for example to stop sending graphic output to the Graph window. This statement is not required, but is often recommended to save compute resources.

❸ The ODS RTF statement tells SAS to start formatting any output which follows as an RTF document, named "report_1.rtf", and to save the document in the ~**&JES**\ods_output folder.

❹ The ODS RTF CLOSE statement tells SAS to stop sending output to the RTF document, and the ODS LISTING statement tells SAS to resume sending output to the LISTING destination.

When you run this code, the report_1.rtf file, shown in Display 7-1, is saved to your ~**\&JES**\ods_output folder.

Display 7-1: The report_1.rtf File Opened in Microsoft Word

Resistance Data by Vendor

Vendor	N	M_Res	N_Fail	P_Fail
ChiTronix	90	20.78	16	0.18
Duality	57	15.06	11	0.19
Empirical	45	15.31	8	0.18

Table 7-1 gives a partial list of the destinations available in ODS. Consult the SAS online documentation for the complete list. The available destinations include SAS formatted, as well as third-party formatted destinations. The SAS formatted destinations are used to create SAS objects that can be used in further SAS programming steps. The third-party formatted destinations are used to convert SAS output to various industry-standard formats, such as RTF, PDF, HTML, and XML, in order to create documents that are easily shared with others and published.

Table 7-1: Selected ODS Destinations

Destination	Type of Output
SAS Formatted	
LISTING	The LISTING destination directs SAS output to the normal default locations, for example the Output or Graph windows that you have been using to view SAS output in the previous chapters.
OUTPUT	The OUTPUT destination creates SAS data sets when certain SAS procedures are run. (See Section 7.5.2.)
DOCUMENT	The DOCUMENT destination is used to capture SAS output objects in a kind of meta-document which can then be reprocessed to send all or part of the output to any of the third-party formatted destinations.
Third-Party Formatted	
RTF	Rich Text Format files, which can be opened by MS Word and other word processing applications. (See Section 7.2.)
PDF	PDF files which can be read by Adobe Reader. (See Section 7.3.)
PS	PostScript printer format.
PCL	PCL files for printing to HP LaserJet printers.
HTML	HTML 4.0 output with embedded stylesheets, which can be opened in any Web browser. (See Section 7.4.)
CHTML	Compact HTML that does not use style information.
HTML3	HTML 3.2 output.
HTMLCSS	HTML output with cascading style sheets.
PHTML	HTML which uses only 12 style elements and no class attributes.
IMODE	Minimal HTML for use with mobile phones.
CSVALL	Columns of data separated by commas, including titles and footnotes, easily imported by spreadsheets. (See Section 3.2.2.)
MSOffice2K	HTML easily read by Microsoft Excel. (See Section 3.2.3.)
ExcelXP	XML easily read by Microsoft Excel. (See Section 3.4.1.)
XHTML	XHTML which is the format slated to replace HTML.

7.1.1 The Sample Report

The examples in Sections 7.2, 7.3, and 7.4 show how to publish a simple four-page quality report in RTF, PDF, and HTML format. The RTF version of the report is shown in Display 7-2. This section lists the code used to create the tables and graphs that make up the report. The examples in the following sections use %INCLUDE statements to run each of these code files when creating the reports.

```
/*=== JES\sample_code\ch_7\vendors_1.sas ===*/  ❶
TITLE1 "Resistance Data by Vendor 4Q 2008";
PROC PRINT DATA=JES.Results_Tab_2(WHERE=(_TYPE_=1)) NOOBS LABEL;
  VAR Vendor N M_Res N_Fail P_Fail;
RUN;

/*=== JES\sample_code\ch_7\vendors_2.sas ===*/  ❷
TITLE1 "Resistance by Vendor - 4Q 2008";
PROC REPORT DATA=JES.Results_Tab(WHERE=(Mon>="2008-10")) NOWINDOWS;
  COLUMN Vendor Mon M_Res R_L R_U N_Fail;
  DEFINE Vendor/ORDER;
  DEFINE Mon /ORDER;
  DEFINE M_Res / ANALYSIS WIDTH=10;
  DEFINE R_L / ANALYSIS WIDTH=10;
  DEFINE R_U / ANALYSIS WIDTH=10;
  DEFINE N_Fail / ANALYSIS;
RUN;

/*=== JES\sample_code\ch_7\vendors_3.sas ===*/  ❸
TITLE1 HEIGHT=5 "Mean Resistance by Vendor and Month";
(SYMBOL and AXIS statements omitted)
PROC GPLOT DATA =JES.Results_Tab;
  PLOT M_Res*Mon=Vendor / NAME="Vend_3"
  HAXIS=AXIS1 VAXIS=AXIS2 VREF=12.5 22.5;
  Format M_Res 3.0;
RUN; QUIT;

/*=== JES\sample_code\ch_7\chitronix_1.sas ===*/  ❹
TITLE1 "Delay vs Resistance - ChiTronix - Q4 2008";
(SYMBOL and AXIS statements omitted)
PROC GPLOT DATA=JES.Results_Q4(WHERE=(Vendor="ChiTronix"))
  PLOT Delay*Resistance=Mon / NAME="Chi_1"
  HAXIS=AXIS1 VAXIS=AXIS2
  LHREF=2 LVREF=2 HREF=12.5 22.5 VREF=150 250;
  FORMAT Resistance 8.0;
RUN; QUIT;

/*=== JES\sample_code\ch_7\chitronix_2.sas ===*/  ❺
TITLE HEIGHT=5 "Resistance - ChiTronix - 4Q 2008";
PROC UNIVARIATE DATA=JES.Results_Q4(WHERE=(Vendor="ChiTronix"));
  VAR Resistance;
  HISTOGRAM Resistance / Normal(MU=EST SIGMA=EST)HREF=12.5 22.5;
  INSET N MEAN STD SKEWNESS KURTOSIS NORMAL(AD ADPVAL)
  / HEIGHT=2.5 FORMAT = 5.3 POSITION=NW;
RUN;
```

❶ vendors_1.sas uses PROC PRINT to create the first table shown on Page 1 of the report.

❷ vendors_2.sas uses PROC REPORT to create the second table shown on Page 1 of the report.

❸ vendors_3.sas uses PROC GPLOT to create the plot shown on Page 1 of the report.

❹ chitronix_1.sas uses PROC GPLOT to create the first plot on Page 2 of the report. Similar code files, duality_1.sas and empirical_1.sas, not shown here, create the first plot on Pages 2 and 3.

❺ chitronix_2.sas uses PROC UNIVARIATE to create the second plot on Page 2 of the report. Similar code files, duality_2.sas and empirical_2.sas, not shown here, create the second plot on Pages 2 and 3.

Display 7-2: Sample Report in RTF, Opened in MS Word

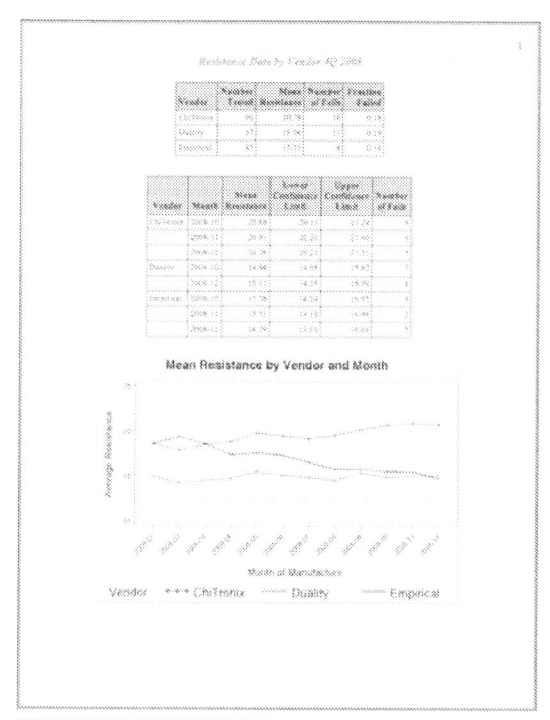

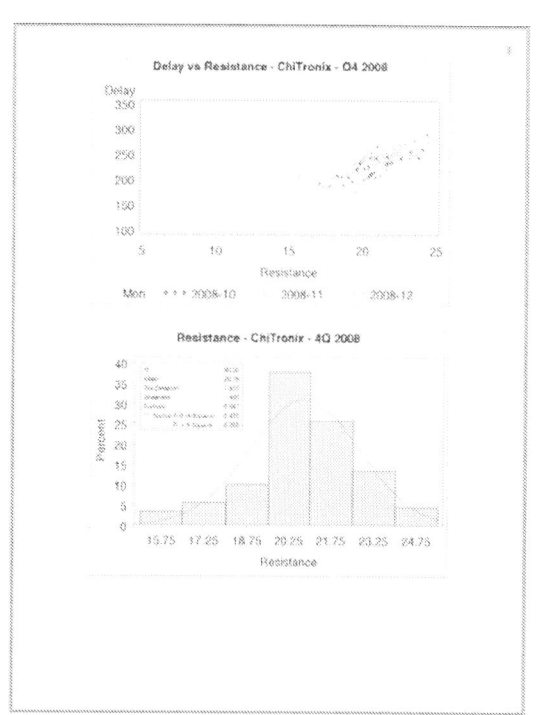

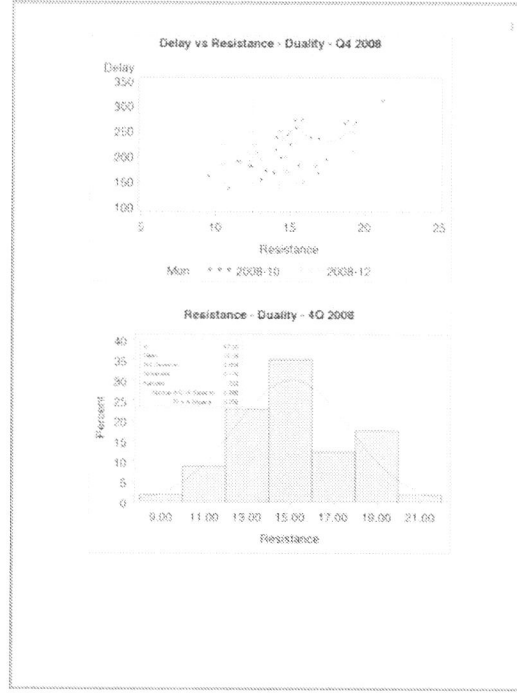

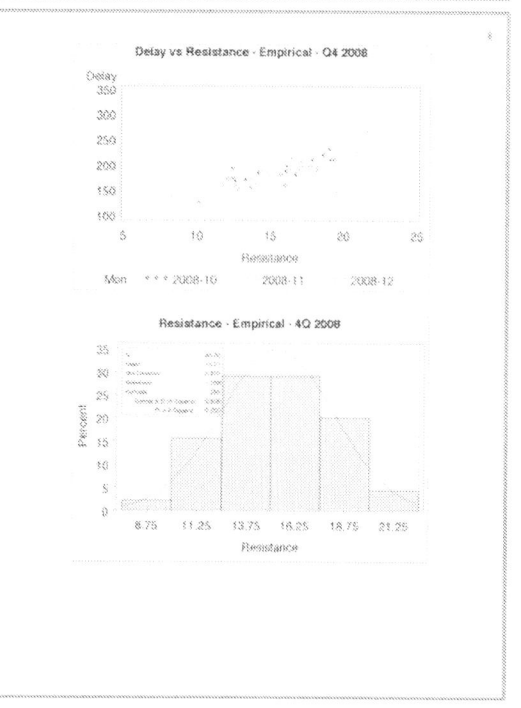

7.2 Publishing Your Report in RTF

This section shows how to publish your sample report to the Rich Text Format, or RTF, destination. RTF documents can be read by Microsoft Word and other word processing software, so this is a good way to get your output into a text document. The code for this section is in ~\JES\sample_code\ch_7\report_rtf.sas.

```
GOPTIONS RESET=ALL BORDER;   ❶
OPTIONS NODATE;
GOPTIONS RESET=ALL BORDER FTEXT='Helvetica' FTITLE='Helvetica/Bold';
GOPTIONS GUNIT=PCT HTEXT=5 DEVICE=SASEMF XMAX=6IN YMAX=4IN;
ODS RTF STARTPAGE=NEVER FILE="&JES.ods_output/report.rtf";   ❷
%INCLUDE "&JES.sample_code/ch_7/vendors_1.sas";   ❸
%INCLUDE "&JES.sample_code/ch_7/vendors_2.sas";
%INCLUDE "&JES.sample_code/ch_7/vendors_3.sas";
%INCLUDE "&JES.sample_code/ch_7/chitronix_1.sas";
ODS RTF EXCLUDE Moments BasicMeasures TestsForLocation Quantiles   ❹
        ExtremeObs ParameterEstimates FitQuantiles GoodnessOfFit;
%INCLUDE "&JES.sample_code/ch_7/chitronix_2.sas";

%INCLUDE "&JES.sample_code/ch_7/duality_1.sas";
ODS RTF EXCLUDE Moments BasicMeasures TestsForLocation Quantiles
        ExtremeObs ParameterEstimates FitQuantiles GoodnessOfFit;
%INCLUDE "&JES.sample_code/ch_7/duality_2.sas";

%INCLUDE "&JES.sample_code/ch_7/empirical_1.sas";
ODS RTF EXCLUDE Moments BasicMeasures TestsForLocation Quantiles
        ExtremeObs ParameterEstimates FitQuantiles GoodnessOfFit;
%INCLUDE "&JES.sample_code/ch_7/empirical_2.sas";

ODS RTF CLOSE;   ❺
```

❶ The first few lines of code set some graphic options. The DEVICE=SASEMF option is recommended for graphic output to the RTF destination. The XMAX and YMAX options specify the size of the graphs to be created.

❷ The ODS RTF statement tells SAS to start formatting any output which follows as an RTF document. The STARTPAGE=NEVER option is used to keep the plot on the same page as the table. Without this statement, the output from each procedure would start on a new page. The FILE = "&JES.ods_output/report.rtf" option tells SAS to store the result as a file named report.rtf in the ~/JES/ods_output directory.

❸ The %INCLUDE statements run the code files shown in Section 7.1.1 to create the tables and graphs.

❹ The ODS RTF EXCLUDE statement suppresses some of the PROC UNIVARIATE output that is not needed for the report. The use of ODS EXCLUDE statements is explained in Section 7.5.3.

❺ The ODS RTF CLOSE statement tells SAS to stop sending output to the RTF document.

After you have run this code, go to your ~/**JES**/ods_output directory and open the report.rtf document in a text editor. Display 7-3 shows the first page of the report opened in MS Word, and Display 7-2 shows the full report. As you can see, it is very easy to send your SAS output to an RTF document. However, you will probably also want to control various aspects of the resulting document, such as page numbers, footnotes, line breaks, etc. All of this, and more, can be done, but it will take some more effort. See the papers by Tong, Hadden, Delaney, Gupta, and McNeill, and the books by Haworth and Gupta, listed in the chapter summary for detailed information on how to control the appearance of your RTF output.

Display 7-3: Page 1 of the Sample Report in RTF, Opened in MS Word

Note that the title of the first table has become a page header, and the title of the second table is not shown. You can use an ODS TEXT = statement to insert text anywhere on the page, for example, to replace the missing title. Consult the SAS online documentation for details.

218 *Just Enough SAS: A Quick-Start Guide to SAS for Engineers*

7.3 Publishing Your Report in PDF

This section shows how to publish your sample report to the PDF destination. PDF documents can be read by Adobe Reader, which is available as a free download from Adobe. Practically anyone with a computer already has Adobe Reader, so this is a good way to distribute a soft copy of your reports in a format that anyone can read and print. This example is very similar to the RTF example in Section 7.2, except that the STARTPAGE, HORIGIN, and VORIGIN options are used to control page breaks and the position of graphs on the page. The code for this section is in ~\JES\sample_code\ch_7\report_pdf.sas.

```
GOPTIONS RESET=ALL BORDER;
OPTIONS NODATE ORIENTATION=PORTRAIT;   ❶
GOPTIONS RESET=ALL BORDER FTEXT='Helvetica' FTITLE='Helvetica/Bold';
GOPTIONS GUNIT=PCT HTEXT=5 DEVICE=SASPRTC;   ❷
ODS LISTING CLOSE;
ODS PDF FILE="&JES.ods_output/report.pdf" STARTPAGE=NEVER PDFTOC=1;   ❸
GOPTIONS HORIGIN=1IN VORIGIN=0.5IN HSIZE=6IN VSIZE=4IN;   ❹
%INCLUDE "&JES.sample_code/ch_7/vendors_3.sas";
%INCLUDE "&JES.sample_code/ch_7/vendors_1.sas";
%INCLUDE "&JES.sample_code/ch_7/vendors_2.sas";

ODS PDF STARTPAGE=YES;   ❺
ODS PDF STARTPAGE=NOW;
GOPTIONS HORIGIN=1IN VORIGIN=5.0IN HSIZE=6IN VSIZE=4IN;
ODS PROCLABEL="Delay vs Resistance - ChiTronix";   ❻
%INCLUDE "&JES.sample_code/ch_7/chitronix_1.sas";
ODS PDF STARTPAGE=NEVER;   ❼
GOPTIONS HORIGIN=1IN VORIGIN=0.5IN HSIZE=6IN VSIZE=4IN;
ODS PDF EXCLUDE Moments BasicMeasures TestsForLocation Quantiles
        ExtremeObs ParameterEstimates FitQuantiles GoodnessOfFit;
ODS PROCLABEL="Distribution of Resistance - ChiTronix";
%INCLUDE "&JES.sample_code/ch_7/chitronix_2.sas";
  (Similar code for Duality and Empirical Pages Omitted)

ODS PDF CLOSE;   ❽
ODS LISTING;
```

❶ The ORIENTATION=PORTRAIT option specifies the orientation of the PDF file to be created. You can also use ORIENTATION=LANDSCAPE.

❷ The DEVICE=SASPRTC option is recommended for graphic output to the PDF destination. Note that the ODS LISTING CLOSE statement is important in this example because, otherwise, the DEVICE = SASPRTC option causes each graph to also be sent directly to your default printer, which is probably not what you want.

❸ The ODS PDF statement tells SAS to start formatting any output that follows as a PDF document. The FILE = "&JES.ods_output/report.pdf" option tells SAS to store the result as a file named report.pdf in the ~/JES/ods_output directory. The STARTPAGE=NEVER option is used to prevent a page break after each procedure. By default, your PDF document includes a fully expanded table of contents. The PDFTOC=1 specifies that the tables of contents be expanded to the first level only. Use the NOTOC option to suppress the table of contents.

❹ The HORIGIN and VORIGIN options specify the horizontal and vertical position of the lower left corner of your graph, relative to the lower left corner of the page, and the HSIZE and VSIZE options specify the size of the graph.

❺ The ODS STARTPAGE=YES and ODS PDF STARTPAGE=NOW statements force a page break. Without these statements, the graphs would all be overlaid on the same page. Note that these statements were not required in the ODS RTF example (Section 7.2).

❻ The ODS PROCLABEL statement specifies the text used to label the output in the table of contents. The default label is the name of the procedure, for example "The GPLOT Procedure" for the first item.

❼ The ODS PDF STARTPAGE=NEVER statement is required to prevent a page break after the first plot.

❽ The ODS PDF CLOSE statement tells SAS to stop sending output to the PDF document. The ODS LISTING statement causes future output to be sent to the LISTING destination.

After you have run this code, go to your ~/JES/ods_output directory and open the report.pdf document with Adobe Reader. It has the same four pages as the RTF document shown in Display 7-2. Page 2 of the PDF document, together with the table of contents, is shown in Display 7-4.

Display 7-4: Sample Report Created with ODS PDF, Opened in Adobe Reader

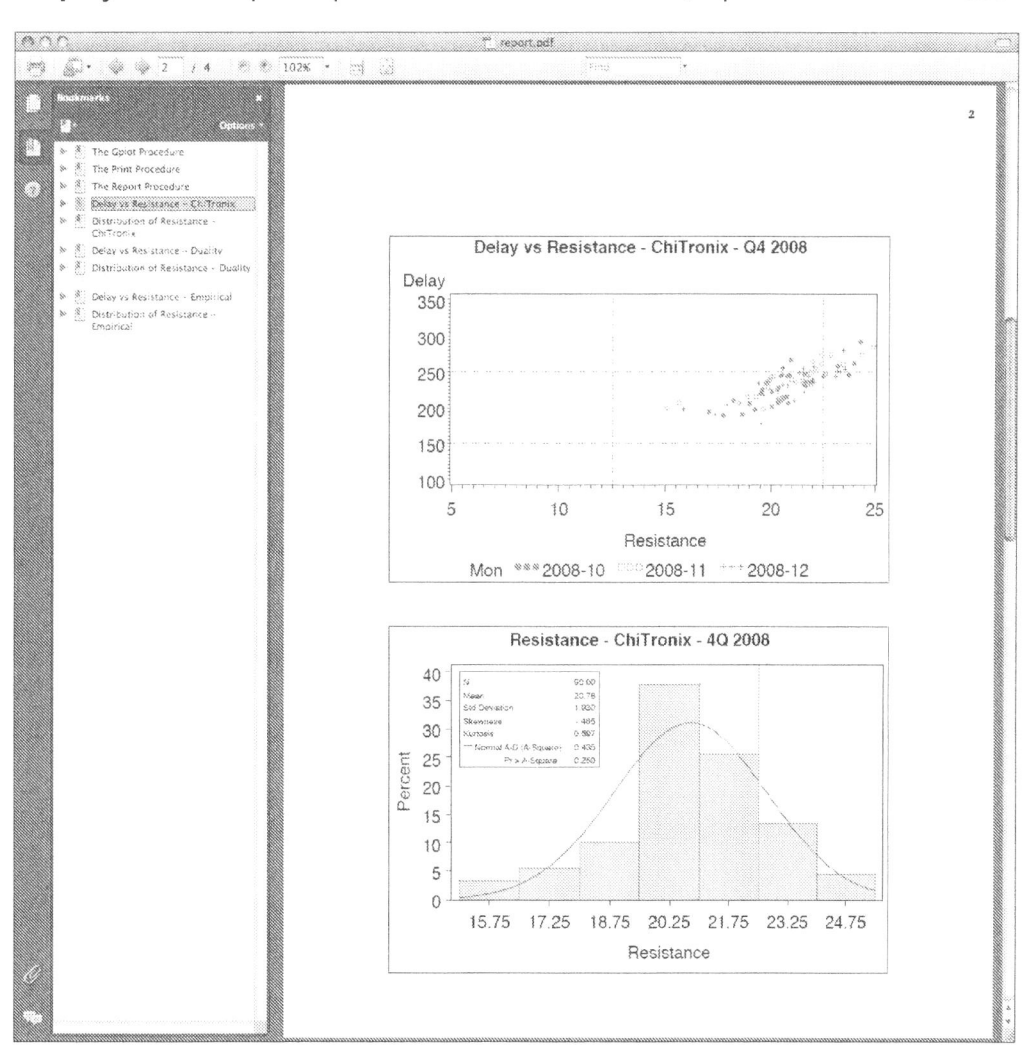

See the papers by Delaney, Gupta, and McNeill, and the books by Haworth and Gupta, listed in the "Chapter Summary" section for detailed information on how to control the appearance of your PDF output.

7.4 Publishing Your Report to the Web

This section shows how to publish your sample report to the HTML destination. If you place your HTML documents in a public Web site, or a site which is internal to your organization, then your results are immediately available to anyone who needs them. This is a very quick and efficient way to distribute your results to a wide audience. The code for this section is in ~\JES\sample_code\ch_7\report_html.sas.

```
PROC DATASETS MEMTYPE=CAT NOLIST; DELETE HTML; RUN; QUIT; ❶
GOPTIONS RESET=ALL BORDER;
OPTIONS NODATE;
GOPTIONS RESET=ALL BORDER FTEXT='Helvetica' FTITLE='Helvetica/Bold';
GOPTIONS GUNIT=PCT DEVICE=PNG HTEXT=4 HTITLE=5 HSIZE=9IN VSIZE=6IN; ❷

ODS HTML PATH="&JES.ods_output/page1" ❸
    (URL=NONE) ❹
    BODY="report.html" ❺
    STYLE=MINIMAL; ❻
    %INCLUDE "&JES.sample_code/ch_7/vendors_1.sas";
    %INCLUDE "&JES.sample_code/ch_7/vendors_2.sas";
    %INCLUDE "&JES.sample_code/ch_7/vendors_3.sas";

    %INCLUDE "&JES.sample_code/ch_7/chitronix_1.sas";
    ODS HTML EXCLUDE Moments BasicMeasures TestsForLocation Quantiles
            ExtremeObs ParameterEstimates FitQuantiles GoodnessOfFit;
    %INCLUDE "&JES.sample_code/ch_7/chitronix_2.sas";
    %INCLUDE "&JES.sample_code/ch_7/duality_1.sas";
    ODS HTML EXCLUDE Moments BasicMeasures TestsForLocation Quantiles
            ExtremeObs ParameterEstimates FitQuantiles GoodnessOfFit;
    %INCLUDE "&JES.sample_code/ch_7/duality_2.sas";
    %INCLUDE "&JES.sample_code/ch_7/empirical_1.sas";
    ODS HTML EXCLUDE Moments BasicMeasures TestsForLocation Quantiles
            ExtremeObs ParameterEstimates FitQuantiles GoodnessOfFit;
    %INCLUDE "&JES.sample_code/ch_7/empirical_2.sas";
ODS HTML CLOSE; ❼
```

❶ The PROC DATASETS statement deletes any graphs in the WORK.HTML catalog, which contains any graphs previously created with ODS HTML. This is done for the same reason that the WORK.GSEG catalog was deleted in the examples in Chapter 6, so that the names of any newly created graphic files are predictable. See the discussion of graph names in Section 6.2.

❷ The DEVICE=PNG option is recommended for graphic output to the HTML destination.

❸ The ODS HTML statement tells SAS to start formatting any output that follows as an HTML document. The PATH="&JES.ods_output/page1" option tells SAS to store the results in the ~/**JES**/ods_output/page1 directory.

❹ The (URL=NONE) option tells SAS to use relative references in any hyperlinks in the HTML code it writes. Without this option, absolute references will be used, and the report.html file may not be able to find the PNG files that you create if, for example, you move the page1 folder to a public Web site.

❺ The BODY="report.html" option specifies the name of the HTML file to be created.

❻ The STYLE=MINIMAL option specifies the style to be used for the Web pages, including font types, background and foreground colors, etc. This option is not required, but is used here to produce output that is easily read when reproduced in this book. See Section 7.6 for a discussion of ODS styles.

❼ The ODS HTML CLOSE statement tells SAS to stop sending output to the HTML document.

After you run this code, go to your ~\JES\ods_output\page1 directory and you will find the report.html file along with seven PNG files containing the graphs for the report. Display 7-5 shows the report.html file opened in a browser window. If you scroll down you will see all the same graphs shown in Display 7-2.

Display 7-5: Sample Report Created with ODS HTML

Vendor	Number Tested	Mean Resistance	Number of Fails	Fraction Failed
ChiTronix	90	20.78	16	0.18
Duality	57	15.06	11	0.19
Empirical	45	15.31	8	0.18

Resistance Data by Vendor - 4Q 2008

Resistance by Vendor - 4Q 2008

Vendor	Month	Mean Resistance	Lower Confidence Limit	Upper Confidence Limit	Number of Fails
ChiTronix	2008-10	20.68	20.11	21.24	5
	2008-11	20.91	20.21	21.60	6
	2008-12	20.76	20.21	21.31	5
Duality	2008-10	14.94	14.05	15.82	7
	2008-12	15.17	14.35	15.99	4
Empirical	2008-10	15.58	14.24	16.93	3
	2008-11	15.55	14.16	16.94	2
	2008-12	14.79	13.56	16.03	3

The following sections show how to publish the same report using frames, or as separate pages with HTML links to move from one page to another, and how to add traffic lighting to the table created with PROC REPORT.

7.4.1 Putting Your ODS HTML Output in Frames

If you have several tables and graphs in the same Web page, it might be a good idea to include a table of contents frame to simplify navigation of the page. This can be done with very little effort as shown in the sample code below.

```
ODS HTML PATH="&JES.ods_output/page2" (URL=NONE)
   BODY="report_body.html"  ❶
   CONTENTS="report_toc.html"  ❷
   FRAME = "report_frame.html"  ❸
STYLE=MINIMAL;
   %INCLUDE "&JES.sample_code/ch_7/vendor_1.sas";
   %INCLUDE "&JES.sample_code/ch_7/vendor_2.sas";
   %INCLUDE "&JES.sample_code/ch_7/vendor_3.sas";

   ODS PROCLABEL="Delay vs Resistance - ChiTronix";  ❹
   %INCLUDE "&JES.sample_code/ch_7/chitronix_1.sas";
   ODS HTML EXCLUDE Moments BasicMeasures TestsForLocation Quantiles
         ExtremeObs ParameterEstimates FitQuantiles GoodnessOfFit;
   ODS PROCLABEL="Distribution of Resistance - ChiTronix";
   %INCLUDE "&JES.sample_code/ch_7/chitronix_2.sas";

(Similar code for Duality and Empirical omitted)

ODS HTML CLOSE;
```

❶ The BODY option specifies the name of the file that will contain your tables and graphs.

❷ The CONTENTS option specifies the name for the table of contents file.

❸ The FRAME option specifies the name for the main HTML page that will reference the BODY and CONTENTS files.

❹ The ODS PROCLABEL statement specifies the text used to label the output in the table of contents. The default label is the name of the procedure, for example "Print" for the first item.

After you run this code you will find three HTML files in your ~\JES\ods_output\page2 directory. The report_frame.html file is the main page which references two other HTML files: report_toc.html and report_body.html. The report_body.html file is identical to the report.html file created in the previous example. Display 7-6 shows report_frame.html opened in a browser window. You can see the various tables and graphs either by scrolling the frame on the right or clicking one of the items in the table of contents frame on the left.

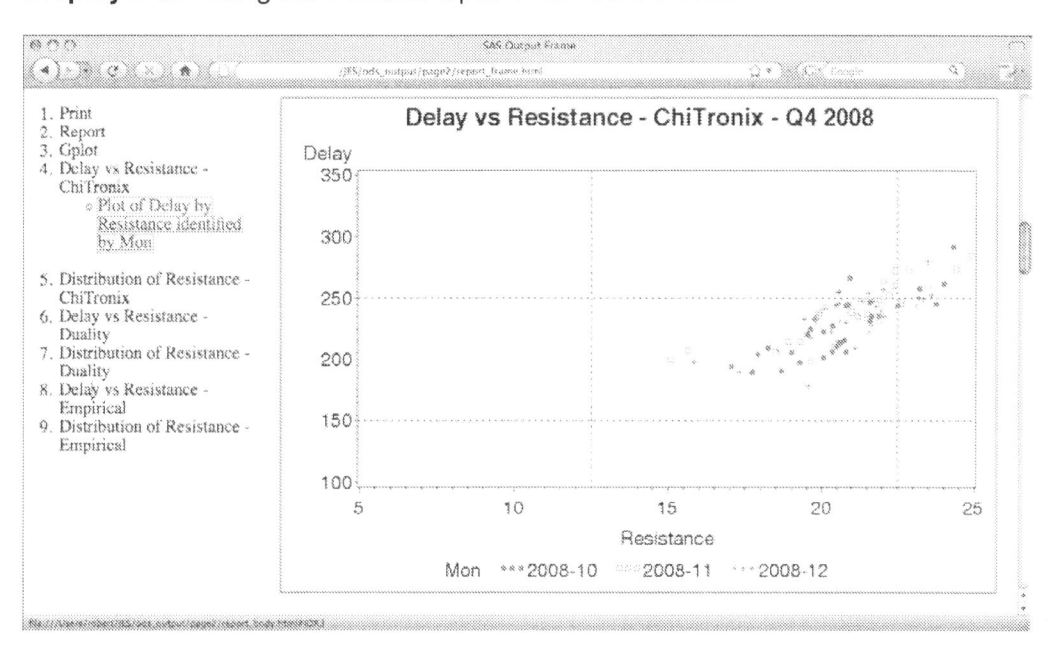

Display 7-6: Using the FRAME Option with ODS HTML

In some cases it is better to have the various output items in different HTML files, but still accessible in the same frame structure. Your next bit of sample code shows how to do this using the NEWFILE option.

```
ODS HTML PATH="&JES.ods_output/page3" (URL=NONE)
  BODY="report_body.html"
  CONTENTS="report_toc.html"
  FRAME = "report_frame.html"
  STYLE=MINIMAL
  NEWFILE=PAGE; ❶
  (other lines as in the previous example)
ODS HTML CLOSE;
```

❶ The NEWFILE=PAGE option starts a new body file for each page of output. In this example, that means separate pages for each of the two tables and seven graphs in the report.

After you run this code the ~\JES\ods_output\page3 folder will contain nine body files: report_body.html, report_body1.html, report_body2.html, etc., in addition to the report_frame.html and report_toc.html files, and seven PNG files. If you open the new report_frame.html in a browser, it will look similar to Display 7-6, except that the frame on the right will contain only the one table or graph selected in the table of contents frame on the left. This is especially convenient if you want to print one of your output items without printing all of them.

7.4.2 Adding Hyperlinks to Your Web Pages

One of the most useful features of HTML is the ability to embed hyperlinks that initiate an action when clicked on, for example open a new page or download a file. ODS makes it very easy to embed such links in your Web pages. The example in this section creates a main Web page similar to Page 1 in Display 7-2, with links to three other Web pages, one for each vendor, similar to Pages 2, 3, and 4 in Display 7-2. The example illustrates five different ways to embed hyperlinks.

1. Add a link to a title using the LINK option of the TITLE statement.

2. Add links to PROC PRINT output by writing the appropriate HTML code into a variable in the data set to be printed.

3. Add links to PROC REPORT output using the "URL" definition in a COMPUTE BLOCK.

4. Add links to the plot points in a graph created by PROC GPLOT using the HTML option.

5. Add links to the legend in a graph created by PROC GPLOT using the HTML_LEGEND option.

The first bit of code creates a Web page for each vendor, just to have some pages to link to. The code shown here creates the ChiTronix.html page. Similar code in report_html.sas, not shown here, creates the Duality.html and Empirical.html pages in the page4 folder.

```
ODS HTML PATH="&JES.ods_output/page4" (URL=NONE)
    BODY="ChiTronix.html" (TITLE="ChiTronix") STYLE=MINIMAL;  ❶
%INCLUDE "&JES.sample_code/ch_7/chitronix_1.sas";
ODS HTML EXCLUDE Moments BasicMeasures TestsForLocation Quantiles
        ExtremeObs ParameterEstimates FitQuantiles GoodnessOfFit;
%INCLUDE "&JES.sample_code/ch_7/chitronix_2.sas";
ODS HTML CLOSE;
```

❶ The (TITLE="ChiTronix") option will replace the default "SAS Output" with "ChiTronix" in the title of your Web page. This is not required, but if you have several open tabs in your browser, it will make it easier to navigate.

Before you create the main page with hyperlinks, some preparation is required. You need to add some variables to your data sets, and then modify your code to take advantage of the new variables. The first step is to use the CATS function (see Section 2.4.3) to add columns to the data sets that SAS will use to create the links.

```
DATA Results_Tab_2; SET JES.Results_Tab_2;
    Print_Link = CATS('<A HREF="',Vendor,'.html"
        TARGET="_blank">',Vendor,'</A>');
    LABEL Print_Link="Vendor";
RUN;
DATA Results_Tab; SET JES.Results_Tab;
    Report_Link=CATS(Vendor,'.html');
    Point_Link=CATS('HREF="',Vendor,'.html"');
    Legend_Link=CATS('HREF="',Vendor,'.html" TARGET="_blank"');
RUN;
```

The first DATA step creates the Results_Tab_2 data set from JES.Results_Tab_2 by adding the Print_Link variable, which will be used to create links in the PROC PRINT output. The LABEL option is used to assign the label "Vendor" to the new variable. This is done so that when the data set is printed, the column heading for the link will be "Vendor" instead of "Print_Link".

The second DATA step creates the Results_Tab data set from JES.Results_Tab by adding the Report_Link, Point_Link, and Legend_Link variables. Report_Link will be used to add hyperlinks to the PROC REPORT output, and Point_Link and Legend _Link will be used to add hyperlinks to the points and legend of the PROC GPLOT output.

```
Results_Tab_2    (selected columns)

 Vendor                                 Print_Link

ChiTronix       <A HREF="ChiTronix.html"     TARGET="_blank">ChiTronix</A>
Duality         <A HREF="Duality.html"       TARGET="_blank">Duality</A>
Empirical       <A HREF="Empirical.html"     TARGET="_blank">Empirical</A>
```

The Print_Link variable contains standard HTML syntax that a browser will interpret as a hyperlink. HREF="ChiTronix.html" specifies the page to be opened and TARGET="_blank" specifies that the page should be opened in a new tab, or a new window, depending on your browser settings. Without the TARGET="_blank" the new page would replace the current page.

```
Results_Tab     (selected rows and columns)

 Vendor         Mon        Report_Link            Point_Link

ChiTronix       2008-01    ChiTronix.html         HREF="ChiTronix.html"
ChiTronix       2008-02    ChiTronix.html         HREF="ChiTronix.html"
ChiTronix       2008-03    ChiTronix.html         HREF="ChiTronix.html"
......
Duality         2008-01    Duality.html           HREF="Duality.html"
Duality         2008-02    Duality.html           HREF="Duality.html"
Duality         2008-03    Duality.html           HREF="Duality.html"
......
Empirical       2008-01    Empirical.html         HREF="Empirical.html"
Empirical       2008-02    Empirical.html         HREF="Empirical.html"
Empirical       2008-03    Empirical.html         HREF="Empirical.html"

                Legend_Link

HREF="ChiTronix.html" TARGET="_blank"
HREF="ChiTronix.html" TARGET="_blank"
HREF="ChiTronix.html" TARGET="_blank"
......
HREF="Duality.html" TARGET="_blank"
HREF="Duality.html" TARGET="_blank"
HREF="Duality.html" TARGET="_blank"
......
HREF="Empirical.html" TARGET="_blank"
HREF="Empirical.html" TARGET="_blank"
HREF="Empirical.html" TARGET="_blank"
```

The Point_Link and Legend_Link variables are similar to Print_Link, but some of the HTML syntax is left out as this will be inserted by ODS HTML. The TARGET="_blank" statement is used for Legend_Link but not for Point_Link just to illustrate the different effects. The Report_Link variable leaves even more of the HTML syntax to be filled in by ODS.

Next you need to modify your code to take advantage of the new variables. The code files vendor_1a.sas, vendor_2a.sas, and vendor_3a.sas, shown here include the required modifications to the original vendor_1.sas, vendor_2.sas, and vendor_3.sas files shown in Section 7.1.1.

The vendors_1a.sas file creates the first table on the page, with a hyperlink in the title and in each row of the table.

```
/*=== JES\sample_code\ch_7\vendors_1a.sas ===*/
TITLE1 "Resistance Data by Vendor - 4Q 2008";
TITLE2 LINK="../report.pdf" "Download Report"; ❶
PROC PRINT DATA=Results_Tab_2(WHERE=(_TYPE_=1)) NOOBS LABEL;
   VAR Print_Link N M_Res N_Fail P_Fail; ❷
RUN;
```

❶ The LINK option of the TITLE2 statement specifies the hyperlink, "../report.pdf" and the title text to be printed, "Download Report". This will create a link to the report.pdf file created in Section 7.3. The "../" is interpreted by the browser as referring to the folder one level higher than the current folder, i.e., to the ~\JES\ods_output folder, where report.pdf is located, instead of the ~\JES\ods_output\page4 folder.

❷ The VAR statement specifies Print_Link to be the first variable printed. The LABEL option causes the label, "Vendor", instead of the variable name, "Print_Link", to be printed as the column header.

The vendors_2a.sas file creates the second table on the page, with hyperlinks for each vendor name in the first column. This example uses a COMPUTE block and a CALL DEFINE routine that are not explained in the brief introduction to PROC REPORT in Section 5.4. The syntax of a simple COMPUTE block has the form

```
COMPUTE report_item;
CALL DEFINE(columnid, attribute, attribute_value);
ENDCOMP;
```

where *report-item* is one of the columns in the report, *columnid* is a column identifier, *attribute* is some attribute of that column, and *attribute_value* is the value to be assigned to *attribute* for the columns identified by *columnid*. A detailed discussion of COMPUTE blocks and CALL DEFINE routines is beyond the scope of this book, but this simple example should be enough to get you started adding hyperlinks to your own PROC REPORT output. The example in Section 7.4.3 shows how to use a COMPUTE block with CALL DEFINE to add traffic lighting to the same report. For complete details on the use of PROC REPORT, see the book by Carpenter listed in the "Chapter Summary" section.

```
/*=== JES\sample_code\ch_7\vendors_2a.sas ===*/
TITLE1 "Resistance by Vendor - 4Q 2008";
PROC REPORT DATA=Results_Tab(WHERE=(Mon>="2008-10")) NOWINDOWS;
   COLUMN Vendor Mon M_Res R_L R_U N_Fail Report_Link; ❶
   DEFINE Vendor/ORDER;
   DEFINE Mon   /ORDER;
   DEFINE M_Res / ANALYSIS;
   DEFINE R_L / ANALYSIS;
   DEFINE R_U / ANALYSIS;
   DEFINE N_Fail / ANALYSIS;
   DEFINE Report_Link / DISPLAY NOPRINT; ❷
   COMPUTE Report_Link; ❸
       CALL DEFINE("_C1_", "URL", Report_Link);
   ENDCOMP;
RUN;
```

❶ Report_Link is added to the COLUMN statement so that it can be used to define links.

❷ A DEFINE statement is used with the NOPRINT option to specify that Report_Link should not be printed in the report.

❸ This COMPUTE block uses the value of one variable in the COLUMN list, Report_link, to modify the attributes of another column in the report, Vendor. This is done with a CALL DEFINE statement that specifies that the variable printed in the first column, "_C1_", should have a URL attribute with the value contained in the Report_Link variable.

The vendors_3a.sas file creates the graph on the main page, with hyperlinks in each plot point and each item in the legend.

```
/*=== JES\sample_code\ch_7\vendors_3a.sas ===*/
TITLE1 HEIGHT=5 "Mean Resistance by Vendor";
SYMBOL1 VALUE=dot    HEIGHT=2 COLOR=green  WIDTH=2 INTERPOL=JOIN;
SYMBOL2 VALUE=square HEIGHT=2 COLOR=red    WIDTH=2 INTERPOL=JOIN;
SYMBOL3 VALUE=plus   HEIGHT=2 COLOR=blue   WIDTH=2 INTERPOL=JOIN;
AXIS1 LABEL=(JUSTIFY=Center HEIGHT=4  FONT='Arial' COLOR=Black "Month
of Manufacture")
    VALUE = (ANGLE=45 ROTATE=0 HEIGHT=3 FONT='Arial')
    OFFSET= (5, 5);
AXIS2 LABEL =(H=4 A=90  FONT='Arial' 'Average Resistance')
    VALUE = (H=3 FONT='Arial')
    ORDER = (10 TO 25 BY 5)
    MINOR = (N=4);
PROC GPLOT DATA =Results_Tab;
  PLOT M_Res*Mon=Vendor / NAME="F6_14_"
  HAXIS=AXIS1 VAXIS=AXIS2 VREF=12.5 22.5
  HTML=Point_Link  ❶
  HTML_LEGEND=Legend_Link;  ❷
  Format M_Res 3.0;
RUN; QUIT;
```

❶ The HTML option specifies that each plot point be a hyperlink with the value specified by the Point_Link variable.

❷ The HTML_LEGEND option specifies that each item in the legend be a hyperlink with the value specified by the Legend_Link variable.

Finally, the next bit of code in the report_html.sas file creates the main page, report_w_links.html, by including these three code files.

```
ODS HTML PATH="&JES.ods_output/page4" (URL=NONE)
  BODY="report_w_links.html" (TITLE="All Vendors")
  STYLE=MINIMAL;
  %INCLUDE "&JES.sample_code/ch_7/vendors_1a.sas";
  %INCLUDE "&JES.sample_code/ch_7/vendors_2a.sas";
  %INCLUDE "&JES.sample_code/ch_7/vendors_3a.sas";
ODS HTML CLOSE;
```

After you run this code, your ~\JES\ods_output\page4 folder will contain four HTML files: ChiTronix.html, Duality.html, Empirical.html, and report_w_links.html, as well as the seven PNG files containing the graphs. Displays 7-7 and 7-8 show the top and bottom parts of the report_w_links.html page, with arrows indicating the five types of hyperlinks on the page.

Display 7-7: The report_w_links.html Page—Top of the Page

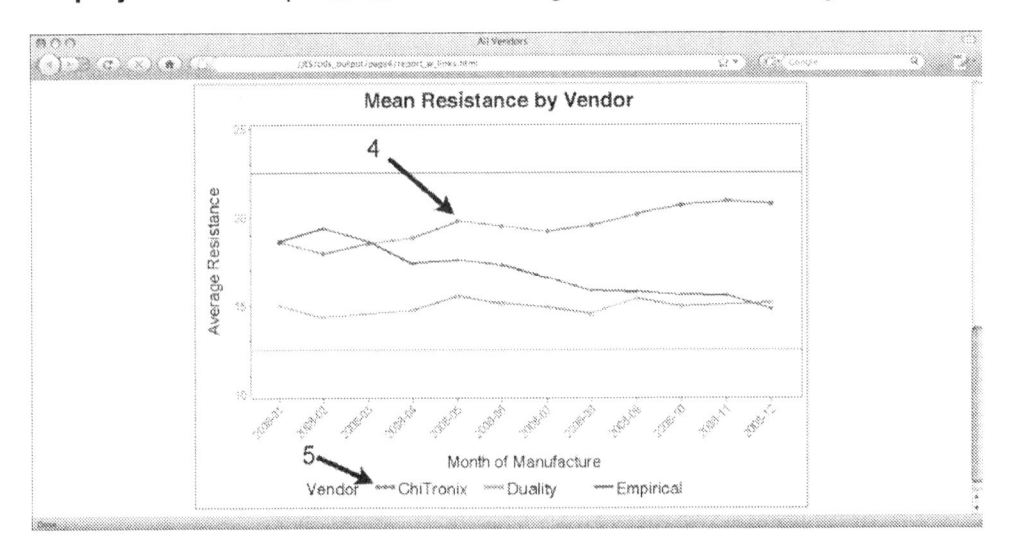

Display 7-8: The report_w_links.html Page—Bottom of the Page

The five arrows in Displays 7-7 and 7-8 indicate the five different types of hyperlink on the report_w_links.html page.

1. If you click on "Download Report" in the second line of the title, the report.pdf file that you created in Section 7.3 will either be opened or downloaded, depending on your browser settings.

2. If you click on one of the vendor names in the first table, the corresponding vendor page, e.g., Duality.html, will open in a separate tab, or in a separate window, depending on your browser settings.

3. If you click on one of the vendor names in the second table, the corresponding vendor page will open in same window, replacing report_w_links.html. You cannot use TARGET="_blank" with PROC REPORT to force the new page to open in a new tab or window, but the example in Section 7.4.3 shows a different way to accomplish this. Click on the browser's Back button to return to the main page.

4. If you click on any of the plot points in the graph, the corresponding vendor page will open in same window, replacing report_w_links.html. If you prefer that the page open in a separate tab or window, redefine the Point_Link variable to include the TARGET="_blank" statement using the same syntax that was used for Legend_Link in the code at the beginning of this section. Click on the browser's Back button to return to the main page.

5. If you click on one of the symbols next to a vendor name in the legend, the corresponding vendor page, e.g. Duality.html, will open in a separate tab, or in a separate window, depending on your browser settings.

If you click on the links in the first table, or the legend symbols, for each vendor in turn, your page should look like Display 7-9 if your browser is set to open a new page in a new tab. Note the names on each tab which make it easy to navigate among the pages.

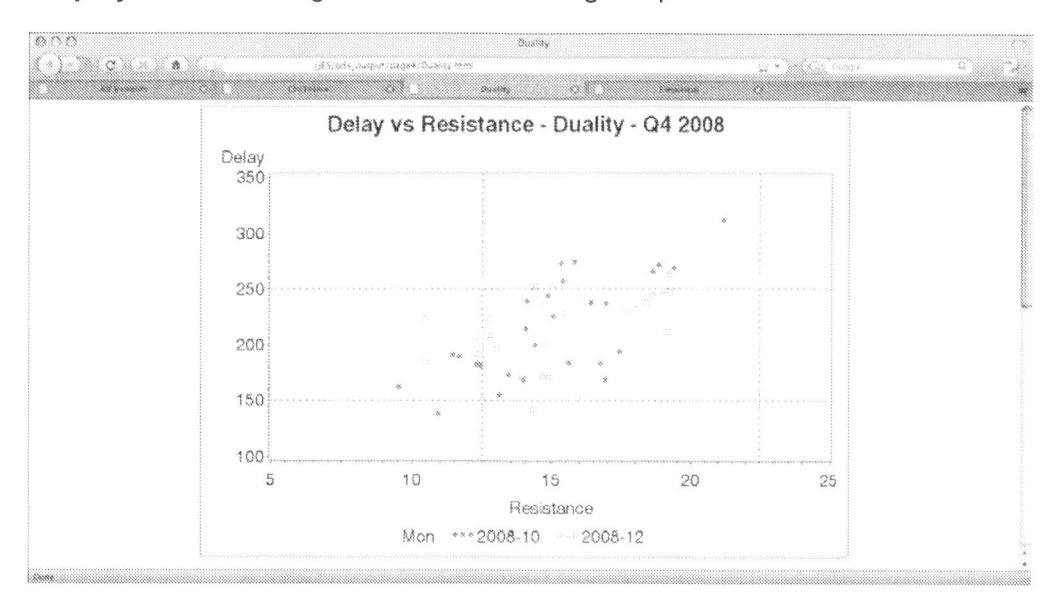

Display 7-9: Main Page with All Vendor Pages Opened

7.4.3 Adding Traffic Lighting to PROC REPORT

It is often useful to add "traffic lighting" to your tables so that the viewer can tell at a glance which cells represent data which is good, marginal, or bad relative to some metric. For example, you might use green for groups that passed a test, red for those that failed, and yellow for those with marginal test results. This section shows how to add traffic lighting to the table created with PROC REPORT in the example from the previous sections. The criteria to be used are

- Green: Number of Fails <=2
- Yellow: Number of Fails =5
- Red: Number of Fails >=6

These are not particularly realistic criteria for this data set, but were chosen so that each color would be used, and to illustrate three different methods for assigning the colors:

- Highlight a cell based on the value in that cell.
- Highlight a cell based on the value in another cell in the same row.
- Highlight an entire row based on the value in one cell of that row.

The vendors_2b.sas code file is the same as vendors_2a.sas, except that a COMPUTE block is added to specify the desired traffic lighting.

```
/*=== JES\sample_code\ch_7\vendors_2b.sas ===*/
TITLE1 "Resistance by Vendor - 4Q 2008";
PROC REPORT DATA=Results_Tab(WHERE=(Mon>="2008-10")) NOWINDOWS;
   COLUMN Vendor Mon M_Res R_L R_U N_Fail Report_Link;
   DEFINE Vendor/ORDER;
   DEFINE Mon   /ORDER;
   DEFINE M_Res / ANALYSIS;
   DEFINE R_L / ANALYSIS;
   DEFINE R_U / ANALYSIS;
   DEFINE N_Fail / ANALYSIS;
   DEFINE Report_Link / DISPLAY NOPRINT;
   COMPUTE Report_Link;
       CALL DEFINE("_C1_", "URL", Report_Link);
   ENDCOMP;
   COMPUTE N_Fail; ❶
       IF N_Fail.SUM >=6 THEN DO;
         CALL DEFINE(_COL_, "STYLE", "STYLE=[BACKGROUND=Salmon]"); ❷
       END;
       IF N_Fail.SUM=5 THEN DO;
         CALL DEFINE("_C2_", "STYLE", "STYLE=[BACKGROUND=Gold]"); ❸
       END;
       IF N_Fail.SUM <= 2 THEN DO;
         CALL DEFINE(_ROW_, "STYLE", "STYLE=[BACKGROUND=PaleGreen]"); ❹
       END;
   ENDCOMP;
RUN;
```

❶ The second COMPUTE block uses the value of the N_Fail variable to specify the background color of certain cells. Note that the value of N_Fail is specified as a compound name "N_Fail.SUM" in each of the comparisons. The general form of a compound name is *variablename.statistic*, where *statistic* is the statistic used to calculate the report item. In this example, SUM is the default statistic, and is equal to the variable value. Consult the SAS online documentation or the references in the "Chapter Summary" section for further details.

❷ If the value of N_Fail is greater than or equal to 6, the background color is set to Salmon for the column specified as _COL_, which refers to the variable being defined in the COMPUTE block, i.e., N_Fail. Note that Salmon, Gold, and PaleGreen are used here because Red and Green are dark, and make the cell text difficult to read on the Web page, while Yellow is very light and difficult to see when the page is printed in black and white, as in Display 7-10.

❸ If the value of N_Fail is equal to 5, the background color is set to Gold for the column specified as "_C2_", which refers to the second variable specified in the COLUMN statement, which is Mon in this example. Use the "_Cn_" notation to specify properties of a cell other than the one named in the COMPUTE statement. Note that "_Cn_" requires quotation marks while _COL_ does not.

❹ If the value of N_Fail is less than or equal to 2, the background color is set to PaleGreen for the columns specified as _ROW_, which means the entire row.

The final bit of code in report_html.sas uses vendors_2b.sas to create a new version of the main page, with the desired traffic lighting.

```
ODS HTML PATH="&JES.ods_output/page4" (URL=NONE)
  BODY="report_w_links_2.html" (TITLE="All Vendors")
  STYLE=MINIMAL
  HEADTEXT="<base TARGET=_BLANK>"; ❶
  %INCLUDE "&JES.sample_code/ch_7/vendors_1a.sas";
  %INCLUDE "&JES.sample_code/ch_7/vendors_2b.sas";
  %INCLUDE "&JES.sample_code/ch_7/vendors_3a.sas";
ODS HTML CLOSE;
```

❶ The HEADTEXT option causes <base TARGET=_BLANK> to be written in the header of the HTML file being created. This sets the default value of TARGET for all links, so the links in the PROC REPORT output, as well as the links in the plot points of the GPLOT output, will now open pages in new tabs or new windows.

Display 7-10 shows the PROC REPORT table in the new report_w_links_2.html page. In a real application you would probably want to apply the highlighting consistently to the same column or columns. Note that the report item used to control the traffic lighting does not have to be included in the report. You could, for example, define a variable named Color in your data set to be "Red", "Yellow", or "Green" based on any appropriate criteria. Then you could use the Color variable to control the cell backgrounds, but exclude it from the report using the NOPRINT option, as was done with the Report_Link variable in the example in Section 7.4.2.

Display 7-10: PROC REPORT Output with Traffic Lighting

Vendor	Month	Mean Resistance	Lower Confidence Limit	Upper Confidence Limit	Number of Fails
ChiTronix	2008-10	20.68	20.11	21.24	5
	2008-11	20.91	20.21	21.60	6
	2008-12	20.76	20.21	21.31	5
Duality	2008-10	14.94	14.05	15.82	7
	2008-12	15.17	14.35	15.99	4
Empirical	2008-10	15.58	14.24	16.93	3
	2008-11	15.55	14.16	16.94	2
	2008-12	14.79	13.56	16.03	3

Resistance by Vendor - 4Q 2008

7.5 Using ODS to Save and Select Procedure Output

An ODS *output object* is an object that contains both the results of a DATA step or PROC step and information about how to format the results. For example, the first code sample in Section 5.7 uses PROC UNIVARIATE to create five output objects that are displayed as five tables in the Output window. Later examples produce additional tables as well as graphic output objects, including a histogram and a probability plot. This section explains how to

- Save selected output objects as SAS data sets.
- Select which of the various output objects should be included in your RTF, PDF, or HTML reports.

7.5.1 Using ODS TRACE to Identify Output Objects

Before you can save or select the output of SAS procedures, you need to know the names of the various objects that are available to you. You can use ODS TRACE to generate a list of the objects created by a procedure. This example uses ODS TRACE to identify the output objects created by PROC UNIVARIATE. The code for this section is in ~\JES \sample_code\ch_7\output_example.sas.

```
GOPTIONS RESET=ALL GUNIT=PCT HTEXT=4 FTEXT='Arial' BORDER;
ODS TRACE ON;  ❶
TITLE HEIGHT=5 "Distribution of Resistance - All Vendors";
PROC UNIVARIATE DATA=JES.Results_Q4;
  VAR Resistance;
  HISTOGRAM Resistance / NORMAL(MU=est SIGMA=est);
  INSET N MEAN STD SKEWNESS KURTOSIS NORMAL(AD ADPVAL)
  / HEIGHT=2.5 FORMAT = 5.3 POSITION=NW;
RUN;
ODS TRACE OFF;  ❷
```

❶ The ODS TRACE ON statement tells SAS to start sending lines to the Log window that identify the output objects created.

❷ The ODS TRACE OFF statement tells SAS to stop sending these lines to the Log window.

When you run this code, nine output objects are created, including eight tables and one graph. You will find the tables in the Output window, and the graph in the Graph window. The first five tables and the histogram are the same as shown in Section 5.7. The ODS TRACE ON statement causes the name of each output object to be written to the Log window in the order in which it is created. Table 7-2 shows the names of the output objects that correspond to the first five tables in the Output window.

Table 7-2: Output Object Names

Table in the Output Window	Name of the Output Object
Moments	Moments
Basic Statistical Measures	BasicMeasures
Tests for Location: Mu0=0	TestsForLocation
Quantiles (Definition 5)	Quantiles
Extreme Observations	ExtremeObs

```
Output Added:
-------------
Name:          Moments
Label:         Moments
Template:      base.univariate.Moments
Path:          Univariate.Resistance.Moments
-------------
Output Added:
-------------
Name:          BasicMeasures
Label:         Basic Measures of Location and Variability
Template:      base.univariate.Measures
Path:          Univariate.Resistance.BasicMeasures
-------------
Output Added:
-------------
Name:          TestsForLocation
Label:         Tests For Location
Template:      base.univariate.Location
Path:          Univariate.Resistance.TestsForLocation
-------------
Output Added:
-------------
Name:          Quantiles
Label:         Quantiles
Template:      base.univariate.Quantiles
Path:          Univariate.Resistance.Quantiles
-------------
Output Added:
-------------
Name:          ExtremeObs
Label:         Extreme Observations
Template:      base.univariate.ExtObs
Path:          Univariate.Resistance.ExtremeObs
-------------
Output Added:
-------------
Name:          Univar
Data Name:     GRSEG
Path:          Univariate.Resistance.Univar
-------------
Output Added:
-------------
Name:          ParameterEstimates
Label:         Parameter Estimates
Template:      base.univariate.FitParms
Path:          Univariate.Resistance.FittedDistributions.Normal.ParameterEstimates
-------------
Output Added:
-------------
Name:          GoodnessOfFit
Label:         Goodness of Fit
Template:      base.univariate.FitGood
Path:          Univariate.Resistance.FittedDistributions.Normal.GoodnessOfFit
-------------
Output Added:
-------------
Name:          FitQuantiles
Label:         Quantiles
Template:      base.univariate.FitQuant
Path:          Univariate.Resistance.FittedDistributions.Normal.FitQuantiles
-------------
```

7.5.2 Using ODS OUTPUT to Save Output to Data Sets

After using ODS TRACE to discover the names of procedure output, you can use ODS OUTPUT to save tabular output to SAS data sets.

```
ODS OUTPUT ❶
  Moments         = WORK.Moments
  BasicMeasures   = WORK.Basic
  FitQuantiles    = Work.Quant;
PROC UNIVARIATE DATA= JES.Results_Q4;
  VAR Resistance;
  HISTOGRAM Resistance / Normal(MU=est SIGMA=est);
  INSET N MEAN STD SKEWNESS KURTOSIS NORMAL(AD ADPVAL)
  / HEIGHT=2.5 FORMAT = 5.3 POSITION=NW;
RUN;
```

❶ The ODS OUTPUT statement tells SAS which objects you want to save, and what data set names to use. For example, Moments = WORK.Moments means that the object named Moments will be saved as a new data set, also called Moments, in the WORK library. The output names must agree with the table names generated by ODS TRACE, but the new data set names can be any valid data set names, for example Quant instead of FitQuantiles.

After you run this code, you will find three new data sets, Moments, Basic and Quant, in your WORK library. If you compare the Moments and Basic data sets to the tables in your Output window (also shown in the example in Section 5.7), you will see that SAS has reformatted the tables to look like data sets.

```
Moments

VarName        Label1              cValue1          nValue1

Resistance     N                       192          192.000000
Resistance     Mean              17.7986361          17.798636
Resistance     Std Deviation      3.68851639          3.688516
Resistance     Skewness          -0.2804873          -0.280487
Resistance     Uncorrected SS    63422.5417              63423
Resistance     Coeff Variation   20.7235901          20.723590

Label2              cValue2          nValue2

Sum Weights              192          192.000000
Sum Observations    3417.33812       3417.338122
Variance            13.6051531        13.605153
Kurtosis            -0.8639843        -0.863984
Corrected SS        2598.58425       2598.584248
Std Error Mean      0.26619574         0.266196
```

- The analysis variable, Resistance, is stored in the VarName variable.
- The first six parameters are stored in the Label1, cValue1, and nValue1 variables:
  - Label1 is the parameter name.
  - cValue1 is a character variable containing the parameter value.
  - nValue1 is a numeric variable containing the parameter value.
- The next six parameters are stored in the Label2, cValue2, and nValue2 variables:
  - Label2 is the parameter name.
  - cValue2 is a character variable containing the parameter value.
  - nNalue2 is a numeric variable containing the parameter value.

```
Basic
VarName      LocMeasure  LocValue  VarMeasure             VarValue
Resistance   Mean        17.79864  Std Deviation           3.68852
Resistance   Median      18.43911  Variance               13.60515
Resistance   Mode           .      Range                  16.67093
Resistance                         Interquartile Range     6.00199
```

- The analysis variable, Resistance, is stored in the VarName variable.
- The location parameters are stored as two separate variables: LocMeasure and LocValue.
- The variability parameters are stored as two separate variables: VarMeasure and VarValue.

```
Quant
VarName     Histogram  Distribution  Percent  ObsQuantile  EstQuantile
Resistance     1         Normal        1.0      9.52923      9.21786
Resistance     1         Normal        5.0     11.94936     11.73157
Resistance     1         Normal       10.0     12.53766     13.07161
Resistance     1         Normal       25.0     14.77605     15.31077
Resistance     1         Normal       50.0     18.43911     17.79864
Resistance     1         Normal       75.0     20.77805     20.28650
Resistance     1         Normal       90.0     22.05705     22.52566
Resistance     1         Normal       95.0     23.22808     23.86571
Resistance     1         Normal       99.0     24.40082     26.37941
```

- The analysis variable, Resistance, is stored in the VarName variable.
- The distribution that was fit to the data, Normal, is stored in the Distribution variable.
- The percent and observed and estimated quantile values are stored in the Percent, ObsQuantile, and EstQuantile variables, respectively.

The Quant data set might be used in a further analysis step, for example, to create a plot of the observed vs. estimated quantiles. The Moments and Basic data sets can be used to capture parameters of interest, which might be used to annotate some future output or to determine the course of further analysis steps. For example, you might want to include the Mean and Standard Deviation values in a TITLE statement, or to perform an analysis step only if the hypothesis of Normality was not rejected.

Section 11.2.2 shows two different ways to create macro variables from variables in a data set, using either PROC SQL or CALL SYMPUT in a DATA step. The subject will be treated in more detail in Chapter 11, but it will be useful to include a simple example here to show how to extract parameters from data sets like Moments or Basic.

```
PROC SQL NOPRINT;
   SELECT nValue2 INTO :Kurt
   FROM Moments
   WHERE Label2 = "Kurtosis";
QUIT;
%PUT Kurtosis = &Kurt;
```

The SELECT statement selects the value of nValue2, from the data set Moments, corresponding to the row where the value of Label2 is equal to "Kurtosis," and puts the result into the macro variable &Kurt. The %PUT statement writes the new macro variable to the Log window just so you can confirm that the kurtosis value has been captured correctly.

7.5.3 Using ODS SELECT and EXCLUDE to Customize Your Report

If a procedure creates more output than you want to include in your report, you can use an ODS SELECT statement or an ODS EXCLUDE statement to get exactly the output you want, as follows:

1. Run your code with ODS TRACE ON, as in Section 7.5.1, to identify the names of the output objects.

2. Review the output in the Output and Graph windows to decide which output to include in your report.

3. Rerun your code using ODS SELECT to select the objects that you want, or ODS EXCLUDE to exclude the objects that you don't want. ODS SELECT does not work for graphic objects, such as the histogram created by PROC UNIVARIATE, so you will have to use ODS EXCLUDE to produce a report that includes graphic output, while excluding other output.

The next two examples create two Web pages with different content from the same PROC UNIVARIATE code. The first step, using ODS TRACE to find the names of the output objects, was done in Section 7.5.1, so you can refer to that section to see the names used in the SELECT and EXCLUDE statements here.

```
ODS HTML PATH="&JES.ods_output/page5" (URL=NONE)
  BODY="select.html" STYLE=MINIMAL;
ODS HTML SELECT BasicMeasures GoodnessOfFit; ❶
GOPTIONS RESET=ALL GUNIT=PCT HTEXT=4 FTEXT='Arial';
TITLE HEIGHT=5 "Distribution of Resistance - All Vendors";
PROC UNIVARIATE DATA= JES.Results_Q4;
  VAR Resistance;
  HISTOGRAM Resistance / Normal(MU=est SIGMA=est);
  INSET N MEAN STD SKEWNESS KURTOSIS NORMAL(AD ADPVAL)
  / HEIGHT=2.5 FORMAT = 5.3 POSITION=NW;
RUN;
ODS HTML CLOSE;
```

❶ The ODS HTML SELECT statement specifies that only the BasicMeasures and GoodnessOfFit tables will be created. The resulting Web page, select.html, is shown in Display 7-11.

```
ODS HTML PATH="&JES.ods_output/page6" (URL=NONE)
  BODY="exclude.html" STYLE=MINIMAL;
ODS HTML EXCLUDE Moments BasicMeasures TestsForLocation Quantiles
        ExtremeObs ParameterEstimates GoodnessOfFit FitQuantiles; ❶
GOPTIONS RESET=ALL GUNIT=PCT HTEXT=4 FTEXT='Arial';
TITLE HEIGHT=5 "Distribution of Resistance - All Vendors";
PROC UNIVARIATE DATA= JES.Results_Q4;
  VAR Resistance;
  HISTOGRAM Resistance / Normal(MU=est SIGMA=est);
  INSET N MEAN STD SKEWNESS KURTOSIS NORMAL(AD ADPVAL)
  / HEIGHT=2.5 FORMAT = 5.3 POSITION=NW;
RUN;
ODS HTML CLOSE;
```

❶ The ODS HTML EXCLUDE statement lists the names of all output objects except Univar, so only Univar, the histogram, is sent to the HTML destination. The resulting Web page, exclude.html, is shown in Display 7-12.

The ODS SELECT and ODS EXCLUDE statements work the same way for the RTF and PDF destinations. This example uses exactly the same EXCLUDE statement used in Sections 7.2, 7.3, and 7.4 to include the histogram from PROC UNIVARIATE, while excluding the tabular output, in the RTF, PDF, and HTML versions of the sample report.

Display 7-11: The select.html Page Created Using ODS HTML SELECT

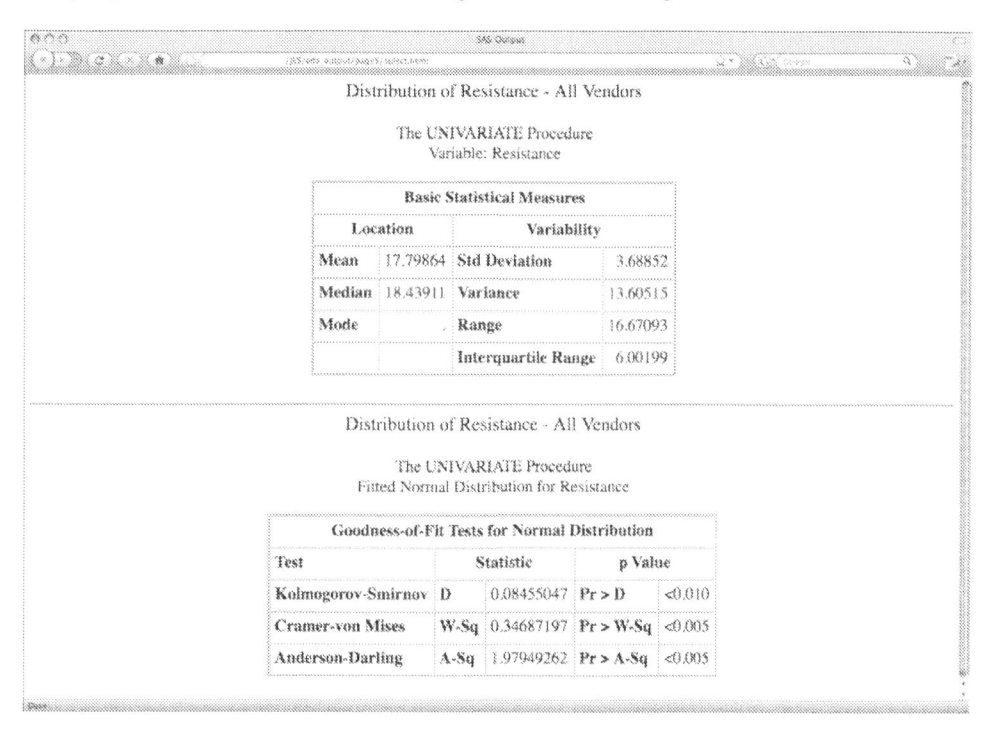

Display 7-12: The exclude.html Page Created Using ODS HTML EXCLUDE

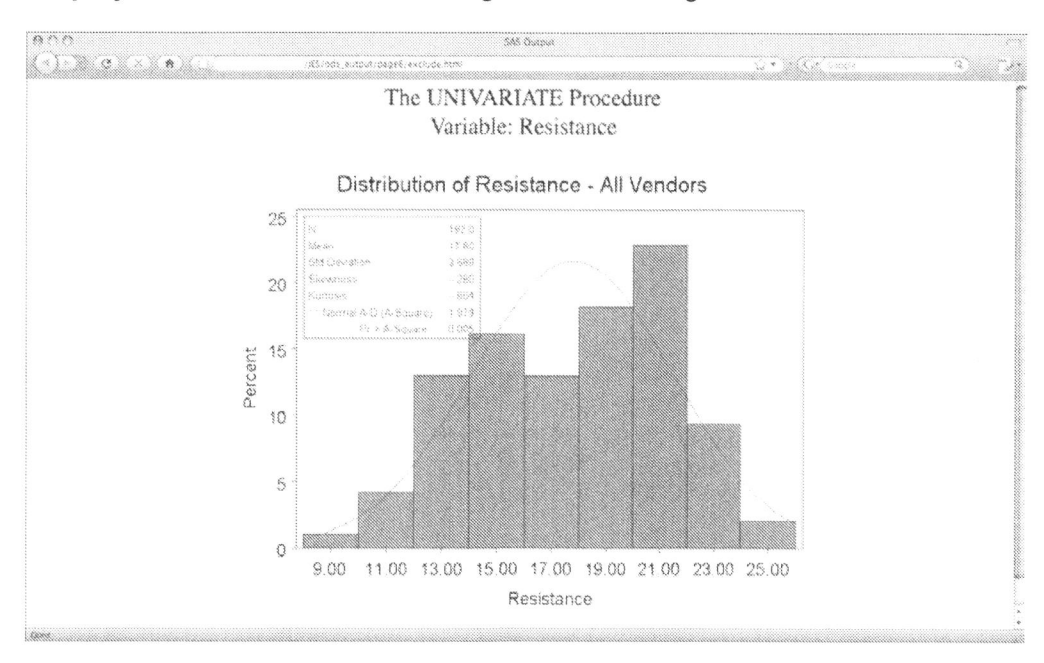

7.6 More Than Enough

This section includes some more advanced techniques for enhancing your Web pages by writing some of your own HTML code, selecting one of the prepackaged styles provided by SAS to control the appearance of your pages, or even creating your own style.

7.6.1 Including Your Own HTML Code

You might sometimes want to customize the pages you create with ODS HTML by including some of your own HTML code. For example, you might want to create an HTML table and place the output created by ODS HTML within the cells of the table. The difficulty is that, by default, ODS HTML writes opening and closing tags at the beginning and end of each HTML file it creates. But you can get around this by using the NO_TOP_MATTER and NO_BOTTOM_MATTER options to suppress the writing of the opening and closing tags.

This example creates an HTML table, and then uses ODS HTML to place a graph within one of the cells of the table. The code for this section is in ~\JES \sample_code\ch_7\notobo.sas.

```
PROC DATASETS MEMTYPE=CAT NOLIST; DELETE HTML; RUN; QUIT;
OPTIONS NODATE;
GOPTIONS RESET=ALL BORDER FTEXT='Helvetica' FTITLE='Helvetica/Bold';
GOPTIONS GUNIT=PCT HTEXT=4 HTITLE=5 HSIZE=9IN VSIZE=6IN;

/* Write your own HTML header code, and start a table */
FILENAME noTopBot "&JES.ods_output/page7/noTopBot.html";   ❶
DATA _NULL_;  FILE noTopBot;   ❷
  PUT "<HTML><HEAD><TITLE>noTopBot Example</TITLE>";
  PUT '</HEAD><BODY><TABLE BORDER=2 ALIGN=CENTER>';
  PUT '<TR><TD ALIGN=CENTER><H1>2008 Quality Report</H1>';
  PUT '</TD></TR><TR><TD>';
RUN;

/* Let SAS write the intermediate HTML code lines */
FILENAME noTopBot   "&JES.ods_output/page7/noTopBot.html" MOD;   ❸
ODS HTML PATH="&JES.ods_output/page7" (URL=NONE)
  BODY=noTopBot(NO_TOP_MATTER NO_BOTTOM_MATTER) STYLE=MINIMAL;   ❹
  %INCLUDE "&JES.sample_code\ch_7\vendors_3.sas";
ODS HTML CLOSE;

/* Write the HTML closing lines   */
FILENAME noTopBot   "&JES.ods_output/page7/noTopBot.html" MOD;   ❺
DATA _NULL_;  FILE NoTopBot;
  PUT '</TD></TR><TR><TD>';
  PUT 'For further information contact
      <a href="mailto:John.Doe@BigCorp.com">
      John.Doe@BigCorp.com</a>';
  PUT '</TD></TABLE></CENTER>';
  PUT '</BODY></HTML>';
RUN;
```

❶ The FILENAME statement defines noTopBot to be a reference to the file noTopBot.html in the ~\JES\ods_output\page7 directory.

❷ The DATA _NULL_ statement writes four lines of HTML code to the noTopBot.html file. This creates some minimal header code for an HTML page, creates a table, writes '2008 Quality Report' in the first row, and starts a new row of the table.

❸ The second FILENAME statement opens the noTopBot file for writing. The MOD option means that new lines will be appended to, and not overwrite, the original file.

❹ The BODY is specified by the file reference noTopBot instead by a filename as in the previous examples. The NO_TOP_MATTER and NO_BOTTOM_MATTER options are used to prevent SAS from writing the opening and closing tags to the HTML code file.

❺ The third FILENAME statement opens the noTopBot file again for writing. The second DATA _NULL_ statement writes four more lines to the noTopBot file. This adds another row to the table, writes the contact information, closes the table, and then writes the ending of the HTML file.

Display 7-13: ODS HTML Output Embedded in an HTML Table

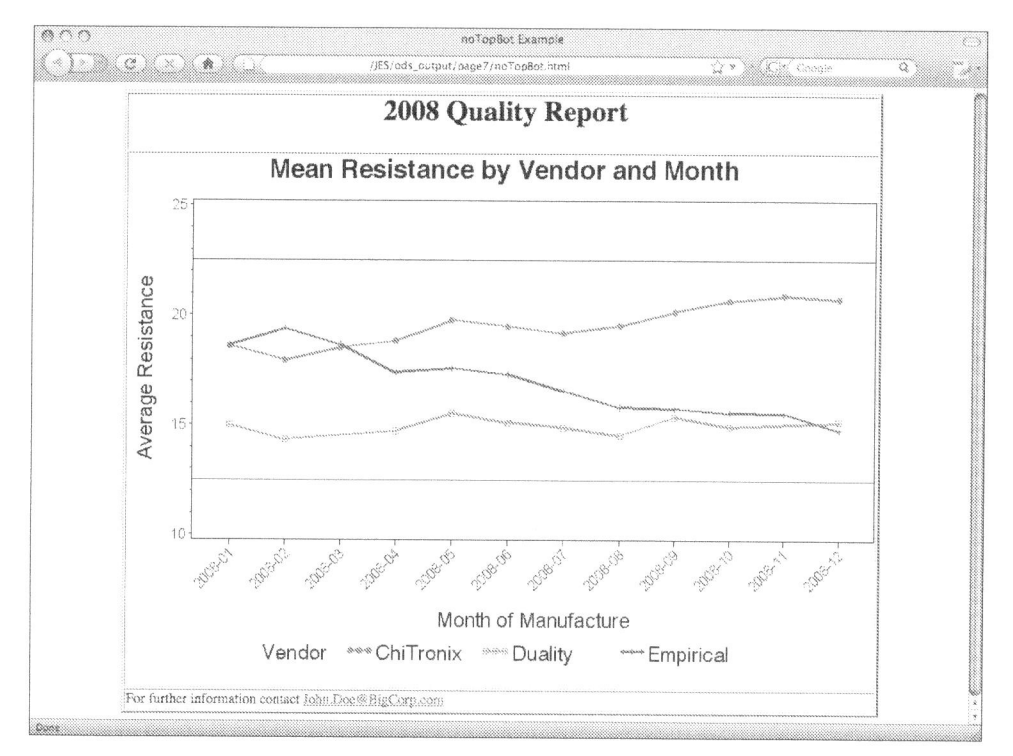

The ODS HTML output is embedded in the second row of the table created by the DATA _NULL_ steps.

Note that the STYLE=MINIMAL statement had no effect as this would have caused style formatting to be written with the top matter, but the NO_TOP_MATTER option prevented this from happening.

7.6.2 Selecting a Style for Your ODS Output

The appearance of the ODS objects you create is governed by a *style* definition that specifies the foreground and background color, font face, font size, and other characteristics to be used in each of the different parts of your document—titles, body text, table column headers, etc. The style can be one of the many prepackaged styles provided by SAS, or one that you create yourself. You select a style by using the STYLE option, as was done with the STYLE=MINIMAL option in Section 7.4. If you do not explicitly select a style, then SAS will choose a default style appropriate to the destination, for example the PRINTER style for ODS PDF or the RTF style for ODS RTF. The code in this section creates a Web page that lists all of the prepackaged SAS styles, and it shows a sample report in each style.

This code was adapted from code in the paper "The Output Delivery System (ODS) from Scratch" by Kevin Smith that is listed in the "Chapter Summary" section. The changes made to the code (highlighted in gray) were to add "&JES.styles/" to redirect the output to your ~/**JES**/styles folder, and to have each style sample open in a new tab with an appropriate title. The code for this section is in ~**JES**\sample_code\ch_7\Smith_Styles.sas.

```
/* This program prints a sample report in HTML, PDF, and RTF */
/* in every style that is specified in ODS path. */
ods listing close;
proc template;
define table vstyle;
column libname memname style links;
define links;
   header = 'Samples';
   compute as '<a href="' || trim(style) || '.html">HTML</a> ' ||
              '<a href="' || trim(style) || '.pdf">PDF</a> ' ||
              '<a href="' || trim(style) || '.rtf">RTF</a>';
end;
end;
run;
/* Print index of all styles. */
ods html file="&JES.styles/index.html" (TITLE="Styles")
    headtext="<base target=_blank>";
data _null_;
   set sashelp.vstyle;
   file print ods=(template='vstyle');
   put _ods_;
run;
ods html close;
%macro generateods();
   ods html file=" &JES.styles/&style..html"
       (TITLE="&style") style=&style;
   ods pdf file=" &JES.styles/&style..pdf" style=&style;
   ods rtf file=" &JES.styles/&style..rtf" style=&style;
   proc contents data=sashelp.class; run;  ❶
   ods rtf close;
   ods pdf close;
   ods html close;
%mend;
/* Print a sample of each style.*/
data _null_;
   set sashelp.vstyle;
   call symput('style', trim(style));
   call execute('%generateods');
run;
ods listing;
```

❶ Note that the PROC CONTENTS line is only there to create the sample output to be displayed in each style. You can replace this line with your own SAS code to see the different styles applied to the output you are interested in.

After you run this code, use your browser to open the index.html file in your ~\JES\styles folder. Your page will look like Display 7-14, except that you will not have the first style, JES. The JES style is discussed in Section 7.6.3, "Creating Your Own Style."

Display 7-14: The index.html Page Created by Smith_Styles.sas

Library Name	Member Name	Style Name	Samples
SASUSER	TEMPLAT	Jes	HTML PDF RTF
SASHELP	TMPLMST	Base.Template.Style	HTML PDF RTF
SASHELP	TMPLMST	Styles.Analysis	HTML PDF RTF
SASHELP	TMPLMST	Styles.Astronomy	HTML PDF RTF
SASHELP	TMPLMST	Styles.Banker	HTML PDF RTF
SASHELP	TMPLMST	Styles.BarrettsBlue	HTML PDF RTF
SASHELP	TMPLMST	Styles.Beige	HTML PDF RTF
SASHELP	TMPLMST	Styles.Brick	HTML PDF RTF
SASHELP	TMPLMST	Styles.Brown	HTML PDF RTF
SASHELP	TMPLMST	Styles.Curve	HTML PDF RTF
SASHELP	TMPLMST	Styles.D3d	HTML PDF RTF

You can use this page to browse through the styles and select the one you like best. Click on one of the HTML, PDF, or RTF links to see a sample of the style in the corresponding format. For example, if you click on HTML in the Styles.D3d line, the page shown in Display 7-15 will open.

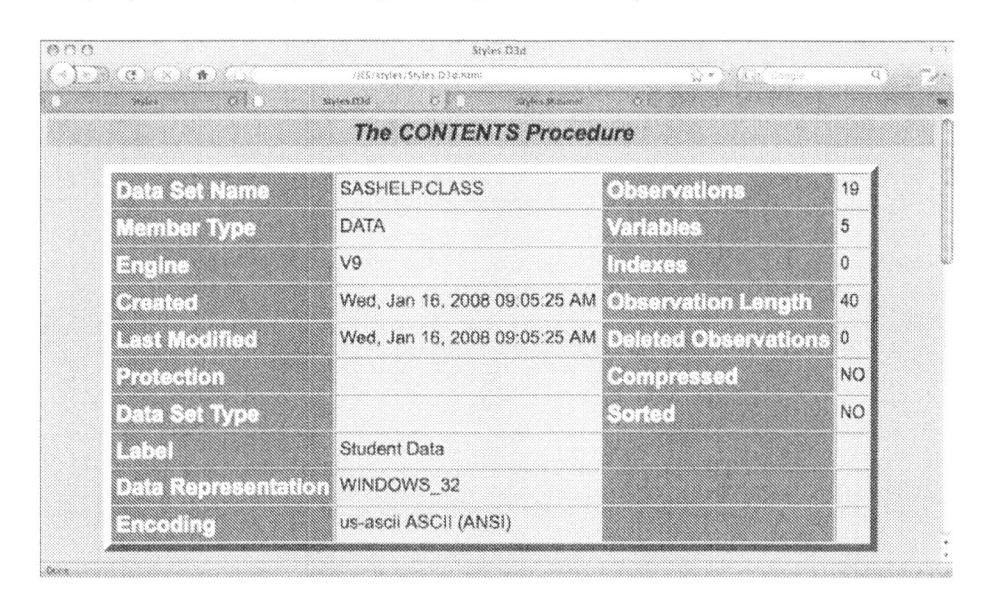

Display 7-15: Sample Page Using the D3D Style

The CONTENTS Procedure

Data Set Name	SASHELP.CLASS	Observations	19
Member Type	DATA	Variables	5
Engine	V9	Indexes	0
Created	Wed, Jan 16, 2008 09:05:25 AM	Observation Length	40
Last Modified	Wed, Jan 16, 2008 09:05:25 AM	Deleted Observations	0
Protection		Compressed	NO
Data Set Type		Sorted	NO
Label	Student Data		
Data Representation	WINDOWS_32		
Encoding	us-ascii ASCII (ANSI)		

242 *Just Enough SAS: A Quick-Start Guide to SAS for Engineers*

7.6.3 Creating Your Own STYLE

A SAS style consists of hundreds of instructions that tell SAS which colors, font types, etc. to use for each element of a document. Changing just a few of these instructions can produce a dramatic difference in the look of your output, but figuring out which elements to change to get the effects that you want is not easy. This section includes a simple example of how to create a new style by changing selected elements of an existing style. The point is to give you a quick look at how it's done, but not to attempt to explain the details. For that you can refer to the references listed in the "Chapter Summary" section, including *Output Delivery System: The Basics* by Lauren Haworth, which includes several chapters on styles, as well as much more detail on the other topics covered in this chapter. The code for this section is in ~\JES \sample_code\ch_7\jes_style.sas.

I like the 3-D look of the D3D style in my Web pages, but I don't like the colors, so I created my own style, JES, using the code below. The JES style is the same as the D3D style with a few color changes.

```
PROC TEMPLATE;  ❶
  DEFINE STYLE JES;
  PARENT=STYLES.D3D;
  STYLE Color_List from Color_List /  ❷
       'FGA'  = White    /* Title */
       'FGA3' = DarkBlue /* Table Headers */
       'FGB2' = DarkBlue /* Links */
       'BGA'  = cx666699 /* Background */
       'BGA2' = White    /* Table Cell Background */
       'BGA3' = GRAYB0   /* TOC and Table Header Background */;
    STYLE Colors from Colors/  ❸
       'datafg' = Color_List('FGA3')  /* Table cell foreground */  ;
  STYLE GraphColors from GraphColors/
       'gwalls' = White;     /* Background of the plot area */
  CLASS  GraphBackground from GraphBackground/ /*Background of the area
                                                  around the plot */
       BackgroundColor=Color_List('BGA3');
  END;  ❹
RUN;
```

❶ The first line begins PROC TEMPLATE, which is used to create and modify styles. The DEFINE STYLE JES; statement specifies that a new style, JES, is to be created. The PARENT=STYLES.D3D; statement specifies that the new style will be the same as the D3D style, except for the properties that are specified in the lines to follow.

❷ The STYLE Color_List .../ statement replaces some of the colors in the Color_List of the D3D style. For example, the first line specifies that the color named 'FGA' will be replaced by White. The colors in the color list are used as a palette to specify the color of various elements later in the style definition. Color names are arbitrary, except that foreground colors begin with 'FG' and background colors begin with 'BG'.

❸ The second STYLE statement sets the color of the style element named 'datafg' to be the value of Color_List('FGA3'), which was set to DarkBlue in the first STYLE statement. The comments indicate the part of the output that uses each of the specified colors. For example, the color assigned to 'datafg' is used as the foreground color in table cells. Note, however, that these comments are not present in the definition of the D3D style, and it took quite a bit of trial and error to discover which elements had to be changed to achieve the desired color scheme.

❹ The END statement ends the DEFINE statement, and the RUN statement ends PROC TEMPLATE.

After you have run this code, your new style, JES, will be available for use in your ODS output. This style definition will persist after you close your SAS session, so you can use it in future SAS sessions without redefining it.

This example uses the STYLE=JES option to create a Web page similar to the examples in Section 7.4, but using the JES style.

```
OPTIONS NODATE;
GOPTIONS RESET=ALL BORDER FTEXT='Helvetica' FTITLE='Helvetica/Bold';
GOPTIONS GUNIT=PCT HTEXT=4 HTITLE=5 HSIZE=9IN VSIZE=6IN;
ODS LISTING CLOSE;
  AXIS1 OFFSET=(5,5);
  SYMBOL1 COLOR=Red   HEIGHT=3 VALUE=Square   I=JOIN;
  SYMBOL2 COLOR=Blue  HEIGHT=3 VALUE=Triangle I=JOIN;
  SYMBOL3 COLOR=Green HEIGHT=3 VALUE=Circle   I=JOIN;
ODS HTML PATH="&JES.ods_output/page4" (URL=NONE)
  BODY="Jes_Body.html"
  CONTENTS="Jes_TOC.html"
  FRAME = "Jes_Frame.html"
  STYLE=JES;
  %INCLUDE "&JES.sample_code/ch_7/vendors_1a.sas";
  %INCLUDE "&JES.sample_code/ch_7/vendors_2b.sas";
  %INCLUDE "&JES.sample_code/ch_7/vendors_3a.sas";
ODS HTML CLOSE;
ODS LISTING;
```

When you run this code, the Jes_Frame.html page shown in Display 7-16 is created in your ~/**JES**/ods_output/page4 folder.

Display 7-16: Web Page Created Using the JES Style

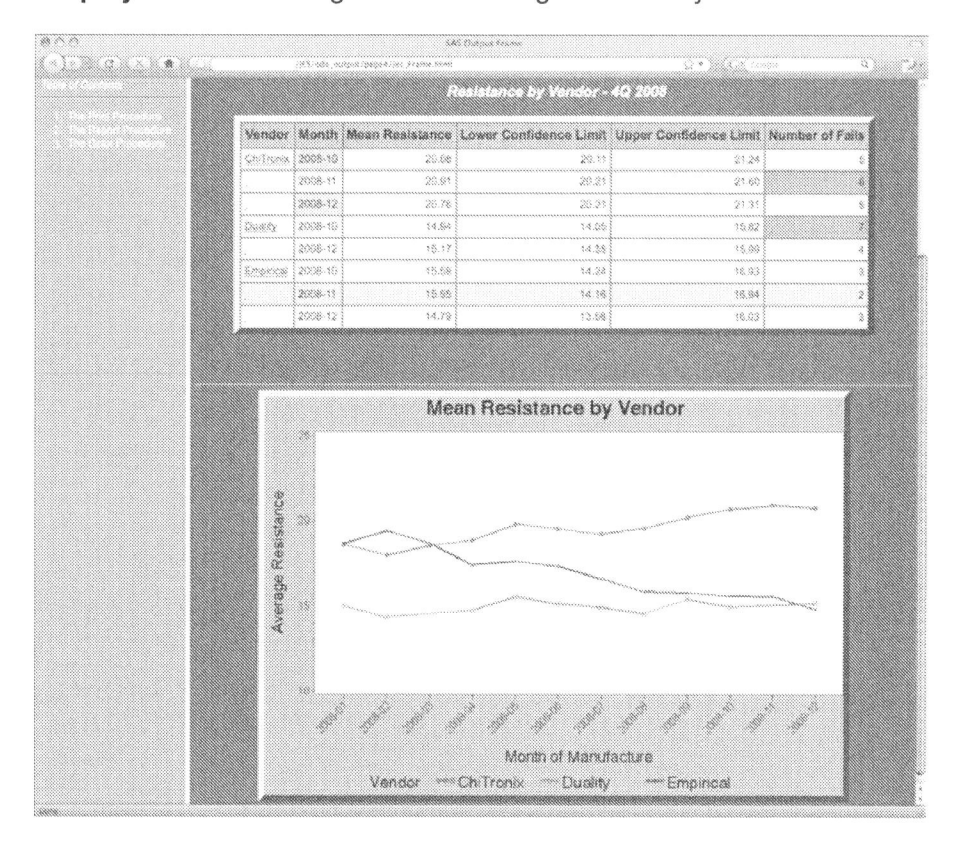

7.7 Chapter Summary

7.7.1 Recap

After finishing this chapter you should know how to

- Publish your SAS output in RTF, PDF, or HTML format.
- Add HTML hyperlinks to your PROC PRINT, PROC REPORT, and PROC GPLOT output.
- Add traffic lighting to PROC REPORT output created with ODS HTML.
- Use ODS TRACE to identify the names of tables created during SAS procedures.
- Use ODS OUTPUT to save tables created by SAS procedures as data sets.
- Use ODS SELECT and EXCLUDE to select which procedure output is included in your documents.
- Customize your output by selecting a SAS style, or creating your own style.

7.7.2 For More Information

Books

Carpenter, Art. 2007. *Carpenter's Complete Guide to the SAS® Report Procedure*. Cary, NC: SAS Institute Inc.

Delwiche, Lora D., and Susan J. Slaughter. 2003. *The Little SAS® Book: A Primer, Third Edition*. Cary, NC: SAS Institute Inc.

Gupta, Sunil. 2003. *Quick Results with the Output Delivery System*. Cary, NC: SAS Institute Inc.

Haworth, Lauren E. 2001. *Output Delivery System: The Basics*. Cary, NC: SAS Institute Inc.

This chapter shows you how to use ODS HTML to create static Web pages. The following two books discuss the use of SAS/IntrNet to create dynamic Web pages which can run SAS processes and create new reports and graphs in response to user input. With SAS/IntrNet you could, for example, allow a visitor to your Web page to select only the vendors, failure modes, and date range he or she is interested in, and then create a report including only the specified information. SAS/IntrNet is a separate SAS product requiring an additional license fee.

Henderson, Don. 2007. *Building Web Applications with SAS/IntrNet®: A Guide to the Application Dispatcher*. Cary NC: SAS Institute Inc.

Pratter, Frederick E. 2006. *Web Development with SAS® by Example, Second Edition*. Cary, NC: SAS Institute Inc.

SAS Conference Papers

Delaney, Kevin P. 2004. "Multiple Graphs on One page, the Easy Way (PDF) and the Hard Way (RTF)." *Proceedings of the Twenty-Eighth Annual SAS Users Group International Conference*. Cary, NC: SAS Institute Inc. Paper 71-28. **[094-28.pdf]**

Gebhart, Eric. 2007. "ODS Markup, Tagsets, and Styles! Taming ODS Styles and Tagsets." *Proceedings of the SAS Global Forum 2007 Conference*. Cary, NC: SAS Institute Inc. Paper 225-2007. **[225-2007.pdf]**

Gupta, Sunil K. 2001. "Using Styles and Templates to Customize SAS® ODS Output."
Proceedings of the Twenty-Sixth Annual SAS Users Group International Conference.
Cary, NC: SAS Institute Inc. Paper 1-26. **[p001-26.pdf]**

Gupta, Sunil. 2008. "SAS® ODS Technology for Today's Decision Makers." *Proceedings of the SAS Global Forum 2008 Conference.* Cary, NC: SAS Institute Inc. Paper 193-2008. **[193-2008.pdf]**

Hadden, Louise. 2005. "PROC TABULATE and ODS RTF: The Perfect Fit for Complex Tables." *Proceedings of the Thirtieth Annual SAS Users Group International Conference.* Cary, NC: SAS Institute Inc. Paper 091-30. **[091-30.pdf]**

Hadden, Louise. 2006. "STOP! WAIT! GO! See What Traffic-Lighting Can Do For You." *Proceedings of the Thirty-First Annual SAS Users Group International Conference.* Cary, NC: SAS Institute Inc. Paper 142-31. **[142-31.pdf]**

Haworth, Lauren. 2001. "ODS for PRINT, REPORT, and TABULATE." *Proceedings of the Twenty-Sixth Annual SAS Users Group International Conference.* Cary, NC: SAS Institute Inc. Paper 3-26. **[p003-26.pdf]**

Haworth, Lauren. 2003. "SAS® with Style: Creating Your Own ODS Style Template." *Proceedings of the Twenty-Eighth Annual SAS Users Group International Conference.* Cary, NC: SAS Institute Inc. Paper 195-28. **[195-28.pdf]**

Haworth, Lauren. 2004. "Introduction to ODS." *Proceedings of the Twenty-Ninth Annual SAS Users Group International Conference.* Cary, NC: SAS Institute Inc. Paper 245-29. **[245-29.pdf]**

Haworth, Lauren. 2004. "SAS® with Style: Creating Your Own ODS Style Template for RTF Output." *Proceedings of the Twenty-Ninth Annual SAS Users Group International Conference.* Cary, NC: SAS Institute Inc. Paper 125-29. **[125-29.pdf]**

Haworth, Lauren. 2005. "SAS with Style: Creating Your Own ODS Style Template for PDF Output." *Proceedings of the Thirtieth Annual SAS Users Group International Conference.* Cary, NC: SAS Institute Inc. Paper 132-30. **[132-30.pdf]**

Haworth, Lauren. 2006. "PROC TEMPLATE: The Basics." *Proceedings of the Thirty-First Annual SAS Users Group International Conference.* Cary, NC: SAS Institute Inc. Paper 112-31. **[112-31.pdf]**

Lund, Pete. 2008. "PDF Can be Pretty Darn Fancy: Tips and Tricks for the ODS PDF Destination." *Proceedings of the SAS Global Forum 2008 Conference.* Cary, NC: SAS Institute Inc. Paper 033-2008. **[033-2008.pdf]**

McNeill, Sandy. 2001. "Changes & Enhancements for ODS by Example (through Version 8.2)." *Proceedings of the Twenty-Sixth Annual SAS Users Group International Conference.* Cary, NC: SAS Institute Inc. Paper 2-26. **[p002-26.pdf]**

Pratter, Frederick E. 2007. "Using the SAS® Output Delivery System and PROC TEMPLATE to Create XHTML Files." *Proceedings of the SAS Global Forum 2007 Conference.* Cary, NC: SAS Institute Inc. Paper 118-2007. **[118-2007.pdf]**

Smith, Kevin D. 2007. "The Output Delivery System (ODS) from Scratch." *Proceedings of the SAS Global Forum 2007 Conference.* Cary, NC: SAS Institute Inc. Paper 219-2007. **[219-2007.pdf]**

Smith, Kevin D. 2007. "PROC TEMPLATE Tables from Scratch." *Proceedings of the SAS Global Forum 2007 Conference.* Cary, NC: SAS Institute Inc. Paper 221-2007. **[221-2007.pdf]**

Tong, Cindy. 2003. "ODS RTF: Practical Tips." *Proceedings of the Sixteenth Annual NorthEast SAS Users Group Conference.* Washington, DC. Paper at007. **[at007.pdf]**

7.7.3 Exercises

- Create a new sample report which includes a trend plot and histograms for Delay as well as Resistance, and publish your report in RTF, PDF, and HTML formats.
- In the PROC REPORT table, set the background color to red if the mean Delay is greater than 225.
- Use ODS CSVALL (see Section 3.2.2) to create CSV files containing the raw data from the JES.Results data set for each combination of Vendor and Month. Add hyperlinks to the trend plots which cause the corresponding CSV file to be downloaded when a point is clicked on.
- Edit the JES style so that the background is green (cx669966) instead of blue (cx666699).

Chapter 8

Plotting Your Data with ODS Graphics

- 8.1 Introduction 248
- 8.2 ODS Statistical Graphics 250
- 8.3 PROC SGPLOT 252
- 8.4 PROC SGPANEL 296
- 8.5 PROC SGSCATTER 298
- 8.6 More Than Enough 304
- 8.7 Chapter Summary 306

8.1 Introduction

ODS Statistical Graphics, or ODS Graphics, which was experimental in SAS 9.1 and production in SAS 9.2, provides you with a significantly enhanced capability to create graphical output for your analyses and reports. Many statistical procedures have been modified to automatically create appropriate plots to accompany the tabular output. This saves you the work of saving the output from the statistical procedure to a data set, and then using SAS/GRAPH to create graphs to aid in the understanding and communication of your results. In addition, there are several new statistical graphics, or "SG", procedures in SAS/GRAPH that take advantage of the ODS Graphics functionality. These procedures enable you to create graphs similar to those illustrated in Chapter 6, with less effort, and to create new plot types not previously available in SAS/GRAPH. The procedures that use ODS Graphics ensure a consistent and attractive look for your graphs, which saves you the trial-and-error process of selecting font types and sizes, colors, legend options, etc., to get your graphs to look just right. But you can also customize your graphs using procedure options, ODS styles, the ODS Graphics Editor, and the Graph Template Language. The paper "Getting Started with ODS Statistical Graphics in SAS® 9.2" by Robert Rodriguez, which is listed in the "Chapter Summary" section, provides a good overview of ODS Graphics.

8.1.1 Statistical Procedures Using ODS Graphics

ODS Graphics provides the ability to automatically generate graphs when certain statistical procedures are run. Table 8-1 shows some of the procedures in Base SAS, SAS/STAT, and SAS/QC that support ODS Graphics. Procedures will continue to be added in new releases of SAS, so check the SAS online documentation for the current list. Section 8.2 includes an example illustrating the use of ODS Graphics with PROC LIFETEST.

8.1.2 New SG Procedures in SAS/GRAPH

There are four new statistical graphics (SG) procedures in SAS 9.2 that you can use to create general plots not tied to any particular statistical procedure. These new procedures use ODS Graphics functionality, and so they produce graphs which are consistent in behavior and appearance with graphs produced by the statistical procedures. For a detailed explanation of the SG procedures, see the paper "Effective Graphics Made Simple Using SAS/GRAPH® SG Procedures" by Dan Heath.

8.1.2.1 PROC SGPLOT
The SGPLOT procedure can be used to create many of the same plots that are available with PROC GPLOT and PROC GCHART. The SGPLOT code is generally easier to write, and the appearance of the resulting output is generally more consistent and attractive than what you would get without a lot more effort using GPLOT or GCHART. SGPLOT is discussed in Section 8.3.

8.1.2.2 PROC SGPANEL
The SGPANEL procedure creates an array, or panel, of similar plots, one for each value of a classification variable. SGPANEL is discussed in Section 8.4.

8.1.2.3 PROC SGSCATTER
The SGSCATTER procedure creates an array of scatter plots, one for each pair from a list of variables. SGSCATTER is discussed in Section 8.5.

8.1.2.4 PROC SGRENDER

The SGRENDER procedure can be used to create more complex graphic displays than can be created with the above procedures. You must first use the Graph Template Language to create a template for the desired plot layout, and then use SGRENDER with the template to create the desired graphic. A detailed discussion of SGRENDER is beyond the scope of this book, but a brief example is given in Section 8.6.

8.1.3 ODS Styles

Section 7.6 includes a brief introduction to the use of ODS styles to control the appearance of your output. Starting with SAS 9.2, the same concept is applied to SAS/GRAPH graphical output. In addition to controlling the fonts and colors of your tabular output, styles now also control line types and colors, symbol types and colors, graph title fonts, and other aspects of your graphical output. The use of ODS Styles to control graphics output will not be discussed here, but you can find an excellent introduction to the subject in "Using ODS Styles with SAS/GRAPH®" by Jeff Cartier and Dan Heath, which is listed in the "Chapter Summary" section.

8.1.4 Graph Template Language

Graph templates are used to control the layout and appearance of the graphs produced by procedures that support ODS Graphics. These templates are written in the Graph Template Language (GTL). SAS provides a default template for every graph, and so you do not need to know anything about graph templates or the GTL in order to create graphs. The exception is PROC SGRENDER, which requires that you explicitly define your own graphics template and then apply it to your data to create the desired graphical output. An example of the use of GTL with PROC SGRENDER is given in Section 8.6. See also the paper "Introduction to the Graph Template Language" by Sanjay Matange which is listed in the "Chapter Summary" section.

Table 8-1: Selected Procedures Supporting ODS Graphics in SAS 9.2

Base SAS					
CORR	FREQ	UNIVARIATE			
SAS/STAT					
ANOVA	BOXPLOT	CALIS	CLUSTER	CORRESP	FACTOR
FREQ	GAM	GENMOD	GLM	GLIMMIX	GLMSELECT
KDE	KRIGE2D	LIFEREG	LIFETEST	LOESS	LOGISTIC
MCMC	MDS	MI	MIXED	MULTTEST	NPAR1WAY
PHREG	PLS	PRINCOMP	PRINQUAL	PROBIT	QUANTREG
REG	ROBUSTREG	RSREG	SEQDESIGN	SEQTEST	SIM2D
TCALIS	TRANSREG	TTEST	VARIOGRAM		
SAS/QC					
ANOM	CAPABILITY	CUSUM	MACONTROL	PARETO	RELIABILITY
SHEWHART					

The data sets used in this chapter are:

- JES.Results, which is the data set containing the 2008 test results for each vendor. (See Section 5.1.1.)
- JES.Results_Q4, which is a subset of JES.Results containing data for only the last three months of 2008.
- JES.Results_Tab, which is a PROC MEANS summary of the JES.Results data set. (See Section 6.1.1.)
- JES.LifeTest, which is a data set containing the results of an accelerated life test. (See Section 10.1.3.)

8.2 ODS Statistical Graphics

The example in this section uses PROC LIFETEST to illustrate the kind of graphical output that is available from statistical procedures with very little effort using ODS Statistical Graphics. Chapters 9 and 10 include examples using ODS Graphics with other statistical procedures. The code for this section is in ~\JES \sample_code\ch_8 \lifetest.sas.

This code uses ODS Graphics to plot the results of a life test.

```
ODS LISTING GPATH="&JES.SG/S_8_2" IMAGE_DPI=300;   ❶
ODS GRAPHICS ON / RESET IMAGENAME="F8_1_" IMAGEFMT=PNG;   ❷

PROC LIFETEST DATA=JES.LifeTest;   ❸
  TIME TestTime*Censor(-1);
  STRATA Temp_C;
RUN;

ODS GRAPHICS OFF;   ❹
```

❶ The ODS LISTING statement specifies that the output be sent to the LISTING destination. The GPATH option specifies that any image files created be saved in the **~JES/SG/S_8_2** folder. The IMAGE_DPI option specifies that the images be created at 300 DPI. If you omit the IMAGE_DPI option, images will be saved at the default DPI for the destination, which is 100 DPI in this case.

❷ The ODS GRAPHICS ON statement turns on ODS Graphics. The RESET option resets all ODS Graphics global options to their default values. The IMAGENAME option specifies the name of the first image file to be created. If the IMAGENAME option is omitted, a default name will be used, in this case "SurvivalPlot.png". The RESET option ensures that the specified name is used, even if an image file with the same name was previously created. You can use RESET=INDEX to achieve the same effect without resetting any other ODS Graphics options. The IMAGEFMT option specifies that the image files have the PNG format. If this option is omitted, images will be saved in the default file type for the destination.

❸ PROC LIFETEST computes survival curves from the test data for each vendor. The syntax of PROC LIFETEST will be explained in Section 10.4.

❹ ODS GRAPHICS OFF; turns off ODS Graphics.

After you run this code, the output from PROC LIFETEST will be listed in the Results window, as shown in Display 8-1, and you will have a new file, F8_1_.png, in your **~\JES\SG\S_8_2** folder. You can view the graph produced by PROC LIFETEST either by double-clicking the **Survival Curves** icon in the Results window, or by double-clicking the image file in your SG folder. In either case, the graph will open in the application, e.g., Adobe Photoshop, that you have designated to handle PNG files. The survival curve plot is shown in Figure 8-1.

Display 8-1: Results Window After Running PROC LIFETEST

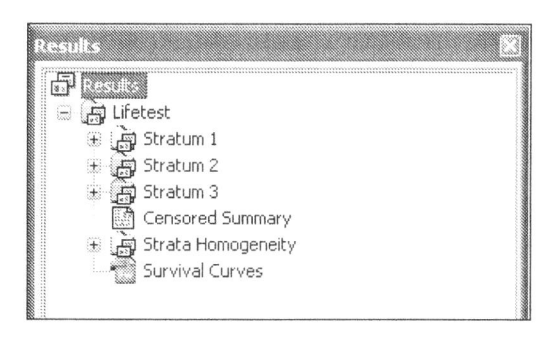

Figure 8-1: Graph Created by PROC LIFETEST

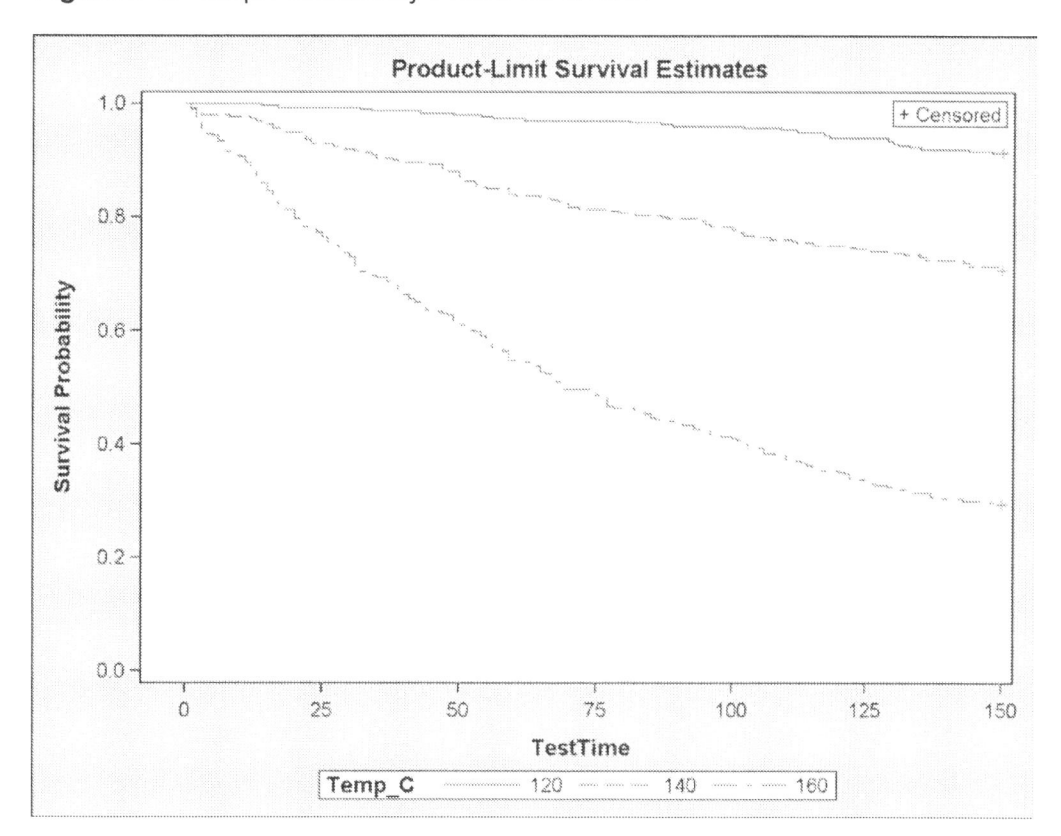

Note that appropriate titles, legends, axes, line types, etc. have been applied without any of the effort required by the methods described in Chapter 6.

8.3 PROC SGPLOT

The SGPLOT procedure can be used to create many of the same plots available with PROC GPLOT and PROC GCHART, and includes additional functionality not available in either of these procedures. The SGPLOT code is generally easier to write, and the appearance of the graphs is controlled by styles, which is easier than specifying all the graphical options yourself, and tends to provide a more consistent look to your graphical output.

The statements available with PROC SGPLOT are shown in Table 8-2. The first five groups of statements in the table list the different types of plots which can be drawn, and the last two groups contain statements that you can use to control the reference lines, legends, and axes of the plots. The classification of the plot types, "Basic X-Y Plots" etc., is my own, and is somewhat arbitrary. Note that HLINE, VLINE, HBAR, VBAR, and PBSPLINE have been included in more than one category corresponding to different ways that they can be used. The subsections of Section 8.3 cover these topics:

- 8.3.1 The plot types created by SGPLOT, using the statements in Table 8-2.
- 8.3.2 How to plot multiple lines on the same or different graphs.
- 8.3.3 Options you can use to customize the appearance of your plots.
- 8.3.4 Some examples of more complicated graphs you can create with SGPLOT.
- 8.3.5 The use of the URL option to add hyperlinks to your HTML output.

The following example shows how to save output to the LATEX, RTF, and HTML destinations.

```
ODS _ALL_ CLOSE;  ❶
ODS LATEX PATH="&JES.SG/S_8_2" BODY="Lifetest.tex" IMAGE_DPI=300;  ❷
ODS RTF   PATH ="&JES.SG/S_8_2" BODY="Lifetest.rtf" IMAGE_DPI=300;  ❸
ODS HTML  PATH="&JES.SG/S_8_2" (URL=NONE) BODY="Lifetest.html";  ❹
ODS GRAPHICS ON / RESET IMAGENAME="F8_1_";  ❺

PROC LIFETEST DATA=JES.LifeTest;
   TIME TestTime*Censor(-1);
   STRATA Temp_C;
RUN;

ODS GRAPHICS OFF;
ODS _ALL_ CLOSE;  ❻
ODS LISTING;
```

❶ The ODS _ALL_ CLOSE statement closes any ODS destinations that might be open.

❷ The ODS LATEX statement causes output to be sent to the LATEX destination. This is a convenient way to create high-resolution (300 DPI) image files in postscript format. The LATEX destination was used to create most of the figures in this book.

❸ The ODS RTF statement causes output to be sent to the RTF destination. The resulting RTF file provides a convenient way to store a number of related images that can be copied and pasted into other documents, such as PowerPoint presentations.

❹ The ODS HTML statement causes output to be sent to the HTML destination. If you select **Tools → Options → Preferences → Results**, and check the **View results as they are generated** box and **View results using Internal Browser**, then the HTML version of your results will open automatically when you run the sample code. This is a good way to see the results very quickly. The examples in the rest of this chapter use only the HTML destination. Note: If you are running SAS for UNIX, you may have to start the Remote Browser Server executable, rbrowser, in order to see your results in a browser window. Consult the SAS online documentation, or SAS Technical Support, if you need help with this.

❺ The IMAGENAME statement specifies the name to be used for image files. The names used in the examples correspond to the figures in this book, for example "F8_1_" for Figure 8-1. This makes it easy to match up the image files, code samples, and the figures.

❻ The ODS _ALL_ CLOSE statement closes all open ODS destinations, including the LISTING destination. The next line opens the LISTING destination again so that it is available for any future output.

Table 8-2: SGPLOT Statements

Statement	Graph Type
Basic X-Y Plots	
SERIES	Line plot of Y vs. X.
SCATTER	Scatter plot of Y vs. X.
STEP	Step function plot of Y vs. X.
PBSPLINE	Smooth spline plot of Y vs. X. This is a fit of the data, which may not pass through each point.
NEEDLE	Lines from a baseline value on the Y axis to each corresponding (X,Y) value.
Limit Plots	
DOT	A single summary point (FREQ, SUM, or MEAN) for each value of a classification variable, with optional limit lines.
HLINE	Horizontal line plot with optional limit lines.
VLINE	Vertical line plot with optional limit lines.
HBAR	Horizontal bar chart with optional limit lines.
VBAR	Vertical bar chart with optional limit lines.
Bar Charts	
HBAR	Horizontal bar chart.
VBAR	Vertical bar chart.
HLINE	Horizontal line plot which can be overlaid on an HBAR plot (DOT plots can be overlaid on an HBAR or HLINE as well).
VLINE	Vertical line plot which can be overlaid on a VBAR plot.
Distribution Plots	
HISTOGRAM	Histogram of the data points.
DENSITY	Normal or kernel density estimate fitted to the data points.
HBOX	Horizontal Box-and-Whisker plot.
VBOX	Vertical Box-and-Whisker plot.
Data Fit Plots	
ELLIPSE	Confidence ellipse overlaid on a SCATTER plot.
LOESS	LOESS curve fit to a SCATTER plot.
REG	Regression fit to a SCATTER plot.
PBSPLINE	Penalized b Spline fit to a SCATTER plot.
BAND	Plots a region defined by UPPER and LOWER variables provided by the user. Does not fit the region to the data.
Legends and Reference Lines	
REFLINE	Add reference lines to various plots.
KEYLEGEND	Add legend to a plot.
INSET	Add an inset to a plot.
Plot Axes	
XAXIS	Control the attributes of the X AXIS.
YAXIS	Control the attributes of the Y AXIS.
XAXIS2	Control the attributes of the second X AXIS.
YAXIS2	Control the attributes of the second Y AXIS.

8.3.1 Plot Types

This section gives a brief overview of the plot types that are available within PROC SGPLOT, grouped according to the classification in Table 8-2.

8.3.1.1 Basic X-Y Plots

This code illustrates five types of basic X-Y plots that can be created using different statements with SGPLOT. These are similar to the plots created in Chapter 6 using the PLOT and PLOT2 statements with PROC GPLOT. The code for this section is in ~\JES \sample_code\ch_8 \basic.sas.

```
ODS HTML PATH="&JES.SG/S_8_3" (URL=NONE) BODY="Basic.html";
ODS GRAPHICS ON / RESET IMAGENAME="F8_2_";

TITLE "SERIES";
PROC SGPLOT DATA=JES.Results_Tab(WHERE=(Vendor="ChiTronix"));
   SERIES Y=M_Res X=Mon; ❶
RUN;

TITLE "SCATTER";
PROC SGPLOT DATA=JES.Results_Tab(WHERE=(Vendor="ChiTronix"));
   SCATTER Y=M_Res X=Mon; ❷
RUN;

TITLE "STEP";
PROC SGPLOT DATA=JES.Results_Tab(WHERE=(Vendor="ChiTronix"));
   STEP Y=M_Res X=Mon; ❸
RUN;

TITLE "NEEDLE";
PROC SGPLOT DATA=JES.Results_Tab(WHERE=(Vendor="ChiTronix"));
   NEEDLE  Y=M_Res X=Mon / BASELINE=20; ❹
RUN;

TITLE "PBSPLINE";
PROC SGPLOT DATA=JES.Results_Tab(WHERE=(Vendor="ChiTronix"));
   PBSPLINE  Y=M_Res X=Month; ❺
RUN;

ODS GRAPHICS OFF;
ODS HTML CLOSE;
```

The five PROC SGPLOT statements are used to create five different types of plot from the same data. The variables to be plotted are specified as Y=M_Res X=Mon rather than with the M_Res*Mon syntax used by PROC GPLOT.

❶ The SERIES statement results in the plot points being joined by straight-line segments.

❷ The SCATTER statement plots only the points, with no connecting lines.

❸ The STEP statement connects the points with a step function.

❹ The NEEDLE statement draws vertical lines from the BASELINE to each point. If BASELINE is not specified, the lines are drawn from the X axis.

❺ The PBSPLINE statement fits a penalized B-spline to the points. In this example, PBSPLINE draws a smooth curve going through each point. If the points cannot be joined by a simple curve, PBSPLINE draws a smooth fit line through the scatter of points, as illustrated in Section 8.3.1.5.

Figure 8-2: Basic X-Y Plots

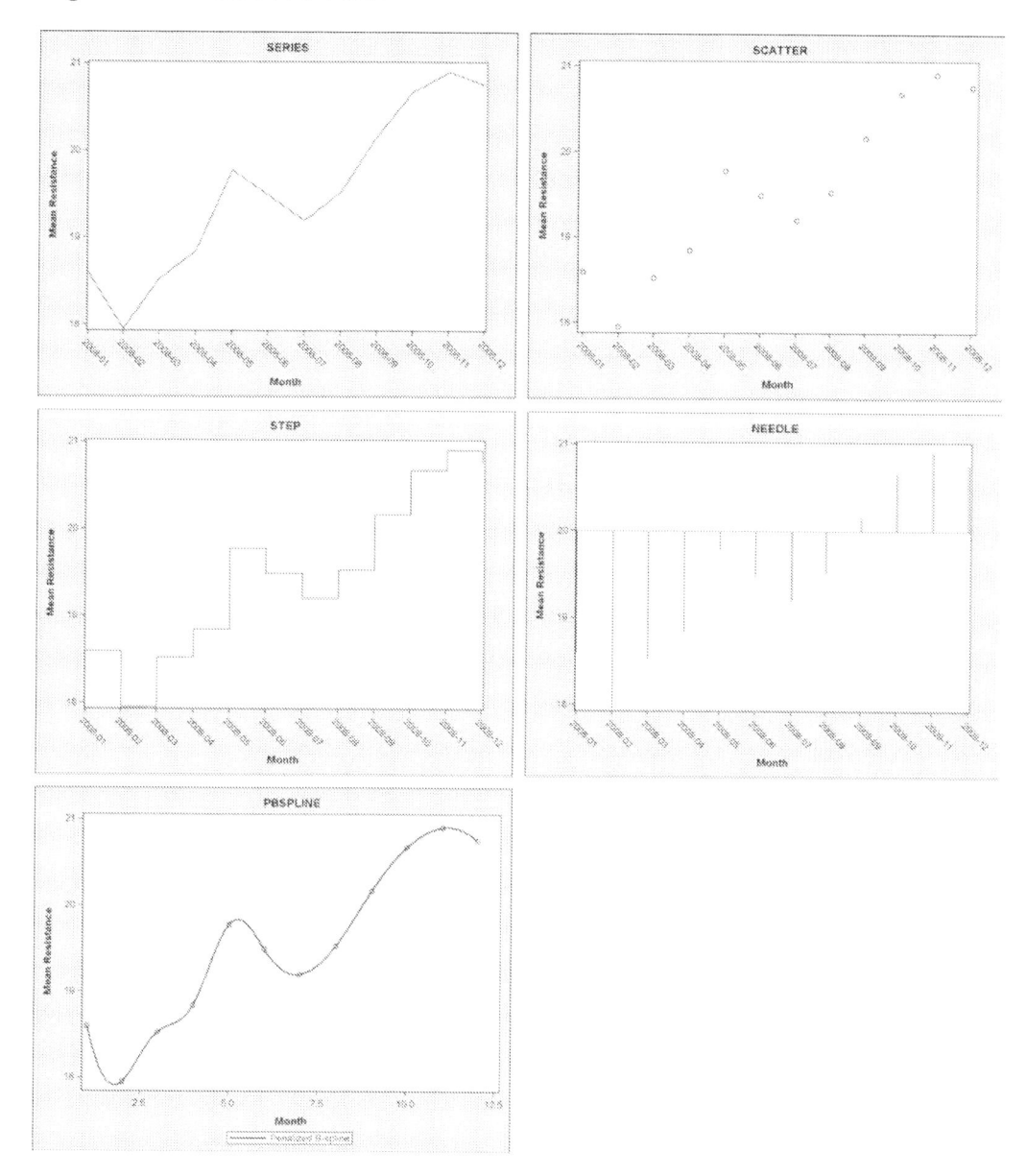

8.3.1.2 Limit Plots

The examples in this section illustrate the DOT, VLINE, and HLINE statements, which optionally include limit lines for each plot point, using the options shown in Table 8-3. You can also add limit lines to plots created with HBAR and VBAR, as shown in Section 8.3.1.3. The code for this section is in ~\JES \sample_code\ch_8 \limits.sas.

Table 8-3: Selected Options for the DOT, HLINE, VLINE, HBAR, and VBAR Statements

Option	Meaning	Possible Values
RESPONSE	Specifies the numeric response variable for the plot.	Any numeric variable in the data set
STAT	Specifies the statistic to be plotted.	FREQ, MEAN, SUM
LIMITS	Add upper and or lower limit lines. Only valid with STAT=MEAN.	BOTH, UPPER, LOWER
LIMITSTAT	Specify the statistic for the limit lines.	CLM, STDDEV, STDERR
ALPHA	[1-Confidence Level] for confidence limits if LIMITSTAT=CLM.	A number between 0 and 1. The default is .05 (95% confidence).
NUMSTD	Number of standard units for the limit lines if LIMITSTAT=STDDEV or STDERR.	Any positive number. The default is 1.

```
ODS HTML PATH="&JES.SG/S_8_3" (URL=NONE) BODY="Limits.html";
ODS GRAPHICS ON / RESET IMAGENAME="F8_3_";
TITLE "DOT";
PROC SGPLOT DATA=JES.Results;
   DOT Vendor /RESPONSE=Resistance STAT=MEAN;   ❶
RUN;

TITLE "DOT with Confidence Limits";
PROC SGPLOT DATA=JES.Results;
   DOT Vendor /RESPONSE=Resistance
     STAT=MEAN LIMITS=BOTH LIMITSTAT=CLM ALPHA=0.10;   ❷
RUN;

TITLE "HLINE";
PROC SGPLOT DATA=JES.Results;
   HLINE Vendor / RESPONSE=Resistance STAT=MEAN;   ❸
RUN;

TITLE "HLINE with Confidence Limits";
PROC SGPLOT DATA=JES.Results;
    HLINE Vendor / RESPONSE=Resistance
      STAT=MEAN LIMITS=BOTH LIMITSTAT=CLM ALPHA=0.10;   ❹
RUN;

TITLE "VLINE";
PROC SGPLOT DATA=JES.Results;
   VLINE Vendor / RESPONSE=Resistance STAT=MEAN;
RUN;

TITLE "VLINE with Confidence Limits";
PROC SGPLOT DATA=JES.Results;
   VLINE Vendor / RESPONSE=Resistance
     STAT=MEAN LIMITS=BOTH LIMITSTAT=CLM ALPHA=0.10;
RUN;

ODS GRAPHICS OFF;
ODS HTML CLOSE;
```

❶ The DOT statement creates a dot plot of the specified statistic, MEAN, of the specified response variable, Resistance, for each value of the variable Vendor.

❷ The second DOT statement uses the LIMITS, LIMITSTAT, and ALPHA options to add 90% confidence limits on the mean resistance.

❸ The HLINE statement creates a horizontal line plot of the specified statistic, MEAN, of the specified response variable, Resistance, for each value of the variable Vendor.

❹ The second HLINE statement uses the LIMITS, LIMITSTAT, and ALPHA options to add 90% confidence limits on the mean resistance.

Figure 8-3: Limit Plots

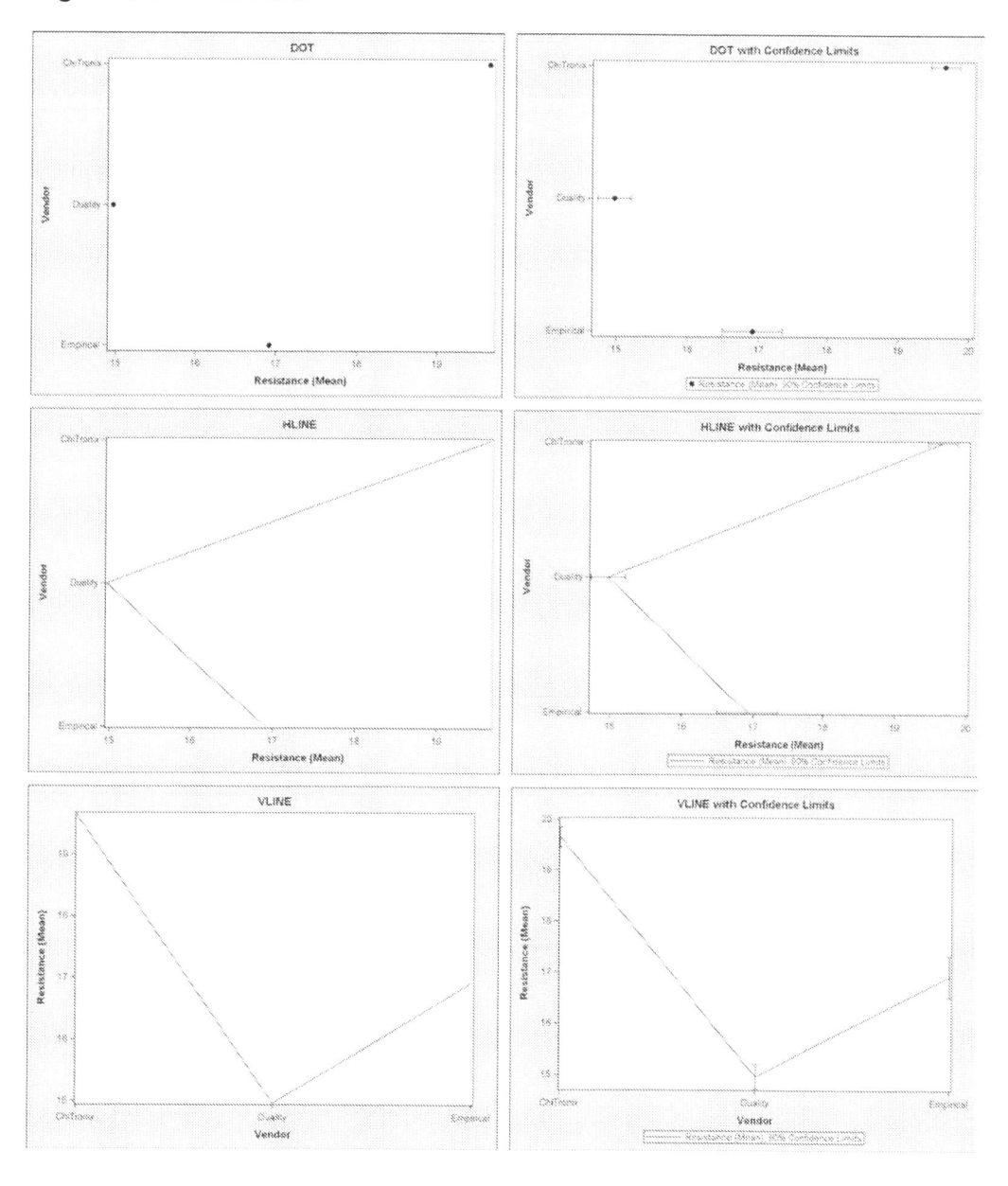

8.3.1.3 Bar Charts

This section illustrates the use of VBAR and HBAR to create vertical and horizontal bar charts, with optional limit lines, and the use of HLINE and VLINE to overlay line plots on the bar charts. Table 8-3 lists some of the VBAR and HBAR options used in these examples. The code for this section is in ~\JES \sample_code\ch_8 \bar.sas.

```
ODS HTML PATH="&JES.SG/S_8_3" (URL=NONE) BODY="Bar.html";
ODS GRAPHICS ON / RESET IMAGENAME="F8_5_";

TITLE "VBAR";
PROC SGPLOT DATA=JES.Results;
   VBAR Vendor /RESPONSE=Resistance STAT=MEAN;    ❶
RUN;

TITLE "HBAR";
PROC SGPLOT DATA=JES.Results;
   HBAR Vendor /RESPONSE=Resistance STAT=MEAN;
RUN;

TITLE "VBAR with Confidence Limits";
PROC SGPLOT DATA=JES.Results;
   VBAR Vendor / RESPONSE=Resistance STAT=MEAN
      LIMITS=BOTH LIMITSTAT=CLM ALPHA=0.10;    ❷
RUN;

TITLE "HBAR with Confidence Limits";
PROC SGPLOT DATA=JES.Results;
   HBAR Vendor / RESPONSE=Resistance STAT=MEAN
      LIMITS=BOTH LIMITSTAT=CLM ALPHA=0.10;
RUN;

TITLE "VBAR and VLINE";
PROC SGPLOT DATA=JES.Results;
   VBAR  Vendor /RESPONSE=Defects STAT=MEAN;    ❸
   VLINE Vendor /RESPONSE=Fail STAT=MEAN;
RUN;

TITLE "HBAR and HLINE";
PROC SGPLOT DATA=JES.Results;
   HBAR  Vendor /RESPONSE=Defects STAT=MEAN;
   HLINE Vendor /RESPONSE=Fail STAT=MEAN;
RUN; QUIT;

ODS GRAPHICS OFF;
ODS HTML CLOSE;
```

❶ The VBAR statement creates a vertical bar plot of the variable Vendor. The RESPONSE and STAT options specify the response variable, Resistance, and the statistic to be plotted, MEAN.

❷ The second VBAR statement uses the LIMITS, LIMITSTAT, and ALPHA options to add 90% confidence limits on the mean resistance.

❸ The VLINE statement can be used with the VBAR statement to create a line plot and a bar chart on the same graph. In this example, the mean number of defects per unit is plotted as a bar chart, and the mean number failing the resistance specification is plotted as a line.

Figure 8-4: Bar Charts

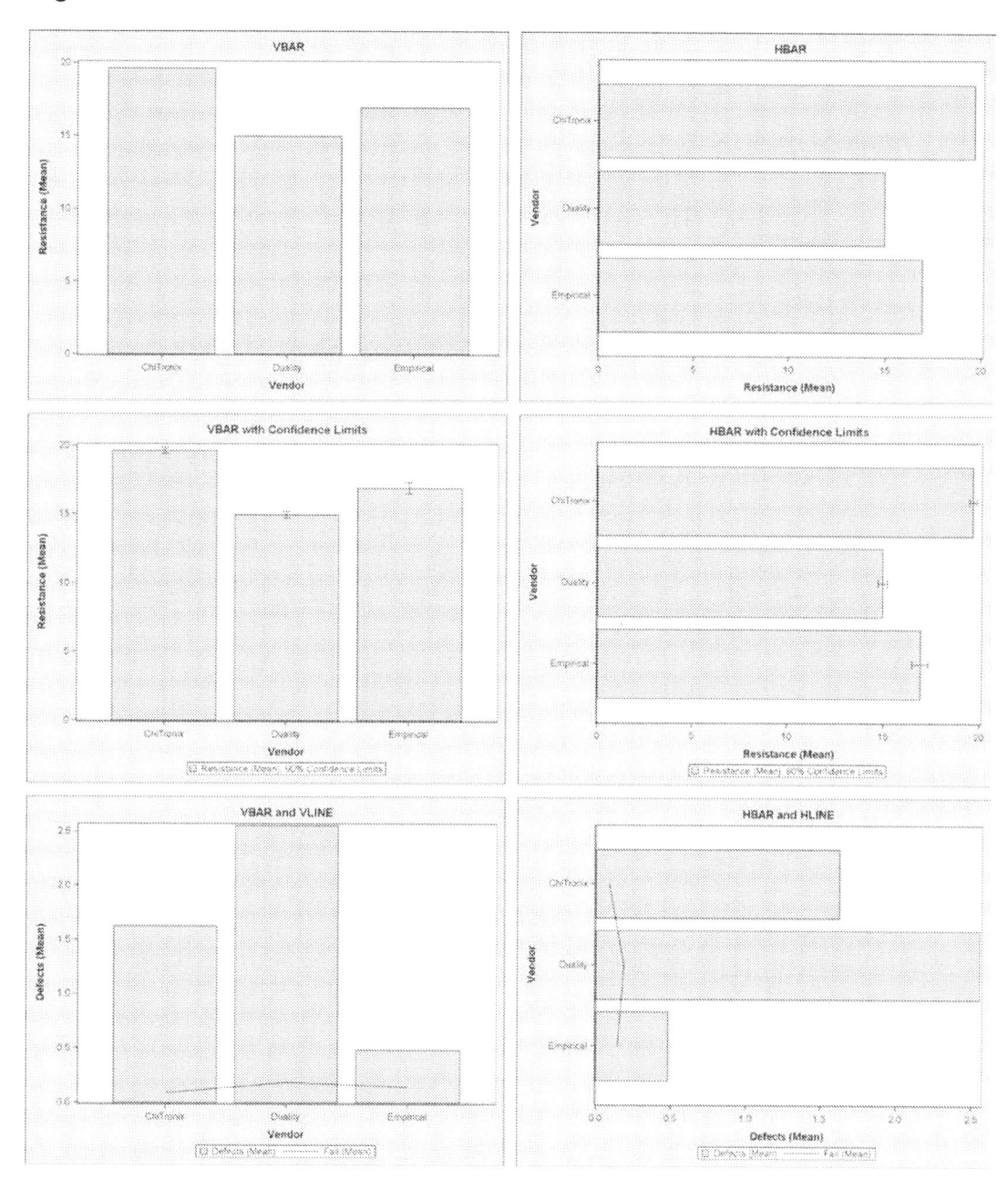

8.3.1.4 Distribution Plots

The plots in this section provide various ways to summarize sample distributions. The code for this section is in ~\JES \sample_code\ch_8 \distribution.sas.

```
ODS HTML PATH="&JES.SG/S_8_3" (URL=NONE) BODY="Distribution.html";
ODS GRAPHICS ON / RESET IMAGENAME="F8_4_";
TITLE "VBOX";
PROC SGPLOT DATA=JES.Results;
   VBOX Resistance / CATEGORY=Vendor;  ❶
RUN;

TITLE "HBOX";
PROC SGPLOT DATA=JES.Results;
   HBOX Resistance / CATEGORY=Vendor;  ❷
RUN;

TITLE "HISTOGRAM";
PROC SGPLOT DATA=JES.Results;
   HISTOGRAM Resistance;  ❸
RUN;

TITLE "DENSITY";
PROC SGPLOT DATA=JES.Results;
   DENSITY Resistance;  ❹
RUN;

TITLE "HISTOGRAM and DENSITY";
PROC SGPLOT DATA=JES.Results;
   HISTOGRAM Resistance;  ❺
   DENSITY Resistance;
RUN;

ODS GRAPHICS OFF;
ODS HTML CLOSE;
```

❶ The VBOX statement specifies a vertical box plot of the Resistance variable for each value of the CATEGORY variable Vendor.

❷ The HBOX statement specifies a horizontal box plot of the Resistance variable for each value of the CATEGORY variable Vendor. The VBOX and HBOX statements are discussed further in Section 8.3.4.2.

❸ The HISTOGRAM statement creates a histogram of the values of the Resistance variable.

❹ The DENSITY statement plots a normal density curve fitted to values of the Resistance variable. You can use the TYPE=KERNEL option to fit a kernel density estimate instead.

❺ The HISTOGRAM and DENSITY statements can be used together to overlay a histogram and fitted density curve on the same graph. This type of plot is discussed in more detail in Section 8.3.4.1.

Figure 8-5: Distribution Plots

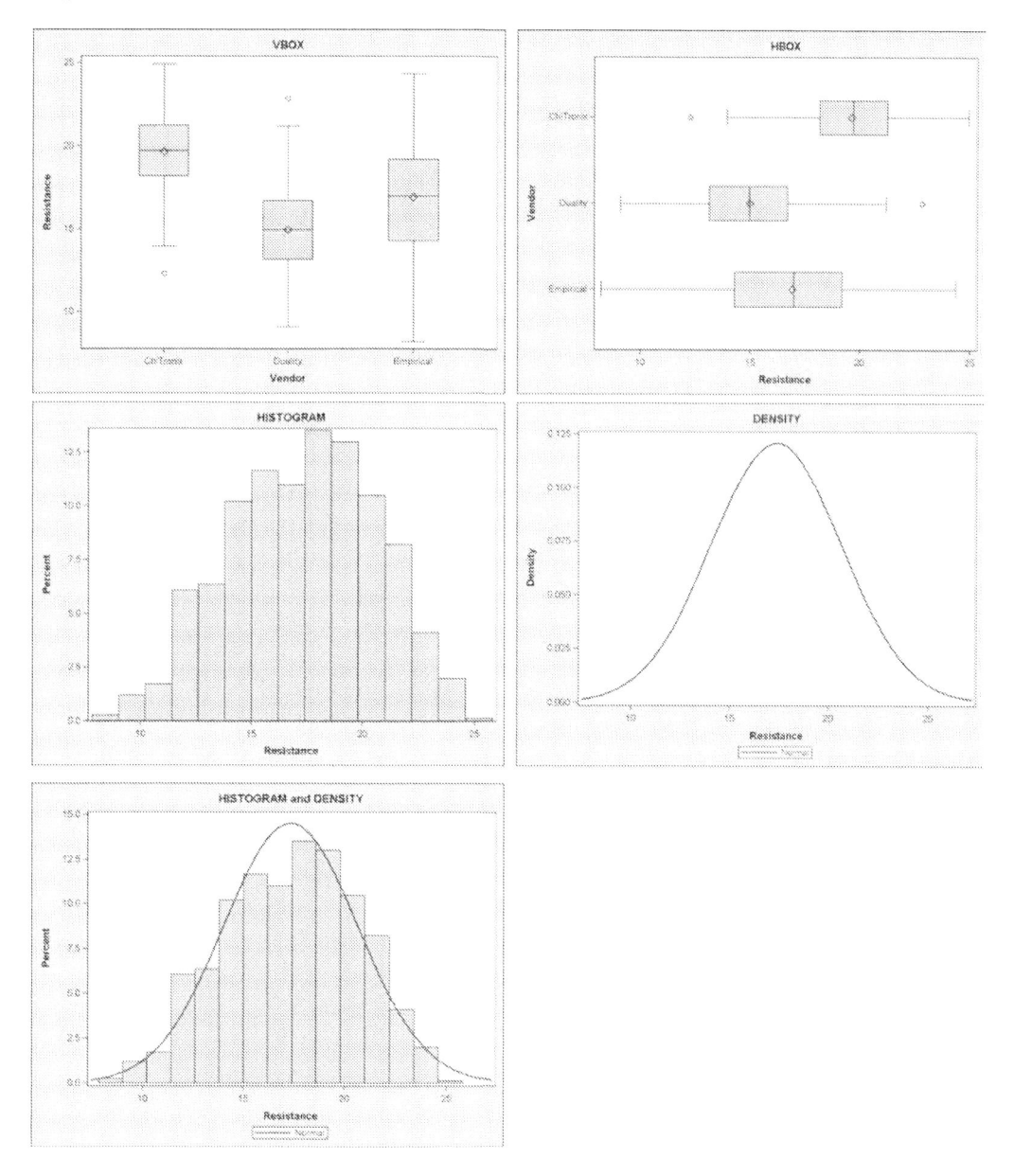

8.3.1.5 Data Fit Plots

This section illustrates some plots that you can use to examine the relationship between two numeric variables. These plots use complex statistical methods to create fitted lines and confidence and prediction limits and ellipses. Consult the SAS online documentation for an explanation of the details of each method. The code for this section is in ~\JES \sample_code\ch_8 \fit.sas.

```
ODS HTML PATH="&JES.SG/S_8_3" (URL=NONE) BODY="Fit.html";
ODS GRAPHICS ON / RESET IMAGENAME="F8_6_";
TITLE "PBSPLINE";
PROC SGPLOT DATA=JES.Results;
   PBSPLINE Y=Delay X=Resistance / CLM CLI DEGREE=3;  ❶
RUN; QUIT;
TITLE "LOESS";
PROC SGPLOT DATA=JES.Results;
   LOESS Y=Delay X=Resistance / CLM INTERPOLATION=CUBIC;  ❷
RUN; QUIT;

TITLE "REG";
PROC SGPLOT DATA=JES.Results;
   REG Y=Delay X=Resistance / CLM CLI DEGREE=3;  ❸
RUN; QUIT;

TITLE "ELLIPSE";
PROC SGPLOT DATA=JES.Results;
   ELLIPSE Y=Delay X=Resistance/TYPE=MEAN FILL FILLATTRS=(COLOR=RED);  ❹
   ELLIPSE Y=Delay X=Resistance /TYPE=PREDICTED;
   SCATTER Y=Delay X=Resistance;
RUN; QUIT;

DATA Results; SET JES.Results;  ❺
    D_Low  = 125 + Resistance + .2*Resistance**2 - 50;
    D_High = 125 + Resistance + .2*Resistance**2 + 50;
RUN;

TITLE "BAND";
PROC SGPLOT DATA=Results;
   BAND X=Resistance UPPER=D_High LOWER=D_Low;  ❻
   SCATTER Y=Delay X=Resistance;
RUN; QUIT;

ODS GRAPHICS OFF;
ODS HTML CLOSE;
```

❶ The PBSPLINE statement fits a penalized B-spline to the sample (X, Y) data points. The DEGREE=3 option specifies the degree of the spline transformation. The CLM and CLI options request confidence limits and prediction limits.

❷ The LOESS statement fits a loess curve to the sample (X,Y) data points. The INTERPOLATION=CUBIC option specifies the degree of the interpolating polynomials. The CLM option requests confidence limits on the mean.

❸ The REG statement fits a regression line to the sample (X, Y) data points. The DEGREE=3 option specifies the degree of the fitted polynomial. The CLM and CLI options request confidence limits and prediction limits.

❹ The first ELLIPSE statement creates a bivariate normal confidence ellipse for the population mean Resistance and Delay. The FILL and FILLATTRS options are used to fill the ellipse because it is very small and difficult to see among the points of the scatter plot. The second

ELLIPSE statement uses the TYPE=PREDICTED option to get a prediction ellipse for a new observation. The SCATTER statement is not required, but is included to show the same points as on the other plots.

❺ The DATA statement adds the D_Low and D_High variables to the Results data set in order to illustrate the BAND statement. These variables were created just to illustrate the syntax of the BAND statement. In practical application, you might have theoretical limits on Delay as a function of Resistance, and use a BAND statement to test the validity of the theoretical relationship, or to detect outliers.

❻ The BAND statement creates a shaded area between the values of the UPPER and LOWER variables. The SCATTER statement is not required, but is included to show the same points as on the other plots.

Figure 8-6: Data Fit Plots

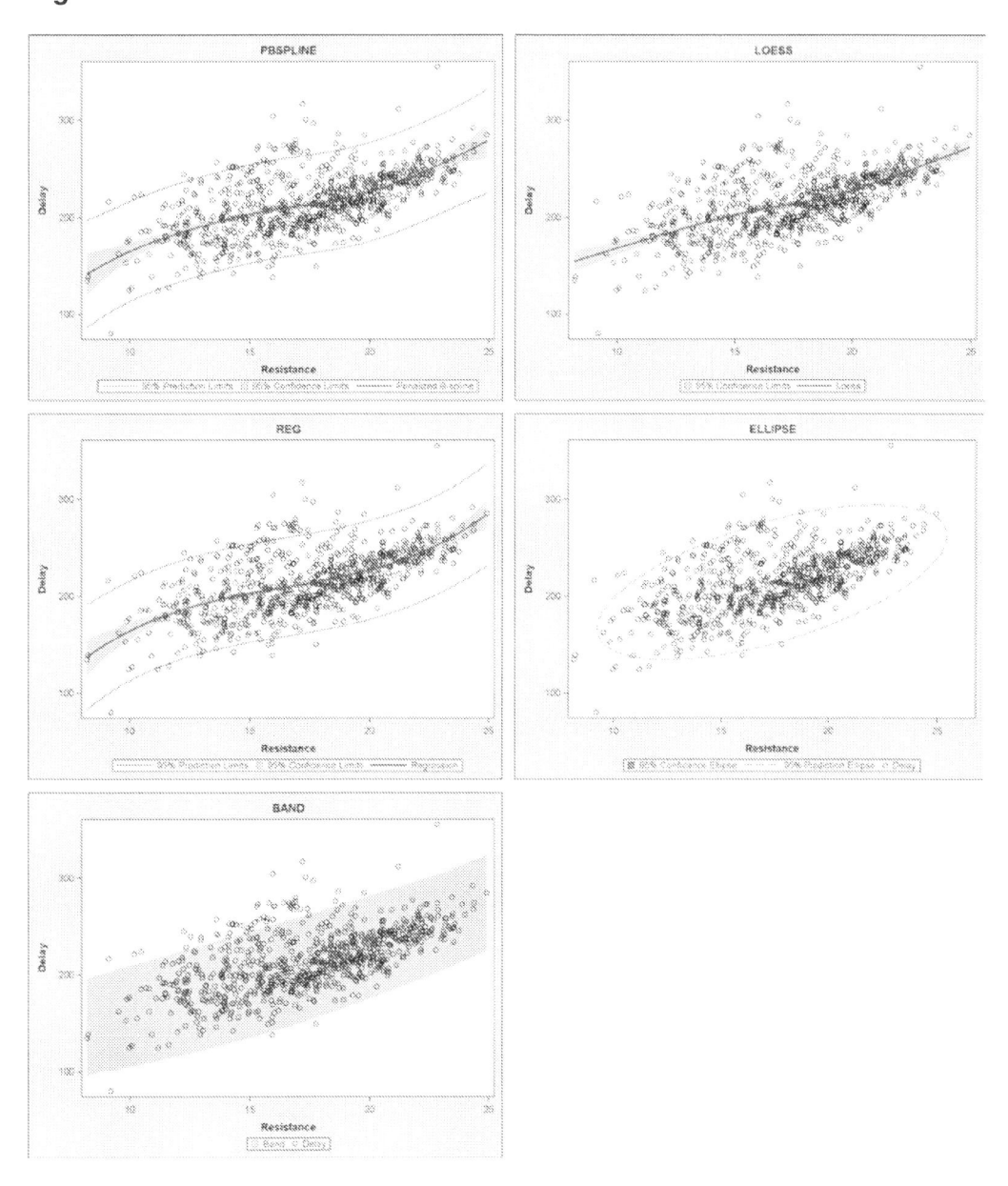

8.3.2 Plotting Multiple Lines on the Same Graph

In the previous section there were some examples of putting multiple plots on the same graph, for example by using an ELLIPSE statement and a SCATTER statement in the same invocation of PROC SGPLOT. In this section we discuss:

- how to use the GROUP or BY option to create multiple plots with a single plot statement
- how to use multiple plot statements within SGPLOT
- how to specify a second X or Y axis for your plots

8.3.2.1 Using the BY and GROUP Options

It is often necessary to create a separate plot for each distinct value of a variable in the data set. You can use the BY statement to create each plot on a separate graph, or the GROUP option to overlay all the plots on a single graph. The code for this section is in ~\JES \sample_code\ch_8 \by_group.sas.

```
ODS HTML PATH="&JES.SG/S_8_3" (URL=NONE) BODY="By_Group.html";
ODS GRAPHICS ON / RESET IMAGENAME="F8_7_";

TITLE "Mean Resistance by Month";
PROC SGPLOT DATA=JES.Results_Tab; BY Vendor;   ❶
  SERIES Y=M_Res X=Mon;
RUN;

ODS GRAPHICS / RESET IMAGENAME="F8_8_";
PROC SGPLOT DATA=JES.Results_Tab;
  SERIES Y=M_Res X=Mon / GROUP=Vendor;   ❷
RUN;

ODS GRAPHICS OFF;
ODS HTML CLOSE;
```

❶ The first PROC SGPLOT uses a BY statement to create a separate plot for each distinct value of MU, with each plot on a separate graph. A BY statement can be used with any of the SGPLOT plot statements. The three plots created by this code are shown in Figure 8-7. Note that subtitles containing the value of Vendor have been automatically added to each plot.

❷ The second PROG SGPLOT uses the GROUP option to create a separate plot for each distinct value of Vendor, with all plots overlaid on the same graph. The GROUP option can be used with the SERIES, SCATTER, STEP, PBSPLINE, NEEDLE, DOT, HBAR, VBAR, HLINE, VLINE, BAND, LOESS, and REG statements, but not with the HISTOGRAM, DENSITY, HBOX, VBOX, or ELLIPSE statements. The resulting graph is shown in Figure 8-8. The three different plots are distinguished by different symbols and identified by the legend.

Figure 8-7: Multiple Plots Using a BY Statement

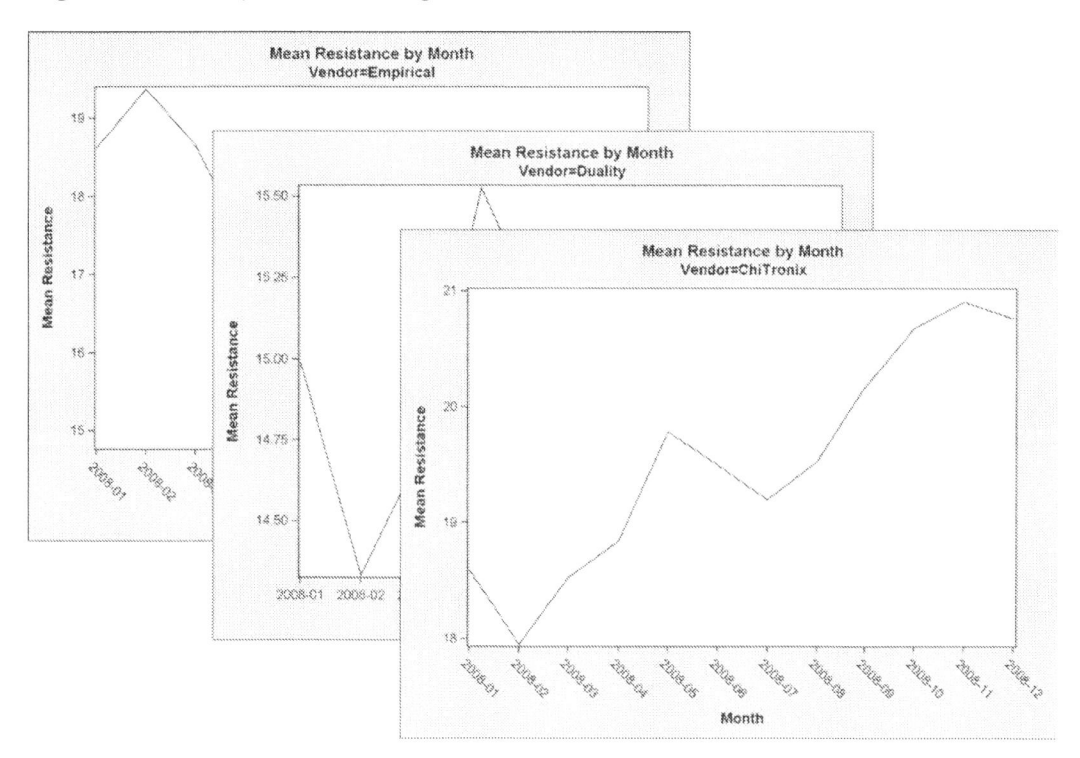

Figure 8-8: Multiple Plots Using the GROUP Option

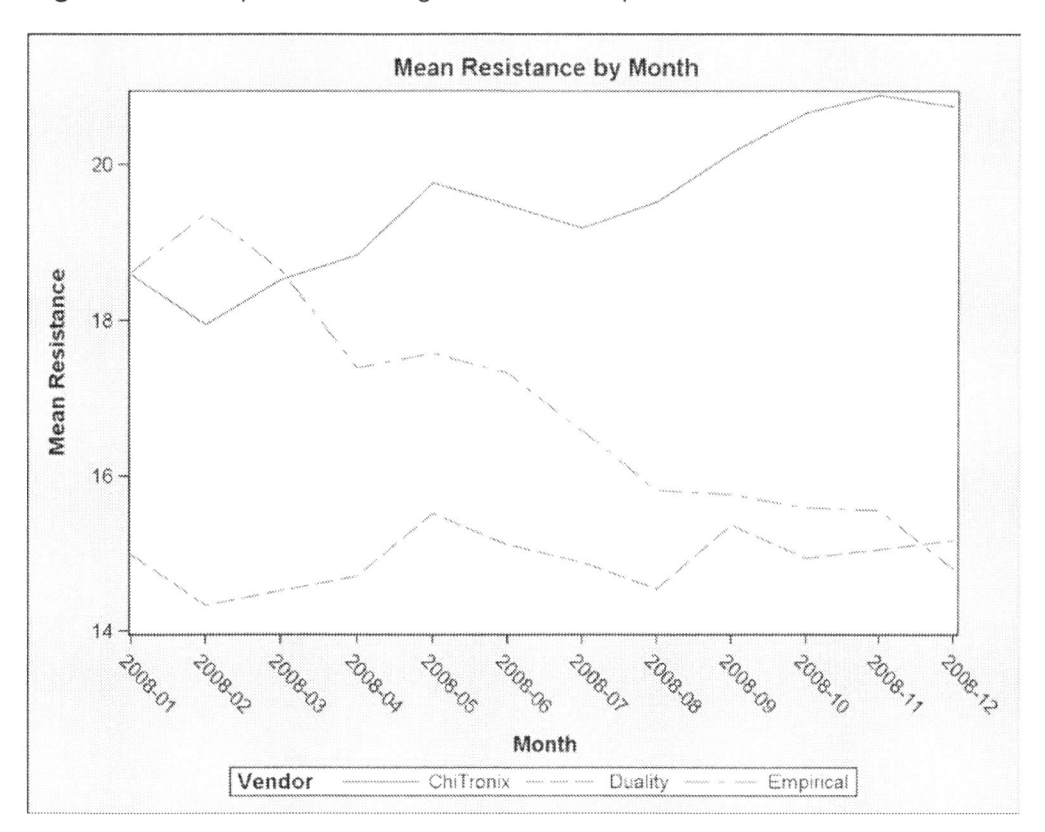

8.3.2.2 Using Multiple Plot Statements

You can use more than one SGPLOT plot statement to create multiple plots in the same graph. You have already seen examples of overlaying a SCATTER plot with a BAND or ELLIPSE plot in Section 8.3.1.5, and a few more examples are provided here. The code for this section is in ~\JES \sample_code\ch_8 \overlay.sas. You can generally overlay different plots types whenever it would make sense to do so, but there are a few restrictions:

- HISTOGRAMs and DENSITY plots may only be combined with each other.
- A VBOX or HBOX plot cannot be used with any other plot type.
- VBAR/VLINE and HBAR/HLINE/DOT plots may only be combined with each other.

```
ODS HTML PATH="&JES.SG/S_8_3" (URL=NONE) BODY="Overlay.html";
ODS GRAPHICS ON / RESET IMAGENAME="F8_9_";

TITLE "Distribution of Resistance";
PROC SGPLOT DATA=JES.Results;  ❶
  HISTOGRAM Resistance;
  DENSITY Resistance / TYPE=NORMAL(MU=15 SIGMA=3);
  DENSITY Resistance / TYPE=KERNEL;
RUN;

ODS GRAPHICS / RESET IMAGENAME="F8_10_";
TITLE "Mean Resistance and Delay by Month for ChiTronix";
PROC SGPLOT DATA=JES.Results_Tab(WHERE=(Vendor="ChiTronix"));  ❷
  SERIES Y=M_Res   X=Mon;
  SERIES Y=M_Del   X=Mon;
RUN;

ODS GRAPHICS OFF;
ODS HTML CLOSE;
```

❶ The first SGPLOT overlays three plots using the resistance values from the JES.Results data set. The HISTOGRAM statement plots a histogram of the resistance values. The first DENSITY statement plots the density of a Normal distribution with mean=15 and Sigma=3. If the (MU=15 SIGMA=3) option is left out, the best fitting normal density, based on the maximum likelihood estimates of the mean and standard deviation, is plotted. The second DENSITY statement plots a kernel density function fit to the resistance data. The plot is shown in Figure 8-9. A legend is automatically added to identify the DENSITY plots. A more detailed discussion of histogram and density plots is given in Section 8.3.4.1.

❷ The second SGPLOT overlays SERIES plots of the mean resistance (M_RES) and the mean delay (M_Del) by month for the ChiTronix units. The plot is shown in Figure 8-10. Note that the legend uses the label of each variable to identify the lines.

The problem with the plot in Figure 8-10 is that the two variables plotted, M_Res and M_Del, are of very different magnitudes. So when they are plotted on the same scale, the M_Res trend is very difficult to see. In Section 8.3.2.3 you will see how to create a more effective graph by plotting each variable on a different Y axis.

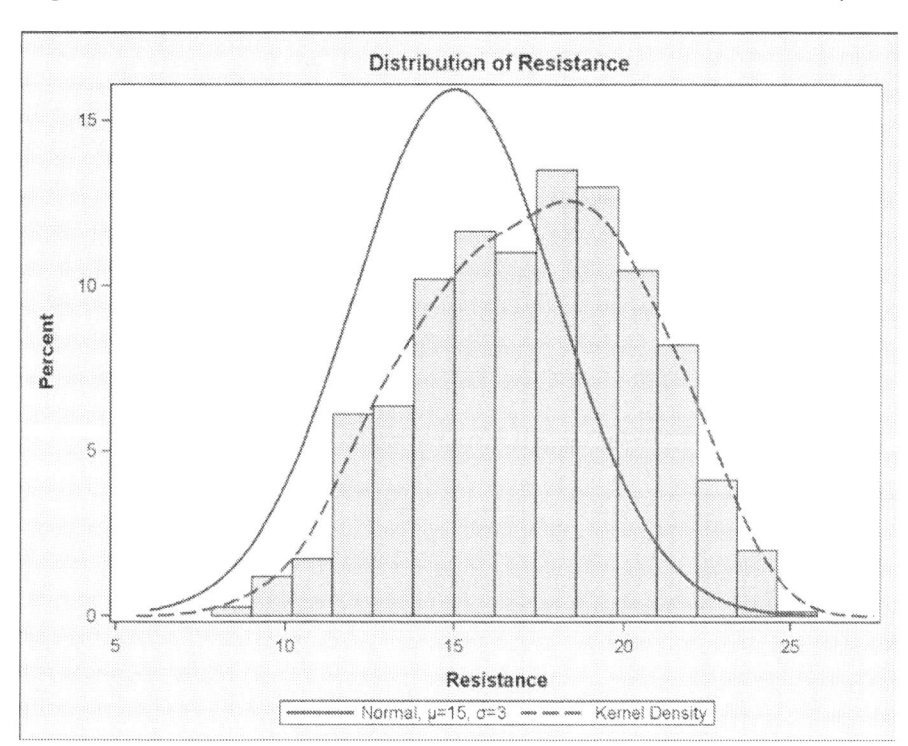

Figure 8-9: HISTOGRAM and DENSITY Plots on the Same Graph

Figure 8-10: Multiple SERIES Plots on the Same Graph

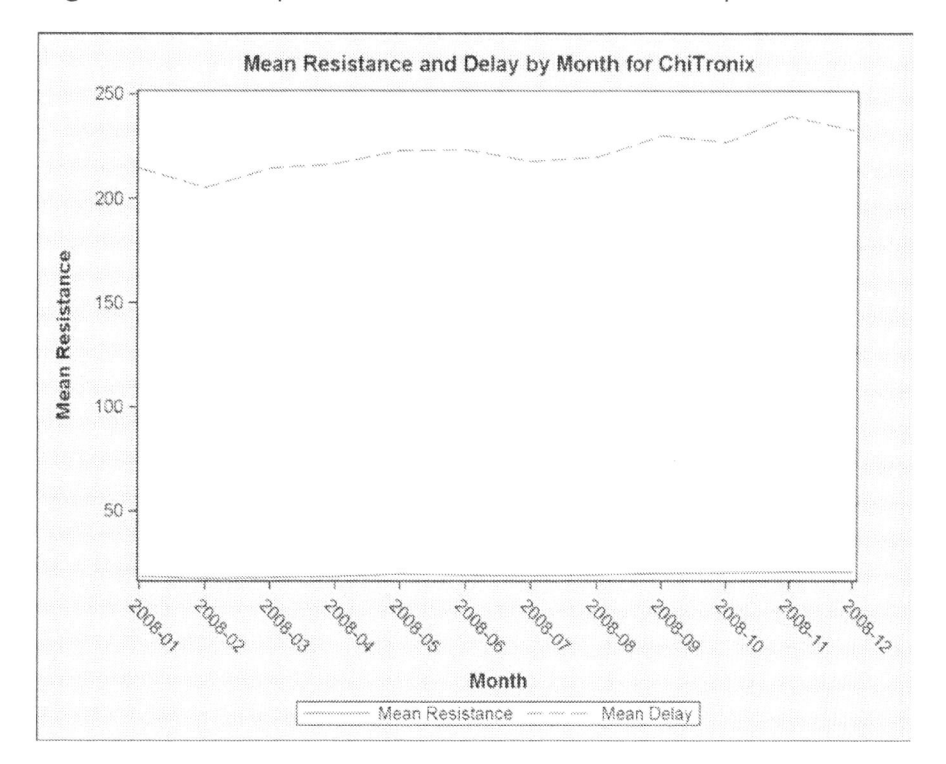

8.3.2.3 Using a Second X or Y Axis

When two different kinds of data are plotted on the same graph, it is often appropriate to choose different Y axis scales for the different plots. A typical example is Figure 8-10 where the Y axis is acceptable for plotting the M_Del variable, but not for plotting the M_Res variable, because the values of M_Res are so much smaller. You can correct this problem by using the Y2AXIS option to create a different Y axis for the M_Del plot. The code for this section is in ~\JES \sample_code\ch_8 \two_axes.sas. The Y2AXIS and X2AXIS options can be used with all of the SGPLOT statements except DOT, HBAR, and VBAR.

```
ODS HTML PATH="&JES.SG/S_8_3" (URL=NONE) BODY="Two_Axes.html";
ODS GRAPHICS ON / RESET IMAGENAME="F8_11_";

TITLE "Mean Resistance and Delay by Month";
PROC SGPLOT DATA=JES.Results_Tab(WHERE=(Vendor="ChiTronix"));
   SERIES Y=M_Res  X=Mon ;
   SERIES Y=M_Del  X=Mon / Y2AXIS; ❶
RUN;
```

❶ The Y2AXIS option specifies that the plot of M_Del will be on a second Y axis. The values of the second Y axis are displayed on the right side of the plot.

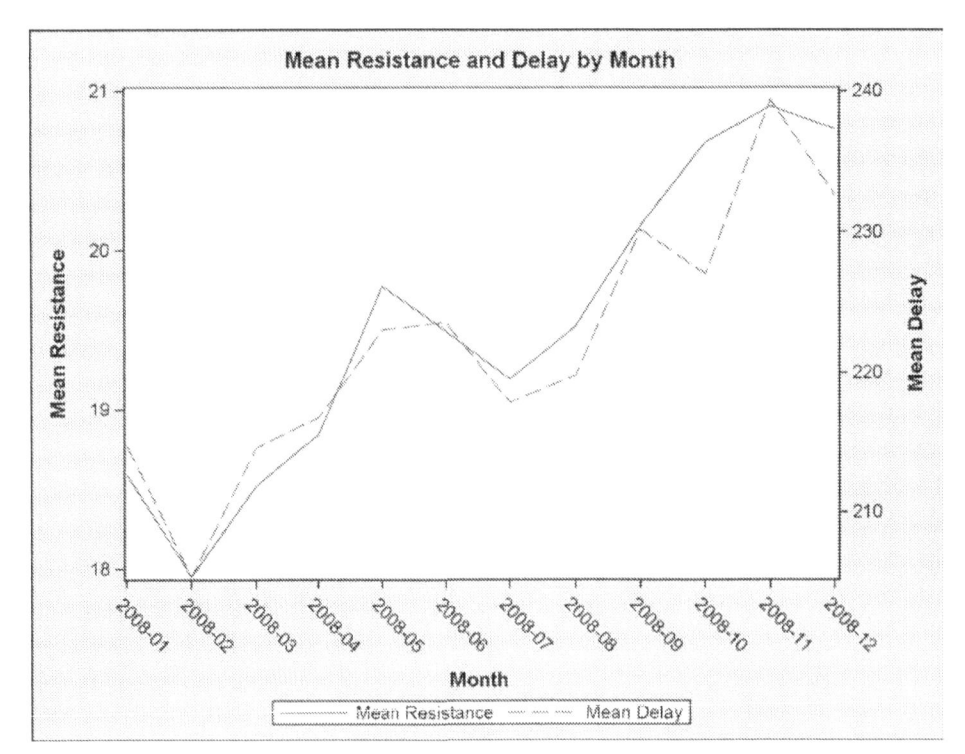

Figure 8-11: Using the Y2AXIS Option

The next example uses the GROUP option together with Y2AXIS to create separate plots of M_Res and M_Del for each vendor, with the M_Res plots using the axis on the left, and the M_Del plots using a second Y axis shown on the right side.

```
ODS GRAPHICS / RESET IMAGENAME="F8_12_";
TITLE "Mean Resistance and Delay by Vendor and Month";
PROC SGPLOT DATA=JES.Results_Tab;
   SERIES Y=M_Res  X=Mon / GROUP=Vendor;
   SERIES Y=M_Del  X=Mon / Y2AXIS GROUP=Vendor;
RUN;

ODS GRAPHICS OFF;
ODS HTML CLOSE;
```

Figure 8-12: Using the Y2AXIS Option with the GROUP Option

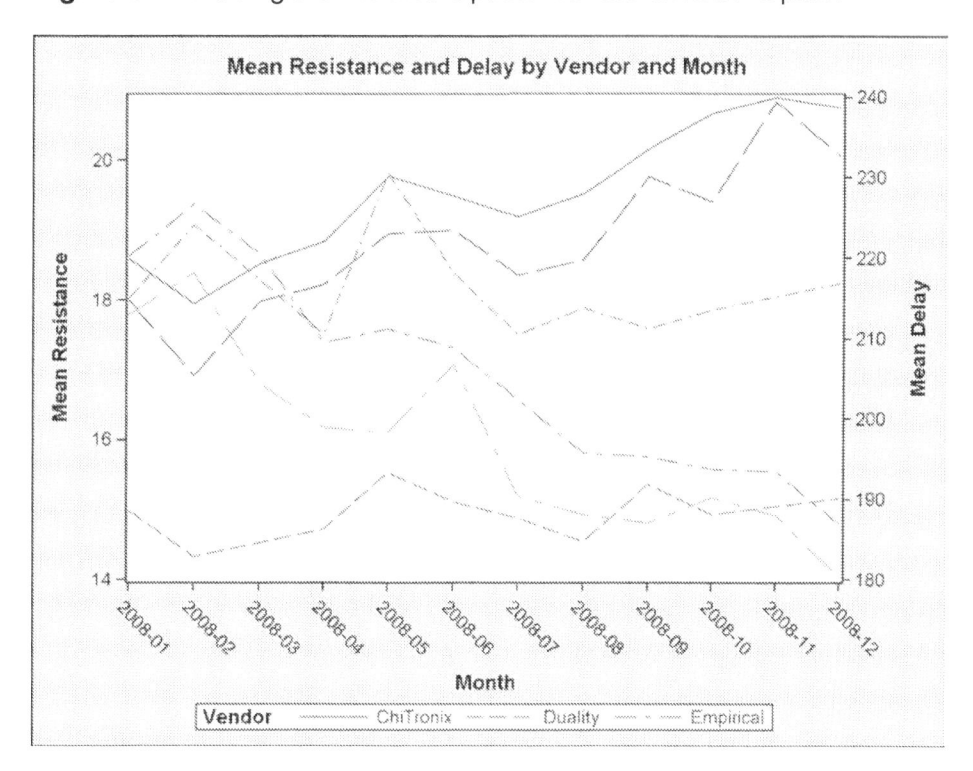

The difficulty with this plot is that the legend gives no indication of which lines are the M_Res plots and which are the M_Del plots. The example in Section 8.3.3.6 shows how to use options to identify the various lines in this plot.

8.3.3 Options to Customize SGPLOT Output

The SGPLOT procedures do not recognize any of the options described in Section 6.3.2 except for the TITLE and FOOTNOTE statements. The code for this section is in ~\JES \sample_code\ch_8 \sgplot_options.sas.

8.3.3.1 SGPLOT Statements and Options

Table 8-2 includes several statements that you can use to customize your SGPLOT output, and Table 8-4 includes many of the options that you can use with the SGPLOT statements. Note that not all options are valid with every plot statement—for example, the FILL option cannot be used with a SERIES statement because there is no area to fill. Consult the SAS online documentation to see which options work with each statement, and what other options are available. The following sections illustrate the use of these options to control the appearance of your graphs.

Table 8-4: Selected Options Used with SGPLOT

Option	Meaning	Possible Values
CYCLEATTRS	Specifies that all plots are drawn with unique attributes, e.g., symbols or line types.	(no value required)
LEGENDLABEL	Specifies a label that identifies the markers in the plot.	Any text string.
LINEATTRS	Specifies the appearance of lines using a style reference and/or one or more of the suboptions: COLOR, PATTERN, THICKNESS.	(COLOR=Red PATTERN=Dot THICKNESS=1)
MARKERS	Add markers (symbols) to each SERIES or fit plot.	(no value required)
NOMARKERS	Remove markers from the SERIES or fit plot.	(no value required)
MARKERATTRS	Specifies the appearance of markers using a style reference and/or one or more of the suboptions: COLOR, SYMBOL, SIZE.	(COLOR=Red SYMBOL=Star SIZE=10)
FILL	Specifies that the bands, histograms, or bars should be filled.	(no value required)
NOFILL	Specifies that the bands, histograms, or bars should not be filled.	(no value required)
FILLATTRS	Specifies the color of the fill.	(COLOR=Green)
OUTLINE	Specifies that the bands, histograms, or bars should be outlined.	(no value required)
NOOUTLINE	Specifies that the bands, histograms, or bars should not be outlined.	(no value required)
TRANSPARENCY	Specifies the degree of transparency for lines, symbols, or area fill.	A number from 0 (opaque) to 1 (transparent).
DATALABEL	Displays a label for each data point.	Any variable in the data set. Default is the Y variable.
CURVELABEL	Specifies a label for plot line.	Any text string. The default is the label of the Y variable.
CURVELABELLOC	Location of the label relative to the graphic area.	INSIDE, OUTSIDE
CURVELABELPOS	Location of the label relative to the X axis.	MIN, MAX, START, END
ALPHA	[1-Confidence Level] for confidence limits.	A number between 0 and 1. The default is .05 (95% confidence level).

8.3.3.2 Titles and Footnotes

TITLE and FOOTNOTE statements work exactly as described in Section 6.3.2.2. The code sample here uses the same TITLE and FOOTNOTE statement as the example in Section 6.3.2.2, and creates a plot, shown in Figure 8-13, which is similar to Figure 6-13.

```
ODS HTML PATH="&JES.SG/S_8_3" (URL=NONE) BODY="sgplot_options.html";
ODS GRAPHICS ON / RESET IMAGENAME="F8_13_";

/*==== Titles and Footnotes =======*/
TITLE1 "Mean Resistance by Vendor";
TITLE2 COLOR= Blue "12 Month Period Beginning Jan. 2008";
FOOTNOTE1 JUSTIFY=LEFT "John Doe" JUSTIFY=RIGHT "BigCorp";
FOOTNOTE2   "Do "
   COLOR=Red UNDERLIN=2 "NOT"
   COLOR=Black UNDERLIN=0 "  Distribute This Report";
PROC SGPLOT DATA =JES.Results_Tab;
   SERIES Y=M_Res X=Mon / GROUP=Vendor;
RUN;
```

Figure 8-13: Using TITLE and FOOTNOTE Statements

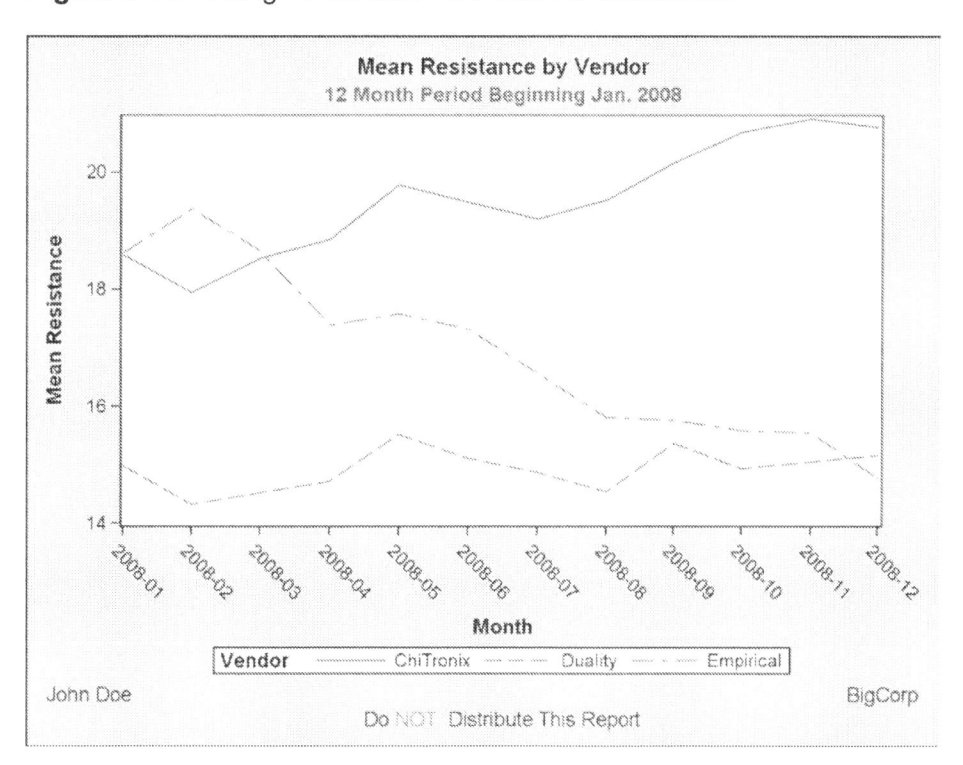

8.3.3.3 Axes

There are four AXIS statements that you can use to specify the range and appearance of each of the axes in your plots. Table 8-5 lists the AXIS statements and some of the more useful AXIS statement options.

Table 8-5: AXIS Statements and Selected AXIS Statement Options

AXIS Statements	
XAXIS	Specifies options for the X axis.
X2AXIS	Specifies options for the X2 axis.
YAXIS	Specifies options for the Y axis.
Y2AXIS	Specifies options for the Y2 axis.
Selected AXIS Statement Options		
Option	Meaning	Some Possible Values
DISPLAY	Specifies which features of the axis are displayed.	ALL, NONE, NOLABEL, NOLINE, NOTICKS, NOVALUES
FITPOLICY	Specifies the method used to fit the tick mark values when there is not enough room.	ROTATE, ROTATETHIN, STAGGER, STAGGERROTATE, STAGGERTHIN, THIN
GRID	Creates gridlines at each tick value.	(no value required)
LABEL	Sets the text used to label the axis.	Any text string, enclosed in quotes.
MIN	Minimum data value for the axis.	Any number
MAX	Maximum data value for the axis.	Any number
REFTICKS	Adds tick marks to the axis opposite from the specified axis.	(no value required)
TYPE	Specifies the type of axis.	DISCRETE, LINEAR, LOG, TIME
VALUES	Values for the ticks on the axis.	(0 TO 20 BY 2)
VALUESHINT	Specifies that the minimum and maximum data values for the axis are determined independently of the tick values specified with the VALUES option. The tick values from the VALUES option are displayed only if they are between the minimum and maximum data values.	(no value required)

The next bit of code in sgplot_options.sas illustrates the use of several of these options.

```
/*===== Axes ==========*/
ODS GRAPHICS / RESET IMAGENAME="F8_14_";
TITLE1 "Mean Resistance and Delay by Vendor";  ❶
FOOTNOTE;
PROC SGPLOT DATA =JES.Results_Tab;
   SERIES Y=M_Res   X=Mon / GROUP=Vendor ;
   SERIES Y=M_Del   X=Mon / GROUP=Vendor Y2AXIS;
   YAXIS   LABEL = 'Mean Resistance'  ❷
        VALUES=(0 TO 30 BY 1)   FITPOLICY=THIN;
   Y2AXIS LABEL = 'Mean Delay' MIN=180 MAX=250 GRID;  ❸
   XAXIS   LABEL="Month of Production"  ❹
        FITPOLICY=ROTATE REFTICKS;
RUN;
```

❶ The TITLE and FOOTNOTE statements create a title and remove the footnotes from the last example.

❷ The YAXIS statement controls the attributes of the Y axis on the left. The LABEL option specifies the text used to label the axis. Note that the label text is automatically rotated 90 degrees. The VALUES option specifies the tick marks for this axis. The FITPOLICY=THIN, which is the default for this axis, specifies that the tick mark values will be "thinned" if there is not enough room to display them all on the axis. In this case, only every other value is displayed.

❸ The Y2AXIS statement controls the attributes of the Y axis on the right. The MIN and MAX options specify the data range of the axis. The GRID option specifies that grid lines should be added at the tick marks.

❹ The XAXIS statement controls the attributes of the X axis. FITPOLICY=ROTATE specifies that the tick mark values will be rotated if there is not enough room to display them all on the axis. Since the axis is displaying character data, this option is not needed here because the values are rotated by default (as for example in Figure 8-13). The REFTICKS option specifies that the tick marks should also be displayed on the opposite axis line, in this case at the top of the plot area.

Figure 8-14: Using XAXIS, YAXIS, and Y2AXIS Statements

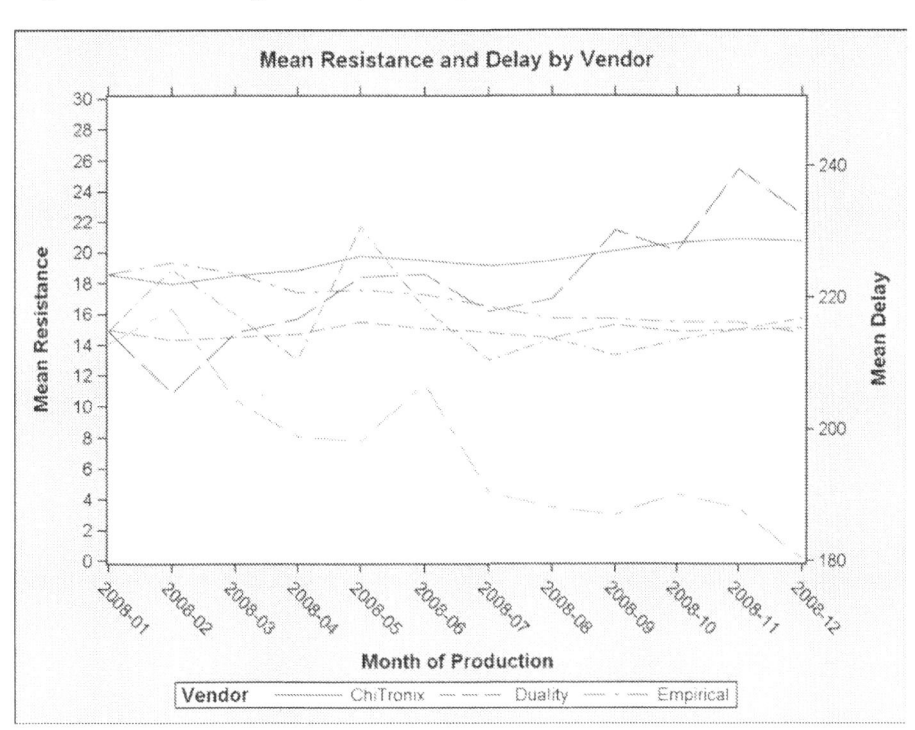

8.3.3.4 Symbols and Lines

You can use the options in Table 8-6 to customize the symbol and line type for each plot on your graph.

Table 8-6: Selected Symbol and Line Options

Option	Meaning	Some Possible Values
CYCLEATTRS	Specifies that all plots are drawn with unique attributes, e.g., symbols or line types.	(no value required)
MARKERS	Adds markers (symbols) to each SERIES or fit plot.	(no value required)
MARKERATTRS	Specifies the appearance of symbols using a style reference and/or the suboptions: COLOR, SYMBOL, SIZE.	(COLOR=Red SYMBOL=Star SIZE=10)
LINEATTRS	Specifies the appearance of lines using a style reference and/or the suboptions: COLOR, PATTERN, THICKNESS.	(COLOR=Red PATTERN=Dot)

```
/*===== SYMBOLs and LINEs =====*/
ODS GRAPHICS / RESET IMAGENAME="F8_15_";
PROC SGPLOT DATA =JES.Results_Tab;
   SERIES Y=M_Res  X=Mon / GROUP=Vendor
   LINEATTRS =(PATTERN=Solid THICKNESS=3);   ❶
   SERIES Y=M_Del  X=Mon / GROUP=Vendor Y2AXIS
   LINEATTRS = (PATTERN=Dash THICKNESS=2)   ❷
      MARKERS
      MARKERATTRS=(COLOR=Red SIZE=10 SYMBOL=Triangle);
   YAXIS  LABEL = 'Mean Resistance'
      VALUES=(0 TO 30 BY 1) FITPOLICY=THIN;
   Y2AXIS LABEL = 'Mean Delay' MIN=180 MAX=250 GRID;
   XAXIS  LABEL = "Month of Production"
      FITPOLICY=ROTATE REFTICKS;
RUN;
```

❶ The first SERIES statement uses the LINEATTRS option to specify that the M_Res plot points should be joined wth a solid line of thickness 3.

❷ The second SERIES statement uses the LINEATTRS, MARKERS, and MARKERATTRS options to specify that the M_Del plot points be marked with red triangles, and be joined with a dashed line of thickness 2. The symbols and line types that you can use with these options are shown in Displays 8-2 and 8-3.

Display 8-2: Symbols to Use with the MARKERATTRS Option

```
List of Marker Symbols

↓  ArrowDown     ▽  HomeDown    ∩  Tilde      ●  CircleFilled
✻  Asterisk      I  Ibeam       △  Triangle   ◆  DiamondFilled
○  Circle        +  Plus        ∪  Union      ⬟  HomeDownFilled
◇  Diamond       □  Square      ×  X          ■  SquareFilled
>  GreaterThan   ☆  Star        Y  Y          ★  StarFilled
#  Hash          T  Tack        Z  Z          ▲  TriangleFilled
```

Display 8-3: Patterns to Use with the LINEATTRS Option

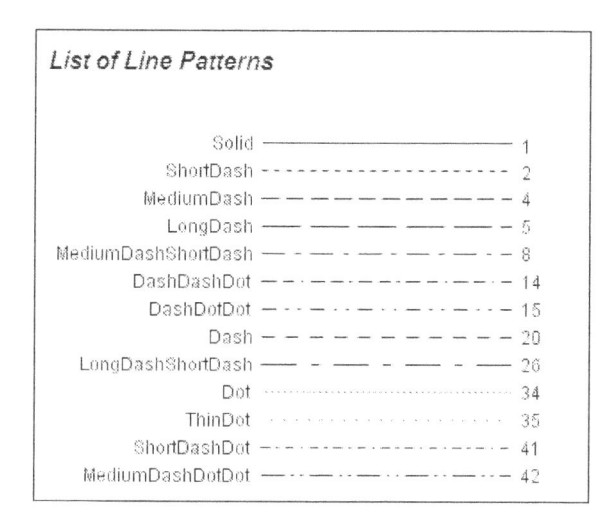

List of Line Patterns

Pattern		Number
Solid	───────	1
ShortDash	- - - - - - -	2
MediumDash	─ ─ ─ ─	4
LongDash	── ── ──	5
MediumDashShortDash	─ - ─ - ─ -	8
DashDashDot	─ ─ · ─ ─ ·	14
DashDotDot	─ · · ─ · ·	15
Dash	─ ─ ─ ─ ─	20
LongDashShortDash	── - ── -	26
Dot	· · · · · ·	34
ThinDot	· · · · · ·	35
ShortDashDot	- · - · - ·	41
MediumDashDotDot	─ · · ─ · ·	42

Figure 8-15: Using the LINEATTRS, MARKERS, and MARKERATTRS Options

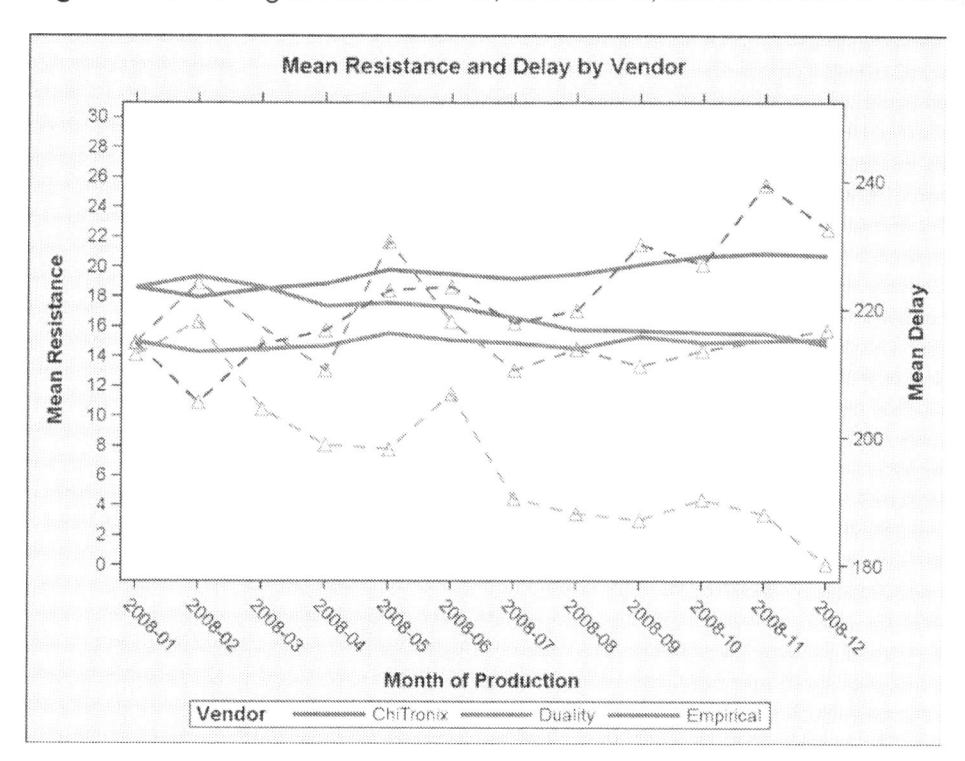

8.3.3.5 Legends

You can identify the plots in your graph using a KEYLEGEND statement as illustrated here, or using the CURVELABEL option as illustrated in the next section. The options for the KEYLEGEND statement are shown in Table 8-7.

Table 8-7: KEYLEGEND Options

Option	Meaning	Some Possible Values
ACROSS	Specifies the number of columns in the legend.	Any integer
DOWN	Specifies the number of rows in the legend.	Any integer
BORDER	Specifies that a border be drawn around the legend.	(no value required)
NOBORDER	Specifies that no border be drawn around the legend.	(no value required)
LOCATION	Specifies the location of the legend relative to the plot area.	OUTSIDE, INSIDE
POSITION	Specifies the position of the legend within the plot.	N, S, E, W, NE, NE, SE, SW, TOP, BOTTOM, RIGHT, LEFT, TOPRIGHT, TOPLEFT, BOTTOMRIGHT, BOTTOMLEFT
TITLE	Specifies a title for the legend.	Any text string

This code uses a KEYLEGEND statement to add a legend to the plot similar to the plot created in Section 8.3.2.3 (Figure 8-11).

```
/*===== LEGENDs ==========*/
ODS GRAPHICS / RESET IMAGENAME="F8_16_";
TITLE1 "Mean Resistance and Delay by Month: ChiTronix";
PROC SGPLOT DATA =JES.Results_Tab(WHERE=(Vendor="ChiTronix"));
   SERIES Y=M_Res  X=Mon / NAME='a'  ❶
   LINEATTRS =(PATTERN=Solid THICKNESS=3)
       LEGENDLABEL="Resistance"; ❷
   SERIES Y=M_Del  X=Mon / NAME='b' Y2AXIS
   LINEATTRS = (PATTERN=Dot THICKNESS=2)
       MARKERS
       MARKERATTRS=(COLOR=Red SIZE=10 SYMBOL=Triangle)
       LEGENDLABEL="Delay";
   YAXIS  LABEL = 'Mean Resistance'
       VALUES=(0 TO 30 BY 1)   FITPOLICY=THIN;
   Y2AXIS LABEL = 'Mean Delay' MIN=180 MAX=250 GRID;
   XAXIS  LABEL = "Month of Production"
       FITPOLICY=ROTATE REFTICKS;
   KEYLEGEND 'a' 'b'  / TITLE='Measurement'  ❸
       ACROSS=1 LOCATION=INSIDE POSITION=NW;
RUN;
```

❶ The NAME option is used in each SERIES statement to identify the plot so that it can be referenced in the KEYLEGEND statement.

❷ The LEGENDLABEL option is used in each SERIES statement to specify the name to be used in the legend. If this option is omitted, the variable label, e.g., 'Mean Resistance' will be used.

❸ The KEYLEGEND statement defines the attributes of the legend. The 'a' 'b' specifies the names of the plot lines to include in the legend. The TITLE option specifies a title for the legend. The LOCATION=INSIDE option specifies that the legend be placed inside of the graphic area. The POSITION=NW option specifies that the legend be placed in the Northwest (NW) corner of the graph.

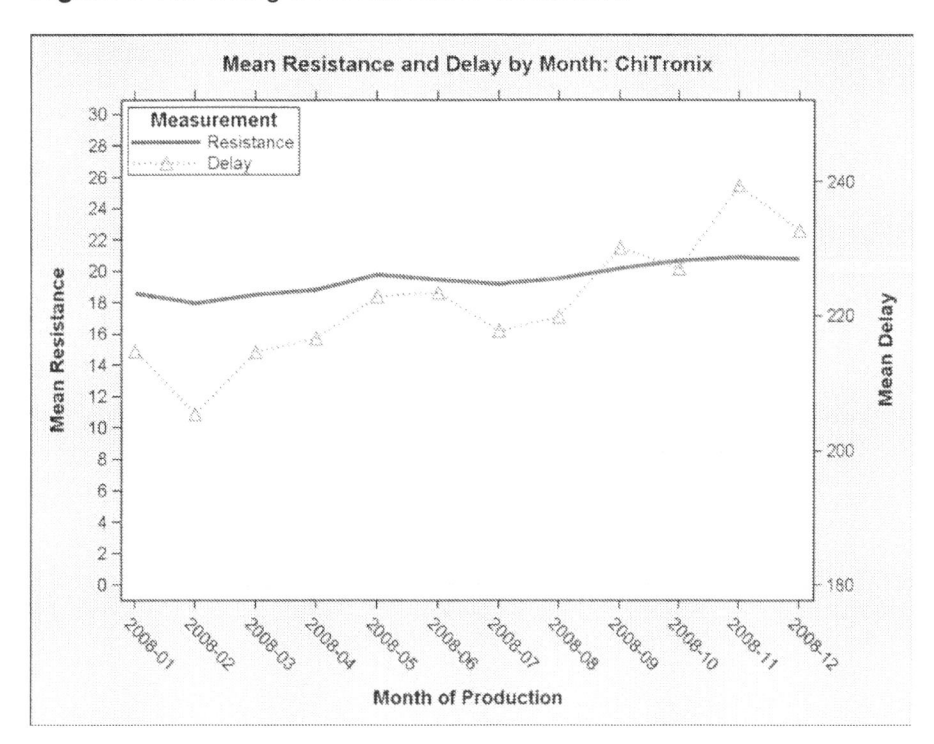

Figure 8-16: Using a KEYLEGEND Statement

8.3.3.6 Curve Labels

Instead of adding a legend to identify the lines in your plot, you can use the CURVELABEL and related options to add a label to each line, either before the first plot point or after the last plot point. The CURVELABEL options are shown in Table 8-8.

Table 8-8: CURVELABEL Options

Option	Meaning	Some Possible Values
CURVELABEL	Specifies the label for plot line.	Any text string. The default is the label of the Y variable or the value of the GROUP variable if the GROUP option is used.
CURVELABELLOC	Specifies whether the labels are placed inside or outside of graphic area.	INSIDE, OUTSIDE
CURVELABELPOS	Specifies whether the labels are placed before the minimum value of the X variable, or after the maximum value.	MIN, MAX, START, END

This code uses CURVELABEL to add labels to the plot from the example in Section 8.3.3.4 (Figure 8-15).

```
/*===== Curve Labels ==========*/
ODS GRAPHICS / RESET IMAGENAME="F8_17_";
TITLE1 "Mean Resistance and Delay by Month and Vendor";
TITLE2 "Solid Lines are Resistance, Dashed Lines are Delay";
PROC SGPLOT DATA =JES.Results_Tab;
   SERIES Y=M_Res  X=Mon / GROUP=Vendor NAME='a'
   LINEATTRS =(PATTERN=Solid THICKNESS=3)
       LEGENDLABEL="Resistance"
       CURVELABEL  ❶
       CURVELABELLOC=INSIDE  ❷
       CURVELABELPOS=MAX;  ❸
   SERIES Y=M_Del  X=Mon /GROUP=Vendor NAME='b' Y2AXIS
   LINEATTRS = (PATTERN=Dot THICKNESS=2)
       MARKERS
       MARKERATTRS=(COLOR=Red SIZE=10 SYMBOL=Triangle)
       LEGENDLABEL="Delay"
       CURVELABEL
       CURVELABELLOC=INSIDE
       CURVELABELPOS=MIN;  ❹
   YAXIS  LABEL = 'Mean Resistance'
       VALUES=(0 TO 30 BY 1)  FITPOLICY=THIN;
   Y2AXIS LABEL = 'Mean Delay' MIN=180 MAX=250 GRID;
   XAXIS  LABEL = "Month of Production"
         FITPOLICY=ROTATE REFTICKS;
   KEYLEGEND 'a' 'b'  / TITLE='Measurement'
         ACROSS=3 LOCATION=INSIDE POSITION=NW;
RUN;
```

❶ The CURVELABEL option adds a label to each plot. You can specify the text to be used for the labels, and if you do not, the default label is used. The default is the Y variable label or, if the GROUP option is used, as in this example, the value of the GROUP variable.

❷ The CURVELABELLOC=INSIDE option specifies that the labels should be inside the graphic area.

❸ The CURVELABELPOS=MAX option specifies that the labels should appear after the last plot point.

❹ For the second SERIES statement, the CURVELABELPOS=MIN option specifies that the labels should appear before the first plot point so that the labels for the Delay plots do not interfere with the labels for the Resistance plots.

The resulting plot is shown in Figure 8-17. Note that white space has been added to the left and right of the plot points to accommodate the labels. A second title statement is used to distinguish the Resistance lines from the Delay lines. The CURVELABEL option works best when there is separation between the plot lines at the maximim or minimum plot points. The line colors can also be used to distinguish the different plot lines in the original plot, but not, of course, in this gray-scale reproduction.

Figure 8-17: Using CURVELABEL Statements

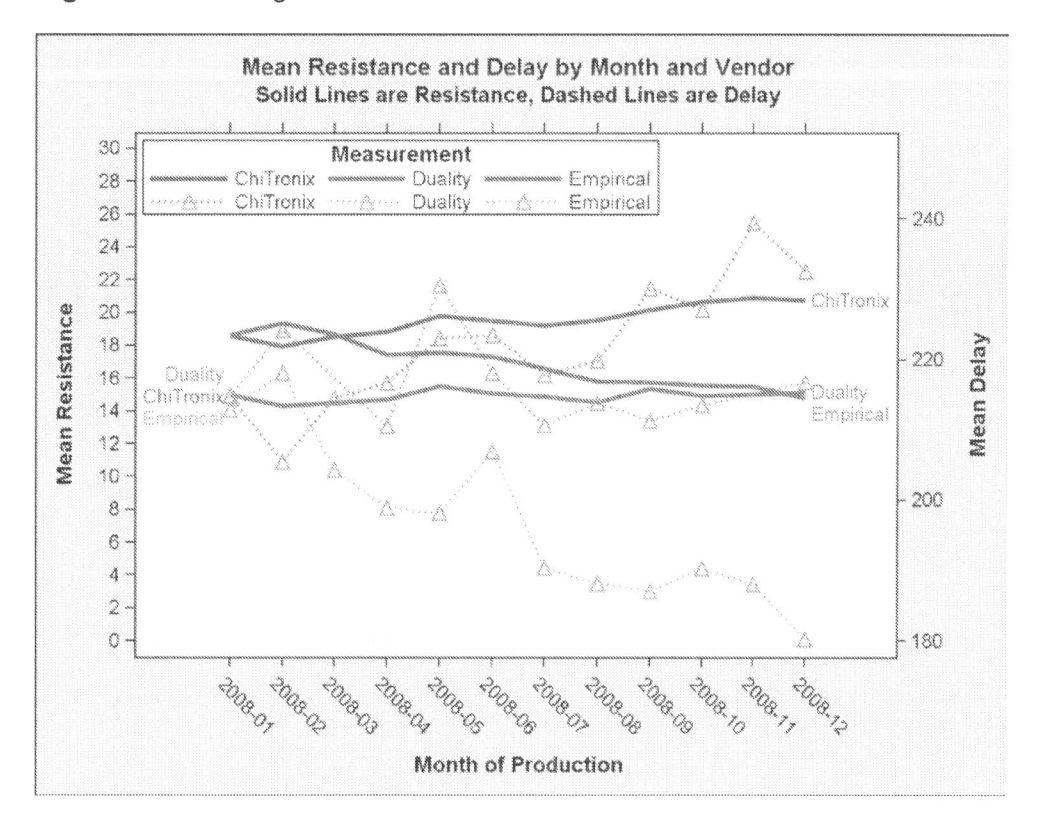

8.3.3.7 Reference Lines

You can use REFLINE statements to add reference lines to one or more of your plot axes. Some of the options used with the REFLINE statement are shown in Table 8-9.

Table 8-9: Selected REFLINE Statement Options

Option	Meaning	Some Possible Values
AXIS	Specifies the axis which contains the reference line values.	X, Y, X2, Y2
LABEL	Specifies the text for the reference line.	Any text. The default is the reference value.
LABELLOC	Location of label—inside or outside of graphic area.	INSIDE, OUTSIDE
LABELPOS	Location of label—min or max value of the axis.	MIN, MAX, START, END
LINEATTRS	Specifies the appearance of the reference line using a style reference and/or one or more of the suboptions: COLOR, PATTERN, THICKNESS.	(COLOR=Red PATTERN=Dot THICKNESS=1)

This example uses REFLINE statements to add the upper and lower specification limits to a scatter plot of Delay vs. Resistance.

```
/*===== Reference Lines ==========*/
ODS GRAPHICS / RESET IMAGENAME="F8_18_";
TITLE1 "Delay vs Resistance by Vendor: Months 10-12";
TITLE2 "Specification Limits: Resistance(12.5-22.5) Delay(150-250)";
PROC SGPLOT DATA=JES.Results_Q4;
  SCATTER Y=Delay X=Resistance / GROUP=Vendor;
  XAXIS  VALUES=(5 TO 25 BY 5);
  YAXIS  VALUES=(100 TO 350 BY 50);
  REFLINE 12.5 22.5 / AXIS=X LABEL=("LSL" "USL") ❶
      LINEATTRS=(COLOR=Red PATTERN=Dash)
      LABELLOC=INSIDE LABELPOS=MIN;
  REFLINE 150 250 / AXIS=Y LABEL=("LSL" "USL") ❷
      LABELLOC=OUTSIDE LABELPOS=MAX;
RUN;
```

❶ The first REFLINE statement adds reference lines at the values 12.5 and 22.5, which are the specification limits for the Resistance variable. The AXIS=X option specifies that the line will be drawn perpendicular to the X axis. The LABEL option specifies the text used to label the reference lines. The LINEATTRS option specifies the color and pattern used for the lines. The LABELLOC=INSIDE option specifies that the label will be drawn inside the graphic area. The LABELPOS=MIN option specifies that the label will be drawn at minimum value of the Y axis.

❷ The second REFLINE statement adds reference lines at the values 150 and 250, which are the specification limits for the Delay variable. The AXIS=Y option specifies that the line will be drawn perpendicular to the Y axis. The LABEL option specifies the text used to label the reference lines. The LABELLOC=OUTSIDE option specifies that the label will be drawn outside the graphic area. The LABELPOS=MAX option specifies that the label will be drawn at the maximum value of the X axis.

The resulting plot is shown in Figure 8-18.

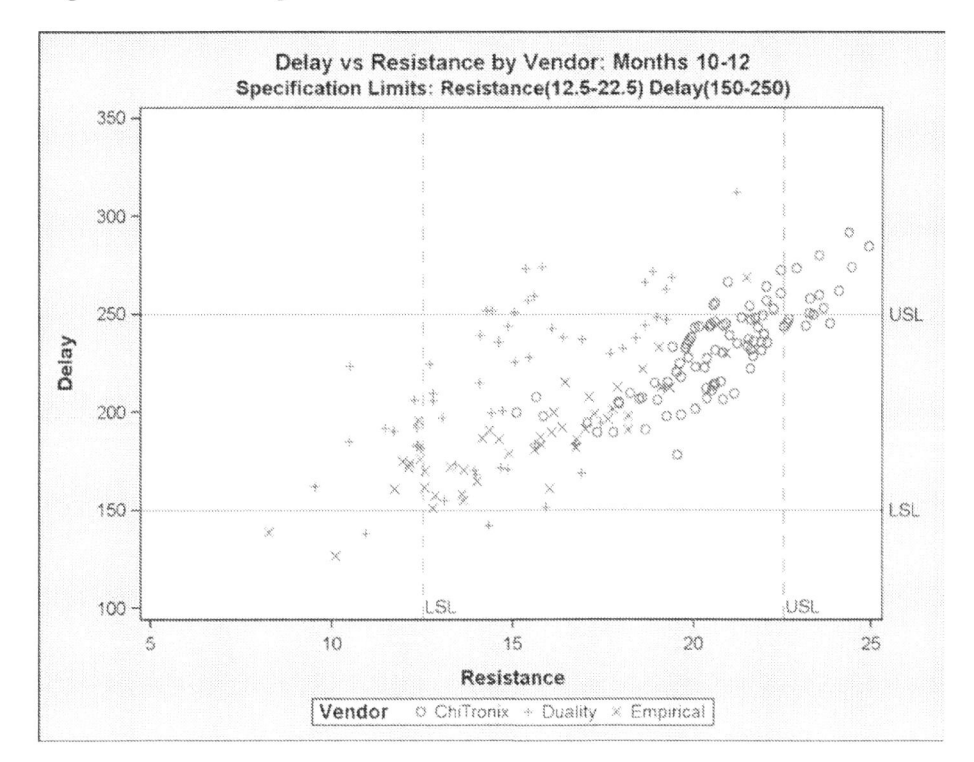

Figure 8-18: Using REFLINE Statements

8.3.3.8 Insets

You can use an INSET statement to add a text box within the area of your plot. Selected options for the INSET statement are shown in Table 8-10.

Table 8-10: Selected INSET Statement Options

Option	Meaning	Some Possible Values
BORDER	Specifies that a border should be drawn around the inset text.	(no value required)
NOBORDER	Specifies that a border should not be drawn around the inset text.	(no value required)
LABELALIGN	Specifies the alignment of labels when you specify label-value pairs.	LEFT, CENTER, RIGHT
VALUEALIGN	Specifies the alignment of values when you specify label-value pairs.	LEFT, CENTER, RIGHT
POSITION	Specifies the position of the inset within the graphic area.	N, S, E, W, NE, SE, NW, SW
TITLE	Specifies a title for the inset.	Any text string

An INSET statement can specify a single text string to be written on one line, or multiple text strings to be written on separate lines. You can also specify a set of label-value pairs, with each pair to be written on a separate line. This code adds an inset table to the plot from Section 8.3.3.7 (Figure 8-18) that includes information about the geographic location of each of the component vendors.

```
/*===== Insets =====*/
ODS GRAPHICS ON / RESET IMAGENAME="F8_19_";
PROC SGPLOT DATA=JES.Results_Q4;
  SCATTER Y=Delay X=Resistance / GROUP=Vendor;
  XAXIS  VALUES=(5 TO 25 BY 5);
  YAXIS  VALUES=(100 TO 350 BY 50);
  REFLINE 12.5 22.5 / AXIS=X LABEL=("LSL" "USL")
      LINEATTRS=(COLOR=Red PATTERN=Dash)
      LABELLOC=INSIDE LABELPOS=MIN;
  REFLINE 150 250 / AXIS=Y LABEL=("LSL" "USL")
      LABELLOC=OUTSIDE LABELPOS=MAX;
  INSET ("ChiTronix"="APAC" "Duality"="EMEA" "Empirical"="AMER") / ❶
  BORDER TITLE="Vendor Locations" ❷
      POSITION=NW ❸
      LABELALIGN=LEFT ❹
      VALUEALIGN=RIGHT;
RUN;

ODS GRAPHICS OFF;
ODS HTML CLOSE;
```

❶ The INSET statement specifies the label-value pairs to be written as an inset.

❷ The BORDER option specifies that a border will be drawn around the text. The TITLE option specifies a title for the inset.

❸ The POSITION=NW option specifies that the inset be placed in the top left (Northwest) of the graphic area.

❹ The LABELALIGN=LEFT option specifies that the labels will be left justified, and the VALUEALIGN=RIGHT option specifies that the values will be right justified. These are the default values for these options, so they do not need not be specified explicitly, but they are included in the code sample to illustrate the syntax.

The resulting plot is shown in Figure 8-19.

Figure 8-19: Using an INSET Statement

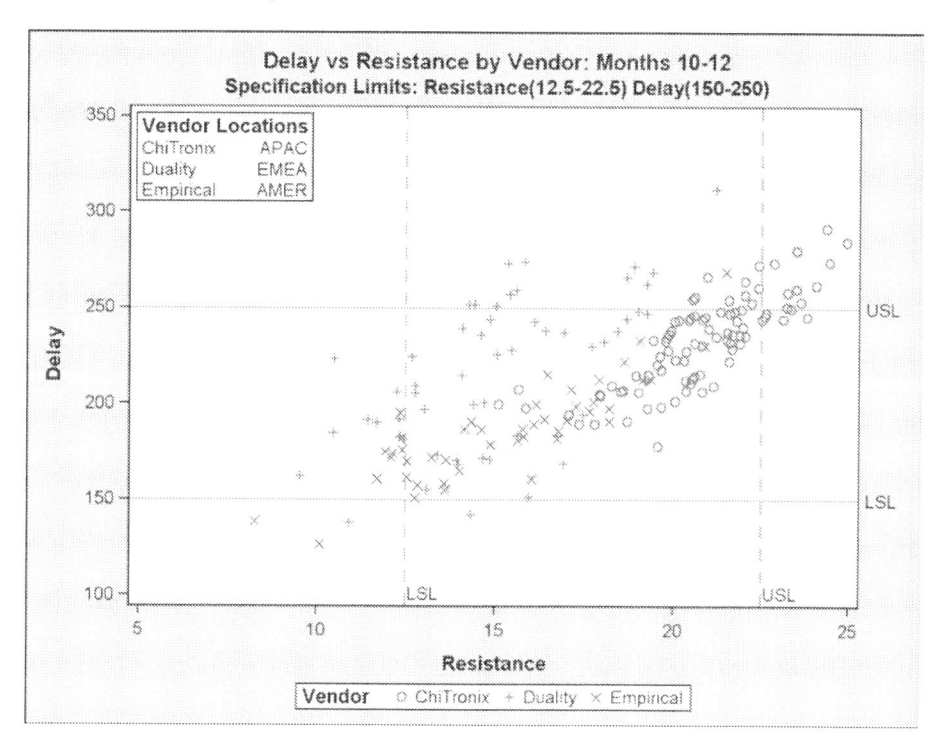

8.3.4 SGPLOT Examples

The previous sections describe all of the plot types available in PROC SGPLOT, how to overlay multiple plots on the same graph, and how to customize the plot symbols, legends, labels, etc. This brief review covers many, but not all, of the options available for each of the plot types. This section includes examples illustrating some of the other options available for the HISTOGRAM, DENSITY, HBOX, and VBOX statements, and an example showing how to plot group confidence limits using HLINE, SERIES, and SCATTER statements.

8.3.4.1 HISTOGRAM and DENSITY

Section 8.3.2.2 includes an example of combining a histogram with density plots on the same graph. This section continues that example, with more detail on the options available for the HISTOGRAM and DENSITY statements. The code for this section is in ~\JES \sample_code\ch_8 \histo_density.sas.

Table 8-11: Selected HISTOGRAM and DENSITY Statement Options

Selected HISTOGRAM Options		
Option	**Meaning**	**Some Possible Values**
BOUNDARY	Specifies whether boundary values are assigned to the lower or upper bin.	LOWER, UPPER
OUTLINE	Specifies that the bars should be outlined.	(no value required)
NOOUTLINE	Specifies that the bars should not be outlined.	(no value required)
FILL	Specifies that the bars should be filled.	(no value required)
NOFILL	Specifies that the bars should not be filled.	(no value required)
FILLATTRS	Specifies the color of the fill.	(COLOR=Green)
SCALE	Specifies the type of axis for the response axis.	COUNT, PERCENT, PROPORTION
Selected DENSITY Options		
TYPE	Specifies the type of density estimate. If TYPE=NORMAL, you can also specify the mean and sigma with the MU and SIGMA suboptions. If you omit these suboptions, SAS will estimate the parameters. See the SAS online documentation for the definition of the C and WEIGHT suboptions of the KERNEL option.	NORMAL, KERNEL NORMAL(MU=0 SIGMA=1) KERNEL(C=2 WEIGHT=QUADRATIC)
LINEATTRS	Specifies the appearance of lines using a style reference and/or one or more of the suboptions: COLOR, PATTERN, THICKNESS.	(COLOR=Red PATTERN=Dot THICKNESS=1)
SCALE	Specifies the type of axis for the response axis.	COUNT, PERCENT, PROPORTION

This code sample illustrates the use of these options.

```
ODS HTML PATH="&JES.SG/S_8_3" (URL=NONE) BODY="histo_density.html";
ODS GRAPHICS ON / RESET IMAGENAME="F8_20_";
TITLE "Distribution of Resistance";
PROC SGPLOT DATA=JES.Results;
   HISTOGRAM Resistance / BOUNDARY=UPPER SCALE=COUNT  ❶
        FILL FILLATTRS=(COLOR=GRAYE0) OUTLINE;  ❷
```

```
        DENSITY Resistance /  ❸
            TYPE=NORMAL(MU=15 SIGMA=3) LEGENDLABEL="Normal(15, 3)"
            LINEATTRS=(COLOR=Red PATTERN=Dot);
        DENSITY Resistance /  ❹
            TYPE=KERNEL LEGENDLABEL="Kernel Fit"
            LINEATTRS=(COLOR=Green PATTERN=Dash) ;
            REFLINE 12.5 22.5 / AXIS=X LABEL=("LSL=12.5" "USL=22.5")
            LINEATTRS=(COLOR=GRAY PATTERN=Dash)
            LABELLOC=OUTSIDE LABELPOS=MAX;
RUN;

ODS GRAPHICS OFF;
ODS HTML CLOSE;
```

❶ The HISTOGRAM statement plots a histogram of the resistance values. The BOUNDARY=UPPER option specifies that any data values which fall on the boundary of a histogram bin will be included in the upper bin. The SCALE=COUNT option specifies that the count should be plotted.

❷ The FILL and FILLATTRS options specify the fill color for the bars. The OUTLINE option specifies that the bars of the histogram should be outlined.

❸ The first DENSITY statement creates a density plot fit to the resistance values. The TYPE=NORMAL option specifies a normal density, with mean and standard deviation specified by the MU=15 SIGMA=3 suboptions. The LEGENDLABEL option specifies the text to be used to identify this curve in the legend. The LINEATTRS option specifies the color and style of this line.

❹ The second DENSITY statement creates a second density plot fit to the resistance values. The TYPE=KERNEL option specifies a kernel density. See the SAS online documentation for options to specify the kernel weight function and bandwidth.

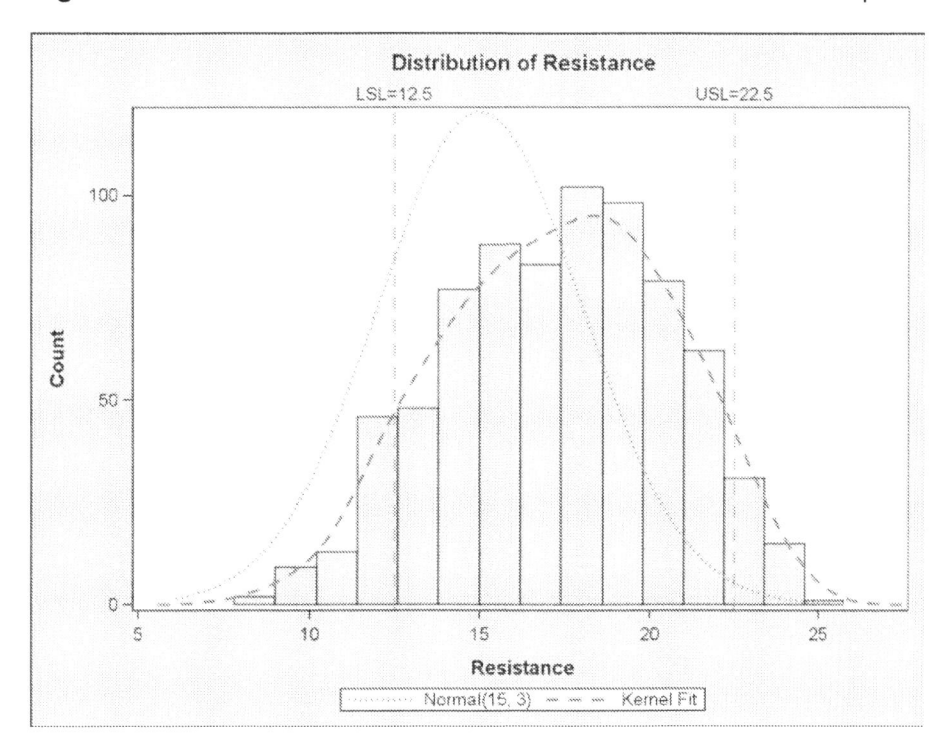

Figure 8-20: HISTOGRAM and DENSITY Plots on the Same Graph

8.3.4.2 HBOX and VBOX

Box plots provide a simple and easily understood graphic display of the distribution of numeric variables. See Section 5.5 for a discussion of box plots and the BOXPLOT procedure, and Sections 5.6 and 6.3.2.7 for alternative ways to create box plots. The HBOX and VBOX statements of PROC SGPLOT were introduced in Section 8.3.1.4. Table 8-12 includes some other options available for these statements. The code for this section is in ~\JES \sample_code\ch_8 \hbox_vbox.sas.

Table 8-12: Selected HBOX and VBOX Statement Options

Option	Meaning	Some Possible Values
BOXWIDTH	Specifies the width of the boxes.	Any number
CATEGORY	Specifies the category variable for the plot.	Any variable in the data set
DATALABEL	Labels the outlier values with the value of the specified variable. If no variable is specified, the response variable is used.	Any variable in the data set
EXTREME	Specifies that the whiskers should extend to the extreme values of the response variable.	(no value required)
SPREAD	Relocates outlier points that have identical values to prevent overlapping.	(no value required)

The code sample illustrates the use of these options.

```
ODS HTML PATH="&JES.SG/S_8_3" (URL=NONE) BODY="hbox_vbox.html";
ODS GRAPHICS ON / RESET IMAGENAME="F8_21_";
TITLE "Box Plot of Resistance by Vendor";
PROC SGPLOT DATA=JES.Results;
   HBOX Resistance / CATEGORY=Vendor
   BOXWIDTH = .75  ❶
   SPREAD  ❷
   DATALABEL=Delay;  ❸
RUN;
ODS GRAPHICS ON / RESET IMAGENAME="F8_22_";
PROC SGPLOT DATA=JES.Results;
   HBOX Resistance / CATEGORY=Vendor
   BOXWIDTH = .25
   EXTREME;  ❹
RUN;
ODS GRAPHICS OFF;
ODS HTML CLOSE;
```

❶ The BOXWIDTH option specifies the relative width of the boxes.

❷ The SPREAD option relocates any outlier points with identical values to prevent overlapping. In this example there are no overlapping outliers, so the option has no effect.

❸ The DATALABEL option specifies that the Delay variable be used to annotate the outlier points. This might be useful to see whether extreme values of Resistance also have extreme Delay values. Any variable in the Results data set could have been used to annotate the points.

❹ In the second PROC SGPLOT, the BOXWIDTH option specifies thinner boxes, and the EXTREME option is used to extend the whiskers to the extreme values of Resistance.

The examples both use the HBOX statement, but of course these options can also be used with the VBOX statement with similar results.

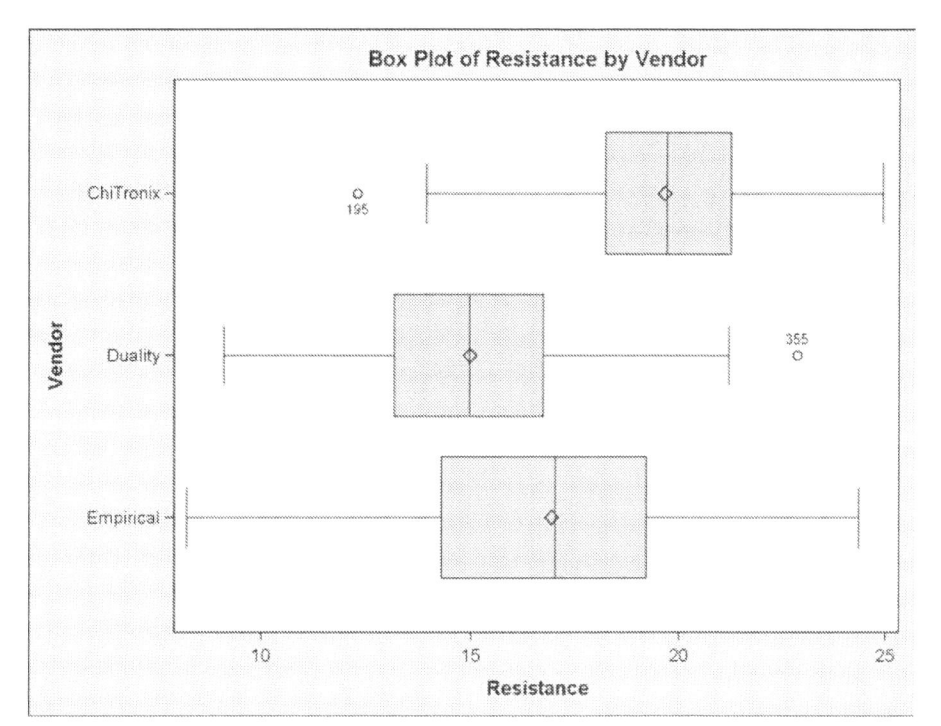

Figure 8-21: Box Plots Created with the HBOX Statement

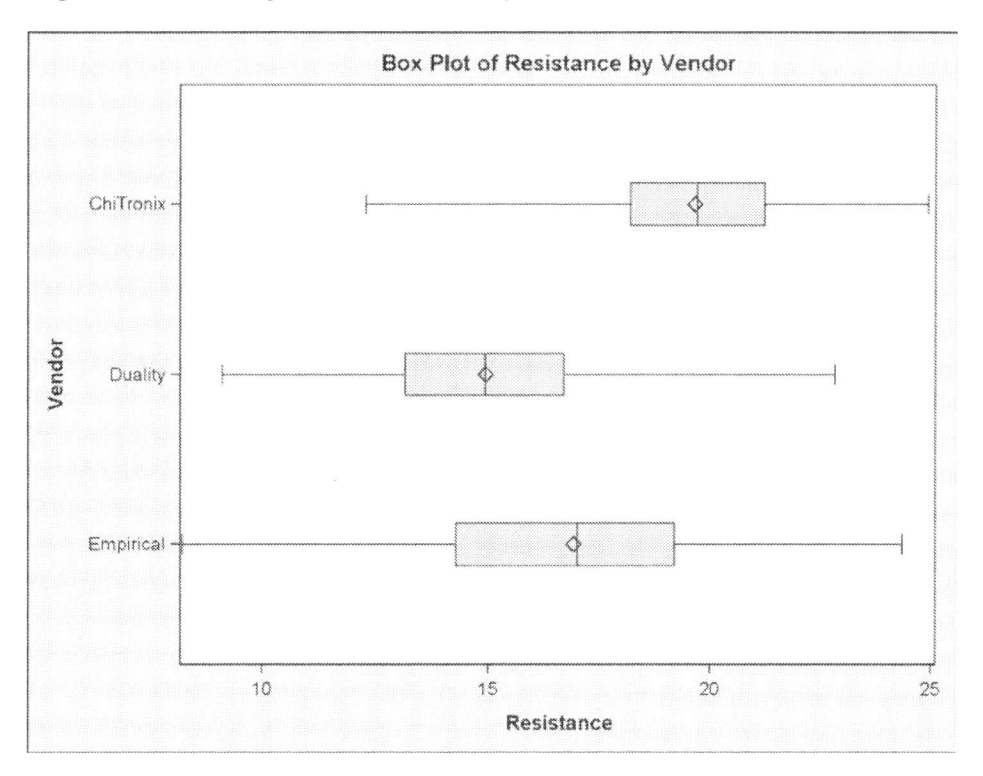

Figure 8-22: Using the EXTREME Option of the HBOX Statement

8.3.4.3 Plotting Group Confidence Limits

The examples in Section 6.3.3 show how to create a plot with confidence limits for each value of a group variable. Doing this with PROC GPLOT requires quite a bit of work. The examples used two DATA steps and one PROC SQL to get a data set in the form required to plot the confidence limits with connecting lines, and then created an annotate data set to label the plot points. The examples in this section show how much easier it is to do the same thing with PROC SGPLOT. The code for this section is in ~\JES \sample_code\ch_8 \conlim_sg.sas.

You can create a plot similar to Figure 6-26 directly from the JES.Results data set using an HLINE statement with PROC SGPLOT.

```
ODS HTML PATH="&JES.SG/S_8_3" (URL=NONE) BODY="conlim_sg.html";

/*===== Normal Confidence Limits with an HLINE Statement =====*/
ODS GRAPHICS ON / RESET IMAGENAME="F8_23_";
TITLE "Resistance by Vendor - Month 1";
PROC SGPLOT DATA=JES.Results(WHERE=(Month=1));
   HLINE Vendor /RESPONSE=Resistance ❶
        STAT=MEAN
        LIMITS=BOTH ❷
        LIMITSTAT=CLM
        ALPHA=0.10
        MARKERS
        DATALABEL; ❸
        XAXIS VALUES=(10 TO 26 BY 2);
REFLINE 12.5 22.5 /
           AXIS=X
           LABEL=("Lower Limit" "Upper Limit")
           LABELLOC=INSIDE LABELPOS=MIN;
RUN;
```

❶ The HLINE statement draws a line plot of Resistance by Vendor. The STAT=MEAN option specifies that the mean resistance be plotted.

❷ The LIMITS=BOTH option specifies both upper and lower limits. The LIMITSTAT=CLM option specifies normal confidence limits on the mean. The ALPHA=0.10 option specifies the confidence limits should be 90% limits.

❸ The DATALABEL option causes the points to be labeled with the values of the calculated response, i.e., the mean of the Resistance variable.

The resulting plot, shown in Figure 8-23, is similar to Figure 6-26, except that Figure 8-23 includes a line joining the means. Note that this plot uses the JES.Results data set, which contains the individual test records, while the example in Section 6.3.3 used JES.Results_Tab, which is a summary of JES.Results created with PROC MEANS. So if you are starting from a set of raw data, such as JES.Results, you need to include the PROC MEANS step in the work required to get such a plot with PROC GPLOT.

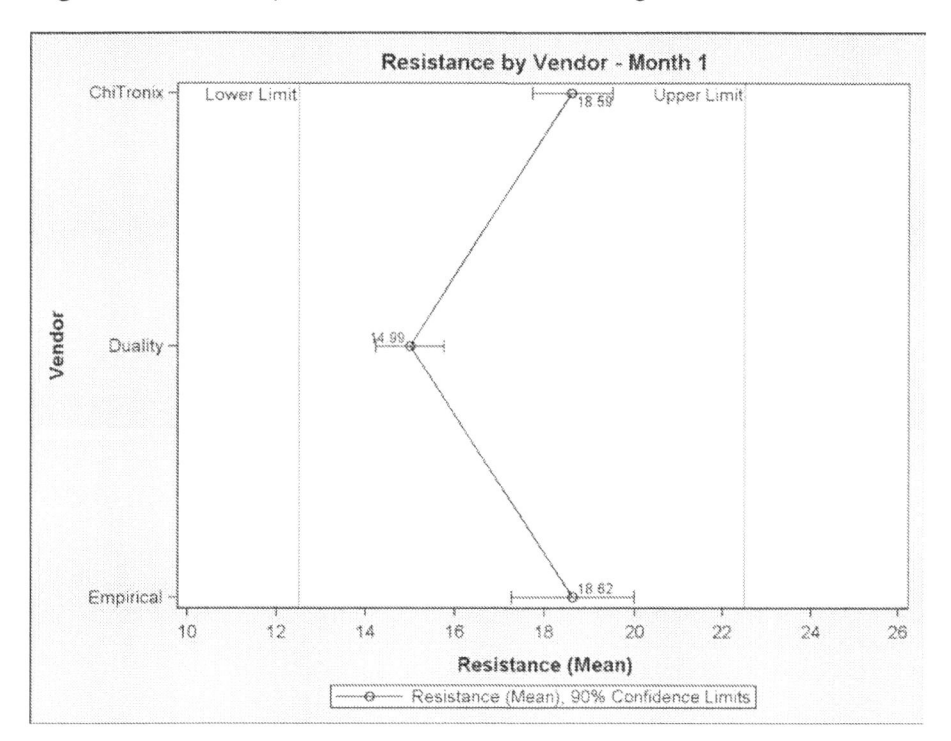

Figure 8-23: Group Confidence Limit Plot Using the HLINE Statement

Note that the point labels added with the DATALABEL option are automatically placed to minimize conflict with other plot elements. The label for Empirical is above the plot point, and the label for ChiTronix is below the plot point, to avoid conflict with the lower and upper borders of the plot area. And each label is placed either left or right of the plot point to avoid conflict with the line joining the means. Achieving the same results with PROC GPLOT would require careful adjustment of the POSITION variable in the annotate data set.

The next example shows how create a similar plot from a summary data set using the SCATTER and SERIES statements.

The previous example used the JES.Results data set, containing all the individual test records, to create a group confidence limit plot. In some cases it is more convenient to work from a summary data set, such as JES.Results_Tab, which contains the mean values and confidence limits for the Resistance variable. (See Section 6.1.1 for the code that was used to create JES.Results_Tab from JES.Results.) The next example shows how to create a plot like Figure 8-23 from the summary data set, using the SCATTER and SERIES statements.

```
/*===== Normal Confidence Limits with a SCATTER Statement======*/
ODS GRAPHICS / RESET IMAGENAME="F8_24_";
PROC SGPLOT DATA=JES.Results_Tab(WHERE=(Month=1));
   SCATTER Y=Vendor X=M_Res / ❶
       XERRORLOWER=R_L
       XERRORUPPER=R_U
       LEGENDLABEL="Resistance(Mean), 90% Confidence Limits"; ❷
   SERIES Y=Vendor X=M_Res / DATALABEL=M_Res ❸
       LEGENDLABEL=" ";

XAXIS VALUES=(10 TO 26 BY 2);
   REFLINE 12.5 22.5 / AXIS=X LABEL=("Lower Limit" "Upper Limit")
   LABELLOC=INSIDE LABELPOS=MAX;
RUN;
```

❶ The SCATTER statement creates a scatter plot of Vendor vs. Mean Resistance (M_Res). The XERRORLOWER and XERRORUPPER options draw limits at R_L and R_U, which are the 90% confidence limits on M_Res. (See Section 6.1.1.)

❷ The LEGENDLABEL option is used to create a legend similar to that in Figure 8-23.

❸ The SERIES statement joins the mean values for each vendor. The DATALABEL option labels each point with the value of M_Res.

The resulting plot, shown in Figure 8-24, is almost identical to Figure 8-23, but note that the order of the Vendor variable on the Y axis is reversed.

You can use any text you want to label the points by creating a variable that contains the desired labels. This example adds the sample size to the point labels. The result is shown in Figure 8-25.

```
DATA Tab; SET JES.Results_Tab(WHERE=(Month=1)); ❶
   Text=".            N="||TRIM(LEFT(PUT(N,8.0)))||
   " Mean="||TRIM(LEFT(PUT(M_Res,8.2)));
RUN;

ODS GRAPHICS / RESET IMAGENAME="F8_25_";
PROC SGPLOT DATA=Tab;
   SCATTER Y=Vendor X=M_Res / DATALABEL=Text ❷
       XERRORLOWER=R_L
       XERRORUPPER=R_U;
   SERIES Y=Vendor X=M_Res;
   XAXIS VALUES=(10 TO 26 BY 2);
   REFLINE 12.5 22.5 / AXIS=X LABEL=("Lower Limit" "Upper Limit")
       LABELLOC=OUTSIDE LABELPOS=MAX; ❸
RUN;
```

❶ The DATA statement adds a new variable, Text, which contains the desired labels. Some spaces are included at the beginning of Text so the labels will clear the upper confidence limit.

❷ The DATALABEL=Text option specifies that the new variable be used to label the points.

❸ The LABELLOC option is set to OUTSIDE so that the REFLINE labels do not interfere with the point labels.

Figure 8-24: Group Confidence Limit Plot Using SCATTER and SERIES Statements

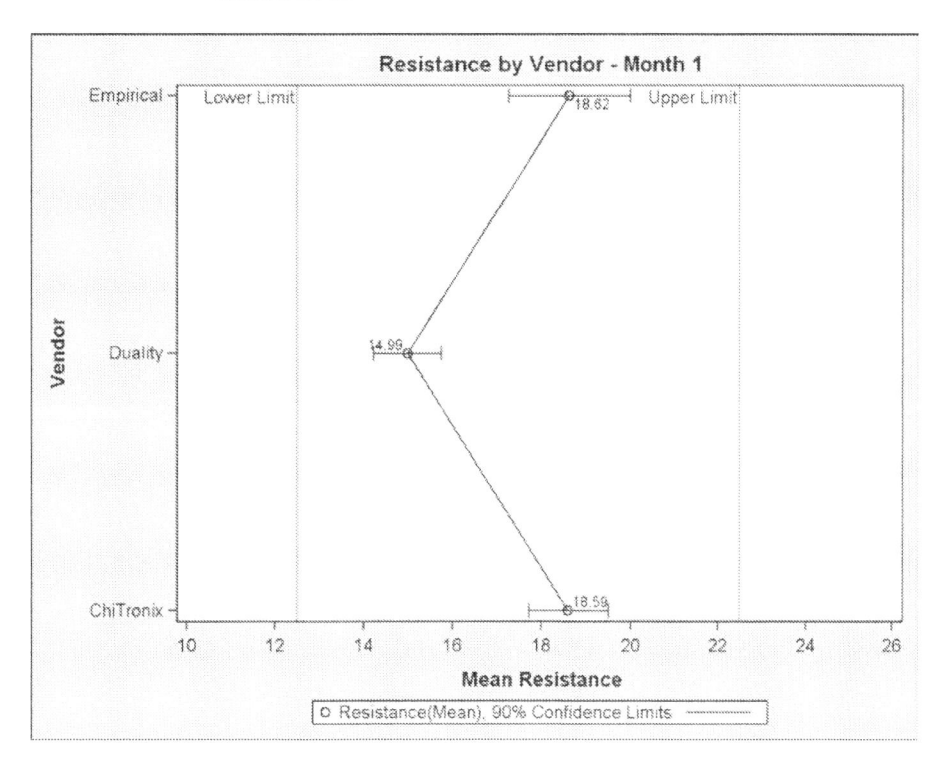

Figure 8-25: Using a Character Variable with the DATALABEL Option to Label the Plot Points

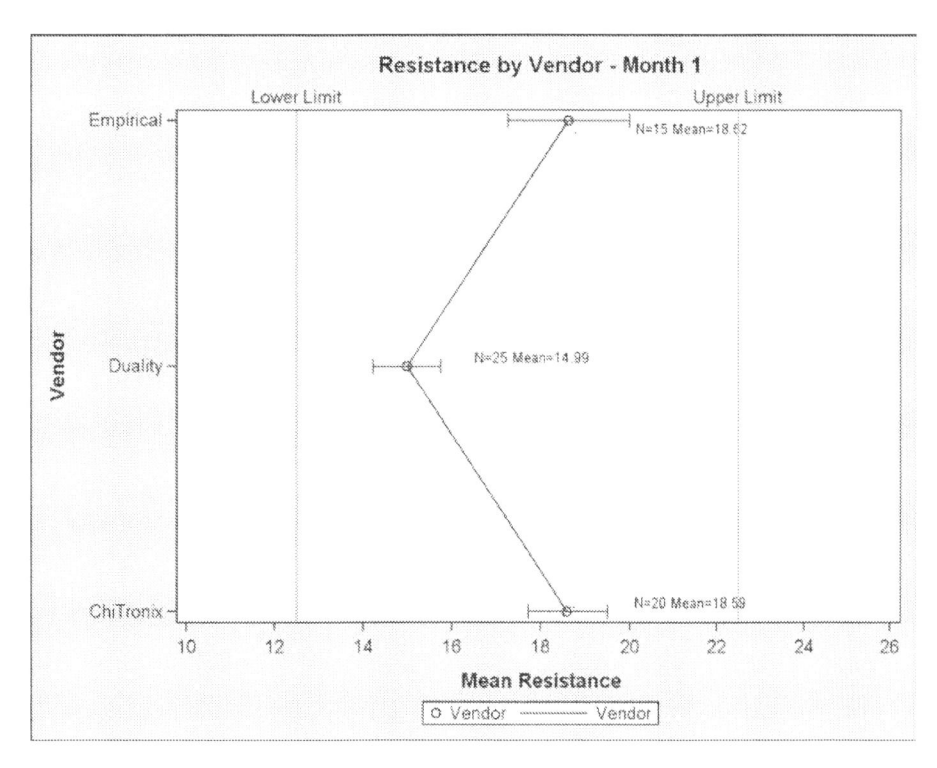

The previous examples created plots with confidence limits based on the normal distribution. In quality and reliability data analysis, you often work with variables that are not normally distributed. For example, the number of units that fail out of N units tested is usually assumed to follow a binomial distribution, and the number of defects found in a wafer, or in a sample of units inspected, is usually assumed to follow a Poisson distribution. The next example shows how to create a group confidence limit plot for Poisson distributed variables. The same method is easily adapted to handle variables assumed to have the binomial distribution.

This example creates a trend plot for defect rate with confidence limits similar to the trend plot that was created with PROC GPLOT in Section 6.3.3.4.

```
/*===== Poisson Confidence Limits with a SCATTER Statement =====*/
DATA Tab_3; SET JES.Results_Tab(WHERE=(Vendor="ChiTronix"));  ❶
   FORMAT LCL UCL 8.3;
   IF N_Def>0 THEN LCL = CINV(.10, 2*N_Def)/(2*N);
   UCL = CINV(.90, 2*(N_Def+1))/(2*N);
   KEEP Mon N N_Def M_Def LCL UCL;
RUN;

ODS GRAPHICS / RESET IMAGENAME="F8_26_";
TITLE "Monthly Defect Rates ChiTronix - 2008";
PROC SGPLOT DATA=Tab_3;
   SCATTER X=Mon Y=M_Def / NAME='a'  ❷
       YERRORLOWER=LCL
       YERRORUPPER=UCL
       MARKERATTRS =(Color=Green SYMBOL=CircleFilled)
       LEGENDLABEL="Mean Defect Rate with 90% Confidence Limits";
   SERIES X=Mon Y=M_Def /  ❸
       DATALABEL=M_Def;
   REFLINE 2 / AXIS=Y  ❹
       LABEL="Spec Limit"
       LABELLOC=INSIDE
       LABELPOS=MIN
       LINEATTRS=(COLOR=Red PATTERN=Dash);
   YAXIS LABEL='Defect Rate' VALUES = (0 TO 3 BY .5);
   XAXIS DISPLAY=(NOLABEL);
   KEYLEGEND 'a' / LOCATION=INSIDE POSITION=N;
RUN;

ODS GRAPHICS OFF;
ODS HTML CLOSE;
```

❶ The DATA step creates the Tab_3 data set containing the upper and lower confidence limit variables, UCL and LCL, based on the assumption that defects have the Poisson distribution. This is the same code used in Section 6.3.3.4, based on the discussion of binomial and Poisson confidence limits in Section 5.2.7. To create a similar plot for a binomial distributed variable, use the expressions involving BETAINV instead of CINV from the code sample in Section 5.2.7.

❷ The SCATTER statement creates a scatter plot of Defect Rate (M_Def) vs. Month. The YERRORLOWER and YERRORUPPER options draw limits at LCL and UCL. The MARKERATTRS option specifies the symbol color and type. The LEGENDLABEL option specifies text to be used in the legend label.

❸ The SERIES statement draws a line connecting the mean defect rates. The DATALABEL option specifies that the values of M_Def be used to label the plot points.

❹ The REFLINE statement specifies the location, label, type, and color of a reference line drawn at the Spec Limit of 2.0.

The resulting plot, shown in Figure 8-26, is very similar to Figure 6-27 created with PROC GPLOT.

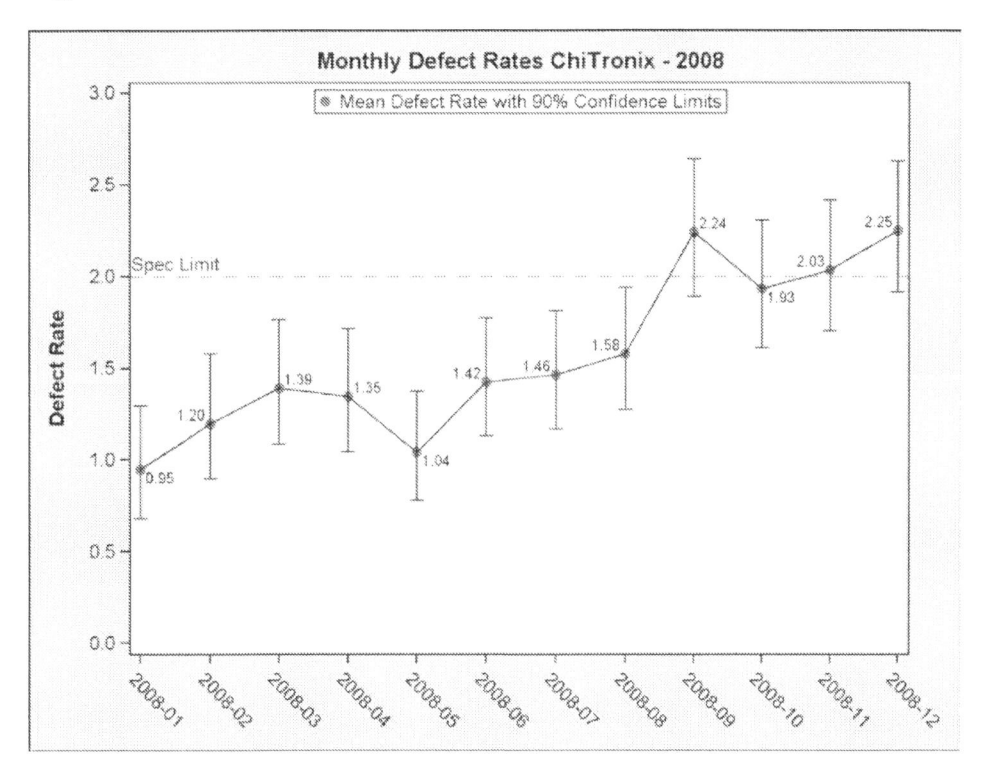

Figure 8-26: Trend Plot with Confidence Limits Based on the Poisson Distribution

The final code sample in this section creates a similar plot from the raw data set, JES.Results, using an HLINE statement. The resulting plot (not shown here) differs from Figure 8-26 in that the confidence limits are based on the assumption that the Defects variable is normally distributed. That is not a good assumption in this example, but might be adequate in cases where the number of defects per unit is greater, and the Poisson distribution is well approximated by a normal distribution.

```
/*===== Confidence Limits Based on the Normal Approximation to the
Poisson =====*/
ODS GRAPHICS ON / RESET IMAGENAME="F8_26a_";
PROC SGPLOT DATA=JES.Results(WHERE=(Vendor="ChiTronix"));
  VLINE Mon /RESPONSE=Defects
       STAT=MEAN
       LIMITS=BOTH
       LIMITSTAT=CLM
       ALPHA=0.10
       MARKERS
       DATALABEL;
     FORMAT Defects 8.2;
     XAXIS VALUES=(0 TO 3 BY .5);
     REFLINE 2 / AXIS=Y
          LABEL="Spec Limit"
          LABELLOC=INSIDE
          LABELPOS=MIN
          LINEATTRS=(COLOR=Red PATTERN=Dash);
RUN; QUIT;
```

8.3.5 Adding Hyperlinks to Plots Created with SGPLOT

Section 7.4.2 showed how to use the HTML option of the PLOT statement to add hyperlinks to graphs created with PROC GPLOT. The SERIES, SCATTER, NEEDLE, STEP, HBAR, and VBAR statements of PROC SGPLOT can use the URL option to do the same thing. The example in this section uses the URL option with a SERIES statement to create a plot similar to the plot in Display 7-8, with links to the individual Vendor pages created in Section 7.4.2. The code for this section is in ~\JES\sample_code\ch_8\hyperlinks.sas.

```
DATA Results_Tab; SET JES.Results_Tab;  ❶
  Point_Link=CATS('../../ods_output/page4/',Vendor,'.html');
RUN;

ODS HTML PATH="&JES.SG/S_8_3" (URL=NONE) BODY="hyperlinks.html";
ODS GRAPHICS ON / RESET IMAGENAME="F8_H_" IMAGEMAP=YES;  ❷

TITLE1 "Mean Resistance by Vendor";
PROC SGPLOT DATA =Results_Tab;
  SERIES Y=M_Res X=Mon / GROUP=Vendor URL=Point_Link  ❸
  LINEATTRS = (PATTERN=SOLID THICKNESS=2)
      MARKERS
      MARKERATTRS=( SIZE=10 );
  YAXIS  LABEL = 'Mean Resistance'
      VALUES=(0 TO 30 BY 1)   FITPOLICY=THIN;
  XAXIS  LABEL="Month of Production"
         FITPOLICY=ROTATE REFTICKS;
RUN;

ODS GRAPHICS OFF;
ODS HTML CLOSE;
```

❶ The DATA step creates the Results_Tab data set by adding the Point_Link variable to JES.Results_Tab. This is similar to the code used to add Point_Link to the Results_Tab data set in Section 7.4.2, but the syntax is slightly different. The 'HREF=' text is not needed because that is added automatically by SAS. Also, the URL starts with '../../ods_output/page4/' so that the link can find the individual Vendor pages, which were saved in the page4 folder when you ran the code in Section 7.4.2.

❷ The IMAGEMAP=YES option adds data tips to the plotted points, and also enables the use of the URL option to attach hyperlinks to the points.

❸ The URL option is used to attach the hyperlink contained in the Point_Link variable to each plot point. This only works if you also specify IMAGEMAP=YES.

```
Results_Tab   (selected columns, first five rows)

Vendor        Mon         M_Res              Point_Link

ChiTronix     2008-01     18.59     ../../ods_output/page4/ChiTronix.html
ChiTronix     2008-02     17.95     ../../ods_output/page4/ChiTronix.html
ChiTronix     2008-03     18.52     ../../ods_output/page4/ChiTronix.html
ChiTronix     2008-04     18.84     ../../ods_output/page4/ChiTronix.html
ChiTronix     2008-05     19.78     ../../ods_output/page4/ChiTronix.html
```

When you run this code, the hyperlinks.html file is created in your **JES\SG\S_8_3** folder. Display 8-4 shows what this page looks like when opened in a Web browser. When you click on the plot points, the corresponding Vendor pages will open in a new tab or window, as shown in Display 7-9.

Display 8-4: Web Page Created Using the URL Option to Add Hyperlinks to the Plot

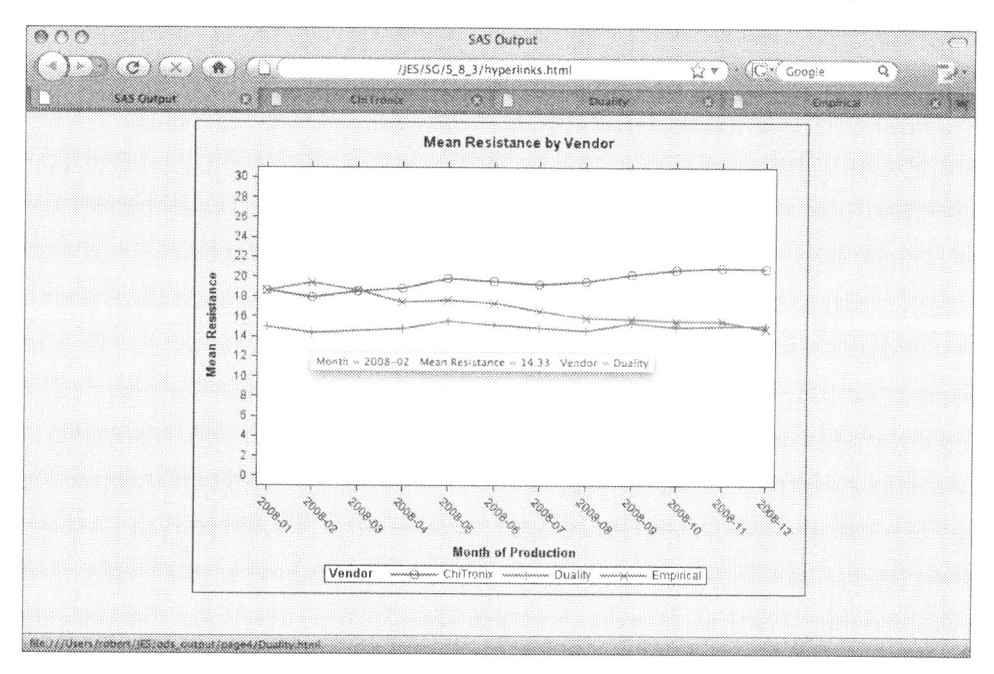

Display 8-4 shows the data tip (or tooltip) information that is displayed when the cursor hovers over the point (Month="2008-02", Vendor="Duality").

8.4 PROC SGPANEL

The SGPANEL procedure creates an array, or panel, of similar plots, one for each value of a classification variable. This is similar to what you would get by using a BY statement with PROC SGPLOT, except that the resulting graphs are assembled into an array of paneled graphs. Note, however, that there are some options supported by SGPLOT that are not supported by SGPANEL, for example CURVELABELLOC. Consult the SAS online documentation to see which options are supported. The code for this section is in ~\JES \sample_code\ch_8 \sgpanel.sas.

To create a panel of plots using any of the plot types described in Section 8.3, you replace the PROC SGPLOT statement with a PROC SGPANEL statement and add a PANELBY statement to specify the classification variable. This example creates a panel of scatter plots of Delay vs. Resistance by Vendor, with one plot for each Month.

```
ODS HTML PATH="&JES.SG/S_8_4" (URL=NONE) BODY="sgpanel.html";
ODS GRAPHICS ON / RESET IMAGENAME="F8_27_";

TITLE1 "Delay vs Resistance by Vendor and Month";
PROC SGPANEL DATA=JES.Results; ❶
   PANELBY Mon / ❷
      ROWS=3 COLUMNS=2 ❸
      UNISCALE=ALL; ❹
      SCATTER Y=Delay X=Resistance / GROUP=Vendor; ❺
RUN;

ODS GRAPHICS OFF;
ODS HTML CLOSE;
```

❶ The PROC SGPANEL statement requests a panel of plots based on the JES.Results data set.

❷ The PANELBY statement specifies that Mon should be the classification variable. You can also use multiple classification variables. For example, you could specify 'PANELBY Mon Vendor /' and leave out the GROUP=Vendor option to create a set of panels with separate scatter plots for each vendor and month.

❸ The ROWS and COLUMNS options specify the grid dimensions. In this case, each panel will have 3 rows and 2 columns, so two panel plots will be created to include all 12 months. If these options are omitted, SAS will make a reasonable guess at the best layout to use.

❹ The UNISCALE=ALL option specifies that the same axes and legends be used for all the plots, which makes it easier to compare the different plots. This is the default value. You can use UNISCALE=ROW or UNISCALE=COLUMN to scale only the row or column axes to be identical.

❺ The SCATTER statement creates a scatter plot of Delay vs. Resistance by Vendor for each value of Month.

Figure 8-27 shows the first of the two panels created by the sample code. The other panel contains the plots for Mon=2008-07 through 2008-12.

Figure 8-27: Panel of Scatter Plots Created with SGPANEL

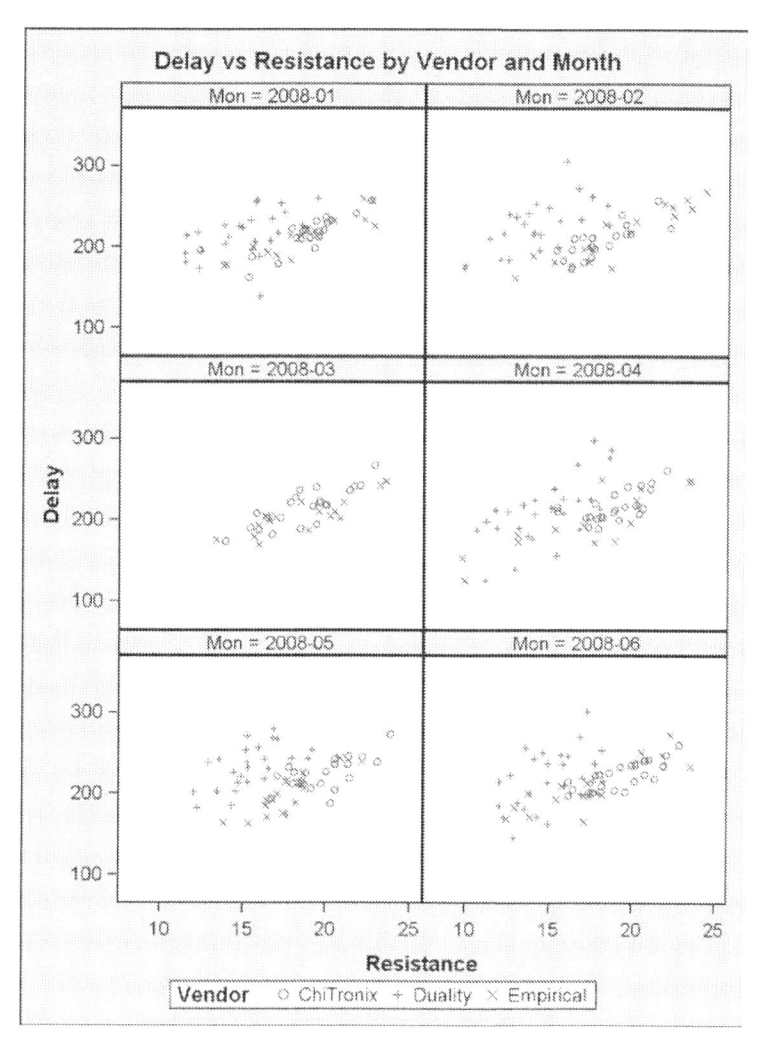

8.5 PROC SGSCATTER

The SGSCATTER procedure creates an array of scatter plots using one of three statements:

- The PLOT statement creates an array of scatter plots with *independent axes*.
- The COMPARE statement creates an array of comparative scatter plots with *shared axes*.
- The MATRIX statement creates a matrix of scatter plots, one for each pair selected from a variable list.

The examples in this section illustrate the use of each of these statements. The code for this section is in ~\JES \sample_code\ch_8 \sgscatter.sas. Selected options for these statements are shown in Table 8-13. Consult the SAS online documentation for other options, as well as other suboptions that can be used to modify the options listed here.

Table 8-13: Selected PLOT, COMPARE, and MATRIX Statement Options

Option	Meaning
DATALABEL	Displays a label for each data point.
ELLIPSE	Adds a confidence or prediction ellipse to the plot.
GROUP	Specifies a classification variable.
LOESS *	Adds a loess fit to the plot.
PBSPLINE *	Adds a fitted penalized B-spline to the plot.
REG *	Adds a regression fit to the plot.
Selected Suboptions for the LOESS, PBSPLINE, and REG Options	
ALPHA	Specifies the confidence level for confidence limits.
CLI **	Creates confidence limits for individual predicted values.
CLM	Creates confidence limits for the predicted mean value.
DEGREE	Specifies the degree of the fit.
* These options can be used with PLOT and COMPARE, but not with MATRIX.	
** This option can be used with PBSPLINE and REG, but not with LOESS.	

The first code sample illustrates the use of the PLOT statement with PROC SGSCATTER.

```
ODS HTML PATH="&JES.SG/S_8_5" (URL=NONE) BODY="sgscatter.html";

ODS GRAPHICS ON / RESET IMAGENAME="F8_28_";
TITLE1 "Delay vs Resistance and Process Temperature vs Date";
PROC SGSCATTER DATA=JES.Results_Q4;
   PLOT Delay*Resistance ProcessTemp*ProcessDate /    ❶
        GROUP=Vendor   ❷
        REG=(DEGREE=2 NOGROUP);   ❸
RUN;
```

❶ The PLOT statement requests scatter plots of Delay vs. Resistance and ProcessTemp vs. ProcessDate.

❷ The GROUP option specifies separate scatter plots for each Vendor on the same graph.

❸ The REG option adds quadratic regression fits to the data. The DEGREE suboption specifies a second-degree polynomial for the fit. The NOGROUP suboption specifies that the fit does not use the group variable from the scatter plot.

The next code sample (not shown here) is the same, but without the NOGROUP suboption of the REG option. The resulting plot, shown in Figure 8-29, has separate fitted polynomials for each Vendor.

Figure 8-28: Array Created Using a PLOT Statement with PROC SGSCATTER

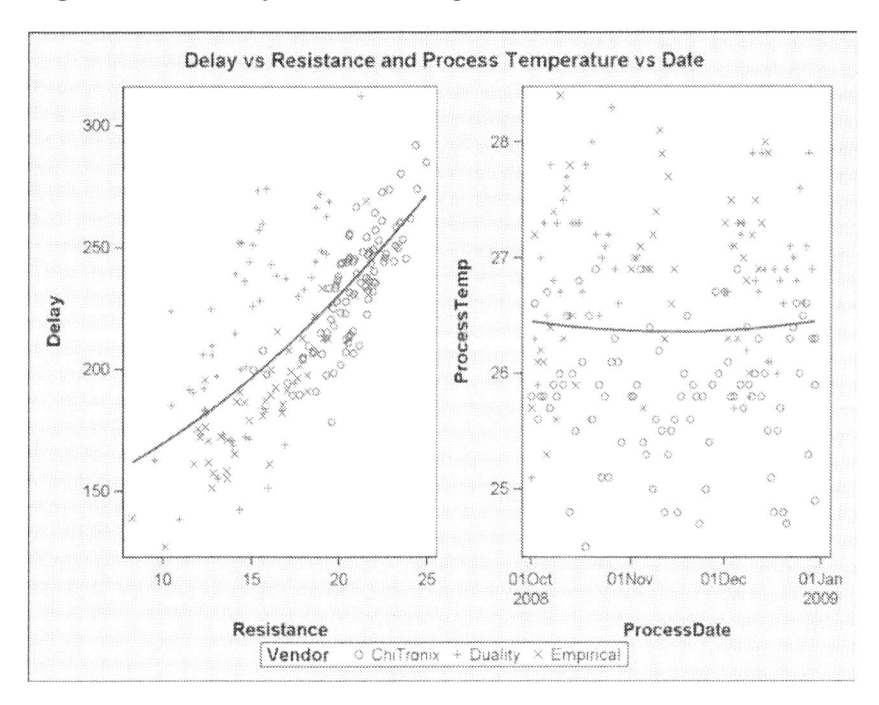

Figure 8-29: Using the REG Option without the NOGROUP Suboption

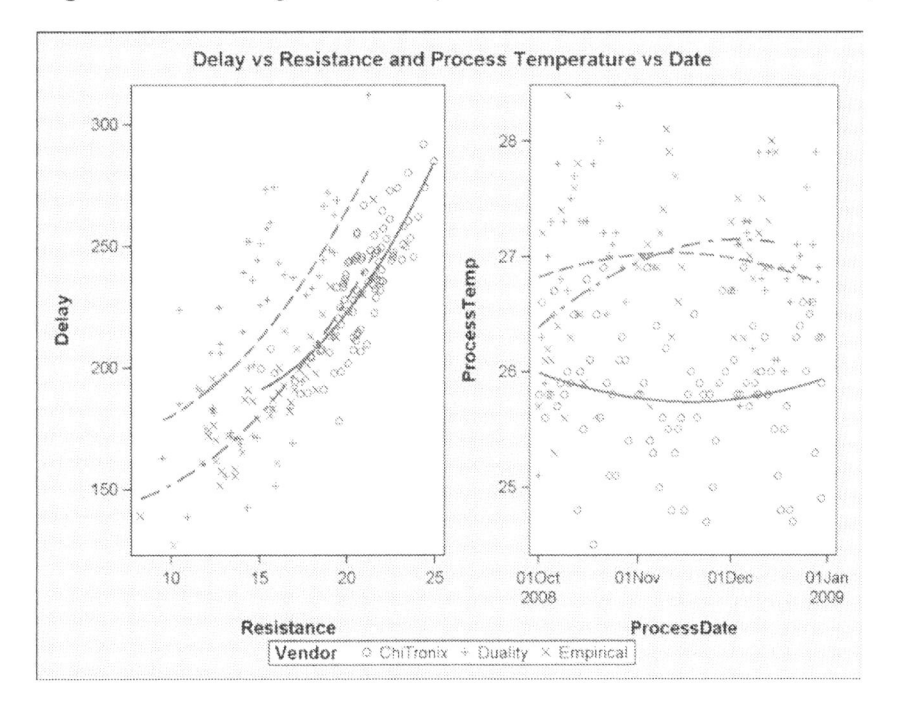

The COMPARE statement of PROC SGSCATTER is similar to the PLOT statement, but uses shared axes for each row and column of scatter plots. The next code sample uses first the PLOT statement and then the COMPARE statement to create scatter plots of Delay and Resistance vs. ProcessDate and ProcessTemp.

```
ODS GRAPHICS / RESET IMAGENAME="F8_30_";
TITLE1 "Resistance and Delay vs Process Date and Temperature";
PROC SGSCATTER DATA=JES.Results_Q4;
   PLOT (Delay Resistance) * (ProcessDate ProcessTemp) / ❶
   GROUP=Vendor
   ELLIPSE=(TYPE=PREDICTED ALPHA=.05); ❷
RUN;

ODS GRAPHICS / RESET IMAGENAME="F8_31_";
TITLE1 "Resistance and Delay vs Process Date and Temperature";
PROC SGSCATTER DATA=JES.Results_Q4;
   COMPARE Y=(Delay Resistance) X=(ProcessDate ProcessTemp) / ❸
   GROUP=Vendor
   ELLIPSE=(TYPE=PREDICTED ALPHA=.05);
RUN;
```

❶ You can use parentheses to specify multiple X or Y variables for the plots.

❷ The ELLIPSE option requests a prediction ellipse (TYPE=PREDICTED) with 95% confidence (ALPHA=.05). Note that there is no option to create separate ellipses for each value of the GROUP variable, Vendor.

❸ Note that the COMPARE statement uses different syntax than the PLOT statement for specifying the X and Y variables.

The plot created by the COMPARE statement (Figure 8-31) makes better use of the available space than the plot created with the PLOT statement (Figure 8-30) because the common axis labels are not repeated for each plot. In general you should use the COMPARE statement if the plots have common axes, and the PLOT statement if they do not, as in Figures 8-28 and 8-29.

Figure 8-30: Using the PLOT Statement with PROC SGSCATTER

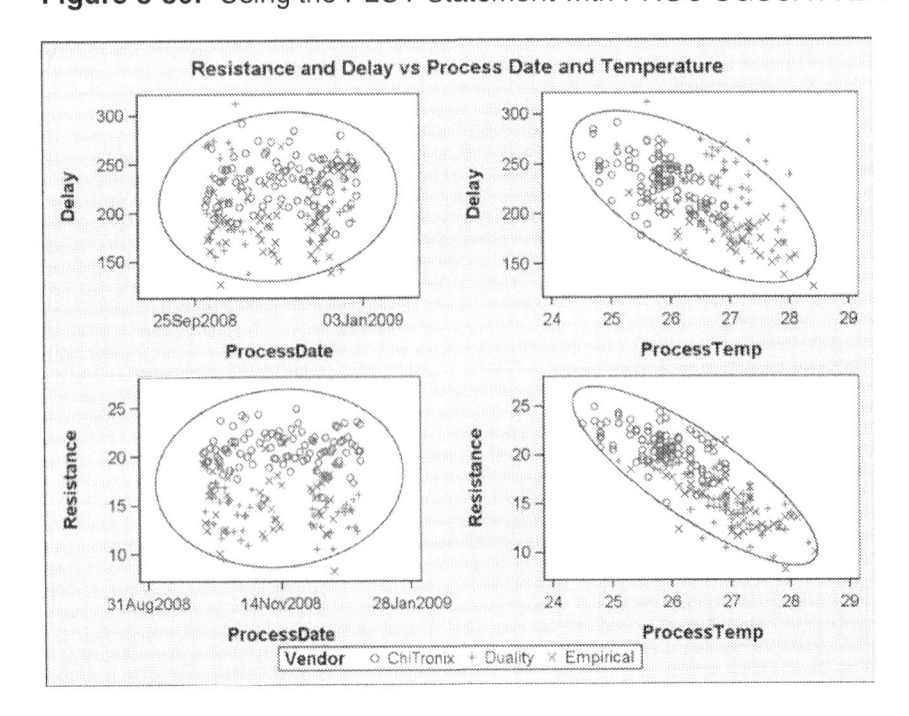

Figure 8-31: Using the COMPARE Statement with PROC SGSCATTER

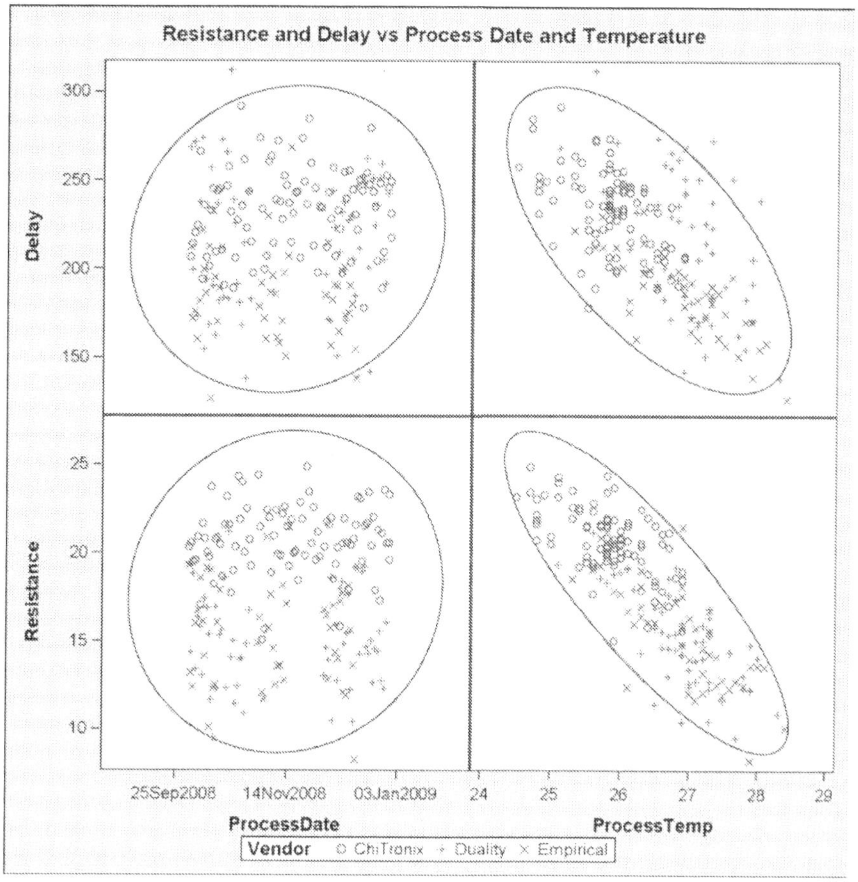

302 *Just Enough SAS: A Quick-Start Guide to SAS for Engineers*

The next example uses a MATRIX statement with PROC SGSCATTER. The MATRIX statement has a DIAGONAL option, which can be used to add plots along the diagonal of the plot matrix. The plot options are HISTOGRAM, NORMAL (for a normal density fit), or KERNEL (for a kernel density fit). If the DIAGONAL option is not used, the variable names are written along the diagonal.

```
ODS GRAPHICS / RESET IMAGENAME="F8_32_";
TITLE1 "Resistance, Delay and Process Temperature";
PROC SGSCATTER DATA=JES.Results_Q4;
   MATRIX Resistance Delay ProcessTemp /  ❶
       GROUP=Vendor
       ELLIPSE=(TYPE=PREDICTED);  ❷
RUN;
```

❶ The MATRIX statement specifies the variable list: Resistance, Delay, and ProcessTemp. Scatter plots are created for each pair of variables from the list. The GROUP option specifies Vendor as the classification variable for the plots.

❷ The ELLIPSE option requests a prediction ellipse for each scatter plot.

Figure 8-32: Array Created Using a MATRIX Statement with PROC SGSCATTER

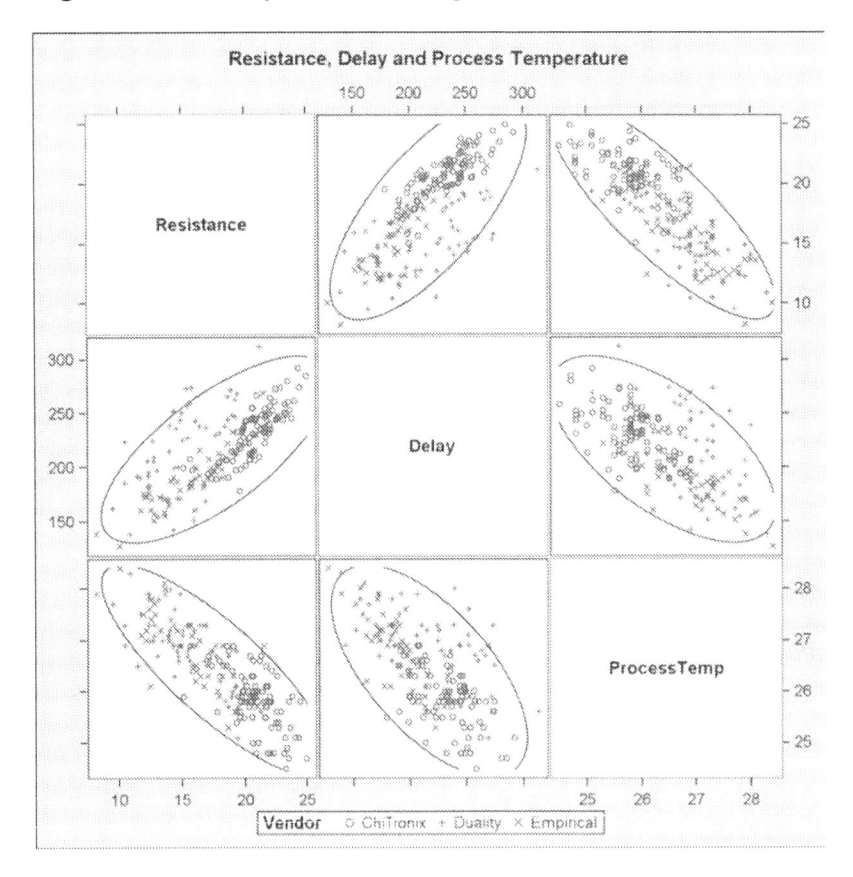

The final example is similar, but uses the DIAGONAL option, and plots a confidence ellipse rather than a prediction ellipse.

```
ODS GRAPHICS / RESET IMAGENAME="F8_33_";
PROC SGSCATTER DATA=JES.Results_Q4;
   MATRIX Resistance Delay ProcessTemp /
       DIAGONAL=(HISTOGRAM NORMAL)  ❶
       GROUP=Vendor
       ELLIPSE=(TYPE=MEAN ALPHA=.01);  ❷
RUN;
```

❶ The DIAGONAL option requests a histogram and normal density fit for each variable.

❷ The ELLIPSE option adds a confidence ellipse for the mean to each scatter plot. The suboptions specify a confidence ellipse (TYPE=MEAN) with 99% confidence (ALPHA=.01). The ellipses are small and difficult to see among the plot points.

Figure 8-33: Using the DIAGONAL Option of the MATRIX Statement

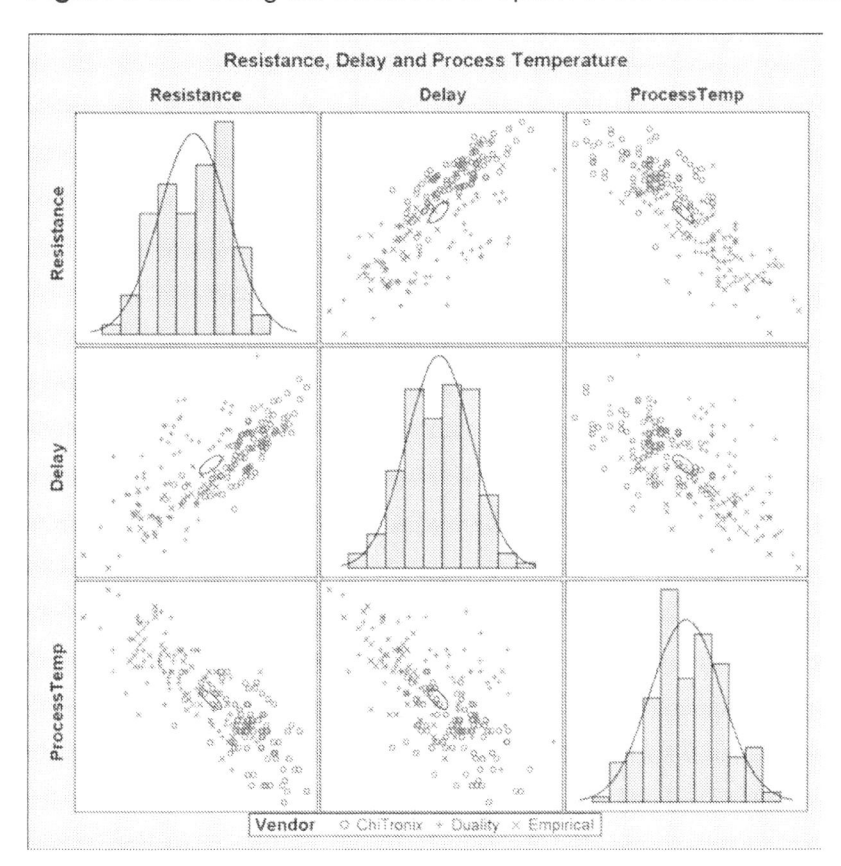

Note that when you use the DIAGONAL statement you lose the axis values and tick marks that are created by the MATRIX statement, as shown in Figure 8-32.

8.6 More Than Enough

This section gives an example using PROC SGRENDER and the Graph Template Language (GTL) to create a custom set of related plots that are easily rerun with different data sets or different variables. A detailed discussion of GTL is beyond the scope of this book, but you can find more information in "Introduction to the Graph Template Language" by Sanjay Matange, which is listed in the "Chapter Summary" section. The code for this section is in ~\JES \sample_code\ch_8 \sgrender.sas. You first use PROC TEMPLATE to create a STATGRAPH template, as illustrated in the first code sample, and then use PROC SGRENDER to create graphs based on the template.

```
PROC TEMPLATE;
  DEFINE STATGRAPH mygraph.hist;  ❶
  MVAR Var1 Var2;
  BEGINGRAPH;
      ENTRYTITLE "Study of " Var1 " Measurements";
      LAYOUT LATTICE / COLUMNS=2 ROWS=2 ROWGUTTER=5px;  ❷
          SCATTERPLOT X=Var1 Y=Var2 / GROUP=Vendor;
          LAYOUT OVERLAY;  ❸
              HISTOGRAM Var1;
              DENSITYPLOT Var1;
          ENDLAYOUT;
          SCATTERPLOT X=Var1 Y=ProcessTemp / GROUP=Vendor;
          BOXPLOT Y=Var1 X=Class / ORIENT=HORIZONTAL;
      ENDLAYOUT;
  ENDGRAPH;
  END;
RUN;
```

❶ The DEFINE statement specifies the type and name of the template to be created. The MVAR statement specifies two macro variables to be provided when PROC SGRENDER is run.

❷ The first LAYOUT statement specifies a 2 by 2 lattice of plots.

❸ The second LAYOUT statement requests an overlay of the specified plot types.

This code creates the mygraph.hist template. The next code sample runs PROC SGRENDER twice to create graphs based on the mygraph.hist template, using first Resistance, and then Delay, as the primary variable. The resulting plots are shown in Figures 8-34 and 8-35.

```
%LET Var1=Resistance; %LET Var2=Delay;
ODS GRAPHICS ON / RESET IMAGENAME="F8_34_";
PROC SGRENDER DATA=JES.Results_Q4 TEMPLATE="mygraph.hist";  ❶
   LABEL Resistance="Resistance" Delay="Delay" Vendor="Vendor";
RUN;
%LET Var1=Delay; %LET Var2=resistance;
ODS GRAPHICS / RESET IMAGENAME="F8_35_";
PROC SGRENDER DATA=JES.Results_Q4 TEMPLATE="mygraph.hist";
   LABEL Resistance="Resistance" Delay="Delay" Vendor="Vendor";
RUN;
```

❶ The PROC SGRENDER statement specifies the data set to be used, and the TEMPLATE option names the STATGRAPH template to be used.

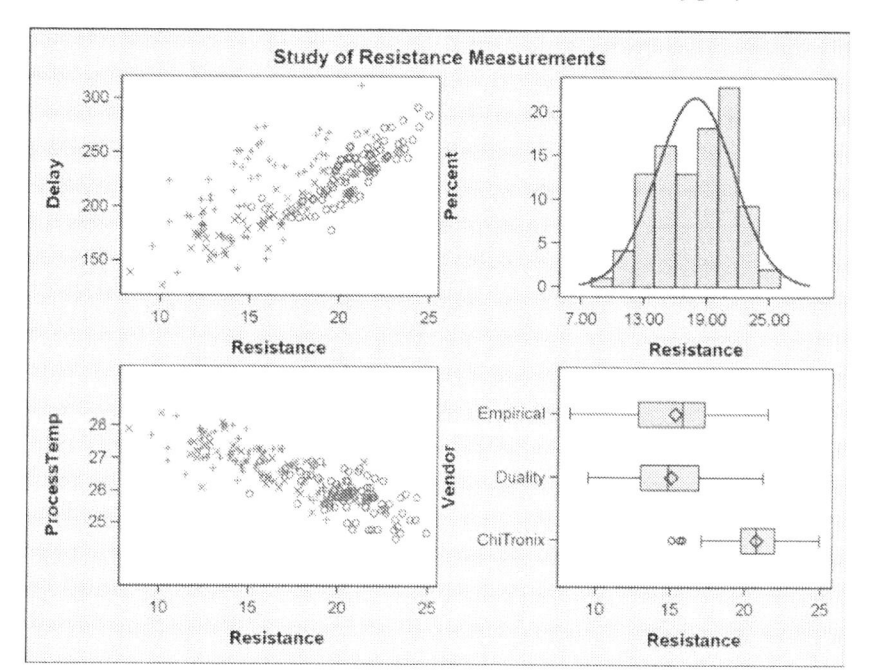

Figure 8-34: Using PROC SGRENDER with the mygraph.hist Template

Figure 8-35: Using PROC SGRENDER with the mygraph.hist Template

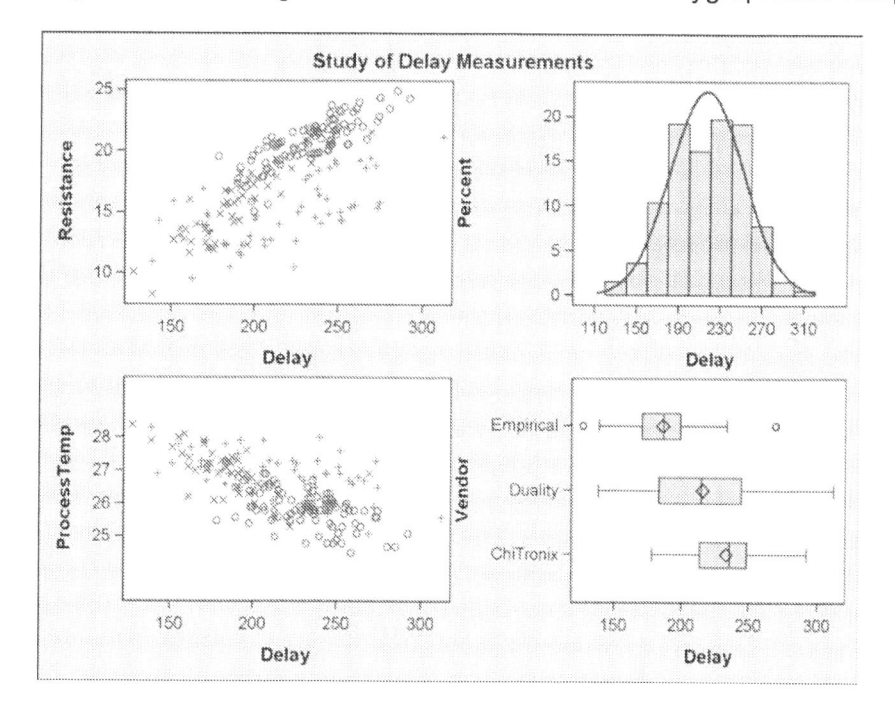

8.7 Chapter Summary

8.7.1 Recap

After finishing this chapter you should know how to

- Use PROC SGPLOT to create a variety of plot types, including:
  - Basic plots using the SERIES, SCATTER, STEP, NEEDLE, and PBSPLINE statements.
  - Limit plot using the DOT, HLINE, VLINE, HBAR, and VBAR statements.
  - Bar charts using the HBAR, VBAR, HLINE, and VLINE statements.
  - Distribution plots using the HISTOGRAM, DENSITY, VBOX, and HBOX statements.
  - Data fit plots using the BAND, ELLIPSE, LOESS, REG, and PBSPLINE statements.
  - Use more than one of the above statements to overlay multiple plots on the same graph.
  - Control the axes on your plots using the XAXIS, YAXIS, XAXIS2, and YAXIS2 statements.
- Use the options available for the statements listed above, as well as the REFLINE, KEYLEGEND, and INSET statements, to customize elements of your graph, including:
  - Titles and footnotes
  - Axis type, range, values, labels, grids, and tick marks
  - Symbol and line type, size, and color
  - Legend content, title, arrangement, and location
  - Curve label content, location, and position
  - Reference line location, color, type, thickness, and label
  - Inset content, location, border, and title
- Use PROC SGPANEL to create a panel of similar plots, one for each value of a classification variable.
- Use PROC SCSCATTER to create an array of scatter plots.

8.7.2 For More Information

Chapter 21 of the SAS/STAT 9.2 documentation, which provides a general overview of ODS Graphics functionality, should be your starting point for learning more.

SAS Conference Papers

Cartier, Jeff, and Dan Heath. 2007. "Using ODS Styles with SAS/GRAPH®." *Proceedings of the SAS Global Forum 2007 Conference.* Cary, NC: SAS Institute Inc. Paper 088-2007. **[088-2007.pdf]**

Heath, Dan. 2007. "SAS/GRAPH® Procedures for Creating Statistical Graphics in Data Analysis." *Proceedings of the SAS Global Forum 2007 Conference.* Cary, NC: SAS Institute Inc. Paper 193-2007. **[193-2007.pdf]**

Heath, Dan. 2008. "Effective Graphics Made Simple Using SAS/GRAPH® SG Procedures." *Proceedings of the SAS Global Forum 2008 Conference.* Cary, NC: SAS Institute Inc. Paper 255-2008. **[255-2008.pdf]**

Matange, Sanjay. 2008. "Introduction to the Graph Template Language." *Proceedings of the SAS Global Forum 2008 Conference.* Cary, NC: SAS Institute Inc. Paper 313-2008. **[gtl.pdf]**

Matange, Sanjay. 2008. "ODS Graphics Editor." *Proceedings of the SAS Global Forum 2008 Conference.* Cary, NC: SAS Institute Inc. Paper 235-2008. **[235-2008.pdf]**

Rodriguez, Robert N. 2008. "Getting Started with ODS Statistical Graphics in SAS® 9.2." *Proceedings of the SAS Global Forum 2008 Conference.* Cary, NC: SAS Institute Inc. Paper 305-2008. **[305-2008.pdf]**

8.7.3 Exercises

Use ODS Graphics to reprogram the sample report of Section 7.1.1.

- Use PROC SGPLOT instead of PROC GPLOT to create the scatter plots.
- Add a page with two plots:
    - Box plots of resistance by Vendor, like Figure 8-21
    - Resistance by Vendor with confidence limits, like Figure 8-25
- For each Vendor, add a page with two plots:
    - A trend plot with confidence limits, like Figure 8-26, but using Resistance instead of Defects
    - A panel plot, like Figure 8-27, but with histograms instead of scatter plots. The panel should have four histograms of the Resistance variable, one for each quarter of 2008.
- Following the methods explained in Section 7.4, use ODS HTML to publish your new report to the Web.
- Use the URL option of the SERIES or SCATTER statement (see Section 8.3.5) to add hyperlinks to the first trend plot, similar to the links added in Section 7.4.2.
    - As in Section 7.4.2, you will first have to create a separate Web page for each Vendor that you can link to.

Chapter 9

Analyzing Quality Data with SAS

9.1 Introduction 310
9.2 Deciding Whether the Process Is In Control 314
9.3 Measuring Process Capability 318
9.4 Monitoring the Process 320
9.5 P-Charts for Fraction Failing 322
9.6 U-Charts for Defects per Unit 324
9.7 More Than Enough 326
9.8 Chapter Summary 328

9.1 Introduction

The terms *quality* and *reliability* are sometimes used interchangeably to refer to the relative "goodness" of manufactured products. In this book the term *quality data* refers to measurements taken on newly manufactured units, and *reliability data* refers to measurements of time to failure. For example, if your new laptop arrives with a dead screen, that's a quality problem, but if it dies after three months of use, that's a reliability problem. Quality data analysis is discussed in this chapter, and reliability data analysis is discussed in Chapter 10.

The three phases of statistical quality improvement are

Phase 1: Bring the process into a state of *statistical control*.

- A process is said to be in *statistical control* if the process is stable and measurements taken on each batch of units can be reasonably assumed to come from the same statistical distribution. The steps to ensure a state of control are
    1. Measure the variable of interest for several sequentially manufactured batches of data.
    2. Run tests for special causes to determine whether the process is in control.
    3. If the process is not in control, attempt to remove any special (or assignable) causes of variation, and then repeat the above steps.
    4. If the process is in control, proceed to Phase 2.

Phase 2: Ensure that the process is *capable* of meeting its specifications.

- A process is said to be *capable* if only an acceptably small fraction of units fail to meet the specification. If the process distribution is approximately normal, process capability is usually measured by a capability index such as C_{pk}, which is defined in Section 9.3. The steps to ensure process capability are
    1. Compute the value of C_{pk} (or other capability index) for your stable process distribution, and compare it to your requirement (e.g. $C_{pk} > 1.33$).
    2. If the C_{pk} does not meet your requirement, improve the distribution by shifting the mean and/or removing any special causes of variation, and then re-measure the process capability.
    3. When the C_{pk} meets your requirement, proceed to Phase 3.

Phase 3: Continually monitor the process to ensure that it remains in control and capable. The steps used to monitor the process are
    1. Take measurements on periodic samples from the production units.
    2. Plot the measurements on a control chart.
    3. Whenever the process is out of control (i.e., fails one or more of the statistical tests), attempt to remove any special causes of variation.

This chapter illustrates some of the SAS procedures that can be used in support of each of these phases.

- PROC SHEWHART is used in the first phase to test for a state of control.
- PROC CAPABILITY is used in the second phase to measure process capability.
- PROC SHEWHART is used again in the third phase to monitor the process.

Table 9-1 lists all of the procedures in SAS/QC software, with a brief description of how each can be used. PROC ANOM is discussed in Section 5.6 and PROC RELIABILITY is discussed in Chapter 10. This chapter focuses on the use of PROC SHEWHART and PROC CAPABILITY to support the implementation of statistical process control.

The examples in this section cover the easy parts of statistical process control—creating the control charts, deciding whether the process is in control and computing the process capability index. The ISHIKAWA, PARETO, OPTEX and FACTEX procedures, as well as the ADX Interface for Design of Experiments, can be used to help with the more difficult tasks—improving the process until it is in control and capable of meeting the specification limits, and then diagnosing and eliminating special causes of variation when the process goes out of control.

Table 9-1: Procedures in SAS/QC Software

\multicolumn{2}{c}{Statistical Quality Measurement and Improvement}
PROC SHEWHART	Creates Shewhart control charts to test whether a process is in control.
PROC CAPABILITY	Compares the distribution of process measurements to specifications and computes C_{pk} and other capability indices.
PROC CUSUM	Creates cumulative sum (cusum) control charts.
PROC MACONTROL	Creates uniformly or exponentially weighted moving average control charts.
\multicolumn{2}{c}{Design of Experiments*}
PROC FACTEX	Constructs orthogonal factorial experimental designs.
PROC OPTEX	Searches for optimal experimental designs.
\multicolumn{2}{l}{* Also consider the SAS ADX Interface for Design of Experiments. This is not a procedure but an interactive environment for creating experimental designs and analyzing the results of the experiments.}
\multicolumn{2}{c}{Other}
PROC ISHIKAWA	Allows interactive creation of Ishikawa, or Fishbone, diagrams to display the factors which affect a particular quality characteristic or problem.
PROC PARETO	Creates Pareto charts showing the relative frequency of different classes of quality-related problems.
PROC ANOM	Compares treatment means and determines which are significantly different from the overall average.
PROC RELIABILITY	Analyzes event time data, for example, product failure or repair times or biological survival and recurrence times.

The examples in this chapter cover only a small fraction of the capabilities of PROC SHEWHART and PROC CAPABILITY. This should be enough to get you started, but please consult the SAS/QC documentation to explore the many other options you can choose to customize the kinds of charts, methods of analysis, and graphical output for these procedures.

9.1.1 Quality Data for the Examples

The JES.First, JES.Second, JES.Third, and JES.Production data sets represent samples taken at the different phases of statistical process control.

- JES.First contains the measurements on the first set of 25 batches of units tested.
    - At this point the process in not in control.
- JES.Second contains the measurements on the second set of 25 batches.
    - At this point the process is in control, but not capable.
- JES.Third contains the measurements on the third set of 25 batches.
    - At this point the process is in control and capable.
- JES.Production contains measurements on samples from the production process.
    - At this point the process is assumed to be in control and capable and is being monitored to detect changes.

The table below shows the first 15 rows of the JES.First data set. The complete data set includes measurements on each of 10 units for each of 25 batches, for a total of 250 rows. The JES.Second, JES.Third, and JES.Production data sets have the same format.

```
JES.First

Batch     Sample     Resistance     Fail     Result      Defects

1         1          15.04          0        Pass        0
1         2          14.87          0        Pass        0
1         3          14.44          0        Pass        2
1         4          14.50          0        Pass        0
1         5          15.39          0        Pass        0
1         6          16.66          0        Pass        0
1         7          10.60          1        Low Res     0
1         8          17.54          0        Pass        0
1         9          16.88          0        Pass        0
1         10         15.69          0        Pass        0
2         1          12.22          1        Low Res     0
2         2          17.07          0        Pass        1
2         3          13.00          0        Pass        0
2         4          20.18          0        Pass        0
2         5          16.54          0        Pass        0
(partial listing)
```

- Each row contains a resistance measurement and a defect count for the corresponding unit.
- If the resistance measurement is outside the specification limits of 12.5-22.5:
    - Result is set to "Low Res" or "High Res".
    - Fail is set to 1.
- A different type of control chart is appropriate for each of these three different types of measurement:
    - *Resistance* is a continuous variable which can be charted with a Mean Chart.
    - *Fail* is a binary variable which can be charted with a P-Chart.
    - *Defects* is a count per unit variable which can be charted with a U-Chart.

PROC SHEWHART can create Mean control charts using raw data (i.e., one row per measurement) such as in the JES.First data set, or summary data containing statistics for each batch. The JES library includes summary data sets created with PROC MEANS for each set of raw data. The JES. First_Sum data set shown here is the summarized version of the JES.First data set.

```
JES.First_Sum

Batch ResistanceN    ResistanceX    ResistanceS    ResistanceR

  1       10            15.16          1.92           6.94
  2       10            15.28          3.68          10.66
  3       10            14.72          2.56           8.45
  4       10            16.80          2.19           7.00
  5       10            13.76          3.91          13.47
  6       10            14.69          1.57           4.67
  7       10            16.04          4.83          13.96
  8       10            15.32          3.11          11.04
  9       10            17.13          3.51          11.31
 10       10            19.42          1.79           5.60
 11       10            18.15          2.90           8.83
 12       10            17.27          3.23          11.06
 13       10            17.16          3.39          11.08
 14       10            16.96          2.81           8.46
 15       10            18.24          2.42           6.56
 16       10            16.97          3.08           9.48
 17       10            17.11          1.92           7.01
 18       10            18.30          2.37           6.39
 19       10            17.66          2.99           9.18
 20       10            18.33          3.93          10.88
 21       10            18.20          3.18           9.55
 22       10            18.80          3.34           8.46
 23       10            18.52          3.27          11.34
 24       10            19.59          3.72          10.73
 25       10            19.99          2.79           8.90
```

In order to use a summary data set instead of the raw data, PROC SHEWHART requires that the variable names be constructed using the analysis or process variable name as the common prefix, followed by a one character code which identifies the statistic. In the example above:

- ResistanceN = the number of resistance measurements in the batch.
- ResistanceX = the mean of the resistance measurements in the batch.
- ResistanceS = the standard deviation of the resistance measurements in the batch.
- ResistanceR = the range of the resistance measurements in the batch.

The raw data sets are used in the examples in Section 9.2, and the summary data sets are used in the examples in Section 9.4.

Sections 9.2–9.4 go through the three phases of statistical quality improvement for the resistance measurements. Sections 9.5 and 9.6 illustrate the use of P-Charts and U-Charts for the Fail and Defects measurements, respectively.

9.2 Deciding Whether the Process Is In Control

In the first phase of the quality improvement process, measurements are taken on sequentially produced batches of units to determine whether the process is in a state of control. The JES.First data set contains this initial set of samples. You can check for control of a continuous variable by creating a Mean Chart, and applying tests for process stability. The code sample uses PROC SHEWHART to create a Mean control chart and perform tests for special causes based on the Western Electric rules.

```
GOPTIONS RESET=ALL;
ODS HTML PATH="&JES.SG/S_9" (URL=NONE) BODY="SPC.html";
ODS GRAPHICS ON / RESET IMAGENAME="F9_1_";
ODS OUTPUT XCHART=Sum_Tab; ❶
PROC SHEWHART DATA=JES.First; ❷
  XCHART Resistance*Batch /STDDEVIATIONS
  TESTS=1 TO 8 ❸
  ZONELABELS TABLEALL;
RUN;
```

❶ The ODS OUTPUT statement requests that summary statistics be saved to the Sum_Tab data set.

❷ The PROC SHEWHART statement calls for analysis of the JES.First data set. The XCHART statement requests a Mean Chart of Resistance, grouped by Batch. The STDDEVIATIONS option specifies that the sample standard deviation, rather than the range, be used to compute the control limits. The limits for the chart can be specified either as sigma limits or probability limits. Sigma limits are computed as a specified number of standard deviations from the mean, while probability limits are set to achieve a specified probability of false alarm when the process is in control. The default is 3 sigma limits, corresponding to a false alarm probability of 0.135%, based on the normal distribution.

❸ The TESTS = 1 TO 8 option requests the 8 standard tests for special causes (see Table 9-2). The ZONELABELS option requests that the A, B, and C zones be displayed on the chart. The TABLEALL option sends a table of batch summary statistics to the Output window.

When you run the code, the Sum_Tab data set and the Mean Chart (Figure 9-1) are created. The Sum_Tab data set (not shown) is similar to the table written to the Output window, but does not include the test descriptions. The chart shows a very clear trend of increasing resistance, and several of the tests are positive, so the process is clearly not in control.

Table 9-2: Tests for Special Causes Using the Western Electric Rules

Index	Pattern Description
1	One point beyond Zone A (outside the control limits)
2	Nine points in a row in Zone C or beyond on one side of the central line
3	Six points in a row steadily increasing
4	Fourteen points in a row alternating up and down
5	Two out of three points in a row in Zone A or beyond
6	Four out of five points in a row in Zone B or beyond
7	Fifteen points in a row in Zone C on either or both sides of the central line
8	Eight points in a row on either or both sides of the central line with no points in Zone C
Zone Definitions	
Zone A	Distance from the center line (mean) is between 2 and 3 sigma
Zone B	Distance from the center line (mean) is between 1 and 2 sigma
Zone C	Distance from the center line (mean) is less than sigma

This is part of the output written to the Output window by the TABLEALL option.

```
Means Chart Summary for Resistance
          Subgroup  --------3 Sigma Limits with n=10 for Mean--------   Special
           Sample     Lower      Subgroup       Average        Upper     Tests
  Batch     Size      Limit        Mean           Mean         Limit    Signaled
    1        10     14.280573    15.161478     17.183435    20.086296
    2        10     14.280573    15.279966     17.183435    20.086296
    3        10     14.280573    14.721389     17.183435    20.086296      5
    4        10     14.280573    16.803839     17.183435    20.086296
    5        10     14.280573    13.758538     17.183435    20.086296    1  6
  (--- Rows for Batches 10-21 Omitted)
   22        10     14.280573    18.797901     17.183435    20.086296      6
   23        10     14.280573    18.521382     17.183435    20.086296
   24        10     14.280573    19.585652     17.183435    20.086296
   25        10     14.280573    19.994593     17.183435    20.086296      5

                           Test Descriptions
             Test 1    One point beyond Zone A (outside control limits)
             Test 2    Nine points in a row on one side of center line
             Test 5    Two out of three points in a row in Zone A or beyond
             Test 6    Four out of five points in a row in Zone B or beyond
```

Figure 9-1: Using the XCHART Statement in PROC SHEWHART

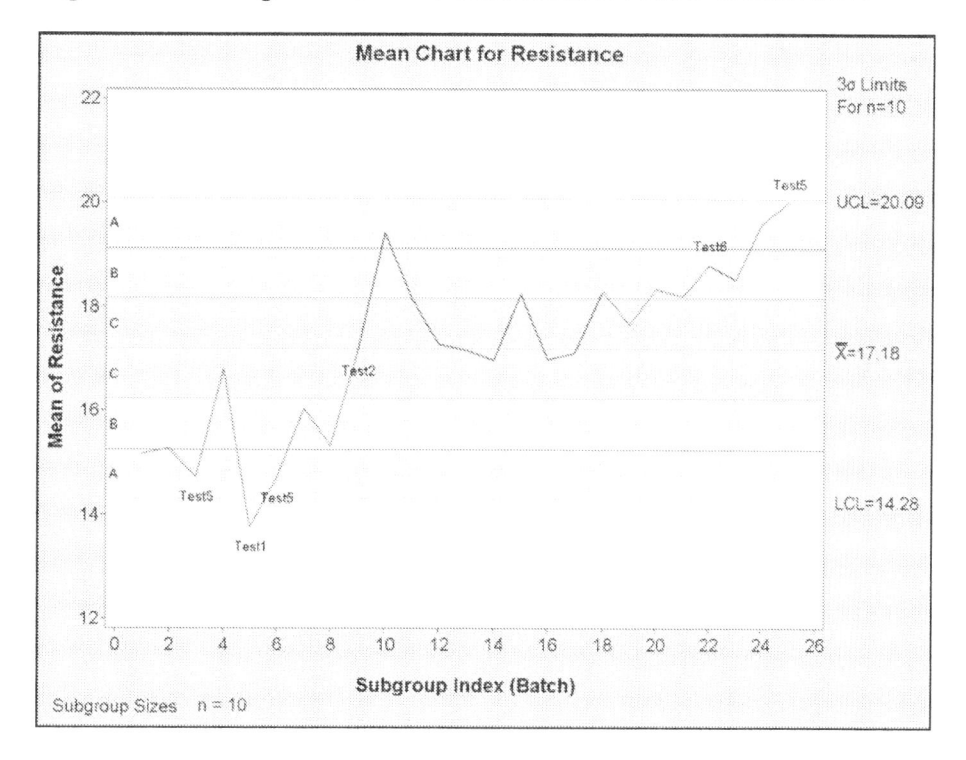

Because the initial process is found to be out of control, action is taken to remove special causes of variation, and then a second set of 25 batches of units are produced with the new process and measured as before. The next bit of code creates a Mean and Standard Deviation Chart from the new data set, JES.Second. This is similar to the code in the previous example, except that an XSCHART statement is used instead of XCHART.

```
/*==== Mean Chart for Second 25 Batches ====*/
ODS GRAPHICS ON / RESET IMAGENAME="F9_2_";
PROC SHEWHART DATA=JES.Second;
   XSCHART Resistance*Batch /
   TESTS=1 TO 8
   TABLEALL ZONELABELS;
RUN;
```

In this case, you can see from the resulting plot (Figure 9-2) that none of the eight tests are positive, so you can assume that the process is in control, with a predictable distribution, and move on to Phase 2.

Next you need to compare this distribution to the product specifications for resistance to see if the process is capable of meeting the specification limits. In Section 9.3, PROC CAPABILITY is used to assess the capability of the process, but first a box plot is created to provide a view of process trend vs. the specifications. You have already seen four ways to create box plots, using PROC BOXPLOT in Section 5.5, PROC ANOM in Section 5.6, PROC GPLOT in Section 6.3.2.7 and PROC SGPLOT in Section 8.3.4.2. The BOXCHART statement of PROC SHEWHART, which creates a box plot superimposed on a Mean Chart, uses many of the same options available for box plots created with PROC ANOM or PROC BOXPLOT.

```
/*==== Box Plots for Second 25 Batches ====*/
ODS GRAPHICS ON / RESET IMAGENAME="F9_3_";
PROC SHEWHART DATA=JES.Second;
   BOXCHART Resistance*Batch / ❶
   BOXCONNECT BOXSTYLE=SCHEMATICID ❷
   VREF=12.5 22.5 VREFLABELS='LSL' 'USL'; ❸
    ID Sample; ❹
RUN;
```

❶ The BOXCHART statement requests a box plot of Resistance by Batch. In a BOXCHART statement, the sample standard deviation is used by default, so the STDDEVIATIONS option is not needed.

❷ The BOXCONNECT option requests a line joining the means for each batch. The BOXSTYLE=SCHEMATIC option requests a schematic box plot in which the outliers are labeled with the value of an ID variable. See Section 5.5 for a discussion of schematic and skeletal box plots.

❸ The VREF and VREFLABELS options add reference lines at the specification limits.

❹ The ID option specifies the variable used to label the outlier points.

The resulting box plot (Figure 9-3) includes the same mean line and control limits as the Mean Chart in Figure 9-2. Note that the control limits apply to the mean values only, and not to any other features of the box plot. It is, however, fair to compare the box plots to the specification limits, LSL and USL. You can see that a large percentage of the units are below the lower specification limit, so the process, though in a state of control, is not yet capable of meeting the specifications. In Section 9.3 PROC CAPABILITY is used to compute the C_{pk} capability index, which provides a more quantitative measure of the process capability.

Figure 9-2: Using the XSCHART Statement in PROC SHEWHART

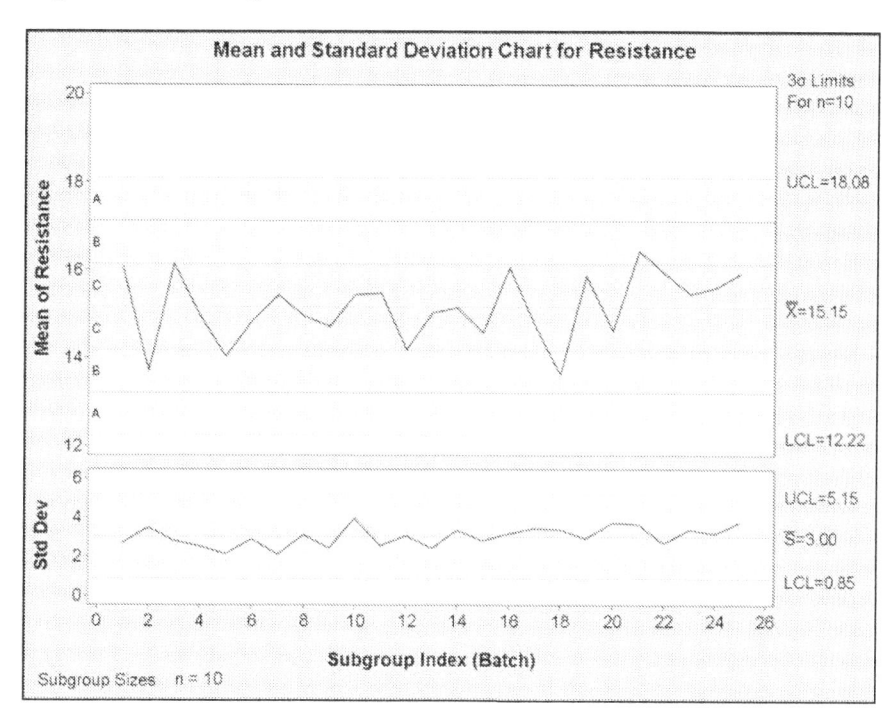

Figure 9-3: Using the BOXCHART Statement in PROC SHEWHART

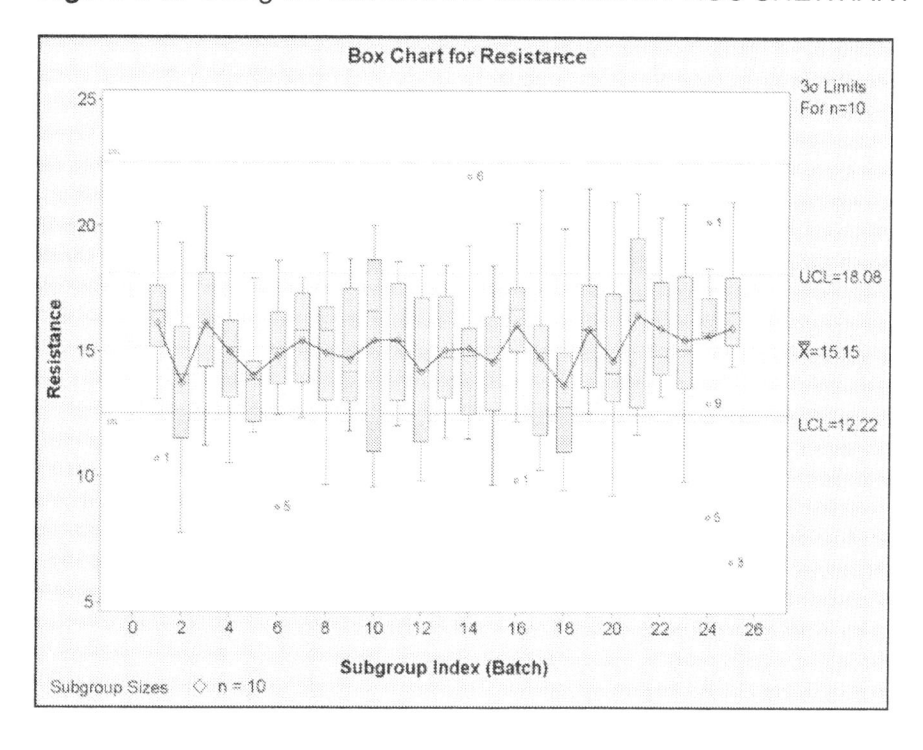

9.3 Measuring Process Capability

When the process is in a state of control, the measurements on each batch are considered to be random samples from the same distribution, so you can therefore use the aggregate data from all batches to estimate the process capability index, C_{pk}, defined by:

$$C_{pk} = \min\left(\frac{USL - \mu}{3\mu}, \frac{\mu - LSL}{3\mu}\right)$$

where μ and σ are the process mean and standard deviation. C_{pk} measures the distance of the process mean, μ, from the nearest specification limit, in units of 3σ. If the measured variable is normally distributed, then a C_{pk} of 1 means that the nearest specification limit is 3σ from the mean, so you can expect 0.135% of the units to be beyond the limit. If $C_{pk} = 1.33$, then the specification limit is 4σ from the mean, and only 0.003% of the units will be beyond the limit. A process is generally considered to be

- *Incapable* if $C_{pk} < 1$
- *Marginally Capable* if $1 < C_{pk} < 1.33$
- *Capable* if $C_{pk} > 1.33$

However, some would recommend putting the boundary at 1.67 (5σ) instead of 1.33 (4σ) in these definitions.

The code below uses PROC CAPABILITY to compute C_{pk} for the second set of 25 batches.

```
/*==== Process Capability - Second 25 Batches ====*/
ODS GRAPHICS ON / RESET IMAGENAME="F9_4_";
PROC CAPABILITY DATA=JES.Second OUTTABLE=Second_Tab;   ❶
   VAR Resistance;   ❷
       SPEC TARGET=17.5 LSL=12.5 USL=22.5   ❸
           CLEFT=Red CRIGHT=Red;
   HISTOGRAM / NORMAL CFILL=GRAYD0 ;   INSET N MEAN STD CPK;   ❹
RUN;
```

❶ The PROC CAPABILITY statement requests that the analysis be run on the JES.Second data set. The OUTTABLE option requests that summary statistics be saved to the Second_Tab data set.

❷ The VAR statement specifies that Resistance is the variable to be analyzed.

❸ The SPEC statement specifies the target value and specification limits for Resistance. The CLEFT and CRIGHT options specify the fill colors to be used for areas outside of the specification limits.

❹ The HISTOGRAM statement requests a histogram of the data. The NORMAL option requests a normal density fit to the histogram. The CFILL option specifies the fill color for the histogram. The INSET statement specifies statistics to be displayed in an inset on the plot.

When you run the code, the Second_Tab data set and the plot shown in Figure 9-4 are created. The Second_Tab data set also includes many other statistics, including skewness, kurtosis, percentiles, and other capability indices.

```
Second_Tab (selected columns)
Obs     _VAR_        _NOBS_     _MEAN_      _STD_     _MEDIAN_    _CPK_
 1     Resistance     250       15.1502    2.98951    15.1932    0.29550
```

Because the C_{pk} is only 0.295, the process is not yet capable, so actions are taken to increase the mean and reduce the variability to improve the C_{pk}. Then a third set of 25 batches of units are produced with the new process and measured as before. Figure 9-5 shows the result of running PROC CAPABILITY on the new data, which is in the JES.Third data set. The process is now marginally capable, with a C_{pk} of 1.1, so you can proceed to Phase 3, monitoring the process.

Figure 9-4: Using PROC CAPABILITY on JES.Second—The Second Set of 25 Batches

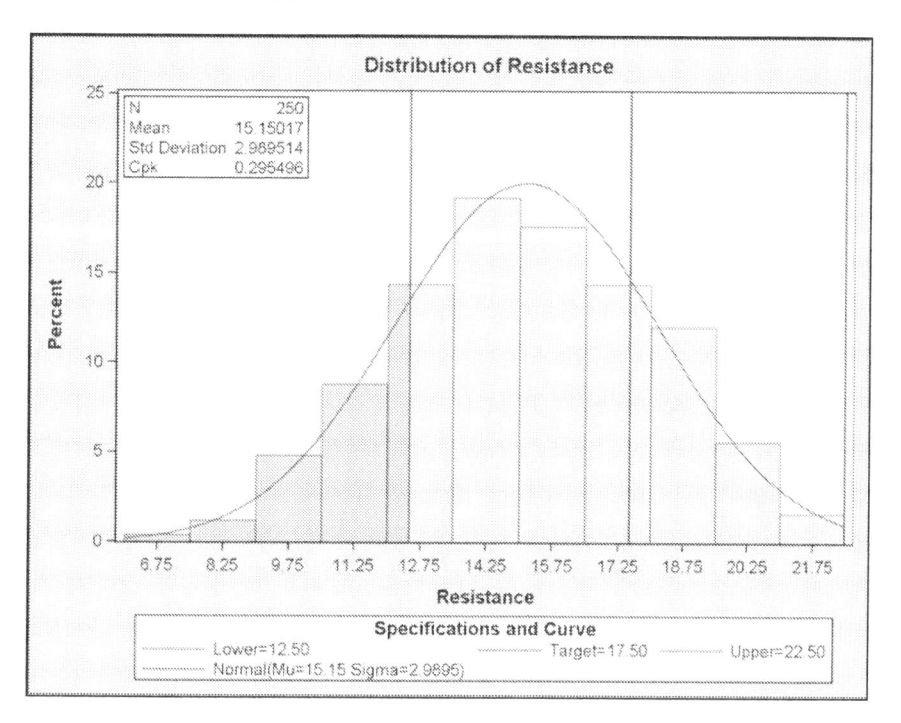

Figure 9-5: Using PROC CAPABILITY on JES.Third—The Third Set of 25 Batches

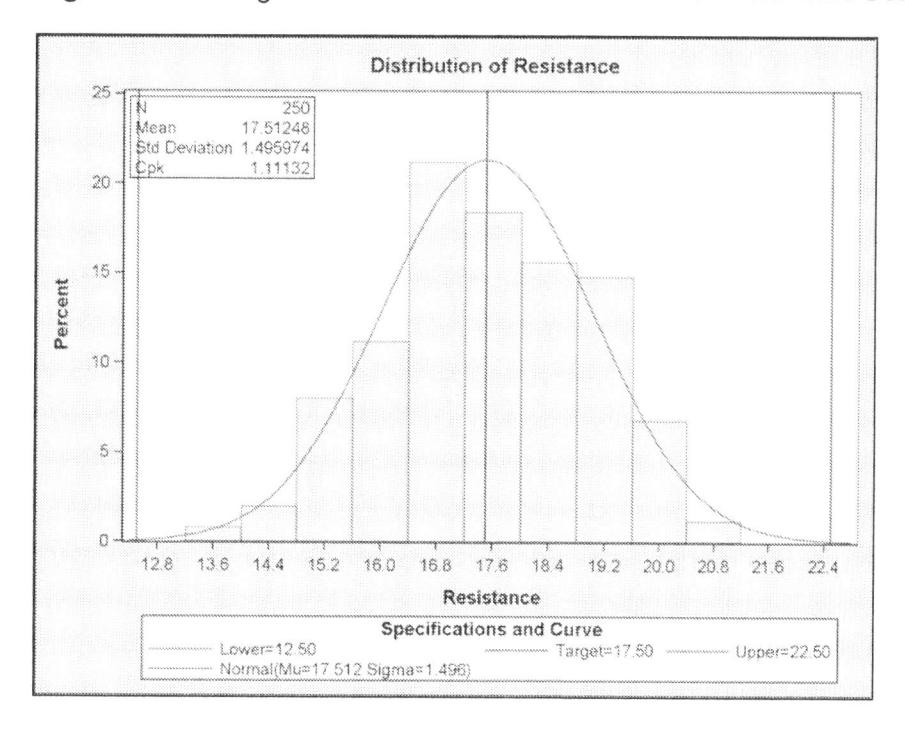

9.4 Monitoring the Process

After the process is in control and at least marginally capable, you need to monitor the process to ensure that it stays that way. The same control charts and tests for special causes that are used to test for a state of control can also be used to detect changes that cause the process to go out of control. You should use the control limits determined during the previous phase to judge the stability of the production process. The next code sample creates a Mean Chart for the JES.Third data set, and saves the control limits to a data set. This example uses the summary data set, JES.Third_Sum, instead of the raw data set, JES.Third, just to illustrate the syntax for running PROC SHEWHART with summary data. The relationship between the raw and summarized data sets is explained in Section 9.1.1.

```
/*==== Mean Chart for Resistance - Third 25 Batches ====*/
ODS GRAPHICS ON / RESET IMAGENAME="F9_6_";
PROC SHEWHART HISTORY=JES.Third_Sum;   ❶
   XCHART Resistance*Batch / STDDEVIATIONS OUTLIMITS=ResLim   ❷
   TESTS=1 TO 8 ZONELABELS TABLEALL;
RUN;
```

❶ This run of PROC SHEWHART uses the summarized data records in the JES.Third_Sum data set. Using the HISTORY= option instead of DATA= tells SAS that this is a summary data set.

❷ Note that in this case, *Resistance* is not the name of a variable in the data set. It is the prefix for the three variable names used to construct the chart, ResistanceN, ResistanceX and ResistanceS, which are described in Section 9.1.1. The OUTLIMITS=ResLim option requests that the computed control limits and other statistics be saved to the ResLim data set. The OUTLIMITS option works exactly the same way if you use raw data instead of summary data.

This table shows selected columns in the ResLim data set, including the statistics used to create the control limits for the production data.

```
ResLim (selected columns)
  _VAR_        _SUBGRP_     _LIMITN_     _LCLX_      _MEAN_      _UCLX_     _STDDEV_
Resistance    Batch           10         16.0712     17.5125     18.9537     1.51922
```

The next bit of code creates a Mean Chart for the production data using the control limits created above.

```
/*==== Mean Chart for Resistance - Production Samples ====*/
ODS GRAPHICS ON / RESET IMAGENAME="F9_7_";
PROC SHEWHART HISTORY=JES.Production_Sum LIMITS=ResLim;   ❶
   XCHART Resistance*Batch /
   TESTS=1 TO 8 ZONELABELS TABLEALL;
RUN;
```

❶ The LIMITS=ResLim option tells SAS to use the control limits from the ResLim data set. By default, the batch means are only plotted for batches with sample size equal to _LIMITN_ in the ResLim data set (all batches in this example), but you can use the ALLN option to force plotting of all means, and the NMARKERS option to use a different symbol to identify batches with different sample sizes.

The resulting plot is shown in Figure 9-7. Note that the control limits are the same as the limits in Figure 9-6, which was created with the first code sample using the JES.Third_Sum data set. You can see that the process is out of control because four out of five points in a row are in Zone B or beyond (Test 6 of Table 9-2). At this point you need to determine the cause of the apparent process shift, and take action to bring the process back into control.

Figure 9-6: Using the HISTORY Option of PROC SHEWHART

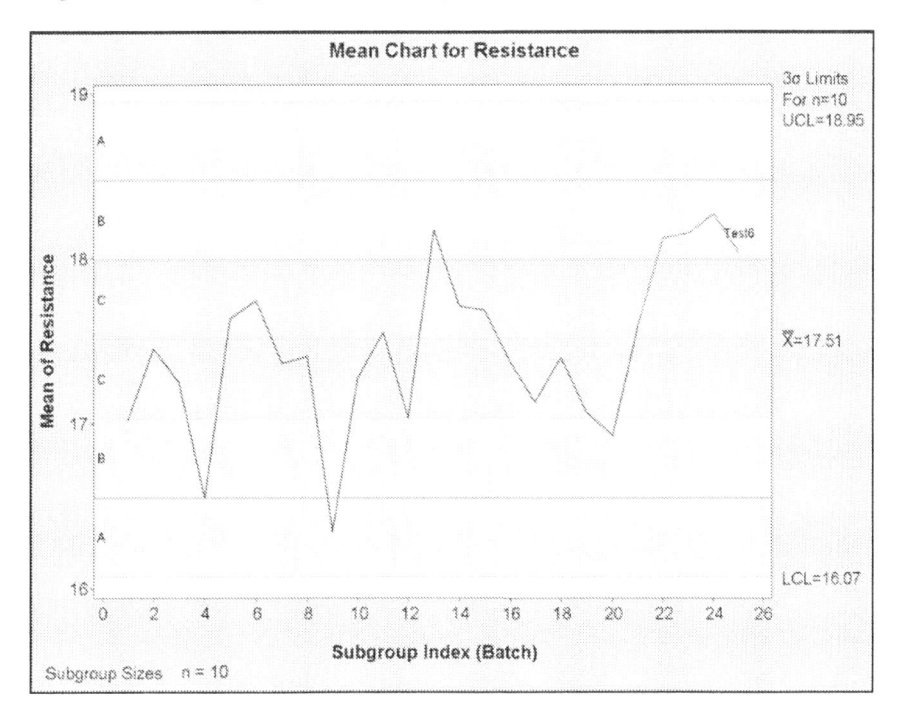

Figure 9-7: Using the LIMITS Option of PROC SHEWHART

9.5 P-Charts for Fraction Failing

The Mean and Standard Deviation Charts discussed in Sections 9.2 and 9.4 are generally appropriate for continuous variables, such as resistance, where the subgroup means can be assumed to follow the normal distribution. But quality data is often captured as the outcome of a pass/fail test, or a defect count for each unit. For these kinds of data, the subgroup means are not well approximated by the normal distribution unless the sample sizes are very large. It is usually more appropriate to use a P-Chart, based on the binomial distribution, for pass/fail data, or a U-Chart, based on the Poisson distribution, for defect per unit data. You can create these charts using the PCHART and UCHART statements of PROC SHEWHART. The PCHART statement is illustrated in this section, and the UCHART statement is illustrated in Section 9.6.

Like the Means Chart, the P-Chart can be used in different ways during the different phases of quality improvement described in Section 9.1. You can use the P-Chart during Phase 1 to test whether the process is in control, and again in Phase 3 to monitor the process. The example in this section is focused on the monitoring phase. Continuing the previous example, the Resistance variable is assumed to be in control, and the process is now being monitored by a simple pass/fail test on a large sample, which is less expensive than the more precise measurement of resistance. The test results are in the Fail variable of the JES.Production_2 data set, which is similar to the JES.First data set shown in Section 9.1.1, except that the sample size for each batch is 1000 instead of 10.

PROC SHEWHART requires that the input data for P-Charts be in summary form, where each observation has summary data for one subgroup, including the number of units tested and the number or proportion of nonconforming units. This code creates the Fail_Tab data set with the required summary data.

```
/*===== P Chart for Fraction Failing - First 25 Batches =====*/
PROC MEANS DATA=JES.Production_2 NWAY NOPRINT;
  CLASS Batch;
  OUTPUT OUT=Fail_Tab N=Num_Test SUM(Fail)=Num_Fail;
RUN;
```

```
Fail_Tab
Batch      _TYPE_     _FREQ_     Num_Test     Num_Fail

  1          1         1000        1000           1
  2          1         1000        1000           1
  3          1         1000        1000           0
  4          1         1000        1000           4
  5          1         1000        1000           3
              (Batches 6-20 Omitted)
 21          1         1000        1000           9
 22          1         1000        1000           8
 23          1         1000        1000           9
 24          1         1000        1000           5
 25          1         1000        1000           8
```

As for the Means Chart (see Section 9.2), the limits for the P-Chart can be specified either as sigma limits or probability limits. Sigma limits are computed as a specified number of standard deviations from the mean, while probability limits are set to achieve a specified probability of false alarm when the process is in control. When the P-Chart is used in Phase 1, the central line is usually computed as the overall average of the data, but in Phase 3 it is more appropriate to specify the known value of P, based on the previous process average. For this example it is assumed that the known process average is 0.1%, and a P-Chart is constructed to monitor the process and detect any changes.

The next code sample uses PROC SHEWHART to create the required P-Chart using probability limits.

```
ODS GRAPHICS ON / RESET IMAGENAME="F9_8_";
PROC SHEWHART DATA=Fail_Tab;
   PCHART Num_Fail*Batch / SUBGROUPN=Num_Test   ❶
   P0 = .001 PSYMBOL=P0   ❷
   ALPHA=.01;   ❸
RUN;
```

❶ The PCHART statement specifies a P-Chart for Num_Fail by Batch. The SUBGROUPN option specifies that the Num_Test variable contains the number of units tested in each batch.

❷ The P0=.001 option sets the known value of P for the chart. If this option is omitted, P is computed as the average fraction failing in the sample. The PSYMBOL option specifies that the central line be labeled 'P_0' to indicate that it is a user-specified value and not the sample average.

❸ The ALPHA=.01 option specifies that the chart limits should be probability limits, with a false alarm probability of 0.01. You can use SIGMAS=K to request K-sigma limits instead. If neither the ALPHA nor SIGMAS options are specified, 3-sigma limits are used by default.

Figure 9-8: Using the PCHART Statement with PROC SHEWHART

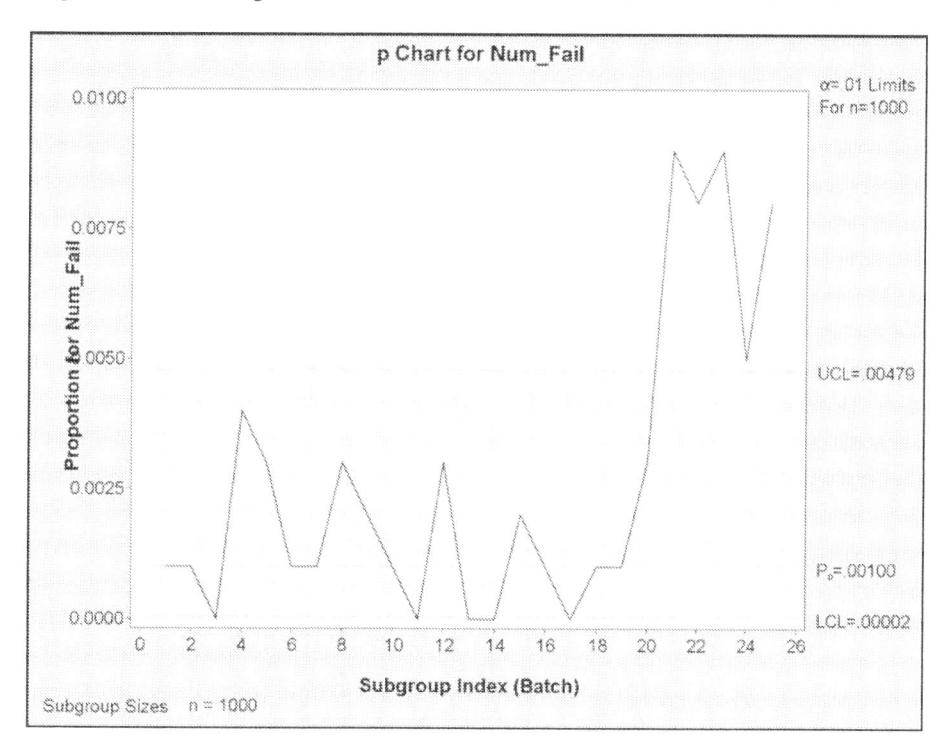

Beginning with Batch 21, several points are above the upper control limit, so the process is clearly out of control and steps should be taken to understand and eliminate the cause of the change.

9.6 U-Charts for Defects per Unit

The U-Chart, which is based on the assumption that the response variable has a Poisson distribution, is usually a good choice when the variable being controlled is the count of defects, or nonconformities, per unit. Because the structure of the data used for P-Charts and U-Charts is so similar—number of units tested and number of failures or defects for each batch—there is sometimes confusion about which chart to use. The answer depends on whether a single unit can fail more than once. If the test is pass/fail, and the number of failures per unit is either zero or one, then the sample data follows the binomial distribution, and the P-Chart should be used. But if a single unit can have more than one defect, for example if the number of defective parts on each circuit board are counted, then the sample data can be represented by the Poisson distribution, and a U-Chart should be used.

Following these rules, the Fail and Defects variables in the JES.First data set (see Section 9.1.1) should use the P-Chart and U-Chart, respectively. The example in this section creates a U-Chart for the Defects variable in the JES.Production_2 data set, which, as noted in Section 9.5, is similar to JES.First but with a batch size of 1000 units. As in Section 9.5, the process is assumed to be in control, with a known defect rate, and the U-Chart is being used to monitor the process.

PROC SHEWHART requires that the input data for U-Charts be in summary form, where each observation has summary data for one subgroup, including the number of units inspected and the total number of defects found on all units in the subgroup. This code sample creates the Defect_Tab data set with the required summary data.

```
/*===== U Chart for Defects per Unit - First 25 Batches =====*/
PROC MEANS DATA=JES.Production_2 NWAY NOPRINT;
  CLASS Batch;
  OUTPUT OUT=Defect_Tab N=Num_Inspect SUM(Defects)=Num_Defects;
RUN;
```

```
Defect_Tab (first 10 rows)
                            Num_         Num_
Batch    _TYPE_    _FREQ_   Inspect      Defects

  1        1       1000     1000         14
  2        1       1000     1000         10
  3        1       1000     1000         10
  4        1       1000     1000         22
  5        1       1000     1000         12
  6        1       1000     1000         10
  7        1       1000     1000         14
  8        1       1000     1000          8
  9        1       1000     1000         10
 10        1       1000     1000          9
```

The limits for the U-Chart can also be specified either as sigma limits, computed as a specified number of standard deviations from the mean, or probability limits, set to achieve a specified probability of false alarm when the process is in control. The central line of the U-Chart can be computed from the sample data, or specified by the user based on requirements or a previously measured process average. For this example it is assumed that the known process average is 0.01 defects per unit, and a U-Chart is constructed to monitor the process and detect any changes.

The code below uses PROC SHEWHART to create a U-Chart using probability limits.

```
ODS GRAPHICS ON / RESET IMAGENAME="F9_9_";
PROC SHEWHART DATA=Defect_Tab;
  UCHART Num_Defects*Batch / SUBGROUPN=Num_Inspect  ❶
  U0=.01  ❷
  USYMBOL=U0
  ALPHA = .01;  ❸
RUN;
```

❶ The UCHART statement specifies a U-Chart for Num_Defects by Batch. The SUBGROUPN specifies that the Num_Inspect variable contains the number of units inspected in each batch.

❷ The U0=.01 option sets the target value of U for the chart. If this option is omitted, U is computed as the average number of defects per unit in the sample. The USYMBOL option specifies that the central line be labeled 'U_0' to indicate that it is a user-specified value and not the sample average.

❸ The ALPHA=.01 option specifies that the chart limits be probability limits, with a false alarm probability of 1%. You can use the SIGMAS=K option to request K-sigma limits instead. If neither the ALPHA nor SIGMAS options are specified, 3-sigma limits are used by default.

Figure 9-9: Using the UCHART Statement with PROC SHEWHART

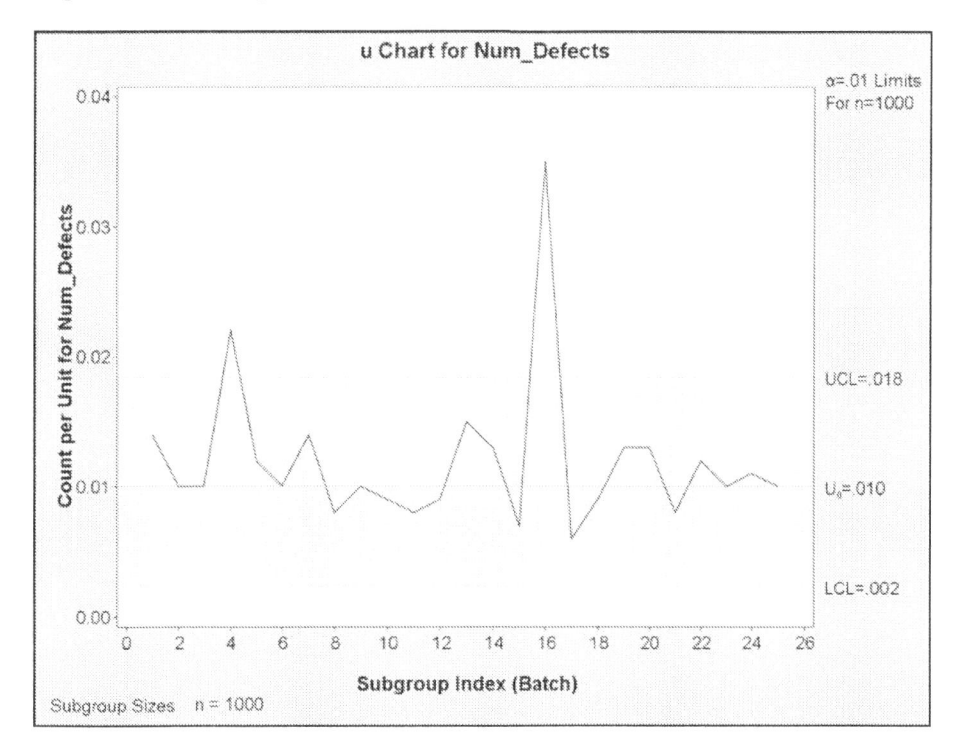

The points for batches 4 and 16 are beyond the control limits, so the process is not in control, and action should be taken to eliminate the causes of variation. Note that it is the *process* that is out of control, not the individual batches, so it is not correct to conclude that the process is "in control again" after batch 16, although this type of reasoning is often used as an excuse for inaction.

9.7 More Than Enough

The Shewhart charts described in the previous sections are the control charts most widely known and used in industry, but SAS/QC also includes procedures for creating other types of control charts. The MACRONTROL procedure can be used to create *Moving Average* charts and *Exponentially Weighted Moving Average* charts, and the CUSUM procedure can be used to create *Cumulative Sum*, or *Cusum*, charts. You can learn more about these procedures from the SAS online documentation. This section includes a very brief introduction to the CUSUM procedure.

Cusum control charts are used for the same purpose as Shewhart control charts in Phase 3 of the quality improvement process—to detect significant process shifts by examining the sample means of a series of subgroups. But instead of comparing each subgroup mean to the control limits independently, a Cusum Chart plots the cumulative sum of the subgroup means, suitably normalized, and tests for a significantly increasing or decreasing trend. The cusum statistic is defined recursively as

$$S_0 = 0$$

$$S_{t+1} = S_t + \frac{\overline{X}_t - \mu_0}{\sigma_0/\sqrt{n}}$$

where μ_0 and σ_0 are the known process mean and standard deviation, and n is the subgroup size. The theoretical advantage of a Cusum Chart over a Means Chart is that it can more quickly detect small shifts in the process mean because each value of the cusum statistic includes data from all previous subgroups. Therefore, the effect of a small shift can accumulate over several samples, and provide enough statistical evidence to trip the alarm before any one subgroup mean has exceeded its control limit. The difficulty, of course, it that the Cusum Chart is more difficult to create, and MUCH more difficult to explain. The CUSUM procedure can at least remove the difficulty in constructing the charts.

This example creates a Cusum Chart for the JES.Production data that can be compared to the Means Chart of the same data created in Section 9.4 and shown in Figure 9-7.

```
ODS GRAPHICS ON / RESET IMAGENAME="F9_10_";
PROC CUSUM DATA=JES.Production;  ❶
  XCHART Resistance*Batch /  ❷
  MU0=17.5 SIGMA0=1.5  ❸
  DELTA=1 ALPHA = .1  ❹
  VAXIS=-12 TO 8 BY 2;  ❺
RUN;
```

❶ The PROC CUSUM statement requests analysis of the JES.Production data set.

❷ The XCHART statement requests a Cusum Chart of Resistance, with subgroups determined by the Batch variable.

❸ The MU0 and SIGMA0 options specify the known process mean and standard deviation. These are the process averages computed in Section 9.4 for the JES.Third data set, and saved to the ResLim data set.

❹ The DELTA option specifies the size of the mean shift to be detected, as a multiple of the standard error. The ALPHA option specifies the probability of a Type 1 error, or false alarm.

❺ The VAXIS option specifies the vertical axis for the plot.

Figure 9-10 shows the graph created by the sample code.

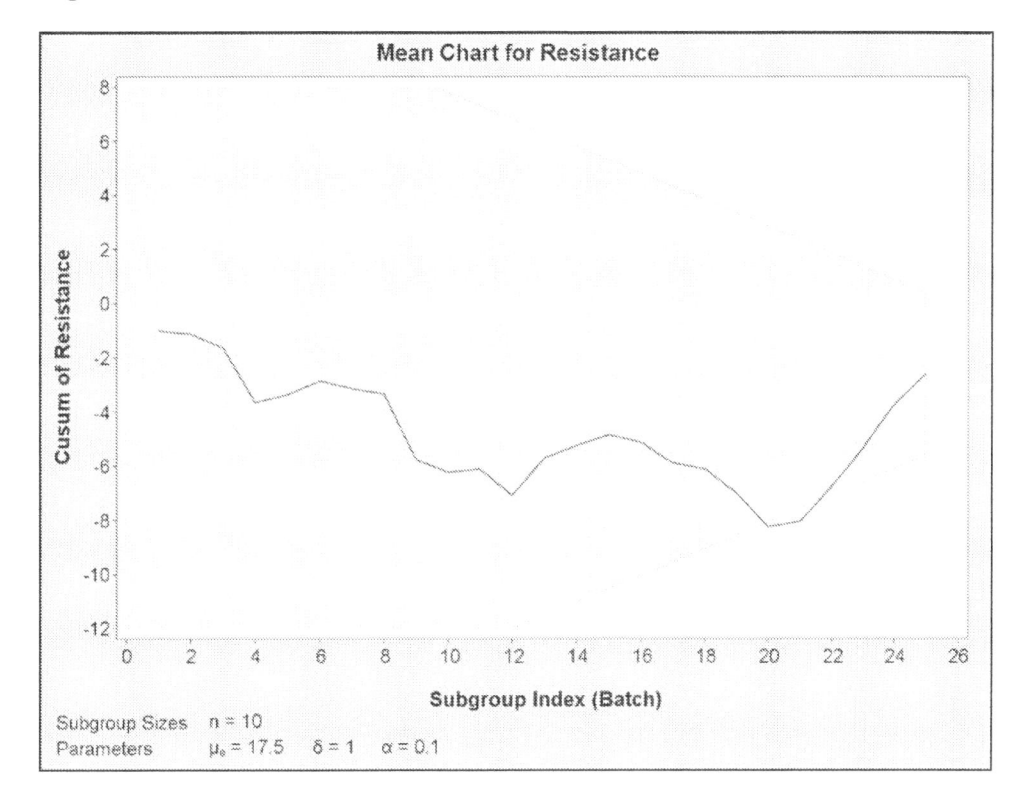

Figure 9-10: Cusum Chart Created with the XCHART Statement of PROC CUSUM

The plotted line is the cusum statistic S(t). The shaded area, or *V-Mask,* is the acceptance or continuation region, and is positioned so that the most recent point (Batch 25) is at the center of the vertical edge on the right. The process is considered to be out of control if any of the plot points are outside the V-Mask. In this example you can see that the points for batches 20 and 21 are outside the V-Mask, so a significant mean shift has been detected. The chart is basically looking backwards from the most recent subgroup to try to detect a trend in the previous several subgroups, and the dimensions of the V-Mask are computed to detect a shift of the size specified by the DELTA option as quickly as possible, with the false alarm probability specified by the ALPHA option.

If you compare this plot with Figure 9-7, you can see that the Cusum Chart has in fact fulfilled its promise by detecting the mean shift before any of the points on the Means Chart exceeded the upper control limit. On the other hand, the Means Chart did trigger an alarm at the same time based on one of the Western Electric rules, because four out of five points in a row are in Zone B or beyond.

9.8 Chapter Summary

9.8.1 Recap

After finishing this chapter you should know how to

- Use PROC SHEWHART to create control charts from SAS data sets, including
  - X-Bar and R (or Sigma) Charts for continuous variables
  - P-Charts for binomial variables, such as fraction failing
  - U-Charts for Poisson variables, such as defect counts
- Use the OUTLIMITS option to save control limits created by PROC SHEWHART.
- Use the LIMITS option to specify the control limits to be used by PROC SHEWHART.
- Use the TESTS option to determine whether the process is in a state of control.
- Use a BOXCHART statement to create box plots of your process data.
- Use PROC CAPABILITY to evaluate the capability of your process.

9.8.2 For More Information

A good place to start is the "Getting Started" sections for PROC CAPABILITY and PROC SHEWHART in the SAS/QC online documentation.

Books

Duncan, Acheson J. 1986. *Quality Control and Industrial Statistics, Fifth Edition*. Homewood, IL.: Irwin.

Grant, Eugene L., and Richard S. Leavenworth. 1988. *Statistical Quality Control, Sixth Edition*. New York: McGraw-Hill.

Montgomery, Douglas C. 1996. *Introduction to Statistical Quality Control, Third Edition*. New York: John Wiley & Sons.

van Dobben de Bruyn, Cornelis S. 1968. *Cumulative Sum Tests: Theory and Practice*. London: Griffin.

Papers

Nelson, Lloyd S. 1984. "The Shewhart Control Chart—Tests for Special Causes." *Journal of Quality Technology,* 16: 237–239.

Nelson, Lloyd S. 1985. "Interpreting Shewhart $\bar{X}$ Control Charts." *Journal of Quality Technology,* 17, 114–116.

9.8.3 Exercises

Your JES library contains data sets JES.Fourth, JES.Fifth and JES.Sixth, which are similar to JES.First, and also the summary data sets JES.Fourth_Sum, JES.Fifth_Sum, and JES.Sixth_Sum, which are similar to JES.First_Sum. For each of these data sets

- Plot a Mean Chart for Resistance to determine whether the process is in control.
  - Do this using the raw data (e.g., JES.Fourth) and also using the summary data (e.g., JES.Fourth_Sum).
- If the process is in control:
  - Use PROC CAPABILITY to estimate the process capability.
  - Rerun the Mean Chart using the OUTLIMITS option to capture the control limits.
  - Rerun the Mean Charts for the other data sets using these control limits.

Chapter 10

Analyzing Reliability Data with SAS

- 10.1 Introduction   332
- 10.2 PROC RELIABILITY   338
- 10.3 PROC LIFEREG   346
- 10.4 PROC LIFETEST   352
- 10.5 PROC PHREG   354
- 10.6 The Reliability of Repairable Units   358
- 10.7 More Than Enough   360
- 10.8 Chapter Summary   362

10.1 Introduction

This chapter will discuss SAS procedures for analyzing the reliability of *non-repairable* and *repairable* units.

Non-repairable units that fail are removed from service and cannot fail again. Typical examples of non-repairable units are memory chips and cell phones. The sample data for non-repairable units is usually right-censored because some of the units have not yet failed at the end of the observation period. Therefore, the observed lifetimes are the times to failure for the units that failed, and the censoring times of the units that did not fail. The objective of the SAS procedures described in this section is to estimate the distribution of T, the time to unit failure, based on the observed lifetimes of a sample of units. The distribution of time to failure can be characterized by any one of the functions F(t), S(t), f(t), h(t), or H(t), shown in Table 10-1. If any of these functions are known, then the others can be computed using the relationships shown in the table.

The distribution function of time to failure, F(t), might be assumed to have a known parametric form, such as the Weibull or Log-Normal distribution, or it might be completely unspecified. You can use the RELIABILITY or LIFEREG procedures to estimate the parameters of F(t) when it has a known parametric form, and the LIFETEST and PHREG procedures to estimate F(t) when the form is unspecified.

Repairable units that fail are repaired and placed into service again, and so might fail multiple times. Typical examples of repairable units are cars and computer systems. The sample data for repairable units is always right-censored because every unit has the possibility of failing as long as it is still in service. The censoring time is either the date the unit is removed from service, or the end of the observation period. The sample data includes the repair times for the units that were repaired, and the censoring times of all units. The reliability functions F(t), S(t), etc. can be used to characterize the distribution of time to first failure of repairable units, but these functions do not account for the second and subsequent failures. One can assume that the time between failures follows the same distribution as the time to first failure, but this assumption is too restrictive in many practical cases. It is more appropriate to use the Mean Cumulative Function, or MCF, to characterize the reliability of repairable units. The MCF, M(t), is defined as the mean number of repairs per unit up to time t. The MCF concept was introduced by Wayne Nelson in the papers cited in the "Chapter Summary" section, which also describe non-parametric methods to compute unbiased estimates and confidence limits for M(t). These methods are available in PROC RELIABILITY.

In some cases there are *explanatory variables* that are believed to influence reliability. For example, it is well known that electrical components tend to fail sooner when operating under stress conditions, such as elevated temperature or increased voltage. In such cases, it is often required to estimate the time to failure distribution as a function of the explanatory variables. The RELIABILITY, LIFEREG, and PHREG procedures have the capability to estimate the effect of explanatory variables on time to failure.

This chapter includes examples of reliability analysis using the RELIABILITY, LIFEREG, LIFETEST, and PHREG procedures. The choice of which procedure to use, and which method of analysis, depends on whether the units are repairable or not, and, for non-repairable units, whether the form of F(t) is known and whether there are explanatory variables to be considered. Table 10-2 summarizes the types of reliability analysis discussed in this chapter, and the SAS procedures that are used for each.

Table 10-1: Reliability Functions for Non-Repairable and Repairable Units

Reliability Functions for Non-Repairable Units (T = Time to Failure)		
Cumulative Distribution Function (CDF)	F(t)	Prob[T <= t]
Survival Function	S(t)	Prob[T>t] = 1-F(t) = exp[-H(t)]
Probability Density Function (PDF)	f(t)	$\lim_{\Delta t \to 0} \frac{\Pr\{t \leq T \leq t+\Delta t\}}{\Delta t} = dF(t)/dt$
Hazard Function	h(t)	$\lim_{\Delta t \to 0} \frac{\Pr\{t \leq T \leq t+\Delta t \mid T > t\}}{\Delta t} = \frac{f(t)}{1-F(t)}$
Cumulative Hazard Function	H(t)	$\int_0^t h(x)dx = -\log(S(t))$
Reliability Function for Repairable Units		
Mean Cumulative Function	M(t)	Mean[Cumulative number of repairs per unit in the time interval (0,t)]

The hazard function, or hazard rate, h(t), is sometimes referred to as the failure rate, but the term hazard is preferred as failure rate can be defined in various ways, while hazard is always defined as in Table 10-1. The Mean Cumulative Function, M(t), for repairable units is analogous to the Cumulative Hazard Function, H(t), for non-repairable units. In particular, if the failure process for repairable units is assumed to follow a Non-Homogeneous Poisson Process (NHPP), with the instantaneous failure rate at age t equal to m(t), independent of the prior repair history of the unit, then

$$m(t) = \lim_{\Delta t \to 0} \frac{\Pr\{\text{fail in}(t, t+\Delta t)\}}{\Delta t}$$

$$M(t) = \int_0^t m(x)dx$$

Table 10-2: Classification of Reliability Analysis Methods

Non-Repairable Units			
Form of F(t)	Explanatory Variables	SAS Procedure	Section
Known	No	RELIABILITY	10.2.1–2
		LIFEREG	10.3.1
Known	Yes	RELIABILITY	10.2.3–4
		LIFEREG	10.3.2
Unspecified	No	LIFETEST	10.4
Unspecified	Yes	PHREG	10.5
Repairable Units			
Form of M(t)	Explanatory Variables	SAS Procedure	Section
Unspecified	No	RELIABILITY	10.6

10.1.1 Parametric Lifetime Distributions

The lifetime distributions that can be fit with PROC RELIABILITY and PROC LIFEREG are Weibull, log-normal, and log-logistic. For each of these distributions, the random variable Y=log(T) belongs to the *location-scale* family of distributions, which means that F(y), the cumulative distribution function of Y, is of the form

$$\Pr\{Y \le y\} = F(y) = G\left(\frac{y-\mu}{\sigma}\right)$$

where

μ = location parameter

σ = scale parameter

G(.) is a cumulative distribution function

Table 10-3 shows the relationship between the distributions of T and log(T) for the distributions that can be used with these procedures.

Table 10-3: Time-to-Failure Distributions and Corresponding Location-Scale Distributions

Distribution of T	Distribution of log(T)
Log-normal	Normal
Weibull	Extreme Value
Log-logistic	Logistic

The examples in this chapter use the Weibull distribution. Table 10-4 shows the functions and parameters which characterize the distribution of T and log(T) for the Weibull and Extreme Value distributions, respectively. It is helpful to understand these relationships because SAS procedures report the parametric estimates in terms of both distributions. For example, if you specify the Weibull distribution, you will get estimates of the Weibull Scale, Weibull Shape, EV Location, and EV Scale.

Table 10-4: Weibull and Extreme Value Distributions

	Weibull Distribution	Extreme Value Distribution
Parameters	Weibull Scale = α = exp(μ) Weibull Shape = β = 1/σ	EV Location = μ = log(α) EV Scale = σ = 1/β
F(t)	$1 - \exp[-(t/\alpha)^\beta]$	$1 - \exp(-\exp[(\log(t)-\mu)/\sigma])$
S(t)	$\exp[-(t/\alpha)^\beta]$	$\exp(-\exp[(\log(t)-\mu)/\sigma])$
f(t)	$(\beta/\alpha)(t/\alpha)^{\beta-1}\exp[-(t/\alpha)^\beta]$	$\left(\frac{1}{\sigma t}\right)\exp\left(\frac{\log(t)-\mu}{\sigma}\right)\exp\left(-\exp\left[\frac{\log(t)-\mu}{\sigma}\right]\right)$
h(t)	$(\beta/\alpha)(t/\alpha)^{\beta-1} = \beta t^{\beta-1}\alpha^{-\beta}$	$(1/\sigma t)\exp[(\log(t)-\mu)/\sigma]$
H(t)	$(t/\alpha)^\beta$	$\exp[(\log(t)-\mu)/\sigma]$

The Weibull Scale parameter, α, is also called the *Characteristic Life*. When t=α, the equation for F(t) reduces to 1 - exp(-1) = 0.632, so the Characteristic Life is the time at which 63% of the units have failed. The Weibull hazard rate is proportional to $t^{\beta-1}$, which is a decreasing, constant, or increasing function of t according as $\beta<1$, $\beta=1$ or $\beta>1$. The Weibull distribution with $\beta=1$ is also called the Exponential, or constant failure rate, distribution.

10.1.2 Acceleration and Proportional Hazards Models

Explanatory variables are continuous variables or classification variables that are assumed to affect the distribution of time to failure. If the form of F(t) is known, the effects of explanatory variables will be represented by an *acceleration model*, which assumes that the effects are multiplicative on the event time, and that the location and scale parameters for the *i*-th unit can be written as

$$\mu_i = \beta_0 + \beta_1 X_{1i} + \ldots + \beta_n X_{ni}$$
$$\sigma_i = \sigma$$

where

$(X_{1i}, X_{2i}, \ldots, X_{ni})$ are the values of the explanatory variables for the *i*-th unit.

$(\beta_0, \beta_1, \ldots, \beta_n)$ and σ are unknown parameters to be estimated from the data.

One of the important uses of acceleration models is to predict component reliability with relatively short test times. For highly reliable components, it is often impractical to test enough units for a long enough time to estimate the time to failure distribution. So the units are tested at high stress conditions, and the data is used to estimate the model parameters. Then the model can be used to predict the time to failure at lower stress conditions by putting the appropriate values of the stress variables, X_{ij}, into the model. The book *Statistical Methods for Reliability Data* by Meeker and Escobar, which is listed in the "Chapter Summary" section, includes a good discussion of several commonly used acceleration models, including the Arrhenius model which is used in the examples in Sections 10.2.4 and 10.3.2. When log(T) has a location-scale distribution, the Arrhenius model can be written as

$$\mu_i = \beta_0 + \beta_1 \left[\frac{1000}{Temp\_C_i + 273.15} \right], \quad \sigma_i = \sigma$$

where $Temp\_C_i$ is the test temperature for the *i*-th unit in degrees Celsius.

If the form of F(t) is unspecified, the effects of explanatory variables can be represented by a *proportional hazards model*, which assumes that the effects are multiplicative on the hazard rate, and that the hazard rate of the *i*-th unit at age t can be written as

$$h_i(t) = h_0(t) \; \exp(\delta_1 X_{1i} + \ldots + \delta_n X_{ni})$$

where

$(X_{1i}, X_{2i}, \ldots, X_{ni})$ are the values of the explanatory variables for the *i*-th unit.

$(\delta_0, \delta_1, \ldots, \delta_n)$ are unknown parameters to be estimated from the data.

$h_0(t)$ is an unknown hazard function to be estimated from the data.

Note that, for a Weibull distribution, an acceleration model is also a proportional hazards model, because the hazard function can be written as

$$h_i(t) = h_0(t) \; \exp\left[(\beta_1/\sigma) X_{1i} + \ldots + (\beta_n/\sigma) X_{ni} \right]$$

10.1.3 Sample Data for Non-Repairable Units

The JES.Lifetest data set is used in the examples in Sections 10.2–10.5 to illustrate the reliability analysis of non-repairable units. This data set represents the results of an accelerated life test performed on 300 units from each of the three vendors, ChiTronix, Duality, and Empirical. One hundred units from each vendor were run for 150 hours at each of three different temperatures, 120, 140, and 160 degrees. The table below shows 18 of the 900 rows in JES.Lifetest, the first two rows of data for each combination of vendor and temperature.

```
JES.Lifetest (partial listing)
  Vendor        SN      TestTime     Censor        Temp_C
ChiTronix      100        150         -1           120
ChiTronix      101        150         -1           120
.....
ChiTronix      500         14          1           140
ChiTronix      501         53          1           140
.....
ChiTronix      800         20          1           160
ChiTronix      801         31          1           160
.....
Duality       1000        150         -1           120
Duality       1001        150         -1           120
.....
Duality       1500         47          1           140
Duality       1501        150         -1           140
.....
Duality       1800          7          1           160
Duality       1801         59          1           160
.....
Empirical     2000        150         -1           120
Empirical     2001        150         -1           120
.....
Empirical     2300         67          1           140
Empirical     2301        150         -1           140
.....
Empirical     2600         15          1           160
Empirical     2601        150         -1           160
```

- SN, the unit serial number, is used to uniquely identify each unit.
- TestTime is equal to the failure time for units which failed, and the time on test, or censoring time, for units which survived.
- Censor is used to identify which values of TestTime are failure times and which are censoring times.
  - Censor = -1 for units which survived until the end of the test.
  - Censor = 1 for units which failed during the test. The fail records are highlighted in the table.
- Temp_C is the temperature at which the unit was tested.

10.1.4 Sample Data for Repairable Units

The JES.Repair data set is used in the example in Section 10.6 to illustrate the reliability analysis of repairable units. This data set represents the repair history for 1000 units from each of the three vendors: ChiTronix, Duality, and Empirical. The JES.Repair data set includes one row for each unit, which specifies the total time in service for that unit, plus one row for each repair event, which specifies the time when the failure occurred.

```
JES.Repair (partial listing - first and last 3 SN for each vendor)

Vendor       SN       Start        Stop         EventDate    EventTime Censor
ChiTronix    1000     26JAN2008    13AUG2008    29JUL2008       185        1
ChiTronix    1000     26JAN2008    13AUG2008    13AUG2008       200       -1
ChiTronix    1001     25JAN2008    13AUG2008    13AUG2008       201       -1
ChiTronix    1002     25JAN2008    13AUG2008    13AUG2008       201       -1
.....
ChiTronix    1997     02APR2007    13AUG2008    27JUL2008       482        1
ChiTronix    1997     02APR2007    13AUG2008    13AUG2008       499       -1
ChiTronix    1998     01APR2007    13AUG2008    13AUG2008       500       -1
ChiTronix    1999     01APR2007    13AUG2008    13AUG2008       500       -1
Duality      5000     26JAN2008    13AUG2008    13AUG2008       200       -1
Duality      5001     25JAN2008    13AUG2008    13AUG2008       201       -1
Duality      5002     25JAN2008    13AUG2008    13AUG2008       201       -1
.....
Duality      5997     02APR2007    13AUG2008    13AUG2008       499       -1
Duality      5998     01APR2007    13AUG2008    13AUG2008       500       -1
Duality      5999     01APR2007    13AUG2008    13AUG2008       500       -1
Empirical    7000     26JAN2008    13AUG2008    13AUG2008       200       -1
Empirical    7001     25JAN2008    13AUG2008    13AUG2008       201       -1
Empirical    7002     25JAN2008    13AUG2008    13AUG2008       201       -1
.....
Empirical    7997     02APR2007    13AUG2008    13AUG2008       499       -1
Empirical    7998     01APR2007    13AUG2008    13AUG2008       500       -1
Empirical    7999     01APR2007    13AUG2008    13AUG2008       500       -1
```

- SN, the unit serial number, is used to uniquely identify each unit.
- Start is the date when the unit was placed into service.
- Stop is the last date for which you have repair data. The Start date varies by unit, but the Stop dates are all equal to August 13, 2008. This kind of data is common when units are sold or installed continuously over time, and all of the repair data is collected on the same date.
- EventDate is the date of failure (and repair) for each failure event, or the Stop date.
- EventTime = EventDate-Start is equal to a failure time or a censoring time, depending on the value of the Censor variable.
- Censor is used to identify which values of EventTime are failure times and which are censoring times.
  - Censor = -1 for records with EventTime equal to the censoring time.
  - Censor = 1 for records with EventTime equal to a failure time.

The date variables, Start, Stop, and EventDate, are not required for the analysis, but are included to help clarify the meaning of the EventTime variable.

338 *Just Enough SAS: A Quick-Start Guide to SAS for Engineers*

10.2 PROC RELIABILITY

This section illustrates the use of PROC RELIABILITY to fit Weibull distributions and acceleration models to life test data. The code for this section is in ~\JES \sample_code\ch_10\reliability.sas.

10.2.1 Fitting a Single Weibull Distribution

The first example fits a single distribution to only the 160° data from the JES.LifeTest data set described in Section 10.1.3. The examples in this chapter continue the practice of sending all results to the HTML destination, as discussed in Section 8.3.

```
GOPTIONS RESET=ALL;
ODS HTML PATH="&JES.SG\S_10_2" (URL=NONE) BODY="Reliability.html";

/*===== Probability Plot for Units tested at 160 =====*/
ODS GRAPHICS ON / RESET IMAGENAME="F10_1_"; ❶
ODS OUTPUT ParmEst=ParmEst_1 PctEst=PctEst_1; ❷
PROC RELIABILITY DATA=JES.LifeTest(WHERE=(Temp_C=160)); ❸
        DISTRIBUTION WEIBULL; ❹
        PROBPLOT TestTime*Censor(-1) / CONFIDENCE=.9; ❺
RUN;
```

❶ The ODS GRAPHICS ON statement requests the ODS Graphics functionality.

❷ The ODS OUTPUT statement requests that parameter and percentile estimates be saved as SAS data sets.

❸ The PROC RELIABILITY statement requests analysis of the rows in JES.LifeTest with Temp_C = 160.

❹ The DISTRIBUTION statement requests a fit to the Weibull Distribution.

❺ The PROBPLOT statement requests a probability plot where TestTime is the name of the life time variable and Censor(-1) specifies that rows with Censor = -1 are censoring times, not failure times. (See Section 10.1.3 for an explanation of censoring and failure times.) The Confidence = .9 option specifies that 90% confidence limits will be included in the plot.

When you run the code, the ParmEst_1 and PctEst_1 data sets are created along with the plot shown in Figure 10-1. The PctEst_1 data set (not shown) contains the estimated percentiles of the fitted distribution along with standard errors and confidence limits. The ParmEst_1 data set contains the estimated parameters of the fitted Extreme Value and equivalent Weibull distributions, along with the standard errors of the estimates and confidence limits. In terms of the notation of Table 10-4

μ = EV Location = 4.7493

σ = EV Scale 1.1202 =

α = Weibull Scale = $\exp(\mu)$ 115.5075 hours

β = Weibull Shape = $1/\sigma$ 0.8927

$F(t) = 1 \; \exp(t/115.51)^{.8927}$

```
ParmEst_1

Parameter          Estimate    Stderr    Lower      Upper

EV Location        4.7493      0.0773    4.6222     4.8764
EV Scale           1.1202      0.0675    1.0146     1.2368
Weibull Scale      115.5075    8.9263    101.7198   131.1639
Weibull Shape      0.8927      0.0538    0.8085     0.9857
```

The estimated Weibull Shape value in ParmEst_1, 0.8927, is slightly less than one, which means that the hazard rate of the fitted distribution is slowly decreasing with unit age.

Figure 10-1: Plot Created by the PROBPLOT Statement of PROC RELIABILITY

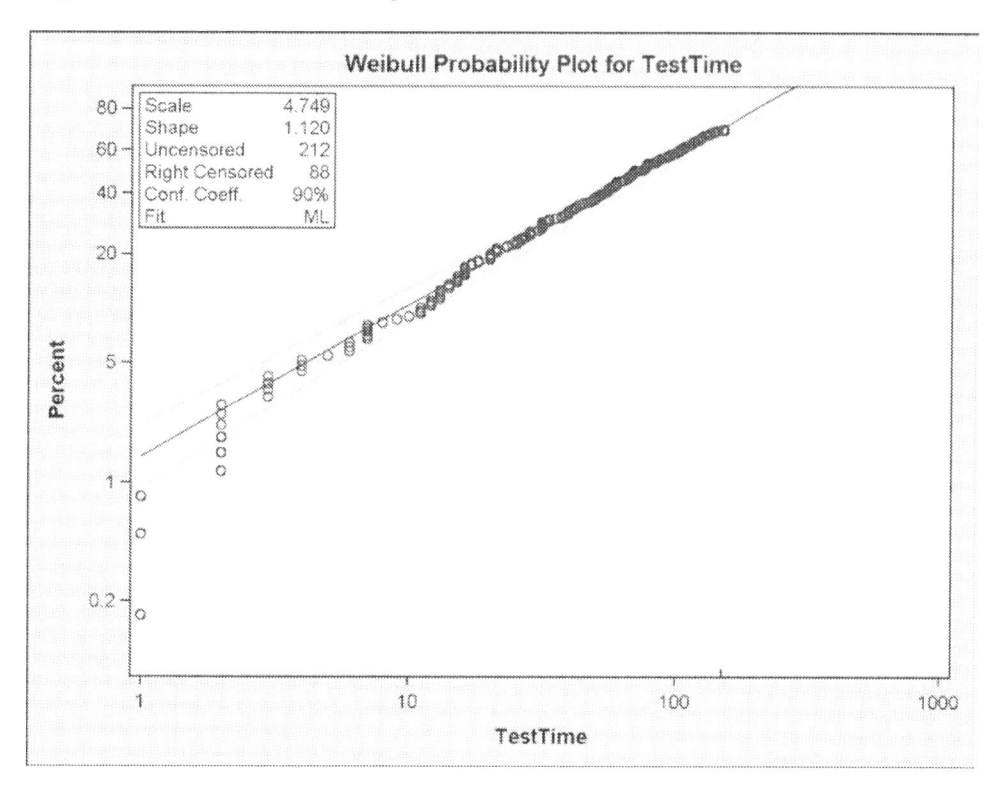

In a Weibull probability plot, the fitted distribution plots as the straight line:

$$\log\left[\log\left(1/[1-F(t)]\right)\right] = \beta \log(t) - \beta \log(\alpha)$$

The slope of the line is equal to β, the Weibull Shape parameter. The plotted circles are the non-parametric estimates of F(t) using, by default, the modified Kaplan-Meier method. See the SAS online documentation for the many options available to modify the default methods for fitting the data and displaying the plot.

The Weibull distribution appears to be a very good fit to the data, because the plot points are close to the fitted line.

10.2.2 Fitting Multiple Weibull Distributions

One of the objectives of an accelerated life test is to understand the relationship of the failure time distribution to stress conditions, such as the elevated temperatures in the JES.LifeTest data set. It is a good idea to start by fitting a distribution to the data taken at each stress condition, and plotting the actual and fitted distributions on the same graph. The next example fits a separate Weibull distribution for each value of Temp_C, and creates a probability plot for each fit on the same graph.

```
/*===== Probability Plots for each temperature on the same
graph=====*/
ODS GRAPHICS ON / RESET IMAGENAME="F10_2_";
ODS OUTPUT ParmEst=ParmEst_2;  ❶
PROC RELIABILITY DATA=JES.LifeTest;  ❷
  DISTRIBUTION WEIBULL;
  PROBPLOT TestTime*Censor(-1)=Temp_C / OVERLAY NOCONF;  ❸
RUN;
```

❶ The ODS OUTPUT statement requests that parameter estimates be saved to the ParmEst_2 data set.

❷ The PROC RELIABILITY statement requests a fit to all of the data in JES.LifeTest.

❸ The PROBPLOT statement requests a separate Weibull fit for each value of the group variable, Temp_C. The Weibull Scale and Shape parameters are fit independently for each value of Temp_C. The OVERLAY option causes all of the plots to be displayed on the same graph. If this option is omitted, three separate graphs are created. The NOCONF option suppresses the plotting of the confidence limits.

```
ParmEst_2

Parameter         Estimate    Stderr      Lower       Upper     Group

EV Location         6.5586    0.3283      5.9150      7.2021     120
EV Scale            0.6445    0.1251      0.4406      0.9427     120
Weibull Scale     705.2591  231.5621    370.5684   1342.2365     120
Weibull Shape       1.5517    0.3011      1.0608      2.2697     120
EV Location         6.2092    0.1800      5.8564      6.5620     140
EV Scale            1.1370    0.1162      0.9306      1.3891     140
Weibull Scale     497.3023   89.5215    349.4557    707.6995     140
Weibull Shape       0.8795    0.0899      0.7199      1.0745     140
EV Location         4.7493    0.0773      4.5979      4.9008     160
EV Scale            1.1202    0.0675      0.9955      1.2605     160
Weibull Scale     115.5075    8.9263     99.2727    134.3972     160
Weibull Shape       0.8927    0.0538      0.7933      1.0045     160
```

The ParmEst_2 data set contains the parameter estimates for the Weibull fit at each temperature. Note that the parameters for Temp_C=160 are exactly the same as the estimates obtained in Section 10.2.1.

The Weibull Scale, or Characteristic Life, parameters are 705 hours, 497 hours, and 116 hours, respectively for the units tested at 120, 140, and 160 degrees. The Weibull Shape parameters for the data taken at 160 and 140 degrees are quite close, but the Weibull Shape for the data taken at 120 degrees is very different. The Shape parameter for 120 degrees, 1.55, is greater than 1, which means that the hazard rate of the fitted distribution is increasing with unit age.

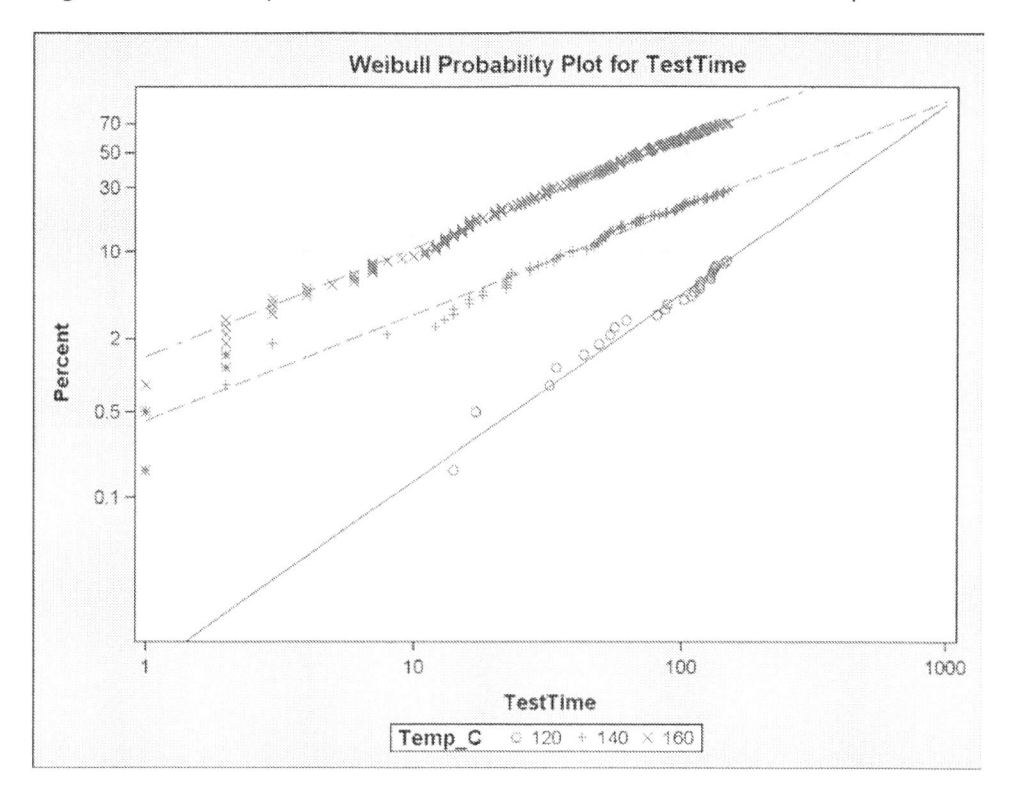

Figure 10-2: Independent Weibull Fits to the Data for Each Temperature

[Weibull Probability Plot for TestTime, showing data for Temp_C 120, 140, and 160]

Figure 10-2 shows the probability plots for the data taken at 120, 140, and 160 degrees. In each case the Weibull appears to be a good fit to the data. The slope of each line is equal to the estimated Weibull Shape parameter, so the fact that the shape parameters for the 140 and 160 degree data are approximately equal also means that the fitted lines are approximately parallel. And the fact that the shape parameter for 120 degree is higher (1.55 vs 0.88 and 0.89) means that the fitted line has a steeper slope.

The next step in the analysis is to try to fit an acceleration model (see Section 10.1.2) that explains the relationship of stress to failure time. A Weibull acceleration model assumes that the EV Location parameter, μ, varies with stress, but the EV Scale parameter, σ, is the same at each stress. This is equivalent to saying that the Weibull Scale parameter, or Characteristic Life, α, varies with stress, but the Weibull Shape parameter, β, is the same at each stress. The fact that the slopes, β, appear to be different for different temperatures suggests that it will be difficult to get a good fitting acceleration model.

10.2.3 Fitting a Weibull Acceleration Model

The next code sample estimates an acceleration model (see Section 10.1.2) by computing the best fitting Weibull distribution for the data taken at each temperature, with the condition that each fit must have the same estimated Weibull Shape parameter. This example uses the Temp_C variable as a classification variable, so the actual numeric temperature values are not used. Section 10.2.4 fits a more conventional acceleration model using the temperature values in an Arrhenius equation.

```
/*===== Plots for each temperature - with a common shape
parameter=====*/
ODS GRAPHICS ON / RESET IMAGENAME="F10_3_";
ODS OUTPUT ParmEst=ParmEst_3;
PROC RELIABILITY DATA=JES.LifeTest;
   DISTRIBUTION WEIBULL;
   MODEL TestTime*Censor(-1)=Temp_C;  ❶
   PROBPLOT TestTime*Censor(-1)=Temp_C / FIT=MODEL  ❷
OVERLAY NOCONF;
RUN;
```

❶ The MODEL statement requests a Weibull regression model fit to the JES.LifeTest data set. The model forces the Weibull Shape parameter to be the same for each value of Temp_C. The Temp_C variable is a classification variable in this fit, so the actual numeric values are not used. You can fit more complex models by including additional independent variables in the MODEL statement, for example:

```
MODEL TestTime*Censor(-1)=Temp_C Voltage Humidity;
```

❷ The PROBPLOT statement includes the FIT=MODEL option which requests that the fit from the previous MODEL statement be plotted.

```
ParmEst_3

Parameter         Estimate      Stderr       Lower        Upper     Group

EV Location         7.6134      0.2046       7.2124       8.0144      120
EV Scale            1.0922      0.0546       0.9903       1.2047      120
Weibull Scale    2025.1141    414.3581    1356.0801    3024.2219      120
Weibull Shape       0.9155      0.0458       0.8301       1.0098      120
EV Location         6.1779      0.1045       5.9731       6.3826      140
EV Scale            1.0922      0.0546       0.9903       1.2047      140
Weibull Scale     481.9623     50.3559     392.7160     591.4901      140
Weibull Shape       0.9155      0.0458       0.8301       1.0098      140
EV Location         4.7424      0.0730       4.5993       4.8855      160
EV Scale            1.0922      0.0546       0.9903       1.2047      160
Weibull Scale     114.7035      8.3747      99.4098     132.3500      160
Weibull Shape       0.9155      0.0458       0.8301       1.0098      160
```

The ParmEst_3 data set contains the parameter estimates for each temperature. The common Weibull Shape parameter is estimated at 0.9155, which means that the hazard rate is decreasing with unit age. Comparing these values to the values obtained in Section 10.2.2, you can see that the fit to the data for 140 and 160 degrees is quite similar, with Characteristic Life values of 482 and 115 hours compared to 497 and 116 and shape parameter 0.92 compared to 0.88 and 0.89. However, the fit to the 120 degree data is quite different, with Characteristic Life of 2025 hours compared to 705 hours, and shape parameter of 0.92 compared to 1.55.

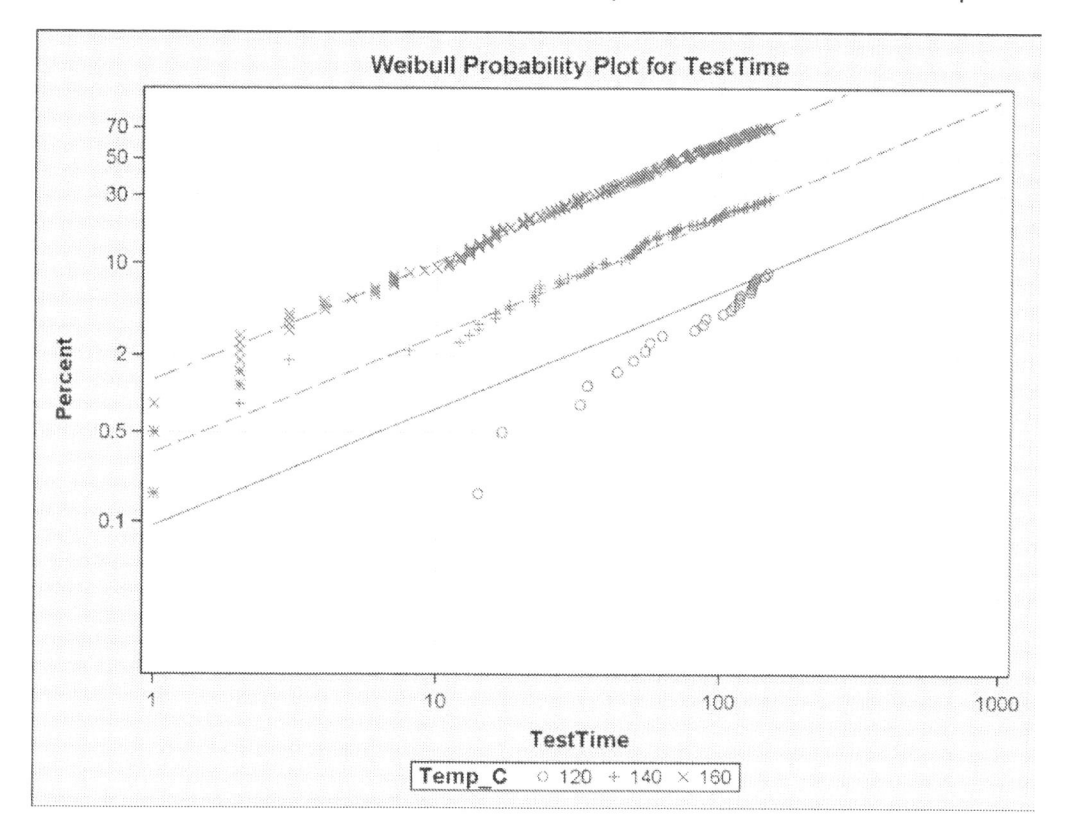

Figure 10-3: Weibull Fits with the Same Shape Parameter for Each Temperature

The fitted lines are all parallel because each is forced to have the same value of the Weibull Shape parameter. This has caused a noticeably poor fit to the data for Temp_C=120, so, if this were real data, you might want to reconsider the model, or confirm that the test conditions were properly controlled.

As noted earlier, the Temp_C variable is used as a classification variable in this example. That is not really an appropriate way to model temperature, and a more reasonable analysis based on the Arrhenius model is given in Section 10.2.4. However, the plot in Figure 10-3 does show you that the plot points for each temperature, which represent the Kaplan-Meier non-parametric estimates of F(t), do not appear to be parallel, so that it will be difficult to get a good fit using Arrhenius or any other model.

In some cases it is appropriate to use a classification variable in an acceleration model, for example to compare reliability among the different vendors at the same stress conditions. An example of this kind of analysis is included in the exercises at the end of the chapter.

10.2.4 Fitting an Arrhenius Acceleration Model

In the example in Section 10.2.3, Temp_C is treated as a classification variable. It is more appropriate to fit an acceleration model, such as the Arrhenius model, to the actual temperature values. For an Arrhenius acceleration model, the location and scale parameters for the *i*-th unit are given by

$$\mu_i = \beta_0 + \beta_1 \left[\frac{1000}{Temp\_C_i + 273.15} \right], \quad \sigma_i = \sigma$$

where $Temp\_C_i$ is the test temperature for the *i*-th unit, and β_0, β_1 and σ are unknown parameters to be estimated from the data. This code fits the Arrhenius model to the LifeTest data.

```
/*===== Fitting an Arrhenius Model to the data =====*/
ODS GRAPHICS ON / RESET IMAGENAME="F10_4_";
ODS OUTPUT ModPrmEst=ModPrmEst_4 ParmEst=ParmEst_4;  ❶
PROC RELIABILITY DATA=JES.LifeTest;
   DISTRIBUTION WEIBULL;
   MODEL TestTime*Censor(-1)=Temp_C/ RELATION=ARRHENIUS;  ❷
   PROBPLOT TestTime*Censor(-1)=Temp_C / FIT=MODEL OVERLAY NOCONF;
RUN;
```

❶ The ODS OUTPUT statement requests that the parameters of the fitted model be saved as SAS data sets. ModPrmEst_4 contains the estimates of the model parameters β_0, β_1 and σ, and ParmEst_4 contains the estimates of μ_i and σ_i for each test condition.

❷ The RELATION=ARRHENIUS option of the MODEL statement specifies that the Arrhenius model is to be fit. The other possible values of RELATION are shown in Table 10-5. In each case, the variable to be fit, Temp_C in the example, is transformed to the value shown in Table 10-5, and then an acceleration model of the form shown in Section 10.1.2 is fit to the transformed variable. Of course you can fit other models by including the appropriate transformed variable in your data set before running PROC RELIABILITY, and using the LINEAR option.

Table 10-5: Possible Values of the RELATION Option

RELATION	Transformed Variable
ARRHENIUS (Nelson parameterization)	1000/(X+273.15)
ARRHENIUS2 (activation energy parameterization)	11606/(X+273.15)
POWER	Log(X) , X>0
LINEAR	X
LOGISTIC	Log(X/(1-X)), 0<X<1

When you run the code, it creates the probability plot shown in Figure 10-4 as well as the ModPrmEst_4 and ParmEst_4 data sets.

```
ModPrmEst_4

Parm              Estimate      Stderr        Lower         Upper

Intercept         -24.0029      2.3063        -28.5231      -19.4827
Temp_C             12.4544      0.9849         10.5240       14.3848
EV Scale            1.0930      0.0547          0.9909        1.2056
Weibull Shape       0.9149      0.0458          0.8295        1.0091
```

The estimated values of β_0 and β_1 are the values of Estimate for "Intercept" and "Temp_C", respectively, so the estimated Weibull Scale parameter at Temp_C can be computed as

$$\text{Weibull Scale} = \exp(\mu_i) = \exp\left(-24.0029 + 12.4544 \times \left[\frac{1000}{Temp\_C + 273.15}\right]\right)$$

```
ParmEst_4
Parameter         Estimate      Stderr       Lower       Upper    Group
EV Location         7.6756      0.2105      7.2630      8.0882      120
EV Scale            1.0930      0.0547      0.9909      1.2056      120
Weibull Scale    2155.1057    453.6309   1426.5864   3255.6601      120
Weibull Shape       0.9149      0.0458      0.8295      1.0091      120
EV Location         6.1421      0.1026      5.9410      6.3432      140
EV Scale            1.0930      0.0547      0.9909      1.2056      140
Weibull Scale     465.0219     47.7158    380.3048    568.6108      140
Weibull Shape       0.9149      0.0458      0.8295      1.0091      140
EV Location         4.7502      0.0731      4.6070      4.8934      160
EV Scale            1.0930      0.0547      0.9909      1.2056      160
Weibull Scale     115.6061      8.4452    100.1842    133.4020      160
Weibull Shape       0.9149      0.0458      0.8295      1.0091      160
```

These Weibull Shape and Scale parameters are not very different from those in the previous example (see ParmEst_3), but in this case the scale parameters can be computed using the equation shown above. The model can be used to predict the distribution of time to failure at another temperature by entering the temperature in this equation to compute the Weibull Scale parameter, and assuming that the Weibull Shape parameter is the same as for the stress test data, 0.9149. Of course, in this particular example, you might not want to trust the model because the fit to the 120 degree data is not very good.

Figure 10-4: Probability Plot Created Using RELATION = ARRHENIUS

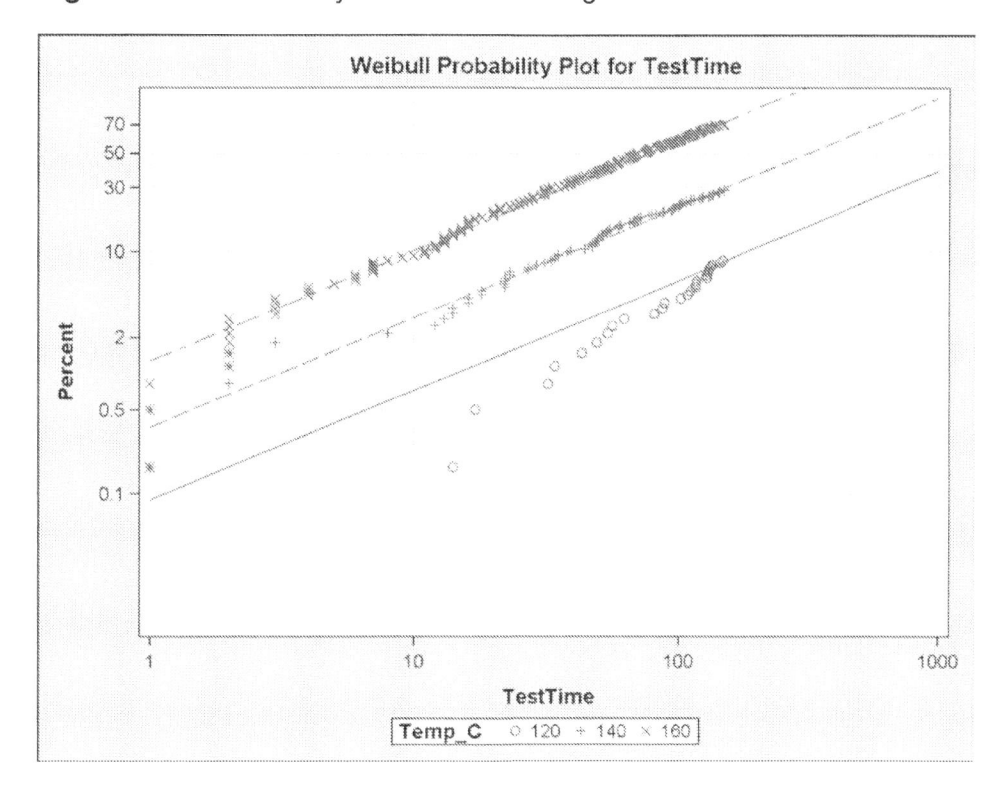

10.3 PROC LIFEREG

This section illustrates the use of PROC LIFEREG to fit Weibull distributions and acceleration models to life test data. The analysis is nearly identical to the examples using PROC RELIABILITY in Section 10.2. Both methods are included because PROC RELIABILITY is in SAS/QC while PROC LIFEREG is in SAS/STAT, and some SAS users will have access to one or the other but not both. The code for this section is in ~\JES\sample_code\ch_10\lifereg.sas.

10.3.1 Fitting a Single Weibull Distribution

The first example fits a Weibull distribution to the 160° data from the LifeTest data set, as in Section 10.2.1.

```
GOPTIONS RESET=ALL;
ODS HTML PATH="&JES.SG\S_10_3" (URL=NONE) BODY="Lifereg.html";

/*===== Fit the 160 degree test results only =====*/
ODS GRAPHICS ON / RESET IMAGENAME="F10_5_";
ODS OUTPUT ParameterEstimates=Estimates_1; ❶
PROC LIFEREG DATA=JES.LifeTest(WHERE=(Temp_C=160)); ❷
 MODEL TestTime*Censor(-1) = Temp_C / ❸
 DISTRIBUTION=WEIBULL;
 PROBPLOT;
 INSET;
 OUTPUT OUT=Fit_1 PREDICTED=Weibull_Scale QUANTILE=.63212055 CDF=CDF; ❹
RUN;
```

❶ The ODS OUTPUT statement requests that the parameter estimates be saved in a SAS data set.

❷ The PROC LIFEREG statement requests an analysis of the rows in LifeTest with Temp_C=160.

❸ The MODEL statement defines the model to be fit. TestTime is the name of the life time variable. Censor(-1) specifies that rows with Censor = -1 are censoring times, not failure times. Temp_C, the independent variable in the model, has the same value for every record, so it is not logically necessary, but LIFEREG requires an independent variable for the regression model. The DISTRIBUTION statement requests a fit to the Weibull distribution. The PROBPLOT statement requests a probability plot of the fitted model. The INSET statement requests that some fit statistics be displayed on the plot.

❹ The OUTPUT statement requests a SAS data set containing the input data and specified statistics. The PREDICTED option requests that a quantile of the fitted distribution be stored as the variable Weibull_Scale. The QUANTILE option requests the .63212055 (= 1/e) quantile which corresponds to the Weibull Scale parameter. If the quantile is not specified, then the median (0.50 quantile) will be reported by default. CDF requests the estimated value of F(t) for each TestTime t.

When you run this code, the Estimates_1 and Fit_1 data sets are created along with the plot shown in Figure 10-5. The Estimates_1 data set contains the estimated model parameters.

```
Estimates_1

                                                                   Prob
Parameter         DF   Estimate   StdErr    LowerCL   UpperCL   ChiSq    ChiSq
Intercept          1    4.7493    0.0773    4.5979    4.9008    3776.96  <.0001
Temp_C             0    0.0000       .         .         .         .        .
Scale              1    1.1202    0.0675    0.9955    1.2605       _        _
Weibull Shape      1    0.8927    0.0538    0.7933    1.0045       _        _
```

The fitted model is defined by the values of the Estimate variable in the Estimates_1 data set as

$$\mu_i = \beta_0 \;\; \beta_1 * \cancel{Temp}\_C_i \;\; 4.\cancel{7}493 = 0 * Temp\_C \;\; 4.7493, \;\; \sigma_i \;\; \sigma \;\; 1.1202$$

which is exactly the same as the fit obtained with PROC RELIABILITY in Section 10.2.1. Note that the "Scale" parameter in Estimates_1 is the EV Scale parameter defined in Table 10-4 of Section 10.1.1. The Weibull Scale parameter can be computed as $\exp(\mu)$, and can also be found as the Weibull_Scale variable in the Fit_1 data set, which was computed as the (1/e) percentile of the fitted distribution.

```
Fit_1 (selected columns)
                                            Weibull_
TestTime   Censor    Temp_C    _PROB_        Scale      CDF

   20        1        160      0.63212      115.507    0.18860
   31        1        160      0.63212      115.507    0.26586
  150       -1        160      0.63212      115.507    0.71712
  150       -1        160      0.63212      115.507    0.71712
  150       -1        160      0.63212      115.507    0.71712
                      (partial listing)
```

This probability plot is almost identical to the plot in Figure 10-1, except for differences in the axis scales, the information contained in the inset, and the default confidence level of 95%. The conclusions are the same: the Weibull distribution is a good fit to the 160 degree life test data.

Figure 10-5: Plot Created by the PROBPLOT Statement of PROC LIFEREG

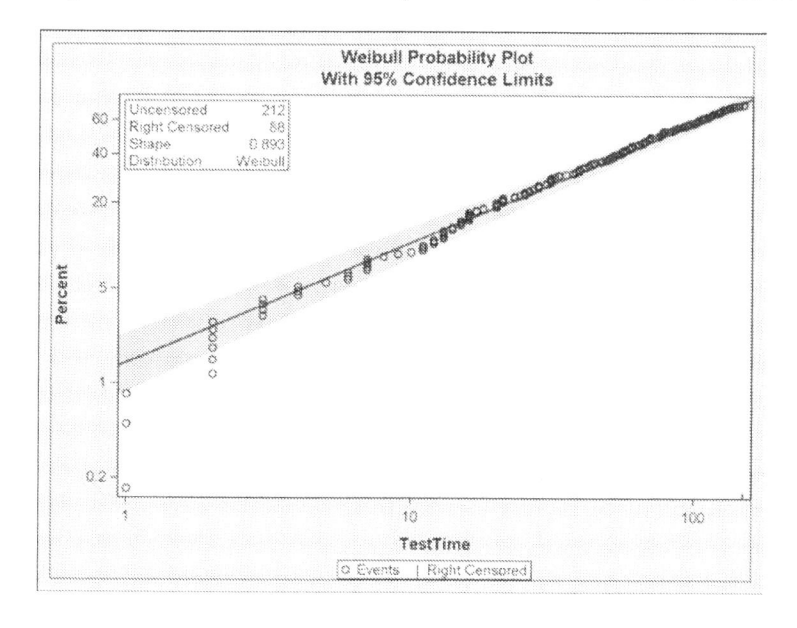

10.3.2 Fitting an Arrhenius Acceleration Model

The next example uses PROC LIFEREG to fit an Arrhenius model to the full LifeTest data set, which is similar to the fit using PROC RELIABILITY in Section 10.2.4.

```
/*===== Fit an Arrhenius Model =====*/
DATA LifeTest_2; SET JES.LifeTest; ❶
  Z = 1000/(Temp_C+273.15);
RUN;
ODS GRAPHICS ON / RESET IMAGENAME="F10_6_";
ODS OUTPUT ParameterEstimates=Estimates_2;
PROC LIFEREG DATA=LifeTest_2(KEEP=Temp_C Z TestTime Censor);
  MODEL TestTime*Censor(-1)= Z / ❷
  DISTRIBUTION=WEIBULL;
  PROBPLOT;
  INSET;
  OUTPUT OUT=Fit_2 PREDICTED=Weibull_Scale QUANTILE= .63212055
CDF=CDF;
RUN;
```

❶ The DATA step adds the variable Z to the JES.LifeTest data set. This is the same as the transformed variable that was specified by the RELATION=ARRHENIUS option used with PROC RELIABILITY in Section 10.2.4.

❷ The MODEL statement specifies Z as the independent variable.

When you run this code, the Estimates_2 and Fit_2 data sets are created along with the plot shown in Figure 10-6. The Estimates_2 data set contains the estimated model parameters and confidence limits on the estimates.

```
Estimates_2
                                                              Prob
Parameter       DF   Estimate   StdErr   LowerCL   UpperCL   ChiSq   ChiSq

Intercept       1    -24.0029   2.3063   -28.5231  -19.4827  108.32  <.0001
Z               1     12.4544   0.9849    10.5240   14.3848  159.89  <.0001
Scale           1      1.0930   0.0547     0.9909    1.2056    _       _
Weibull Shape   1      0.9149   0.0458     0.8295    1.0091    _       _
```

The fitted model is defined by the values of the Estimate variable as

$$\text{Weibull Scale} = \exp(-24.0029 + 12.4544 \times Z), \quad \text{Weibull Shape} = .9149$$

which is the same as the fitted model in Section 10.2.4. The Fit_2 data set includes the variable Scale, which is the estimated Weibull Scale parameter.

The probability plot generated by the PROBPLOT statement (Figure 10-6) requires some explanation.

- The points labeled "Events" are the Kaplan-Meier estimate of F(t), computed from *all* of the rows in LifeTest, and so this is a mixture of the data from all three temperatures.
- The straight line fit and the shaded confidence limit region correspond to the fitted model evaluated for fixed value(s) of the independent variable(s). The default values are
  - the average value of any continuous independent variables (Z in the example)
  - the highest level of any categorical independent variables (none in the example)
- Others values can be set by using an XDATA option. See the SAS online documentation for details.

Therefore, the plotted line corresponds to the average value of Z, so unless temperature has no effect on time to failure, you should not expect the plotted model to be a good fit to the Event points. The example in Section 10.3.3 shows how to use the Fit_2 data set to plot the Weibull fit from this analysis for each value of Temp_C.

```
Fit_2   (partial listing - first 3 rows for each value of temp_C)
                                                    Weibull_
TestTime    Censor      Temp_C       Z        _PROB_    Scale        CDF
   150        -1         120      2.54356    0.63212   2155.11    0.083613
   150        -1         120      2.54356    0.63212   2155.11    0.083613
   150        -1         120      2.54356    0.63212   2155.11    0.083613

    14         1         140      2.42043    0.63212   465.022    0.03975
    53         1         140      2.42043    0.63212   465.022    0.12812
    79         1         140      2.42043    0.63212   465.022    0.17925

    20         1         160      2.30867    0.63212   115.606    0.18197
    31         1         160      2.30867    0.63212   115.606    0.25913
   150        -1         160      2.30867    0.63212   115.606    0.71891
```

Figure 10-6: Plot Created for a Fitted Acceleration Model by PROC LIFEREG

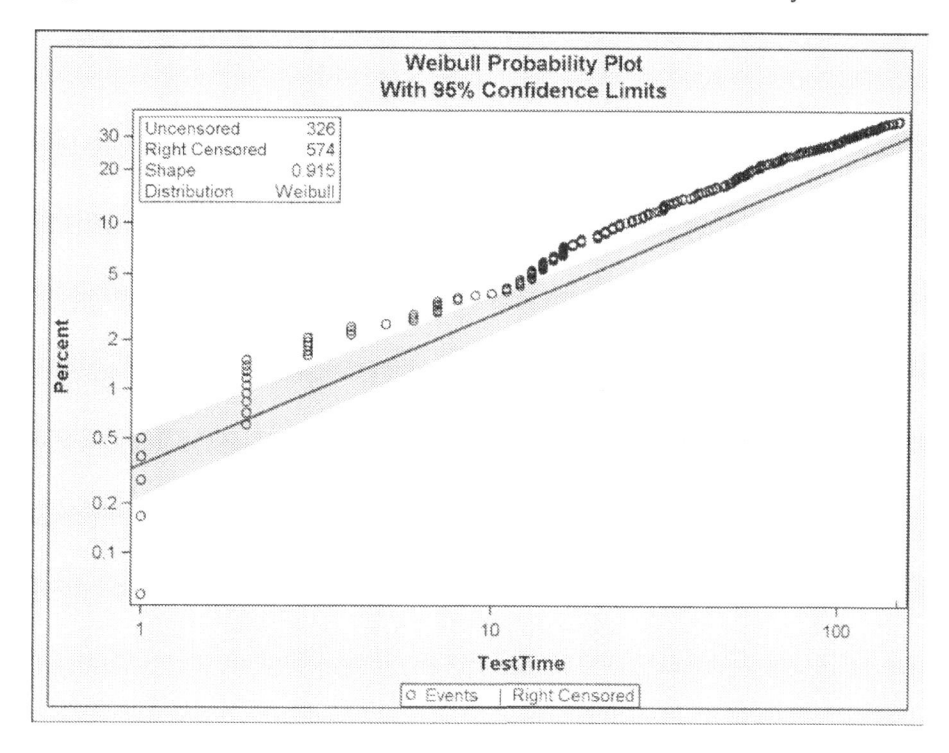

10.3.3 Probability Plots for Each Temperature

You can use the Fit_2 data set created with PROC LIFEREG in the Section 10.3.2 to create probability plots of the fitted Weibull distributions similar to those created using PROC RELIABILITY in Section 10.2.4. As noted in Section 10.2.1, a Weibull probability plot is a plot of

$$\log\left[\log\left(1/[1-F(t)]\right)\right] \text{ vs } \log(t)$$

where the actual data is represented by the Kaplan-Meier (or other non-parametric) estimate of F(t), while the fit is represented by the fitted value of F(t). The fitted value of F(t) has already been calculated as the CDF variable in the Fit_2 data set. If all units have the same test time, as they do in the example, then the Kaplan-Meier estimate of F(t) is equal to

$$F(t_i) = (i - .5)/n$$

where n is the number of units on test and t_i is the time of the i-th failure. This code computes the plot points for the actual and fitted distributions.

```
/*===== Add the Actual and Fitted values of Log(1/(1-F(t)) =====*/
PROC SORT DATA=Fit_2 OUT=Fit_2A; BY Temp_C TestTime; RUN;  ❶
DATA Fit_2A; SET Fit_2A; BY Temp_C TestTime; RETAIN Cum_Fail;  ❷
  IF FIRST.Temp_C=1 THEN DO;
      IF Censor=-1 THEN Cum_Fail=0;
      IF Censor= 1 THEN Cum_Fail=1;
  END;
  IF FIRST.Temp_C=0 THEN DO;
      IF Censor=1 THEN Cum_Fail=Cum_Fail+1;
  END;
  P_Fail= (Cum_Fail-.5)/(300);
  LSA  = LOG(1/(1-P_Fail));  ❸
  LSF  = LOG(1/(1-CDF));
RUN;
```

❶ The PROC SORT sorts the data set in increasing order of the failure times, within each value of Temp_C, to enable counting of the cumulative number of failures at or before each value of TestTime.

❷ The DATA step creates the Fit_2A data set from Fit_2 by adding the variables required for the plot. A RETAIN variable is used to compute Cum_Fail, the cumulative number of failures up until TestTime. The RETAIN statement is explained in Section 2.5. P_Fail is the Kaplan-Meier estimate of F(t_i), computed using the equation shown above, with i=Cum_Fail and t_i = TestTime, and n = 300, which is the sample size for each temperature in the LifeTest data set. The computation of P_Fail would be more complicated if the groups did not all have the same number of units (300) or the units did not all have the same test time (150 hours).

❸ LSA is equal to Log[1/(1-F(t)] for the Kaplan-Meier estimate, and LSF is equal to Log[1/(1-F(t)] for the fitted Weibull distribution.

The next bit of code uses PROC SGPLOT to plot the Kaplan-Meier estimates along with the fitted Weibull distributions for each value of Temp_C. The resulting plot, shown in Figure 10-7, is essentially the same as the plot created using PROC RELIABILITY, Figure 10-4, in Section 10.2.4.

```
/*===== Plot Actual and Fit =====*/
TITLE1 "Arrhenius Fit to Accelerated Life Test Data";
ODS GRAPHICS ON / RESET IMAGENAME="F10_7_";
PROC SGPLOT DATA=Fit_2A;
  SCATTER Y=LSA X=TestTime / GROUP=Temp_C;
  SERIES  Y=LSF X=TestTime / GROUP=Temp_C;
  YAXIS LABEL="Percent Failed" TYPE=LOG; XAXIS TYPE=LOG;
  FORMAT LSA PERCENT8.0;
RUN; QUIT;
```

```
Fit_2A (selected variables)
TestTime  Censor  Temp_C     CDF    Cum_Fail   P_Fail     LSA       LSF
      14       1     120  0.009922         1   0.001667  0.001668  0.009972
      17       1     120  0.011839         2   0.005000  0.005013  0.011910
      32       1     120  0.021020         3   0.008333  0.008368  0.021244
.........
       1       1     140  0.003620         1   0.001667  0.001668  0.003627
       1       1     140  0.003620         2   0.005000  0.005013  0.003627
       2       1     140  0.006814         3   0.008333  0.008368  0.006838
.........
       1       1     160  0.01287          1   0.00167   0.00167   0.01296
       1       1     160  0.01287          2   0.00500   0.00501   0.0129
       1       1     160  0.01287          3   0.00833   0.00837   0.01296

       (partial listing - first 3 rows for each value of Temp_C)
```

Figure 10-7: Probability Plot Created by PROC SGPLOT, Using Results from PROC LIFEREG

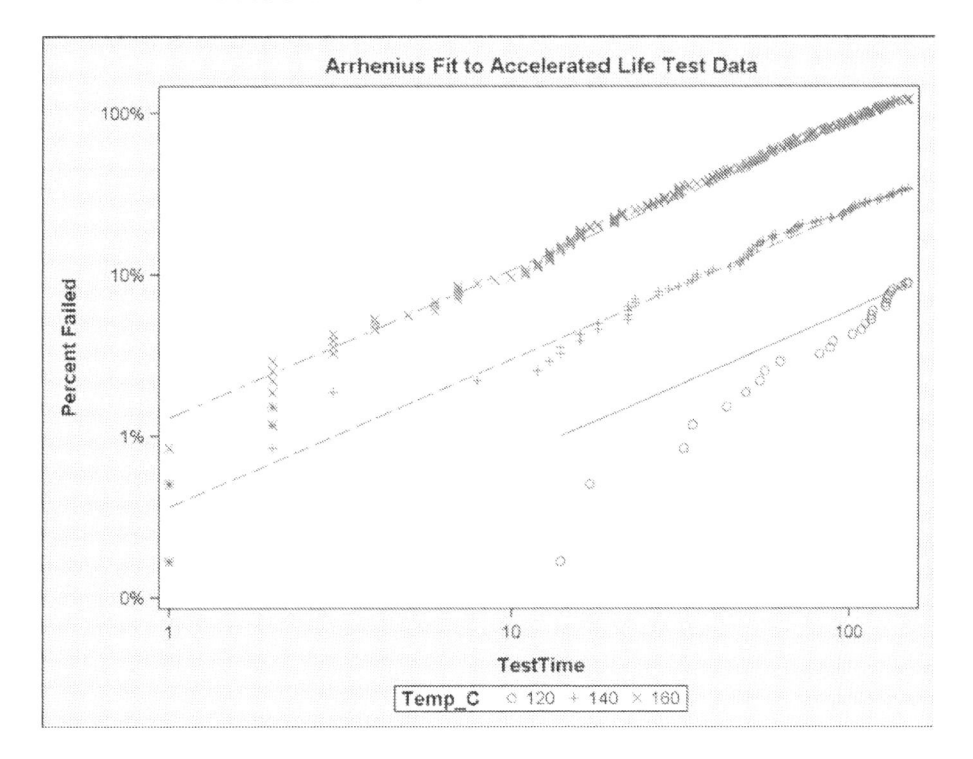

10.4 PROC LIFETEST

You can use PROC LIFETEST to compute non-parametric estimates of the survival function, S(t), using either the Product-Limit (Kaplan-Meier) or the Life-Table method. This section illustrates the use of the Kaplan-Meier method. See the SAS online documentation for a discussion of the Life-Table method. The code for this section is in ~\JES \sample_code\ch_10\lifetest.sas.

The sample code computes the Kaplan-Meier estimates for each value of Temp_C in the JES.LifeTest data set and creates a plot of the estimated survival curves.

```
GOPTIONS RESET=ALL;
ODS HTML PATH="&JES.SG\S_10_4" (URL=NONE) BODY="Lifetest.html";

ODS GRAPHICS ON / RESET IMAGENAME="F10_8_";
ODS OUTPUT HomTests = Tests;   ❶
PROC LIFETEST DATA=JES.LifeTest METHOD=KM ALPHA=.1;   ❷
  TIME TestTime*Censor(-1);   ❸
  STRATA Temp_C;   ❹
  SURVIVAL OUT=Life_Table   ❺
  PLOTS=SURVIVAL(EPB ATRISK STRATA=PANEL);   ❻
RUN;

ODS GRAPHICS OFF;
ODS _ALL_ CLOSE;
ODS LISTING;
```

❶ The ODS OUTPUT statement requests that default test statistics be saved to the Tests data set.

❷ The PROC LIFETEST statement requests an analysis of the data in the JES.LifeTest data set. The METHOD=KM option specifies the Kaplan-Meier method be used. The ALPHA=.1 specifies that 90% confidence limits for S(t) be computed.

❸ The TIME statement specifies that TestTime is the name of the life time variable, and the rows with Censor = -1 are censoring times, not failure times.

❹ The STRATA statement requests a separate estimate of S(t) for each value of Temp_C.

❺ The SURVIVAL statement creates the Life_Table data set containing the results of the analysis.

❻ The PLOTS option requests survival plots of the estimates of S(t). The EPB option requests equal precision confidence bands for S(t). The ATRISK option requests that the number of units at risk at each of several time points be shown on the plot. The STRATA=PANEL option specifies a panel of plots. If this option is left out, the estimated survival curves are overlaid on the same plot.

When you run the code, the Tests and Life_Table data sets, and the plot shown in Figure 10-8, are created.

```
Tests

  Test          ChiSq       DF    ProbChiSq
Log-Rank      315.0423      2       <.0001
Wilcoxon      307.7735      2       <.0001
-2Log(LR)     305.6244      2       <.0001
```

The Tests data set contains the results of three non-parametric tests of the hypothesis that all of the groups defined by the STRATA statement have the same survival curve. In this case the *p*-value for each test is less than .0001, so equality can be rejected at the 0.01% level of significance.

```
Life_Table (partial listing)
          Test
Temp_C    Time    _CENSOR_   SURVIVAL   CONFTYPE   SDF_LCL   SDF_UCL   STRATUM

  120       0        .        1.00000               1.00000   1.00000     1
  120      14        0        0.99667   LOGLOG     0.98285   0.99936     1
  120      17        0        0.99333   LOGLOG     0.97882   0.99791     1
  120      32        0        0.99000   LOGLOG     0.97436   0.99612     1
  120      34        0        0.98667   LOGLOG     0.96991   0.99412     1
  120      43        0        0.98333   LOGLOG     0.96554   0.99198     1
```

The Life_Table data set includes the estimated value of S(t) for each t_i, and the 90% equal precision confidence bands on S(t). The CONFTYPE variable indicates that the LOGLOG method was used to compute the confidence bands. See the SAS online documentation for the other available methods.

Figure 10-8: Survival Probability Plot Created with the PLOTS Statement of PROC LIFETEST

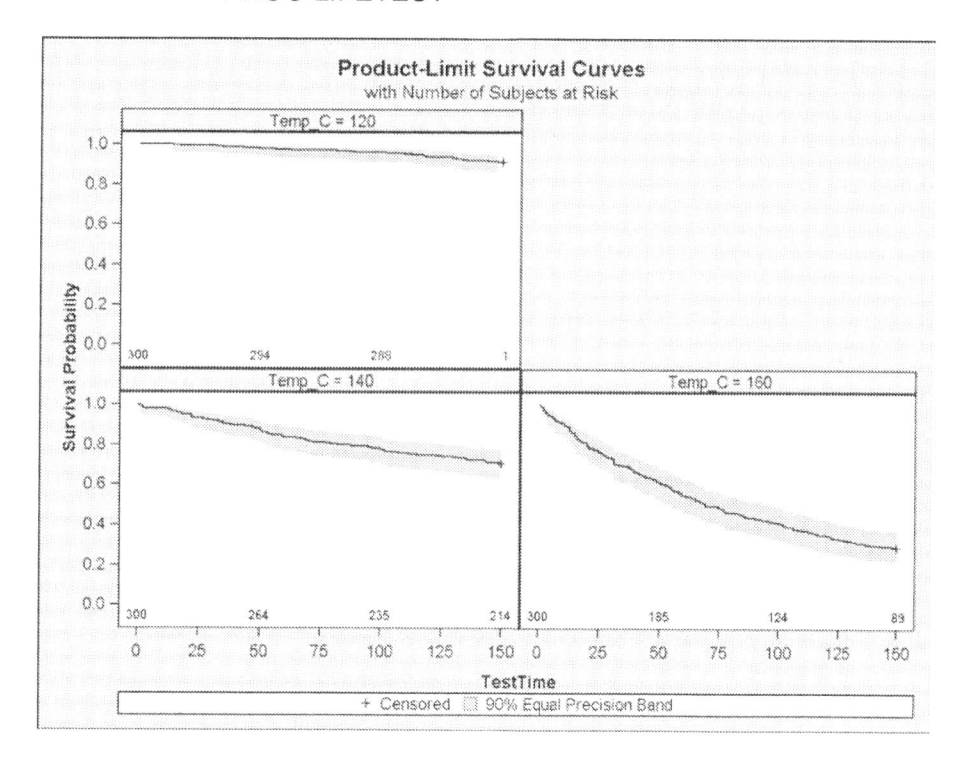

The survival curves provide a good visual summary of the life test data, including the significant effect of temperature on survival. The numbers at the bottom of the plot area indicate the number of units still at risk at various time points, starting with 300 at the beginning of the test, and ending with the number surviving until the end of the test.

10.5 PROC PHREG

Sections 10.2 and 10.3 show how to fit a regression model to censored life test data with explanatory variables when the time to failure has a known type of distribution, Weibull in the examples. It is sometimes preferable to leave the distributional form unspecified, and in these cases you can use PROC PHREG to get regression fit using the proportional hazards model (see Section 10.1.2). The example in this section uses PROC PHREG to compute a regression model similar to the Arrhenius model fitted in Sections 10.2.4 and 10.3.2. The code for this section is in ~\JES\sample_code\ch_10\phreg.sas.

```
DATA LifeTest_2; SET JES.LifeTest;    ❶
  Z = 1000/(Temp_C+273.15);
RUN;
PROC SQL NOPRINT;    ❷
  CREATE TABLE Temp_Values
    AS SELECT DISTINCT Temp_C, Z FROM LifeTest_2;
QUIT;
DATA Nominal; SET Temp_Values(OBS=1);
  Temp_C= 90;   Z = 1000/(Temp_C+273.15);        OUTPUT;
RUN;
DATA Temp_Values; SET Nominal Temp_Values;       RUN;

ODS GRAPHICS ON / RESET IMAGENAME="F10_9_";
ODS OUTPUT ParameterEstimates=ParmEst;    ❸
PROC PHREG DATA=LifeTest_2 PLOT(OVERLAY CL)=SURVIVAL;    ❹
  MODEL TestTime*Censor(-1)= Z;    ❺
  BASELINE OUT=Base SURVIVAL=Surv LOWER=LCL UPPER=UCL    ❻
  COVARIATES=Temp_Values;
RUN;
```

❶ The first DATA step adds the variable Z to the JES.LifeTest data set, as was done for the example in Section 10.3.2, so that it can be used in the regression.

❷ PROC SQL and the following DATA steps are used to create the Temp_Values data set, containing a list of the values of the explanatory variables for which estimates of S(t) are required. The PHREG example uses this data set to request estimated survival curves for each of the test temperatures, and also a predicted survival curve at the use condition of 90 degrees.

```
Temp_Values
Obs     Temp_C        Z
 1         90      2.75368
 2        120      2.54356
 3        140      2.42043
 4        160      2.30867
```

❸ The ODS OUTPUT statement requests that the parameter estimates be saved in the ParmEst data set.

❹ The PROC PHREG statement requests an analysis of the Life_Test_2 data set. The PLOT option requests plots of the survival curves, with confidence limits, overlaid on the same graph.

❺ The MODEL statement specifies Z as the independent variable.

❻ The BASELINE statement requests estimated survival curves for the values in data set specified by the COVARIATES option, Temp_Values. The OUT=Base option specifies the name of the output data set. The SURVIVAL, LOWER, and UPPER options specify the variable names to be used in the Base data set for the survival curves and confidence limits.

When the code is run, the ParmEst and Base data sets and the plot shown in Figure 10-9 are created. The ParmEst data set contains the estimated regression coefficients, the standard error of the estimates, and a chi-squared test of the hypothesis that the coefficient is significantly different from zero.

```
ParmEst
                                                    Prob      Hazard
Parameter    DF      Estimate       StdErr    ChiSq  ChiSq      Ratio
    Z         1     -11.29090      0.76454   218.099  <.0001     0.000
```

The estimated coefficient for the Z variable is -11.29090, so the estimated hazard function for a unit tested at Temp_C is

$$h(t) = h_0(t)\exp(-11.29090 \, Z), \text{ where } Z = 1000/(\text{Temp\_C} + 273.15)$$

You can compare this fit to the Weibull fit to the same data. Using the relationship noted in Section 10.1.2, the fitted model from Sections 10.2.4 and 10.3.2 can be rewritten as a proportional hazards model:

$$h(t) = h_0(t) \times \exp\left[-(\beta_1/\sigma)Z\right] = h_0(t) \times \exp\left[-(12.4544/1.093)Z\right] = h_0(t) \times \exp\left[-11.39 \times Z\right]$$

which agrees very well with the model fitted using PHREG.

Figure 10-9: Survival Probability Plots Created with the PLOT Option of PROC PHREG

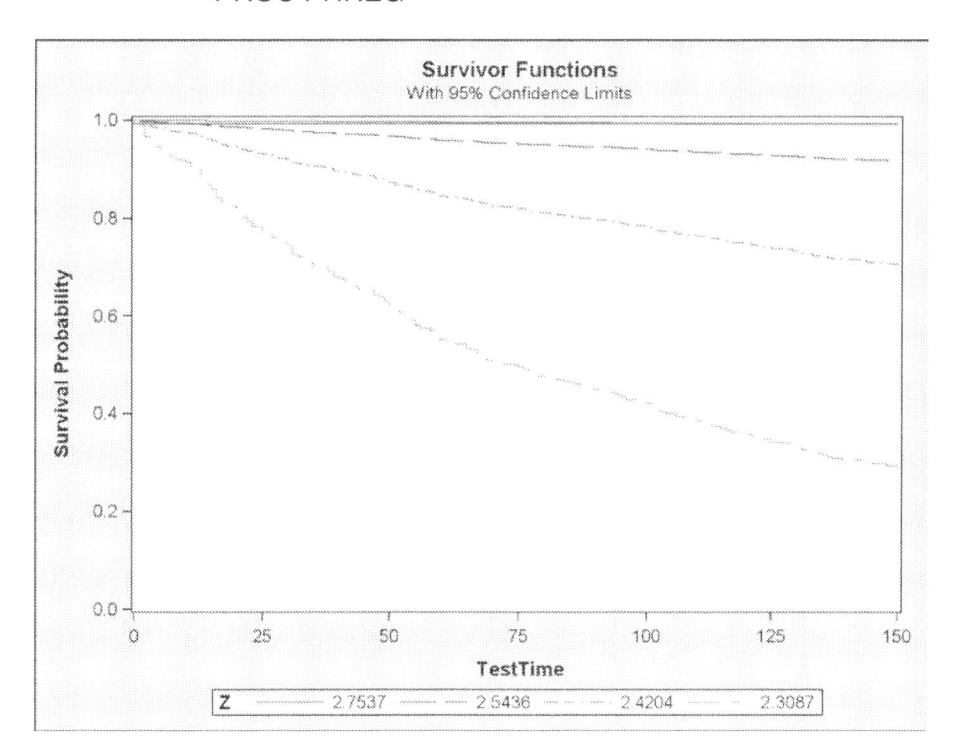

Figure 10-9 includes estimated survival curves for the three test temperatures: 120, 140, and 160, as well as a prediction of what the survival curve would be for units run at 90 degrees. The plots for 120, 140 and 160 degrees are similar to the plots created with PROC LIFETEST (Figure 10-8), but here the curves are constrained to meet the proportional hazards criterion. The predicted curve for 90 degrees is barely noticeable at the top of the plot, but the values are saved in the Base data set, which can be used for further plotting and analysis, as shown in the next example.

The Base data set includes the estimated survival curve, with confidence limits, for each of the Z values specified in Temp_Values and each distinct value of TestTime in the LifeTest_2 data set.

```
Base

Temp_C      Z         TestTime      Surv        LCL         UCL

  90     2.75368          0        1.00000       .           .
  90     2.75368          1        0.99992     0.99983     1.00000
  90     2.75368          2        0.99977     0.99959     0.99996
  90     2.75368          3        0.99969     0.99945     0.99993
  90     2.75368          4        0.99964     0.99937     0.99991
  90     2.75368          5        0.99962     0.99934     0.99990
.....
  90     2.75368        139        0.99225     0.98761     0.99691
  90     2.75368        143        0.99217     0.98749     0.99687
  90     2.75368        144        0.99206     0.98731     0.99682
  90     2.75368        148        0.99198     0.98719     0.99679
  90     2.75368        150        0.99187     0.98701     0.99674
.....

 120     2.54356        139        0.91994     0.89729     0.94315
 120     2.54356        143        0.91918     0.89636     0.94259
 120     2.54356        144        0.91804     0.89496     0.94173
 120     2.54356        148        0.91728     0.89402     0.94115
 120     2.54356        150        0.91613     0.89260     0.94028
.....
 140     2.42043        139        0.71525     0.68107     0.75114
 140     2.42043        143        0.71290     0.67859     0.74893
 140     2.42043        144        0.70936     0.67487     0.74561
 140     2.42043        148        0.70700     0.67238     0.74339
 140     2.42043        150        0.70345     0.66865     0.74006
.....
 160     2.30867        139        0.30615     0.26042     0.35992
 160     2.30867        143        0.30261     0.25704     0.35627
 160     2.30867        144        0.29734     0.25200     0.35083
 160     2.30867        148        0.29386     0.24869     0.34723
 160     2.30867        150        0.28869     0.24378     0.34187

(partial listing - first 5 rows and last 5 rows for each value of Temp_C)
```

The next bit of sample code uses PROC SGPLOT with the Base data set to create a plot of the survival curves and confidence limits for Temp_C equal to 90 and 120 degrees.

```
ODS GRAPHICS ON / RESET IMAGENAME="F10_10_";
TITLE "Estimated Survival Probability at 90 and 120 Degrees";
PROC SGPLOT Data=Base(WHERE=(Temp_C<=120));
   SERIES Y=Surv X=TestTime / GROUP=Temp_C;
   BAND X=TestTime LOWER=LCL UPPER=UCL /GROUP=Z TRANSPARENCY=.5;
   YAXIS VALUES=(.9 TO 1 BY .02);
RUN;
```

Figure 10-10: Estimated Survival Functions Created with the BASELINE Option of PROC PHREG

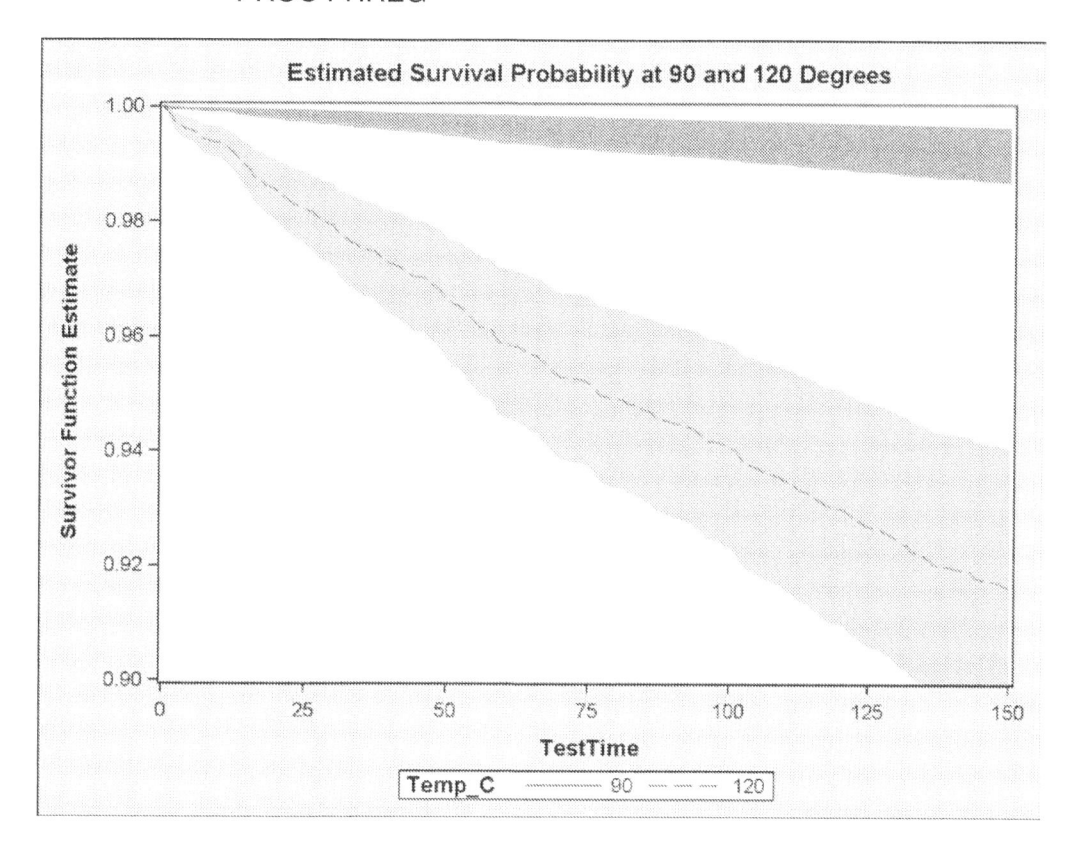

One important difference between the parametric methods available with PROC RELIABILITY and PROC LIFEREG (Sections 10.2 and 10.3) and the non-parametric methods based on PROC LIFETEST and PROC PHREG (Sections 10.4 and 10.5) is that the parametric estimates are easily extrapolated to times, t, beyond the range of the data. For example, if you estimate Weibull Shape and Scale parameters based on test data up to 150 hours, as in the examples in Sections 10.2 and 10.3, then you can use the equations in Table 10-4 to estimate the survival curve for values of t beyond 150 hours. There is not a corresponding method to extrapolate the non-parametric estimates beyond 150 hours.

10.6 The Reliability of Repairable Units

The differences between non-repairable and repairable units are explained in Section 10.1. Sections 10.2–10.5 describe various methods for characterizing the reliability of non-repairable units by estimating the Cumulative Distribution Function, F(t), or the Survival curve, S(t), for such units. This section illustrates the use of PROC RELIABILITY to analyze the reliability of repairable units. As noted in Section 10.1, the reliability of repairable units can be characterized by the Mean Cumulative Function, or MCF, defined as

M(t)=Mean[Cumulative number of repairs per unit in the time interval (0,t)]

The example in this section uses the MCFPLOT statement of PROC RELIABILITY to estimate MCF curves for each Vendor from the JES.Repair data set described in Section 10.1.4. The code for this section is in ~\JES \sample_code\ch_10\mcf.sas.

```
GOPTIONS RESET=ALL;
ODS HTML PATH="&JES.SG\S_10_6" (URL=NONE) BODY="MCF.html";

ODS GRAPHICS ON / RESET IMAGENAME="F10_11_";
ODS OUTPUT MCFEST=MCF_by_Vendor; ❶
TITLE1 HEIGHT=5 "Cumulative Repairs vs Unit Age by Vendor";
PROC RELIABILITY DATA=JES.Repair; ❷
      UNITID SN; ❸
      MCFPLOT EventTime*Censor(-1)=Vendor / ❹
      INTERPOLATE=STEP NOCENPRINT
      OVERLAY;
RUN;
ODS GRAPHICS OFF;
ODS _ALL_ CLOSE;
ODS LISTING;
```

❶ The ODS OUTPUT statement specifies that the MCF curves be saved to the MCF_by_Vendor data set.

❷ The PROC RELIABILITY statement requests analysis of the JES.Repair data set.

❸ The UNITID statement specifies that SN is the variable to be used to identify the units and thus relate the repair times and the censoring time on the same unit.

❹ The MCFPLOT statement requests separate MCF plots for each value of Vendor, where EventTime is the variable that contains the repair and censoring times, and Censor(-1) specifies that rows with Censor = -1 are censoring times, not repair times. The INTERPOLATE=STEP option specifies that the plot points be connected by step functions. The NOCENPRINT option specifies that only repair times, and not censoring times, be included in the output data set. The OVERLAY option specifies that the MCF plots for each Vendor should be displayed on the same graph.

When you run the code, the MCF_by_Vendor data set and the plot in Figure 10-11 are created.

```
MCF_by_Vendor (partial listing - first row and last 2 rows for each vendor)
     Age        Mcf        Std       Lower      Upper       ID      Group
    5.00      0.001      0.001      -0.001      0.003     1424   ChiTronix
 .....
  482.00      0.310      0.027       0.257      0.363     1997   ChiTronix
  495.00      0.362      0.058       0.249      0.476     1986   ChiTronix
    5.00      0.001      0.001      -0.001      0.003     5303     Duality
 .....
  384.00      0.044      0.008       0.029      0.060     5645     Duality
  391.00      0.047      0.008       0.031      0.063     5911     Duality
    1.00      0.001      0.001      -0.001      0.003     7948   Empirical
 .....
  394.00      0.195      0.015       0.165      0.226     7694   Empirical
  409.00      0.199      0.016       0.168      0.229     7699   Empirical
```

Figure 10-11: Using the MCFPLOT Statement of PROC RELIABILITY

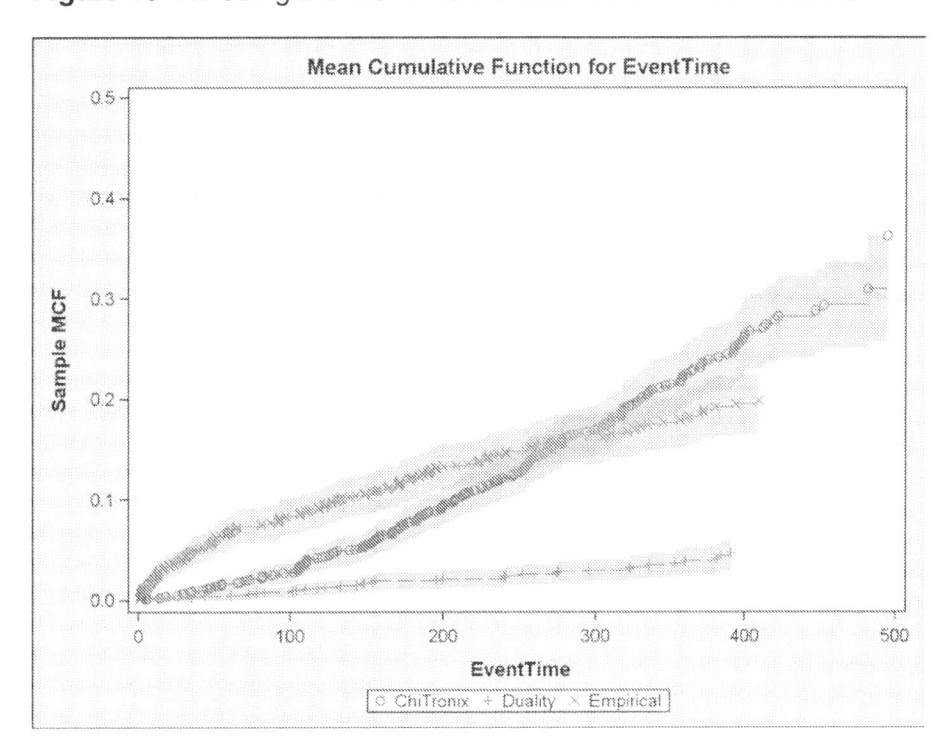

When interpreting MCF plots, it is important to keep in mind that, for any value of EventTime, the cumulative failure rate is equal to the height of the MCF at that point, and the instantaneous failure rate is equal to the slope of the MCF plot at that point. Therefore, lower curves with flatter slopes are better. You can see from Figure 10-11 that

- The Empirical units have a higher failure rate at the beginning, but the rate decreases over time.
- The ChiTronix units have a low failure rate at the beginning, but the rate increases over time.
- The Duality units are consistently better than units from either of the other vendors.

10.7 More Than Enough

Starting with SAS 9.2, the LIFEREG and PHREG procedures can perform Bayesian analysis of reliability data, in addition to the classical analysis described in Sections 10.3 and 10.5. This enables you to incorporate prior information about model parameters into the analysis of new data sets. A detailed discussion of Bayesian methods is way beyond the scope of this book, but this section provides a simple example to get you started if you are interested in the Bayesian approach. You can find more information about these procedures in the SAS online documentation, and the book by Meeker and Escobar, listed in the "Chapter Summary" section, provides a good introduction to the use of Bayesian methods in reliability analysis. The code for this section is in ~\JES \sample_code\ch_10\bayes.sas. This example is similar to the example in Section 10.3.1, but uses a BAYES statement to request Bayesian analysis.

```
GOPTIONS RESET=ALL;
ODS HTML PATH="&JES.SG\S_10_7" (URL=NONE) BODY="Lifereg.html";

/*===== Fit the 160 degree test results only =====*/
ODS GRAPHICS ON / RESET IMAGENAME="F10_12_";
ODS OUTPUT ParameterEstimates=MLE  ❶
           PostSummaries=Summary;
PROC LIFEREG DATA=JES.LifeTest(WHERE=(Temp_C=160));
  MODEL TestTime*Censor(-1) = Temp_C /
  DISTRIBUTION=WEIBULL;
  BAYES SEED=1 WEIBULLSHAPEPRIOR=GAMMA;  ❷
RUN;

ODS GRAPHICS OFF;
ODS _ALL_ CLOSE;
ODS LISTING;
```

❶ The ODS OUTPUT statement requests that the MLE and Summary data sets be created.

❷ The BAYES statement requests Bayesian analysis. The SEED option is included to ensure reproducible results from the Monte Carlo estimation of the posterior distribution. The WEIBULLSHAPEPRIOR option specifies a prior distribution for the Weibull Shape parameter.

```
Summary
Parameter        N       Mean     StdDev      P25       P50       P75
Intercept      10000    4.7529    0.0775    4.6998    4.7518    4.8051
WeibShape      10000    0.8901    0.0537    0.8542    0.8895    0.9260
```

When you run the code, the MLE and Summary data sets are created, along with the diagnostic plots shown in Figures 10-12 and 10-13. The MLE data set contains the same Maximum Likelihood Estimates of the Weibull parameters shown in the Estimates_1 data set in Section 10.3.1. The Summary data set contains the mean, standard deviation, and selected percentiles of the posterior distributions of the Weibull parameters. The Intercept parameter is the EV Location parameter defined in Table 10-4 of Section 10.1.1. The diagnostic plots for each parameter include a trace plot, an autocorrelation plot, and the posterior density of the parameter.

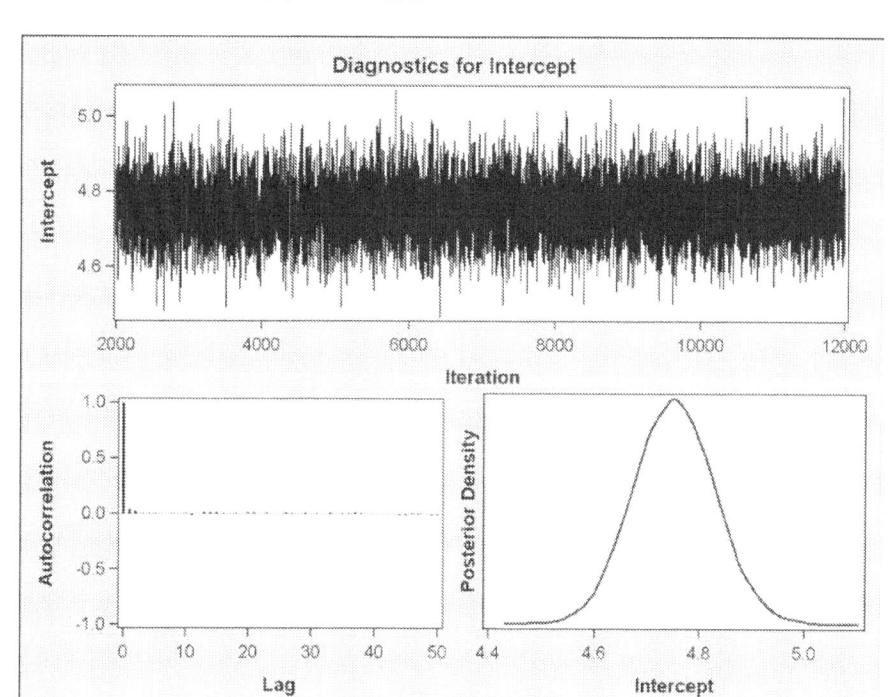

Figure 10-12: Diagnostic Plots Created by the BAYES Statement of PROC LIFEREG

Figure 10-13: Diagnostic Plots Created by the BAYES Statement of PROC LIFEREG

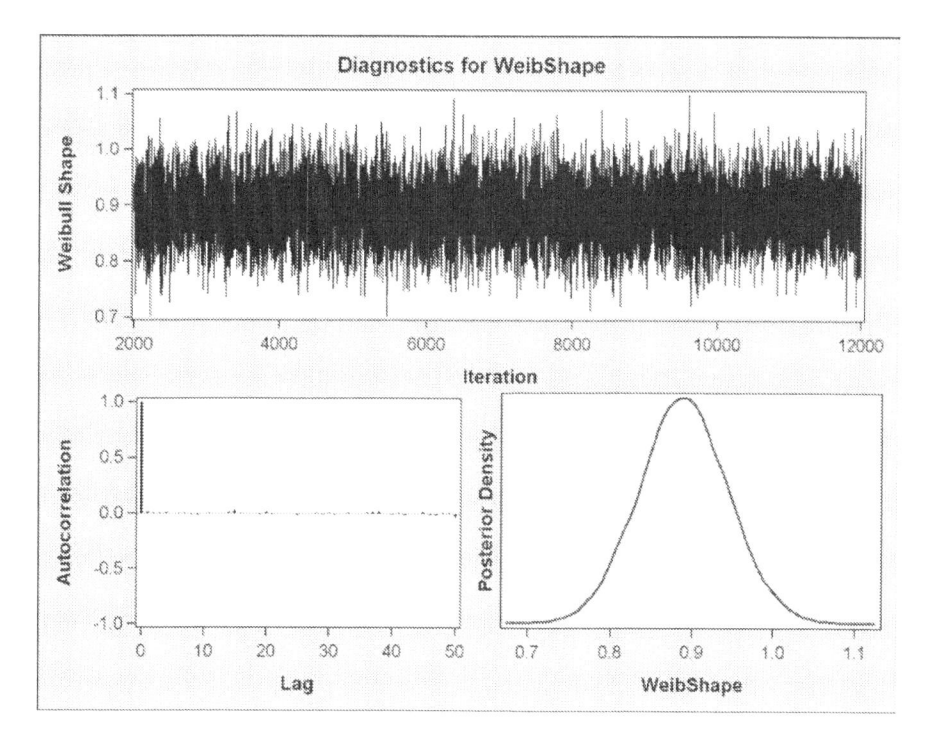

10.8 Chapter Summary

10.8.1 Recap

After finishing this chapter, you should know how to

- Relate the Weibull, log-normal, and log-logistic distributions to the corresponding location-scale distributions: extreme value, normal, and logistic.
- Use PROC RELIABILITY to fit any of these distributions to censored lift test data:
  - Simultaneously fit distributions for each value of a classification variable:
    - with parameters estimated independently
    - with the scale parameter constrained to be the same for each population
  - Fit an acceleration model to life data taken at different stress conditions:
    - using an Arrhenius, power, linear or logistic model to relate life time to stress
- Use PROC LIFEREG to fit the same kinds of models listed above.
- Use PROC LIFETEST to compute a non-parametric estimate of the survival function, S(t), from life test data, based on the Kaplan-Meier or Life-Table method.
- Use PROC PHREG to fit regression models to censored life test data based on the proportional hazards model.
- Use the MCFPLOT statement of PROC RELIABILITY to create a non-parametric estimate of the Mean Cumulative Function.

10.8.2 For More Information

Books

The book by Allison provides a very thorough and detailed explanation of the use of PROC LIFETEST, PROC LIFEREG, and PROC PHREG for survival analysis.

Allison, Paul D., 1995. *Survival Analysis Using the SAS® System: A Practical Guide.* Cary, NC: SAS Institute Inc.

Meeker, William Q., and Luis A. Escobar. 1998. *Statistical Methods for Reliability Data.* New York, NY: John Wiley & Sons.

Nelson, Wayne. 1982. *Applied Life Data Analysis.* New York, NY: John Wiley & Sons.

Tobias, Paul A., and David C. Trindade. 1995. *Applied Reliability, Second Edition.* New York, NY: Van Nostrand Reinhold.

Papers

Refer to the seminal papers on mean cumulative functions by Wayne Nelson.

Nelson, Wayne. 1988. "Graphical Analysis of System Repair Data." *Journal of Quality Technology*, 20 (1): 24–35.

Nelson, Wayne. 1995. "Confidence Limits for Recurrence Data—Applied to Cost or Number of Product Repairs." *Technometrics*, 37: 147–157.

10.8.3 Exercises

The JES.Lifetest data set (see Section 10.1.3) includes test data for each of the three vendors. The examples in this chapter use the data for all vendors to estimate the distribution of time to failure as a function of stress temperature. Use the same data set to explore the differences among the vendors.

- Using the method described in Section 10.2.2, estimate the distribution of time to failure at TempC=160 separately for each vendor.
    - Do the same for Temp_C=120 and 140.
- For each value of Temp_C, use the methods from Section 10.2.3 to fit Weibull distributions with the same shape parameter for each vendor. Is this a good fit for all vendors?
- Using the methods from Section 10.2.4, fit an Arrhenius acceleration model separately for each vendor.
- Repeat each of the steps listed above using PROC LIFEREG instead of PROC RELIABILITY.
- Use PROC LIFETEST to plot survival curves at each temperature separately for each vendor.
- Use PROC PHREG to fit a proportional hazards regression model separately for each vendor.

Chapter 11

SAS Macro Programming

- 11.1 Introduction 366
- 11.2 Macro Variables 368
- 11.3 Macro Programs 378
- 11.4 Utility Macros 390
- 11.5 More Than Enough 404
- 11.6 Chapter Summary 406

11.1 Introduction

You can increase your effectiveness as a SAS programmer by taking advantage of macro variables and programs to create code that is reusable, adaptable, and more easily maintained. The brief example in this section is presented without much detail just to give you an idea of the value of using macro variables and programs in your code. Macro variables and programs are discussed more fully in Section 11.2 and 11.3, respectively.

To get an idea of the value of using macro variables and programs in your code, consider the sample report in Section 7.1.1. That report required three separate programs: chitronix_1.sas, duality_1.sas, and empirical_1.sas, to create the scatter plots for each vendor, and three more programs to create the distribution plots for each vendor. You can use macro variables and/or macro programs to write a single program that creates the required reports for each vendor. The code for this section is in ~\JES\sample_code\ch_11\intro.sas. Here is the code that creates the histogram for ChiTronix in the sample report in Section 7.1.1.

```
GOPTIONS RESET=ALL BORDER;
OPTIONS NODATE;
GOPTIONS RESET=ALL BORDER FTEXT='Helvetica' FTITLE='Helvetica/Bold';
GOPTIONS GUNIT=PCT HTEXT=5 DEVICE=SASEMF XMAX=6IN YMAX=4IN;

/*=== JES\sample_code\ch_7\chitronix_2.sas ===*/
TITLE HEIGHT=5 "Resistance - ChiTronix - 4Q 2008";
PROC UNIVARIATE DATA=JES.Results_Q4(WHERE=(Vendor="ChiTronix"));
  VAR Resistance;
  HISTOGRAM Resistance / Normal(MU=EST SIGMA=EST)HREF=12.5 22.5;
  INSET N MEAN STD SKEWNESS KURTOSIS NORMAL(AD ADPVAL)
  / HEIGHT=2.5 FORMAT = 5.3 POSITION=NW;
RUN;
```

You can use a macro variable to turn this into code that is reusable for each vendor by replacing "ChiTronix" with the macro variable name "&Vendor" in two places in the code.

```
%LET Vendor=Duality; ❶
TITLE HEIGHT=5 "Resistance - &Vendor - 4Q 2008"; ❷
PROC UNIVARIATE DATA=JES.Results_Q4(WHERE=(Vendor="&Vendor"));
  VAR Resistance;
  HISTOGRAM Resistance / Normal(MU=EST SIGMA=EST)HREF=12.5 22.5;
  INSET N MEAN STD SKEWNESS KURTOSIS NORMAL(AD ADPVAL)
  / HEIGHT=2.5 FORMAT = 5.3 POSITION=NW;
RUN;
```

❶ The %LET statement creates a macro variable, &Vendor, with the value "Duality".

❷ The vendor name, ChiTronix in the first code sample, is replaced by &Vendor in the TITLE statement and also in the WHERE clause on the following line.

When you run this code, &Vendor is replaced by its value, Duality, before the statements are executed, and the result is exactly the same as for the duality_2.sas code in Section 7.1.1. So you can generate the histogram for each vendor by running the same code after setting &Vendor to ChiTronix, Duality, and Empirical in turn.

The process is even easier if you use a macro program, as illustrated in the next code sample.

```
%MACRO DistPlot(Vendor); ❶
TITLE HEIGHT=5 "Resistance - &Vendor - 4Q 2008";
PROC UNIVARIATE DATA=JES.Results_Q4(WHERE=(Vendor="&Vendor"));
   VAR Resistance;
   HISTOGRAM Resistance / Normal(MU=EST SIGMA=EST)HREF=12.5 22.5;
   INSET N MEAN STD SKEWNESS KURTOSIS NORMAL(AD ADPVAL)
   / HEIGHT=2.5 FORMAT = 5.3 POSITION=NW;
RUN;
%MEND DistPlot; ❷
%DistPlot(ChiTronix) ❸
%DistPlot(Duality)
%DistPlot(Empirical)
```

❶ The %MACRO statement begins the definition of a new macro program, DistPlot, with one parameter, Vendor.

❷ The %MEND statement ends the definition of the %DispPlot macro. The statements between %MACRO and %MEND are the lines that are executed when the macro is called.

❸ The %DistPlot(ChiTronix) statement causes the %DistPlot macro to be run with &Vendor equal to ChiTronix, producing the same result as the original chitronix_2.sas code. The next two lines run %DistPlot with Vendor equal to Duality and Empirical.

Not only is the macro program less work than creating three separate programs for the three vendors, it is also easier to maintain. For example, if you want to add a footnote to each plot, or remove SKEWNESS and KURTOSIS from the inset, you can make the change in one place instead of three.

Note also that using macro programs is not very difficult. The %DispPlot macro was created by simply wrapping the chitronix_2.sas code between %MACRO and %MEND statements, and replacing "ChiTronix" with "&Vendor". This brief introduction is probably enough to get you started writing such simple macro programs, but there is a lot more you should know in order to take full advantage of the power of macro programming. Section 11.2 shows various ways to create, view, use, and delete macro variables. Section 11.3 provides more details on creating simple macros like %DistPlot, and also introduces the use of looping, conditional logic and branching to create more complex data-driven macros.

11.2 Macro Variables

Macro variables are discussed briefly in Section 1.4, and the &JES macro variable that you defined in Section 1.5 is used in several examples in the previous chapters. This section contains more detailed information on how to create and work with macro variables. The code for this section is in ~\JES\sample_code\ch_11\macro_var.sas.

Here are some basic facts about macro variables:

- Macro variables are text strings—even though they sometimes look like numbers.
- There are two kinds of macro variables:
  - Automatic macro variables are generated by SAS.
  - Global macro variables are generated by the user.
- The names of macro variables:
  - Are case insensitive.
  - Contain at most 32 characters, which must be letters, numbers, or the underscore.
  - Must begin with a letter or an underscore.
- Macro variable names to avoid:
  - You should not create macro variable names beginning with "SYS", "AF", or "DMS" because such names may be used by SAS to create automatic macro variables.
  - Consult the SAS online documentation for a list of reserved words, such as "END", "FILE", "GO", and "IF" that should not be used as macro variable names.
- A macro variable name must be preceded by an ampersand (&) in order to be resolved (i.e., replaced with its value) in SAS statements.

11.2.1 Viewing Macro Variables

You can see all of the currently defined macro variables either by writing them to the Log window, or looking at the SASHELP.VMACRO data set, which is automatically maintained by SAS.

```
%PUT _ALL_;  ❶
DATA MacroVars; SET SASHELP.VMACRO; RUN;  ❷
%PUT _AUTOMATIC_;
%PUT _GLOBAL_;
```

❶ The %PUT _ALL_ statement writes all macro variables to the Log window. To see only the global or automatic variables, use %PUT _GLOBAL_ ; or %PUT _AUTOMATIC_;.

❷ The DATA step copies the SASHELP.VMACRO data set into the MacroVars data set. This data set contains the same information written to the Log window by the %PUT _ALL_; statement. The SASHELP library also contains many other data sets that are automatically generated and maintained by SAS.

The MacroVars data set lists the Scope (GLOBAL or AUTOMATIC), Name, and Value of each macro variable. The first variable in MacroVars is the global variable &JES that is defined by the autoexec.sas code that you created and saved in Section 1.5. The remaining variables are the automatic variables defined by SAS. A few of the more useful automatic variables are highlighted, and are illustrated in the next bit of sample code. Consult the SAS online documentation for the definitions of the automatic variables.

You can use a %PUT statement to write the values of specific macro variables to the Log window.

```
%PUT &JES;
%PUT This SAS Version &SYSVER session started at &SYSTIME on
&SYSDATE9 by user &SYSUSERID running &SYSSCP &SYSSCPL;
```

Chapter 11: SAS Macro Programming **369**

When you run this code, the results of the %PUT statements are written to the Log window.

```
3     %PUT &JES;
c:\JES\
4     %PUT This SAS Version &SYSVER session started at &SYSTIME on
5       &SYSDATE9 by user &SYSUSERID running &SYSSCP &SYSSCPL ;
This SAS Version 9.2 session started at 12:12 on 11OCT2008 by user
Robert running WIN XP_PRO
```

scope	name	offset	value
	MacroVars (selected rows)		
GLOBAL	JES	0	c:\JES\
AUTOMATIC	AFDSID	0	0
(rows omitted)			
AUTOMATIC	SYSDATE	0	11OCT08
AUTOMATIC	SYSDATE9	0	11OCT2008
AUTOMATIC	SYSDAY	0	Saturday
(rows omitted)			
AUTOMATIC	SYSNCPU	0	1
AUTOMATIC	SYSODSPATH	0	SASUSER.TEMPLAT(UPDATE)
AUTOMATIC	SYSPARM	0	SASHELP.TMPLMST(READ)
AUTOMATIC	SYSPBUFF	0	
AUTOMATIC	SYSPROCESSID	0	41D6F023E9BBF7CF4018000000000000
AUTOMATIC	SYSPROCESSNAME	0	DMS Process
AUTOMATIC	SYSPROCNAME	0	DATASTEP
AUTOMATIC	SYSRC	0	0
AUTOMATIC	SYSSCP	0	WIN
AUTOMATIC	SYSSCPL	0	XP_PRO
AUTOMATIC	SYSSITE	0	0070056501
AUTOMATIC	SYSSIZEOFLONG	0	4
AUTOMATIC	SYSSIZEOFUNICODE	0	2
AUTOMATIC	SYSSTARTID	0	
AUTOMATIC	SYSSTARTNAME	0	
(rows omitted)			
AUTOMATIC	SYSTCPIPHOSTNAME	0	machugh
AUTOMATIC	SYSTIME	0	12:12
AUTOMATIC	SYSUSERID	0	Robert
AUTOMATIC	SYSVER	0	9.2
AUTOMATIC	SYSVLONG	0	9.02.01M0P020508
AUTOMATIC	SYSVLONG4	0	9.02.01M0P02052008
AUTOMATIC	SYSWARNINGTEXT	0	

11.2.2 Creating Macro Variables

This section describes three commonly used methods for creating macro variables, using the %LET function, CALL SYMPUT and PROC SQL. In each example the macro variables are written to the Log window with the %PUT statement so that you can see the value of each variable.

11.2.2.1 Using the %LET Function

The %LET function has a very simple syntax:

%LET *variable-name* = *variable-value*;

For example:

```
%LET Var1 = The project lasted for;
%LET Var2 = 99;
%PUT &VAR1 &Var2 Months;
```

When you run this code, the variables &Var1 and &Var2 are created, and their values are written to the Log window.

```
84   %PUT &VAR1 &Var2 Months;
The project lasted for 99 Months
```

Note that the values of &Var1 and &Var2 are substituted before the %PUT statement is run.

11.2.2.2 Using CALL SYMPUT in a DATA Step

A CALL SYMPUT statement can be used during a DATA step to create a macro variable whose value is equal to a variable, or derived from one or more variables, in the data set. The syntax of CALL SYMPUT is

CALL SYMPUT(*variable-name, text*);

where *text* is the name of a variable in the data set or a text expression involving one or more variables in the data set. The next bit of code creates the Tab data set, which is used in the examples.

```
DATA Tab;   FORMAT Test_Date DATE9.;
  DO n=1 TO 90 BY 30;
      Vendor="Duality Logic";
      Test_Date='05JUN2008'd+n;
      Num_Test=1000;
      Num_Fail=ROUND(Num_Test*RANUNI(12345)/50);
      Rate=Num_Fail/Num_Test;
      OUTPUT;
  END;
  DROP N;
RUN;
```

```
Tab
Test_Date      Vendor          Num_Test    Num_Fail    Rate
06JUN2008      Duality Logic   1000        7           0.007
06JUL2008      Duality Logic   1000        15          0.015
05AUG2008      Duality Logic   1000        17          0.017
```

The examples create macro variables from the values in the second row of the Tab data set, using CALL SYMPUT here and PROC SQL in Section 11.2.2.3. The next code sample uses CALL SYMPUT four times to create four macro variables from the Tab data set.

```
DATA _NULL_; SET Tab;  ❶
  IF Test_Date='06JUL2008'd THEN DO;
      CALL SYMPUT('Vend', Vendor);  ❷
      CALL SYMPUT('NTest', Num_Test);  ❸
      CALL SYMPUT('Rate', PUT(Rate, 6.3));  ❹
      CALL SYMPUT('Test_Date', PUT(Test_Date, DATE9.));  ❺
  END;
RUN;
%PUT Vend=[&Vend] NTest=[&NTest] Rate=[&Rate] Test_Date=[&Test_Date];  ❻
```

❶ A DATA _NULL_ step is used to avoid changing Tab or creating a new data set. The IF statement ensures that the following lines are executed only when Test_Date=06JUL2008, so the macro variables will be created from the values in the second row of Tab.

❷ The first CALL SYMPUT writes the value of Vendor to the macro variable &Vend.

❸ The second CALL SYMPUT writes the value of Num_Test to the macro variable &NTest. Num_Test is a numeric variable, so there will be a note in the Log indicating that the numeric value was converted to a character value before being written to &NTest.

❹ The third CALL SYMPUT uses a PUT function (see Section 2.4.3) to convert Rate to a text string, using the 6.3 format, and then defines the macro variable &Rate to be equal to that string.

❺ The fourth CALL SYMPUT writes Test_Date to the variable &Test_Date with the DATE9. format.

❻ The %PUT statement writes the new macro variables to the Log window. Square brackets are printed around the macro variables so that you can see where there are any leading or trailing blanks in the macro variables.

When you run the code, the four macro variables are created, and these lines are written to the Log window. Note that there are four leading blanks in &NTest because Num_Test has an 8.0 format, while "1000" takes only four places.

```
NOTE: Numeric values have been converted to character values at the places given by:
    (Line):(Column).
    102:30

107  %PUT Vend=[&Vend] NTest=[&NTest] Rate=[&Rate] Test_Date=[&Test_Date];
Vend=[Duality Logic] NTest=[    1000] Rate=[ 0.015] Test_Date=[06JUL2008]
```

11.2.2.3 Using PROC SQL

Chapter 4 and Section 5.9 show how to use PROC SQL to manipulate SAS data sets and to extract data from a relational database. You can also use PROC SQL to create macro variables from selected variable values in a data set, or from summary statistics, such as COUNT or MAX, computed from the values in a data set. The syntax of the SELECT statement for creating macro variables is

SELECT Var_1, Var_2,...,Var_n INTO :$Name_1$, :$Name_2$,..., :$Name_n$

where the *Var*s are variables in the data set, or functions of those variables, and the *Name*s are the macro variable names. The next code sample uses PROC SQL with the Tab data set to create the same macro variable created with CALL SYMPUT in Section 11.2.2.2.

```
PROC SQL NOPRINT;  ❶
   SELECT Vendor, Num_Test, Rate, Test_Date  ❷
      INTO :Vend, :NTest, :Rate, :Test_Date  ❸
   FROM Tab  ❹
   WHERE Test_Date='06JUL2008'd;  ❺
QUIT;  ❻
%PUT Vend=[&Vend] NTest=[&NTest] Rate=[&Rate] Test_Date=[&Test_Date];
```

❶ The NOPRINT option disables output to the Output window. It is not required, but it helps to avoid cluttering up the Output window with results you usually don't need.

❷ The SELECT statement specifies the variables to be used to create macro variables. Note that these must be separated by commas.

❸ After the SELECT list, the INTO list specifies the names of the macro variables to be created. Note that each name is preceded by a colon, and the names are separated by commas.

❹ The FROM clause specifies the data set to be used.

❺ The WHERE clause specifies that the row with Test_Date='06JUL2008'd be used to create the variables.

❻ The QUIT; statement ends the SQL procedure.

When you run this code, the variables are created and these lines are written to the Log window. Note that the variables are exactly the same as those created in Section 11.2.2.2.

```
116  %PUT Vend=[&Vend] NTest=[&NTest] Rate=[&Rate] Test_Date=[&Test_Date];
Vend=[Duality Logic] NTest=[    1000] Rate=[  0.015] Test_Date=[06JUL2008]
```

You can use the summary functions of PROC SQL, for example COUNT, MIN, MAX, RANGE, and MEAN, to create macro variables from summary statistics computed for a data set. The next example uses the MAX and COUNT functions to extract the maximum value of Rate and the total number of rows from the Tab data set. Consult the SAS online documentation for the full list of summary functions that you can use with PROC SQL.

```
PROC SQL NOPRINT;
   SELECT MAX(Rate), COUNT(*) INTO :MaxRate, :Num_Row
   FROM Tab;
QUIT;
%PUT MaxRate=[&MaxRate], Num_Row = [&Num_Row];
```

When you run the code, the &MaxRate and &Num_Row variables are created, and these lines are written to the Log window.

```
122  %PUT MaxRate=[&MaxRate], Num_Row = [&Num_Row];
MaxRate=[ 0.017], Num_Row = [    3]
```

You can also use the SEPARATED BY option to select multiple values of the same variable into a single macro variable. This method was used in Section 4.3.7 to create a list of serial numbers to be passed to an SQL query for extracting data from an Oracle database. The next code sample uses the SEPARATED BY option to create a single macro variable containing all the values of Test_Date in the Tab data set.

```
PROC SQL NOPRINT;
   SELECT Test_Date INTO :DateList
   SEPARATED BY " , "
   FROM Tab;
QUIT;
%PUT DateList=[&DateList];
```

Here is the resulting &DateList variable written to the Log window.

```
130  %PUT DateList=[&DateList];
DateList=[06JUN2008 , 06JUL2008 , 05AUG2008]
```

11.2.3 Deleting Macro Variables

When you create a global macro variable, its value remains defined until the end of the SAS session unless you redefine it, using one of the methods described in Section 11.2.2, or delete it. You can use the %SYMDEL function to delete one or more macro variables as shown here.

```
%SYMDEL MaxRate Num_Row;   ❶
%PUT MaxRate=[&MaxRate], Num_Row = [&Num_Row];   ❷
```

❶ The %SYMDEL statement deletes the &MaxRate and &Num_Row variables created in Section 11.2.2.3.

❷ The %PUT function generates warning messages in the Log window because the referenced variables no longer exist. Note that, after the warnings, a line is written to the Log window with "&MaxRate" and "&Num_Row" in place of the missing values.

```
262  %SYMDEL MaxRate Num_Row;
263  %PUT MaxRate=[&MaxRate], Num_Row = [&Num_Row];
WARNING: Apparent symbolic reference MAXRATE not resolved.
WARNING: Apparent symbolic reference NUM_ROW not resolved.
MaxRate=[&MaxRate], Num_Row = [&Num_Row]
```

When running the examples in this chapter, you should delete all global macro variables from time to time, just to be sure that the variables created in one example don't interfere with later examples. But you should not delete the &JES variable because that will still be needed for the %INCLUDE statements in some examples. You can use the %delvars macro program to delete all global macro variables except &JES.

```
%MACRO delvars;
   DATA Vars; SET SASHELP.VMACRO; RUN;
   DATA _NULL_; SET VARS;
       IF Scope='GLOBAL' THEN DO;
           IF Name NE "JES"  AND SUBSTR(Name,1,3) NE "SYS"
           THEN DO;
               CALL EXECUTE('%SYMDEL '||TRIM(LEFT(Name))||';');
           END;
       END;
   RUN;
%MEND delvars;
```

This macro, which is also one of the utility macros listed in Section 11.4, is adapted from an example in Section 13.1.6 of *Carpenter's Complete Guide to the SAS® Macro Language*. You can consult the book for a detailed explanation of the macro but the general idea is to apply the %SYMDEL function to every global variable found in SASHELP.VMACRO (see Section 11.2.1), except for &JES. The code also excludes any macro variable beginning with 'SYS', because %SYMDEL does not work on such variables. As noted in Section 11.2, it is best not to create macro variables beginning with "SYS", "AF", or "DMS" as these are used by SAS to create automatic variables. This code runs %delvars, and then writes all global macro variables to the Log window to verify the results.

```
%INCLUDE "&JES.utility_macros/delvars.sas";
%delvars
%PUT _GLOBAL_;
```

```
294  %PUT _GLOBAL_;
GLOBAL JES c:\JES\
```

11.2.4 Using Macro Variables

This section covers some basic methods for using and transforming macro variables.

11.2.4.1 Using a Period to Delimit a Macro Variable Name

In some cases you need to put a period at the end of a macro variable name so that SAS knows where the variable name ends. For example, if you want to append characters to the end of a macro variable, place a period after the variable name and before the other characters. The period does not appear in the result.

```
%LET Num=1; %LET Char =one;
%LET Cash = M&Char.y;
%PUT Num=[&Num] Char=[&Char] Cash=[&Cash] Cash=[M&Char.y];
```

Another case where a period is required is when you are using a macro variable to identify the name of a SAS library. In the next example, the &theLib variable is defined with the value JES, which is the library name used throughout this book. Then &theLib is used to reference one of the data sets in JES. The first period after &theLib identifies the end of the variable name, and the second period is the separator between library name and member name, so that &theLib..Lifetest resolves to JES.Lifetest.

```
%LET theLib = JES;
DATA Temp; SET &theLib..Lifetest; RUN;
```

```
332  %PUT Num=[&Num] Char=[&Char] Cash=[&Cash] Cash=[M&Char.y];
Num=[1] Char=[one] Cash=[Money] Cash=[Money]

861  %LET theLib = JES;
862  DATA Temp; SET &theLib..Lifetest; RUN;

NOTE: There were 900 observations read from the data set JES.LIFETEST.
```

11.2.4.2 Using Double Quotes to Resolve Macro Variables

A macro variable included in SAS statements will generally be resolved (i.e., replaced with its value) before the statement is run. However, if a macro variable is enclosed in a quoted string, it will be resolved if it is enclosed in double quotes, but not if it is enclosed in single quotes, as illustrated in the next example.

```
TITLE1 "The Number &Num is Spelled &Char";
TITLE2 'The Number &Num is Spelled &Char';
DATA Temp;
   Num=&Num;
   Char="&Char";
   Char2='&Char';
RUN;
PROC PRINT DATA=Temp; RUN;
```

```
The Number 1 is Spelled one
The Number &Num is Spelled &Char

Obs    Num    Char    Char2
 1      1     one     &Char
```

11.2.4.3 Using Macro Functions and Autocall Programs

Macro functions are built-in SAS functions that process one or more arguments and return a result. Autocall programs are macro programs, similar to the macro programs that you can write yourself (see Section 11.3), but written by SAS and made available to you in the autocall library. In Section 11.3.6 you will learn how to add your own macro programs to the autocall library. This section shows how to use some of the SAS supplied macro functions and autocall programs to manipulate your macro variables. The code for this section is in ~\JES\sample_code\ch_11\macro_functions.sas.

Table 11-1 lists a selection of macro functions and autocall programs. Several of these are similar to the functions described in Section 2.4.3, which can be applied to variables in a DATA step.

Table 11-1: Selected Macro Functions and Autocall Programs

Macro Functions	
%UPCASE(*string*)	Converts the characters in *string* to uppercase.
%SCAN(*string, i, char*)	Returns the *i*-th substring of *string*, if *char* is used as a delimiter.
%SUBSTR(*string, k <,n>*)	Returns the substring of *n* characters of *string* starting at the *k*-th character. If *n* is omitted, all characters starting at the *k*-th are returned.
%LENGTH(*string*)	Returns the number of characters in *string*.
%INDEX(*string, substring*)	Location of the first occurrence of *substring* within *string*. Returns zero if *substring* is not found in *string*.
%EVAL(*string*)	Evaluates the numeric value of *string*, using integer arithmetic.
%SYSEVALF(*string*)	Evaluates the numeric value of *string*, using decimal arithmetic.
Autocall Programs	
%LEFT(*string*)	Removes leading blanks from *string*.
%DATATYP(*string*)	Returns the data type, CHAR or NUMERIC, of *string*.

The examples in this section use the macro variables created in Section 11.2.2 to illustrate the use of these functions. The first bit of sample code (not shown here) re-creates these variables and writes them to the Log window.

```
484     %PUT Vend=[&Vend] NTest=[&NTest] Rate=[&Rate] Test_Date=[&Test_Date];
Vend=[Duality Logic] NTest=[    1000] Rate=[   0.015] Test_Date=[06JUL2008]
```

The next bit of code illustrates several of the functions in Table 11-1. In the box below, the code and corresponding results are written on the same line to make it easier to read. This is a compressed view of what you will see in the Log window after you run the code, with each result line added on to the end of the corresponding code line. Most of the examples use macro variables as arguments to the functions, but the arguments can also be simple text strings, as in the last line where the argument is the text string 'qwerty'. Note that the quotes are not used in the function call. The syntax is %UPCASE(qwerty), not %UPCASE('qwerty').

```
-------------------Code---------------------------    ---Result----------
%LET Upper=%UPCASE(&Vend);      %PUT Upper=[&Upper];  Upper=[DUALITY LOGIC]
%LET Last =%SCAN(&Vend, 2, ' '); %PUT Last=[&Last];   Last=[Logic]
%LET V9   =%SUBSTR(&Vend, 1, 9); %PUT V9=[&V9];       V9=[Duality L]
%LET Len  =%LENGTH(&Vend);      %PUT Len=[&Len];      Len=[13]
%LET Loc  =%INDEX(&Vend, Log);  %PUT Loc=[&Loc];      Loc=[9]
%LET VTyp =%DATATYP(&Vend);     %PUT VTyp=[&VTyp];    VTyp=[CHAR]
%LET RTyp =%DATATYP(&Rate);     %PUT RTyp=[&RTyp];    RTyp=[NUMERIC]
%LET NLeft=%LEFT(&NTest);       %PUT NLeft=[&NLeft];  NLeft=[1000]
%PUT %UPCASE(qwerty);                                 QWERTY
```

11.2.4.4 The %EVAL and %SYSEVALF Functions

The %EVAL and %SYSEVALF functions operate on strings which are either arithmetic or logical expressions. The use of these functions with arithmetic expressions is discussed in this section, and their use with logical expressions is discussed in Section 11.3.3.3. An arithmetic expression consists of any combination of numbers and the arithmetic operators, +, -, *, / and ** (exponentiation), which represents a valid calculation according to the usual rules of arithmetic. For example:

```
------------------Code------------------------     ---Result----------
%LET A=%EVAL(4*(3**2)+ (10/5) -6); %PUT A=&A;      A=32
```

The difference between %EVAL and %SYSEVALF is that %EVAL accepts only integers, not decimal values, as input, and rounds the result down to the next smallest integer in the output.

```
------------------Code------------------------     ---Result----------
%LET B=%EVAL(7/4);         %PUT B=&B;              B=1
%LET C=%SYSEVALF(7/4);     %PUT C=&C;              C=1.75
```

If the argument to either of these functions is not a valid arithmetic expression, you will get an error message. %EVAL only accepts integer values, so an expression involving a decimal number is considered invalid. In the next examples, the expression 2*5.5 is invalid for %EVAL because %EVAL does not recognize the decimal number 5.5, but the same expression works with %SYSEVALF. Note, however, that %SYSEVALF does not recognize 06JUL2008 as a number.

```
%LET D=%EVAL(2*5.5);                %PUT D=&D;
ERROR: A character operand was found in the %EVAL function or %IF
condition where a numeric
       operand is required. The condition was: 2*5.5
D=
%LET E=%SYSEVALF(2*5.5);            %PUT E=&E;
E=11
%LET F=%SYSEVALF(06JUL2008+30);     %PUT F=&F;
ERROR: A character operand was found in the %EVAL function or %IF
condition where a numeric
       operand is required. The condition was: 06JUL2008+30
F=
```

The next set of examples illustrates some calculations using the macro variables defined earlier, including an example of the correct way to represent a date as a number in the %SYSEVAL function.

```
------------------Code----------------------        ----Result-------
%LET G=%EVAL(&NTest+300);          %PUT G=[&G];     G=[1300]
%LET H=%SYSEVALF(&NTest+300);      %PUT H=[&H];     H=[1300]
%LET I=%EVAL(&NTest/30);           %PUT I=[&I];     I=[33]
%LET J=%SYSEVALF(&NTest/30);       %PUT J=[&J];     J=[33.3333333333333]
%LET K=%EVAL(&Rate*10);            %PUT K=[&K];     ERROR: A character…
%LET L=%SYSEVALF(&Rate*10);        %PUT L=[&L];     L=[0.15]
%PUT "&Test_Date"d;                                 "06JUL2008"d
%LET M=%EVAL("&Test_Date"d+1);     %PUT M=[&M];     ERROR: A character…
%LET N=%SYSEVALF("&Test_Date"d+1); %PUT N=[&N];     N=[17720]
```

- The calculation of &K fails because &Rate resolves to 0.015 which is not an integer. The ERROR message is the same as in the previous example, but is truncated here to save space.
- The last two examples, use "&Test_Date"d which resolves to "06JUL2008"d, a valid SAS date, equal to 17719. This expression is accepted by %SYSEVALF, but not by %EVAL.

11.3 Macro Programs

As noted in the introduction to this chapter, you can significantly improve your programming efficiency by writing macro programs that can be used over and over again to perform the same data processing steps on different data sets or with different conditions. The examples in this section create a series of successively more complex macros, %Report1 through %Report6, to illustrate how you can create a more efficient, flexible, and bug-free version of a report like the sample report in Section 7.1.1. The sample reports created in this section are even simpler than the report in Section 7.1.1, just a single PROC PRINT of a data set, but the same principles can be applied to more complex reports.

The examples use four data sets: JES.Vendor_List, which is used to automate the creation of reports for each Vendor in the list, and JES.Rates_2006, JES.Rates_2007, and JES.Rates_2008, which contain the data to be used for each report. JES.Rates_2007 is not shown here, but is similar to JES.Rates_2006.

```
JES.Vendor_List
  Vendor      N
  ChiTronix   1
  Duality     2
  Empirical   3
```

```
JES.Rates_2006

                                               Defect_
Year   Qtr        Vendor      N_Test  Defects  Rate

2006   2006_Q1    ChiTronix   2700    134      0.050
2006   2006_Q1    Duality     2700     89      0.033
2006   2006_Q1    Empirical   2700    179      0.066
2006   2006_Q2    ChiTronix   2730    172      0.063
2006   2006_Q2    Duality     2730     69      0.025
2006   2006_Q2    Empirical   2730    205      0.075
2006   2006_Q3    ChiTronix   2760    348      0.126
2006   2006_Q3    Duality     2760     78      0.028
2006   2006_Q3    Empirical   2760    200      0.072
2006   2006_Q4    ChiTronix   2760    164      0.059
2006   2006_Q4    Duality     2760     84      0.030
2006   2006_Q4    Empirical   2760    198      0.072
```

```
JES.Rates_2008

                                               Defect_
Year   Qtr        Vendor      N_Test  Defects  Rate

2008   2008_Q1    ChiTronix   2730     68      0.025
2008   2008_Q1    Empirical   2730    193      0.071
2008   2008_Q2    ChiTronix   2730    286      0.105
2008   2008_Q2    Empirical   2730    206      0.075
2008   2008_Q3    ChiTronix   2760     48      0.017
2008   2008_Q3    Empirical   2760    212      0.077
2008   2008_Q4    ChiTronix   2760     57      0.021
2008   2008_Q4    Empirical   2760    174      0.063
```

11.3.1 Writing a Macro Program

The first example illustrates the basic syntax for defining macro programs. The code for this section is in ~\JES \sample_code\ch_11\macro_programs.sas.

```
%MACRO myMacro;
   %PUT Some SAS code here;
%MEND myMacro;
```

A macro program definition begins with the %MACRO keyword, followed by the name of the macro. The macro definition ends with a % MEND statement. It is not necessary to repeat the macro name in the %MEND statement, but it is a good programming practice, and it will simplify the debugging of your code. Between the %MACRO and %MEND statements you can put any SAS code that you want to run, including the code in any of the examples in this book. When you run this code, the %myMacro program will be defined, but not run. To run your new macro program, type the macro name, including the % sign, and submit the line:

```
%myMacro
```

When you submit this line, the code within the macro definition will be run, writing "Some SAS code here" to the Log window. Note that it is not necessary to put a semicolon at the end of the macro call.

The next code sample defines a macro that creates a simple report from the Rates_2006 data set, and then runs the new macro.

```
%MACRO Report1;  ❶
   TITLE1 "Defect Rate Report for 2006 at ChiTronix";
   DATA Temp; SET JES.Rates_2006;
      IF Vendor="ChiTronix";
   RUN;
   PROC PRINT DATA=Temp NOOBS LABEL;
      VAR Qtr N_Test Defects Defect_Rate;
   RUN;
%MEND Report1;
%Report1  ❷
```

❶ The lines from %MACRO to %MEND define the %Report1 macro. The DATA step selects the ChiTronix data from the JES.Rates_2006 data set, and the PROC PRINT statement prints a simple report.

❷ The final line, %Report1, runs the macro.

When you run the code, this report is written to the Output window.

```
Defect Rate Report for 2006 at ChiTronix

            Number      Number      Defect
Quarter     Tested      Defects     Rate

2006_Q1     2700        134         0.050
2006_Q2     2730        172         0.063
2006_Q3     2760        348         0.126
2006_Q4     2760        164         0.059
```

The example in the next section shows how to make a more useful macro by adding macro parameters that enable you to create a variety of reports with a single macro.

11.3.2 Using Macro Parameters

Your macro programs can be more useful if you include parameters in the definition. The %Report1 macro always creates the same report – the average defect rates for ChiTronix units during 2006. It would be more useful to have a macro that can create a report for any specified vendor and year. You can create such a macro by adding parameters to your macro definition, as shown in the next example.

```
%MACRO Report2(Vendor, Year);   ❶
   TITLE1 "Defect Rate Report for &Year at &Vendor";   ❷
   DATA Temp; SET JES.Rates_&Year;   ❸
       IF Vendor="&Vendor";
   RUN;
   PROC PRINT DATA=Temp NOOBS LABEL;
       VAR Qtr N_Test Defects Defect_Rate;
   RUN;
%MEND Report2;
%Report2(Empirical, 2008)   ❹
```

❶ The %MACRO statement includes two parameters: Vendor and Year. Parameters are entered as an ordered list, separated by commas, and enclosed in parentheses. The parameters are treated like macro variables in the SAS code that follows.

❷ The &Year and &Vendor macro variables are used to customize the TITLE statement.

❸ In the DATA step, &Year is used to specify the data set to be used, and &Vendor is used to specify which lines to select from the data set.

❹ When you run %Report2(Empirical, 2008), a report on the defect rates for Empirical units during 2008, similar to the report in Section 11.3.1, is written to the Output window.

There are two kinds of parameters that you can use in your macro programs: *positional parameters* and *keyword parameters*. *Positional parameters* are entered as a sequence of one or more variable names separated by commas, for example (Vendor, Year) in the example above. *Keyword parameters* are entered as a sequence of name=value definitions, also separated by commas. The next example includes both parameter types.

```
%MACRO Report3(Vendor, Year=2006, Note = Normal Test Process);   ❶
   TITLE1 "Defect Rate Report for &Year at &Vendor";
   TITLE2 "&Note";
   DATA Temp; SET JES.Rates_&Year;
       IF Vendor="&Vendor";
   RUN;
   PROC PRINT DATA=Temp NOOBS LABEL;
       VAR Qtr N_Test Defects Defect_Rate;
   RUN;
%MEND Report3;
%Report3(ChiTronix)   ❷
%Report3(Empirical, Note= Extended Test Process, Year=2007)   ❸
```

❶ The %MACRO statement specifies three parameters: Vendor, Year, and Note. Vendor is a positional parameter. Year is a keyword parameter with a default value of 2006. Note is a keyword parameter with a default value of "Normal Test Process".

❷ The first time the %Report3 macro is run, the value of &Vendor is ChiTronix, and the values of &Year and &Note are not specified, so the default values of 2006 and "Normal Test Process" are used.

❸ The second time the %Report3 macro is run, the value of &Vendor is Empirical, and the values of &Year and &Note are set to 2007 and "Extended Test Process", respectively.

When you run the code, these reports are written to the Output window.

```
Defect Rate Report for 2006 at ChiTronix
Normal Test Process

              Number       Number       Defect
Quarter       Tested       Defects      Rate

2006_Q1       2700         134          0.050
2006_Q2       2730         172          0.063
2006_Q3       2760         348          0.126
2006_Q4       2760         164          0.059
```

```
Defect Rate Report for 2007 at Empirical
Extended Test Process

              Number       Number       Defect
Quarter       Tested       Defects      Rate

2007_Q1       2700         195          0.072
2007_Q2       2730         169          0.062
2007_Q3       2760         212          0.077
2007_Q4       2760         197          0.071
```

There are a few simple rules that govern your use of macro parameters:

- When the macro is defined,
  - all positional parameters must be defined before any keyword parameters
- When the macro is run,
  - all positional parameters must be specified—in the same order that they were listed in the definition
  - keyword parameters are optional, and can be specified in any order, as long as they come after the positional parameters

Keyword parameters are particularly useful in two circumstances:

- If you want to use a default parameter value most of the time, but want to retain the ability to change it when necessary, using a keyword parameter saves you from having to enter the value each time you use the macro.
- If you are using a macro throughout your projects, and later find that you need a macro that is similar but requires an additional parameter, you can add a keyword parameter to the existing macro. This is often more efficient that creating a new macro for the purpose or changing the existing macro references throughout your code.

11.3.3 Looping, Conditional Execution, and Branching

This section shows how to use macro language statements to add do loops, conditional execution, and branching to your macro programs. The code for this section is in ~\JES \sample_code\ch_11 \macro_statements.sas.

11.3.3.1 Macro Language Statements

Macro language statements instruct the SAS macro processor to perform an operation. Some examples are the %LET, %PUT, and %SYMDEL statements used in earlier sections. This section introduces some macro language statements that you can use to control the execution of statements in your macro programs. These statements can only be used in macro programs and not in open code.

11.3.3.2 Using a %DO Statement to Create a Do Loop

This example shows how to use a %DO statement to create three reports at once, one for each vendor in the JES.Vendor_List data set shown at the beginning of Section 11.3.

```
%MACRO Report4(Year);
%DO N = 1 %TO 3 %BY 1;  ❶
  PROC SQL NOPRINT;
      SELECT Vendor into :Vendor
      FROM JES.Vendor_List
      WHERE N=&N;  ❷
  QUIT;
  %LET Vendor=%TRIM(&Vendor);
  DATA Temp; SET JES.Rates_&Year;
      IF Vendor="&Vendor";
  RUN;
  TITLE1 "Defect Rate Report for &Year at &Vendor";
  PROC PRINT DATA=Temp NOOBS LABEL;
      VAR Qtr N_Test Defects Defect_Rate;
  RUN;
%END; /* Completes %DO N = 1 %TO ...; code block */  ❸
%MEND Report4;
%Report4(2006)  ❹
%Report4(2008)
```

❶ The %DO statement starts the do loop. N=1 specifies that the macro variable &N is set to 1 for the first iteration of the loop. %TO 3 specifies that the last value of &N is 3. The %BY 1 option specifies that &N is incremented by 1 for each successive iteration. If the %BY option is omitted, the increment defaults to 1. Therefore, it is not needed in this example, but is included to illustrate the syntax.

❷ The macro variable &N is used in the PROC SQL statement to select each vendor in turn from the JES.Vendor_List data set. The remainder of the code creates a report for the selected vendor.

❸ The %END; statement ends the DO loop.

❹ The %Report4(2006) statement creates three reports similar to those shown in Section 11.3.2. The %Report4(2008) statement creates only two reports, for ChiTronix and Empirical, because the JES. Rates_2008 data set does not include any data for Duality.

You can use logical expressions to modify your reports based on the data being processed. Section 11.3.3.4 shows how to use %IF ... %THEN statements to create an exception report for the case where there is no data for a particular vendor.

11.3.3.3 Logical Expressions

Sections 11.3.3.4 and 11.3.3.5 show how to control the execution of macro code using

%IF *expression* %THEN

statements, where *expression* represents an arithmetic or logical expression. The expression is first evaluated using the %EVAL function (see Table 11-1), so the statement is equivalent to

%IF %EVAL(*expression*) %THEN

The expression is considered to be false if the result of %EVAL(*expression*) is 0 and true otherwise. The use of %EVAL with arithmetic expressions is discussed in Section 11.2.4.4. This section covers logical expressions, which are more commonly used with %IF statements. Table 11-2 lists the logical operators that can be used in constructing logical expressions.

Table 11-2: Operators Used in Logical Expressions

Symbol	Alternate	Example	Meaning
\multicolumn{4}{c}{**Comparison Operators**}
<	LT	A < B	A LT B	A is less than B
<=	LE	A <= B	A LE B	A is less than or equal to B
=	EQ	A = B	A EQ B	A is equal to B
^=	NE	A ^= B	A NE B	A is not equal to B
>=	GE	A >= B	A GE B	A is greater than or equal to B
>	GT	A > B	A GT B	A is greater than B
\multicolumn{4}{c}{**Logical Operators**}
&	AND	A & B	A AND B	Both A and B are true
|	OR	A | B	A OR B	Either A is true or B is true (or both are true)
^	NOT	^ A	NOT A	A is not true

Here are some examples using these operators and some of the macro variables from Section 11.2.2. In each case, the expression is considered to be false if the result is 0, and true otherwise.

```
%PUT Vend=[&Vend] NTest=[&NTest] Rate=[&Rate] Test_Date=[&Test_Date];
Vend=[Duality Logic] NTest=[    1000] Rate=[   0.015] Test_Date=[06JUL2008]

----------Code-------------------------------    ----Result---------------
%PUT %EVAL(&Vend = Duality);                      0
%PUT %EVAL(&Ntest > 50 OR &Vend = Duality);       1
%PUT %EVAL(&Vend EQ Duality Logic);               1
%PUT %EVAL(%UPCASE(&Vend) = DUALITY LOGIC);       1
%PUT %EVAL(&NTest > 50 AND NOT &NTest = 1000);    0
%PUT %EVAL(1 > 9);                                0
%PUT %EVAL(0.1 > .9);                             1
%PUT %SYSEVALF(0.1 > .9);                         0
%PUT %EVAL(2+2);                                  4
```

- In the highlighted example, the result seems to be wrong because 0.1>.9 should be false, but %EVAL does not accept decimal numbers as numbers. When it sees a decimal number it converts all numbers in the expression to character values, and compares the character strings. The character "0" is greater than "." in the character sort order, so the result is 1 (true).

- If you need to compare decimal values, use %SYSEVALF in your %IF statement. For example %IF %SYSEVALF(0.1>.9)… resolves to %IF %EVAL(%SYSEVALF(0.1>.9)…, which resolves to %IF %EVAL(0) …, giving the correct result.

11.3.3.4 Using an %IF Statement for Conditional Execution

The %Report4(2008) statement in Section 11.3.3.2 does not create a report for Duality because there is no Duality data in JES.Rates_2008. The %Report5 macro uses an %IF statement to check whether there is any data for a report, and creates an alternate report if there is no data.

```
%MACRO Report5(Year);
  %DO N = 1 %TO 3 %BY 1;
    PROC SQL NOPRINT;
        SELECT Vendor into :Vendor
        FROM JES.Vendor_List
        WHERE N=&N;
    QUIT;
    %LET Vendor=%TRIM(&Vendor);
    DATA Temp; SET JES.Rates_&Year;
        IF Vendor="&Vendor";
    RUN;
    PROC SQL NOPRINT; ❶
        SELECT COUNT(*) INTO :Num FROM Temp;
    QUIT;
    TITLE1 "Defect Rate Report for &Year at &Vendor";
    %IF &Num>0 %THEN %DO; ❷
        PROC PRINT DATA=Temp NOOBS LABEL;
            VAR Qtr N_Test Defects Defect_Rate;
        RUN;
    %END; /* Ends the %IF ... %THEN; DO; */
    %ELSE %DO;
        DATA Temp; Message= "No Data for &Year at &Vendor"; RUN;
        PROC PRINT DATA=Temp NOOBS; RUN;
    %END; /* Ends the %ELSE; DO; */
  %END; /* Ends the  %DO N = 1 %TO 3 loop */
%MEND Report5;
%Report5(2008) ❸
%Report5(2009) ❹
```

❶ The second PROC SQL puts the number of rows in Temp into the &Num variable.

❷ The %IF statement evaluates the expression &Num>0. If the expression is true, the lines between %THEN %DO; and %END; are executed. If the expression is false, the lines between %ELSE %DO; and %END; are executed. The %ELSE DO; ... %END; block is optional. In some cases you will want to execute a block of code if the expression is true, but do nothing if it is false.

❸ The %Report5(2008) statement creates reports for ChiTronix and Duality similar to the report in Section 11.3.1, and a "no data" report for Duality:

```
Defect Rate Report for 2008 at Duality
          Message
No Data for 2008 at Duality
```

❹ The %Report5(2009) statement generates an ERROR: message in the Log window because you are trying to use the JES.Rates_2009 data set—which doesn't exist. However, it also creates three incorrect reports using the Temp data set created by %Report(2008). If you run this code and don't check the Log, you might incorrectly report the 2008 data as 2009 data.

The next section shows how to check for the existence of a data set before trying to use it.

11.3.3.5 Using a %GOTO Statement to Branch to a Code Line

The Report6 macro tests for the existence of the Rates_&Year data set, and uses a %GOTO statement to branch to the macro label %Done at the end of the program if it does not exist.

```
%MACRO Report6(Year);
%IF %SYSFUNC(EXIST(JES.Rates_&Year))=0 %THEN  ❶
%GOTO Done;  ❷
%DO N = 1 %TO 3 %BY 1;
   PROC SQL NOPRINT;
        SELECT Vendor into :Vendor
        FROM JES.Vendor_List
        WHERE N=&N;
   QUIT;
   %LET Vendor=%TRIM(&Vendor);
   DATA Temp; SET JES.Rates_&Year;
        IF Vendor="&Vendor";
   RUN;
   PROC SQL NOPRINT;
        SELECT COUNT(*) INTO :Num FROM Temp;
   QUIT;
   TITLE1 "Defect Rate Report for &Year at &Vendor";
   %IF &Num>0 %THEN %DO;
        PROC PRINT DATA=Temp NOOBS LABEL;
             VAR Qtr N_Test Defects Defect_Rate;
        RUN;
   %END; /* Ends the %IF ... %THEN; DO; */
   %ELSE %DO;
        DATA Temp; Message= "No Data for &Year at &Vendor"; RUN;
        PROC PRINT DATA=Temp NOOBS; RUN;
   %END; /* Ends the %ELSE; DO; */
%END; /* Ends the  %DO N = 1 %TO 3 loop */
%Done:  ❸
%MEND Report6;
%Report6(2006)
%Report6(2009)
%Report6(2008)
```

❶ The expression %SYSFUNC(EXIST(JES.Rates_&Year)) resolves to true (or 1) if the JES.Rates_&year data set exists, and false (or 0) otherwise. Therefore, the code following %THEN is executed only if the data set *does not* exist. Note that the %DO; … %END; statements used in the previous example are not needed here as there is only one statement to be executed after the %THEN. A detailed explanation of the %SYSFUNC function is beyond the scope of this book, but you should understand the usage in this particular example, so you can use similar statements in your macros.

❷ The %GOTO statement causes execution to switch to the macro label named 'Done'.

❸ The %Done: statement creates the macro label named 'Done'. Note that a colon, not a semicolon, is used at the end of the macro label specification, to let SAS know that this is a label and not a macro program.

In my experience, the most common bugs remaining in production code have to do with trying to use data sets that either don't exist, or exist but have zero rows. You should use the techniques illustrated in %Report6 to avoid these kinds of problems. The %obscnt macro, discussed in Section 11.4.8, checks both conditions as once, and can be used to further simplify your code.

11.3.4 Global and Local Macro Variables

A macro variable can be *global* or *local*. A *local* variable is defined only within a macro program, and its value cannot be accessed outside of the program. A *global* variable is defined globally and can be accessed anywhere in your code. You can use %GLOBAL and %LOCAL statements in your macro programs to control which variables are global and local. The code for this section is in ~\JES\sample_code\ch_11\global_local.sas.

The rules for determining whether a macro is global or local are

- If a variable is defined outside of any macro program, it is global, but it can also be specified as local within a program using a %LOCAL statement.
- If a variable is defined within a macro program, it is local to that program, unless it is made global by a %GLOBAL statement within the macro,
    - or it was previously defined as global outside the macro program. Unless it is specified as local with a %LOCAL statement inside the macro.

The code sample illustrates all of the possibilities.

```
%INCLUDE "&JES.utility_macros/delvars.sas";
%delvars;

%LET One=1; %LET Two=2;
%PUT Before Test: One=&One, Two=&Two, Three=&Three, Four=&Four;
%MACRO Test;
   %LOCAL Two;
   %GLOBAL Four;
   %LET One=1000;
   %LET Two=2000;
   %LET Three = 3000;
   %LET Four=4000;
   %PUT Inside Test: One=&One, Two=&Two, Three=&Three, Four=&Four;
%MEND Test;
%Test;
%PUT After Test: One=&One, Two=&Two, Three=&Three, Four=&Four;
```

It is important to begin by running %delvars (see Section 11.2.3) to make sure the variables are not defined before the code is run. The code includes three %PUT statements to write the values of the macro variables to the Log window before, during, and after the %Test macro is run. The relevant lines from the Log window are shown here.

```
1691 %PUT Before Test: One=&One, Two=&Two, Three=&Three, Four=&Four;
WARNING: Apparent symbolic reference THREE not resolved.
WARNING: Apparent symbolic reference FOUR not resolved.
Before Test: One=1, Two=2, Three=&Three, Four=&Four

Inside Test: One=1000, Two=2000, Three=3000, Four=4000
1702 %PUT After Test: One=&One, Two=&Two, Three=&Three, Four=&Four;
WARNING: Apparent symbolic reference THREE not resolved.
After Test: One=1000, Two=2, Three=&Three, Four=4000
```

- The &One variable is global because it was defined before %Test was run.
  - Therefore, its value changed from 1 to 1000 when the macro was run, and retained the value 1000 after the macro finished.
- The &Two variable is global because it was defined before %Test was run, but it is also specified by the %LOCAL statement to be local within %Test.
  - Therefore, its value is 2000 while in %Test, but reverts to the global value, 2, after %Test is run.
- The &Three variable is local to %Test by default because it was not defined before %Test was run.
  - Therefore, it takes the value 3000 while in %Test, but is undefined after %Test is run.
- The %Four variable is global because of the %GLOBAL statement.
  - Therefore, it retains the value of 4000 after %Test is run.

There is one more rule to be considered: if you run a macro within another macro, a variable local to the outer macro cannot be declared global within the inner macro, as shown in this example.

```
%delvars
%MACRO Test_2;
   %LET Four=400;
   %Test;
%MEND Test_2;
%Test_2;
```

When you run this code, you will see an ERROR: message in the Log window.

ERROR: Attempt to %GLOBAL a name (FOUR) which exists in a local environment.
Inside Test: One=1000, Two=2000, Three=3000, Four=4000

The problem is that when &Four is defined in the %Test_2 macro, it is local to %Test_2 by default, so you can't then declare it to be global within %Test. This is a trap that's pretty easy to fall into. The way out is to make sure that &Four is not local, either by using a %GLOBAL statement within %Test_2, or by defining &Four outside of %Test_2.

Note that a variable local to a macro can be accessed anywhere within that macro, including in another macro embedded within it. So in the example above, the value of the &Four variable is available within the %Test macro.

11.3.5 Using MPRINT, MLOGIC, and SYMBOLGEN

There are three SAS options that you can use to help you understand what happens when your macro programs execute. Each of these options writes a different kind of information to the Log window as your macro runs. These options can be very useful in debugging your macros, or just understanding how they work.

- MPRINT writes the SAS code generated by the macro to the Log window.
- MLOGIC writes details about the execution of each step of the macro to the Log window.
- SYMBOLGEN writes the current value of each macro variable to the Log window.

The sample code reruns the %Report5 macro (see Section 11.3.3.4) with these options turned on. The code for this section is in ~\JES\sample_code\ch_11\mprint.sas.

```
%INCLUDE "&JES.utility_macros\delvars.sas";  ❶
%delvars
OPTIONS MPRINT MLOGIC SYMBOLGEN;  ❷
%Report5(2009)
OPTIONS NOMPRINT NOMLOGIC NOSYMBOLGEN;  ❸
```

❶ The %delvars macro deletes any previously defined macro variables.

❷ The first OPTIONS statement turns on the MPRINT MLOGIC and SYMBOLGEN options. These options are off by default when you start a SAS session. When %Report5 is run, the options cause many additional lines to be written to the Log window.

❸ The second OPTIONS statement turns off these options. It is generally a good idea to turn them off when you're not using them to avoid cluttering the Log window.

Selected lines from the Log window are shown below.

```
MLOGIC(REPORT5):  Beginning execution.
MLOGIC(REPORT5):  Parameter YEAR has value 2009
MLOGIC(REPORT5):  %DO loop beginning; index variable N; start value is 1; stop
     value is 3; by value is 1.
MPRINT(REPORT5):   PROC SQL NOPRINT;
SYMBOLGEN:  Macro variable N resolves to 1
MPRINT(REPORT5):   SELECT Vendor into :Vendor FROM Vendor_List WHERE N=1;
MPRINT(REPORT5):   QUIT;
NOTE: PROCEDURE SQL used (Total process time):
     real time         0.00 seconds
     cpu time          0.00 seconds

MLOGIC(REPORT5):  %LET (variable name is VENDOR)
MLOGIC(TRIM):  Beginning execution.
MLOGIC(TRIM):  This macro was compiled from the autocall file C:\Program
     Files\SAS\SASFoundation\9.2\core\sasmacro\trim.sas
SYMBOLGEN:  Macro variable VENDOR resolves to ChiTronix
MLOGIC(TRIM):  Parameter VALUE has value ChiTronix
```

Note that in addition to the details on the steps of the Report5 macro, the output also includes the steps of the %TRIM macro supplied by SAS, and the location of the autocall library.

11.3.6 Using SASAUTOS to Make Your Macros Accessible

After you define a macro program, it is available during your SAS session, but it must be redefined each time you start a new session before you can use it again. You can redefine your macros with %INCLUDE statements as we have been doing for the %delvars macro.

```
%INCLUDE "&JES.utility_macros/delvars.sas";
```

But it is more convenient to store macros you use frequently in an autocall macro library. Each time you try to run a macro which is not yet defined, SAS will look for the macro name in your autocall library and, if the name is found, define the macro and then run it. This section shows how to include the macros in your ~JES\utility_macros folder in an autocall library. These macros are described in Section 11.4. The code for this section is in ~\JES \sample_code\ch_11 \autocall.sas.

Open the autoexec.sas file that you created in Section 1.5. It should already contain the first two lines shown below. Add the other lines and save the file back to the same location, e.g., c:\Program Files \SAS\SAS 9.2, if that is the location of your autoexec.

```
%LET JES=c:\JES\; /* Use the path to your JES folder */
LIBNAME JES "&JES.sas_data";
FILENAME jesautos "&JES.utility_macros";  ❶
/*=== you may need these lines for UNIX ====  ❷
 %let SASROOT =/(location of your sasautos)/;
 FILENAME SASAUTOS "&SASROOT.sasautos";
=========================================*/
OPTIONS MAUTOSOURCE SASAUTOS=(SASAUTOS, jesautos);  ❸
%check_autocall  ❹
```

❶ The FILENAME statement defines jesautos to be a reference to the ~JES\utility_macros folder.

❷ You might need to uncomment these lines to define your autocall library on a UNIX system.

❸ The OPTIONS statement defines the autocall libraries. The MAUTOSOURCE option turns on the autocall facility. The SASAUTOS option specifies the folders that will be searched for macros. SASAUTOS is a reference to the location of the macros supplied by SAS, such as those shown in Table 11-1. SAS will search for macro names in the order in which the libraries are specified. If you want your own version of a macro (e.g., %LEFT) to take precedence over the SAS version, reverse the order to (jesautos, SASAUTOS) in your autoexec.sas

❹ %check_autocall is a macro in your utility_macros folder that writes a message to the Log.

After you have saved the new version of autoexec.sas, open a new SAS session. If SAS is able to find your macros, the %check_autocall macro runs, and the message "Your Autocall Library is Working" is written to the Log window. If the %check_autocall macro is not found, you will get an error message. When you get this to work correctly, all of the macros in ~JES\utility_macros will be available to you. You can edit these macros, and add your own macros to the folder, and all of these macros will be available to you in any SAS session, without having to use %INCLUDE statements to define them explicitly. But there are two conditions to keep in mind.

- When SAS looks for an autocall macro, it searches on the filenames. Therefore, your macro name must match the name of the file it is in. For example, the %check_autocall macro is in the check_autocall.sas file.
- If a macro is already defined, SAS will not look for it in the autocall library. Therefore, if you edit an autocall macro, the new version might not be available until you explicitly define it, e.g., with a %INCLUDE, or start a new SAS session.

11.4 Utility Macros

This section describes the macro programs that are in your **~JES\utility_macros** folder. If you succeeded in adding jesautos to your autocall library using the method described in Section 11.3.6, then all of these macros will be available to use in your code at any time. You can edit these macros and add your own macros to the folder, and all the latest versions will be available to you each time you start a SAS session. The macros will be described in the order in which they are listed Table 11-3. The code for this section is in ~\JES \utility_macros.

Table 11-3: Macro Programs in the ~JES\utility_macros Folder

Macro	Action
%check_autocall	Writes a message to the Log if your autocall library is found.
%theDate	Creates macro variables &theDate, &theDateTime.
%sasver_os	Creates macro variables containing the SAS version and operating system you are using.
%foot_note(H=3, Note=)	Runs a standardized FOOTNOTE statement.
%my_symbols(H=2, W=1, I=JOIN)	Runs a set of SYMBOL*n* statements with specified symbol height, line width, and line type.
%delvars	Deletes GLOBAL macro variables.
%dups(*dsn*, *variable*)	Finds rows in the *dsn* data set with duplicate values of *variable*.
%obscnt(*dsn*)	Returns the number of rows in the *dsn* data set.
%makedir(*newdir*)	Makes a new directory, *newdir*, if it does not already exist.
%varexist(*dsn*, *varlist*)	Checks whether the names in *varlist* are variables in the *dsn* data set.
%vars_in(*dsn*, *varlist*)	Creates a macro variable containing the names in *varlist* which are variables in *dsn*.
%DBMSlist, %RunQuery, %MakeList	Select records from a relational database which match values found in a SAS data set.
%sas_papers	Creates an HTML page that provides access to the SAS papers included with this book.

The four highlighted macros, %delvars, %obscnt, %makedir, and %varexist, were taken from *Carpenter's Complete Guide to the SAS® Macro Language, Second Edition*. Please refer to this book for a step-by-step explanation of each macro, as well as many other macros you might find useful. The sections describing each macro include the code for defining the macro, followed by some code that runs the macro to illustrate how it can be used. When you open the files in utility_macros you will find the example code in the commented out section at the top. The code in the macros is not explained here. Most are relatively easy to understand, except for those taken from Carpenter's book, which includes methods beyond the scope of this book.

11.4.1 %check_autocall

The %check_autocall macro writes a message to the Log window. You can include this macro in your autoexec.sas to verify that your autocall library is defined, as explained in Section 11.3.6.

```
%MACRO check_autocall;
    %PUT Your Autocall Library is Working!;
%MEND check_autocall;
```

11.4.2 %theDate

This macro creates two macro variables, &theDate and &theDateTime, containing the date and time at the time the macro is run. You can use &theDate or &theDateTime in footnotes to keep track of when various graphs or tables were created. This macro is used by the %foot_note macro in Section 11.4.4.

```
%MACRO theDate;
  %GLOBAL theDate theDateTime;
  DATA _NULL_;
      CALL SYMPUT('theDate', PUT(TODAY(),DATE9.));
      CALL SYMPUT('theDateTime', PUT(DATETIME(),DATETIME20.));
  RUN;
%MEND theDate;
%theDate;
%PUT theDate = [&theDate] theDateTime=[&theDateTime];
```

%PUT theDate = [&theDate] theDateTime=[&theDateTime];
theDate = [05FEB2008] theDateTime=[05FEB2008:16:25:06]

11.4.3 %sasver_os

The %sasver_os macro creates four macro variables:

- &OS is either PC or UNIX depending on where your session is running.
- &Slash is the character used in path descriptions, / for UNIX or \ for Windows.
- &SASver is the SAS version you are running (e.g., 8 or 9).
- &SASnum is the SAS version and release number you are using (e.g., 9_1 or 9_2).

The &OS and &Slash variables are useful if you need to write code to run on both UNIX and Windows. The &SASver and &SASnum variables are useful to correct for the (very few!) cases where code written in one version will not run on a later version, or where you want to take advantage of features in a new version without making the code unusable in an older version. There is an example using the &slash variable with the %makedir macro in Section 11.4.9.

```
%MACRO sasver_os;
  %GLOBAL OS Slash SASver SASnum;
  %LET SASver=%SUBSTR(&SYSVER,1,1);
  %LET SASnum=&SASver._%SUBSTR(&SYSVER,3,1);
  %LET OS=UNIX;
  %IF (&SYSSCP=WIN) %THEN %LET OS = PC;
  %IF "&OS"="PC"    %THEN %LET Slash = \;
  %IF "&OS"="UNIX"  %THEN %LET Slash = /;
%MEND sasver_os;
```

%sasver_os
%PUT OS=[&OS] Slash=[&Slash] SASver=[&SASver] SASnum=[&SASnum];
OS=[PC] Slash=[\] SASver=[9] SASnum=[9_2]

11.4.4 %foot_note

The %foot_note macro runs a FOOTNOTE statement (see Section 6.3.2.2), which includes your name and the date along with an optional note. The H parameter sets the text size. Edit this macro to include your own name. The example for the %my_symbols macro includes the use of %foot_note.

```
%MACRO Foot_Note(H=3, Note=);
   %theDate
   FOOTNOTE HEIGHT=&H JUSTIFY=LEFT "&Note    "
      JUSTIFY=RIGHT "(Your Name Here) &theDate";
%MEND Foot_Note;
```

11.4.5 %my_symbols

The %my_symbols macro runs 34 SYMBOL statements using each of the 34 "Special Symbols for Plotting Data Points" shown in Display 6-3 in Section 6.3.2.6, and a variety of colors. Each statement uses the same symbol height, line width, and line type (e.g., NONE. JOIN, STEPJ, SM40, etc.). You can use this macro to avoid writing several SYMBOL statements each time that you create a new plot with PROC GPLOT. The symbol height and line width and type are parameters, so you set these when you call the macro, but they are the same for all the SYMBOL statements. You can edit the VALUE and COLOR choices to get your own preferred set of symbols.

```
%MACRO my_symbols(H=2, W=1, I=JOIN);

   SYMBOL1    VALUE='PLUS'      HEIGHT=&H    COLOR=green    W=&W I=&I ;
   SYMBOL2    VALUE='x'         HEIGHT=&H    COLOR=red      W=&W I=&I ;
   SYMBOL3    VALUE='star'      HEIGHT=&H    COLOR=blue     W=&W I=&I ;
   SYMBOL4    VALUE='square'    HEIGHT=&H    COLOR=brown    W=&W I=&I ;
   SYMBOL5    VALUE='diamond'   HEIGHT=&H    COLOR=black    W=&W I=&I ;

   SYMBOL6    VALUE='triangle'  HEIGHT=&H    COLOR=brown    W=&W I=&I ;
   SYMBOL7    VALUE='hash'      HEIGHT=&H    COLOR=orange   W=&W I=&I ;
   SYMBOL8    VALUE='Y'         HEIGHT=&H    COLOR=violet   W=&W I=&I ;
   SYMBOL9    VALUE='Z'         HEIGHT=&H    COLOR=cyan     W=&W I=&I ;
   SYMBOL10   VALUE='paw'       HEIGHT=&H    COLOR=violet   W=&W I=&I ;
(lines specifying SYMBOL11 - SYMBOL33 omitted)
   SYMBOL34   VALUE=':'         HEIGHT=&H    COLOR=violet   W=&W I=&I ;
%MEND my_SYMBOLs;
```

This example creates a plot, shown in Figure 11-1, that displays the first 10 symbol types.

```
DATA A; DO Y=1 TO 10;
   X=Y; OUTPUT; X=Y+5; OUTPUT; END;
RUN;
%my_symbols(H=3, W=2, I=JOIN)
%foot_note(Note=Test of my_symbols macro)
TITLE1 H=4 "my_symbols Example";
PROC GPLOT DATA=A;
   PLOT Y*X=Y;
RUN; QUIT;
```

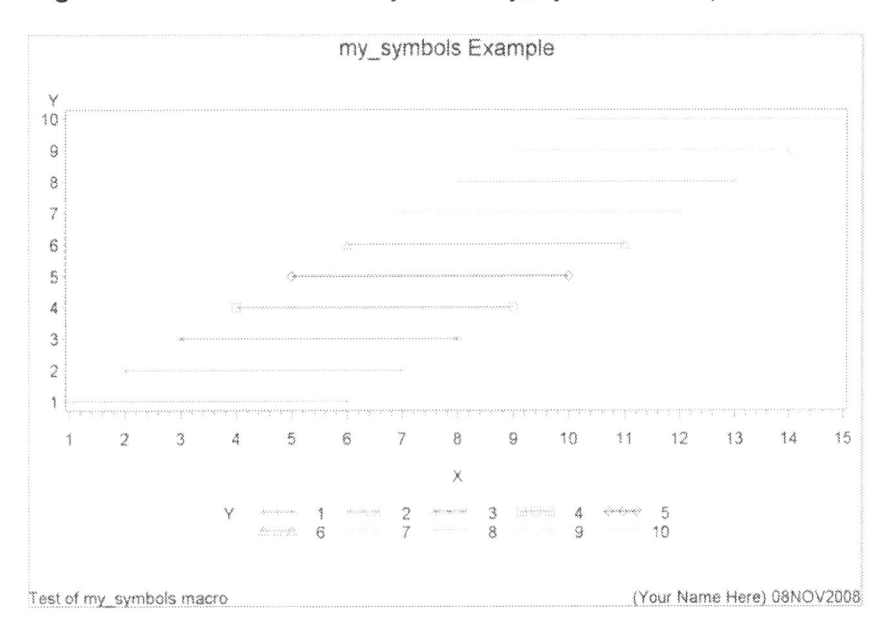

Figure 11-1: Plot Created by the %my_symbols Sample Code

![my_symbols Example plot showing Y vs X with 10 different series]

11.4.6 %delvars

The %delvars macro, which was described in Section 11.2.3, deletes all GLOBAL macro variables, except for &JES and variables beginning with 'SYS'. This macro was adapted from the %delvars macro which appears on p. 357 of *Carpenter's Complete Guide to the SAS® Macro Language, Second Edition*.

```
%MACRO delvars;
   DATA Vars; SET SASHELP.VMACRO; RUN;
   DATA _NULL_; SET VARS;
      IF Scope='GLOBAL' THEN DO;
         IF Name NE "JES"  AND SUBSTR(Name,1,3) NE "SYS" ❶
         THEN DO;
            CALL EXECUTE('%SYMDEL '||TRIM(LEFT(Name))||';');
            END;
         END;
   RUN;
%MEND delvars;
```

❶ This IF..THEN condition was added to the original %delvars macro to avoid deleting the &JES variable, or any variable beginning with "SYS". If you attempt to use the %SYMDEL function with a macro variable beginning with "SYS", an ERROR message is sent to the Log window, and the variable is not deleted.

11.4.7 %dups

The %dups macro takes two parameters

- *dsn* = name of a data set
- *variable* = name of a variable in the *dsn* data set

and creates three data sets

- &*dsn*._DP = all rows of *dsn* which have the same value of *variable* as one or more other rows
- &*dsn*._ND = all rows of *dsn* which have values of *variable* that do not appear in any other row
- &*dsn*._U = a subset of rows of *dsn* which includes one row for each unique value of *variable*

Data analysis methods often depend on the fact that a data set contains no duplicate values of a specific variable. For example, a data set containing installation records might be required to have one and only one record for each unit serial number. Unfortunately, duplicates are often found in the source data, so it's a good idea to check for duplicates before using the data. Section 2.7.2 shows how to use the FIRST and LAST automatic variables to remove duplicate records from a data set. The %Dups macro uses the same method to provide a quick way to detect and eliminate such duplicate records.

This macro is adapted from code on p.110 of *Combining and Modifying SAS® Data Sets: Examples, Version 6, First Edition*. This book does not discuss macro programming, but contains many examples that are good candidates for utility macros.

```
%MACRO dups(dsn, variable);
  %GLOBAL Num_Dups Num_Unique;
  PROC SORT DATA=&dsn; BY &variable; RUN;
  DATA &dsn._DP &dsn._ND &dsn._U;
      SET &dsn; BY &variable;
          IF FIRST.&variable AND LAST.&variable ❶
              THEN OUTPUT &DSN._ND;
          ELSE OUTPUT &DSN._DP;
          IF LAST.&variable THEN OUTPUT &DSN._U;
  RUN;
%MEND Dups;
```

❶ See Section 2.7 for an explanation of how FIRST.*variable* and LAST.*variable* are used to identify unique and duplicate values of *variable*.

This example uses the same data set, JES.Units, as in the example in Section 2.7.

```
DATA Units; SET JES.Units; RUN;
%Dups(Units, SN)
```

When you run the sample code, the Units_DP, Units_ND and Units_U are created. If this were real data, you could use the Units_U data set in the next step of your analysis, because it contains no duplicate values of SN. And you could examine the Units_DP data set to try to understand and eliminate the source of the duplicate records.

```
Units

Obs    SN      Install       Loc

 1    0005    06/21/2006    NY
 2    0007    06/16/2006    CA
 3    0007    06/09/2006    NY
 4    0015    06/09/2006    CA
 5    0015    06/08/2006    CA
 6    0016    06/06/2006    NY
 7    0027    06/05/2006    CA
 8    0035    06/17/2006    CA
 9    0035    06/11/2006    NY
10    0061    06/09/2006    NY
```

```
Units_DP

Obs    SN      Install       Loc

 1    0007    06/16/2006    CA
 2    0007    06/09/2006    NY
 3    0015    06/09/2006    CA
 4    0015    06/08/2006    CA
 5    0035    06/17/2006    CA
 6    0035    06/11/2006    NY
```

```
Units_U

Obs    SN      Install       Loc

 1    0005    06/21/2006    NY
 2    0007    06/09/2006    NY
 3    0015    06/08/2006    CA
 4    0016    06/06/2006    NY
 5    0027    06/05/2006    CA
 6    0035    06/11/2006    NY
 7    0061    06/09/2006    NY
```

```
Units_ND

Obs    SN      Install       Loc

 1    0005    06/21/2006    NY
 2    0016    06/06/2006    NY
 3    0027    06/05/2006    CA
 4    0061    06/09/2006    NY
```

11.4.8 %obscnt

The %obscnt macro is adapted from *Carpenter's Complete Guide to the SAS® Macro Language, Second Edition*, p. 320. This macro takes one parameter, *dsn*, the name of a SAS data set, and returns the number of rows in the data set. If the data set does not exist, zero is returned. This is an example of a *macro function*, which differs from a *macro program* in that it returns a value. The methods used to write a macro so that it returns a variable are beyond the scope of this book. If you are interested in learning how to write macro functions, see the books by Carpenter and Burlew, or the paper by Carpenter, "Macro Functions: How to Make Them—How to Use Them," all of which are listed in the "Chapter Summary" section.

```
%MACRO obscnt(dsn);
   %LOCAL nobs;
   %LET nobs=.;
   %IF %SYSFUNC(EXIST(&dsn)) %THEN %DO;   ❶
       %LET dsnid=%SYSFUNC(OPEN(&dsn));
       %IF &dsnid %THEN %DO;
           %LET nobs=%SYSFUNC(ATTRN(&dsnid, nlobs));
           %LET rc  =%SYSFUNC(CLOSE(&dsnid));
       %END;
       %ELSE %DO;
           %PUT Unable to open &dsn - %SYSFUNC(SYSMSG());
       %END;
   %END;
     %ELSE %DO;
        %LET NOBS=0;
   %END;
   &nobs
%MEND obscnt;
```

❶ This line was added to the original %obscnt macro so that it can be used to simultaneously check whether a data set exists and, if so, to return the number of rows.

You can use the %obscnt macro in your own macro programs in two ways:

- to check whether a data set exists, and has at least one row, before trying to use it
- to get the numbers of rows to use in a DO loop which used each row of a data set

The %Test macro illustrates both of these uses.

```
%MACRO Test;
   %IF %obscnt(JES.Vendor_List)=0 %THEN %GOTO DONE;   ❶
   %DO i=1 %TO %obscnt(JES.Vendor_List);   ❷
       %PUT &i;   ❸
   %END;
%DONE:;
%MEND Test;
%Test
```

❶ In this line, %obscnt is used to branch to the end of the program if the JES.Vendor_List data set does not exist, or has zero rows.

❷ Here %obscnt is used to specify that the limit of the DO loop is the number of rows in the data set.

❸ The %PUT statement doesn't do much, but you could replace it with code that operates on the information on the *i*-th row of JES.Vendor_List. For example, you could create a report for the Vendor in the *i*-th row, as in the %Report6 macro in Section 11.3.3.5.

11.4.9 %makedir

The %makedir macro is taken from *Carpenter's Complete Guide to the SAS® Macro Language, Second Edition*, p. 251. The %makedir macro checks for the existence of a directory, and creates one if it does not already exist.

```
%MACRO makedir(newdir);
   %LET rc=%SYSFUNC(FILEEXIST(&newdir));
   %IF &rc=0 %THEN %DO;
       %SYSEXEC mkdir &newdir;
   %END;
   %LET rc=%SYSFUNC(FILEEXIST(&newdir));  ❶
   %IF &rc=1 %THEN %PUT &newdir exists;
%MEND makedir;
```

❶ This line and the next were added to the original macro to confirm that the new directory exists.

It is often convenient to create separate folders to store different categories of data set, for example a separate folder for each of your product lines. Then you can use LIBNAME statements in your SAS code to reference each of these folders. But each time you add a new product, you need to remember to create the corresponding folder before trying use a LIBNAME statement to store data on the new product. You can use the %makedir macro to automatically create the required folder before you try to use it, as shown in the example.

```
LIBNAME products "&JES.output\products";  ❶
%makedir(&JES.output\products)
LIBNAME products "&JES.output\products";  ❷
%sasver_os  ❸
%makedir(&JES.output&Slash.products)
%makedir(&JES.output&Slash.prod&Slash.gizmo)  ❹
```

❶ This LIBNAME statement fails because the products directory does not exist.

❷ This LIBNAME statement succeeds because the products directory was created by %makedir in the previous line. Note that this example uses "\" in the path specification, so it will not work in a UNIX environment which requires "/" in the path specification.

❸ The %sasver_os macro (see Section 11.4.3) defines the &slash variable to be "\" if you are running SAS for Windows or "/" if you are running SAS for UNIX. The next %makedir uses the &slash variable, and so it will work in either Windows or UNIX.

❹ The third %makedir creates two directories, first "prod" and then "gizmo" within prod. This works in Windows, but not in UNIX because you cannot create a lower-level directory unless the higher-level directory already exists. You would need to use %makedir twice, first to create the "prod" folder, and then to create the "gizmo" folder.

11.4.10 %varexist

The %varexist macro is taken from *Carpenter's Complete Guide to the SAS® Macro Language, Second Edition*, p. 313. This macro takes two parameters

- *dsn* = the name of a data set
- *varlist* = a list of variable names

and returns the value 1 if all names in *varlist* are variables in *dsn*, and 0 otherwise. This macro is useful if you need to check whether the variables you need are present in a data set before trying to use them.

```
%MACRO varexist(dsn, varlist);
  %LOCAL dsid i ok var num rc;
  %LET dsid=%SYSFUNC(OPEN(&dsn, i));
  %LET i=1;
  %LET ok=1;
  %LET var=%SCAN(&varlist, &i);
  %DO %WHILE(&var ne );
      %LET num=%SYSFUNC(VARNUM(&dsid, &var));
      %IF &num=0 %THEN %LET ok=0;
      %LET i=%EVAL(&i+1);
      %LET var=%SCAN(&varlist,&i);
  %END;
  %LET rc=%SYSFUNC(CLOSE(&dsid));
  &ok
%MEND varexist;
```

The example uses %varexist to check for the presence of specified variables in the JES.Results data set (see Section 5.1.1).

```
%PUT %varexist(JES.Results, Vendor Month Resistance);
%PUT %varexist(JES.Results, Vendor Week Resistance);
```

The first %varexist statement returns 1 because Vendor, Month, and Resistance are all variables in the JES.Results data set. The second %varexist statement returns 0 because Week is not a variable in JES.Results.

```
%PUT %varexist(JES.Results, Vendor Month Resistance);
1
%PUT %varexist(JES.Results, Vendor Week Resistance);
0
```

11.4.11 %vars_in

The %vars_in macro, which calls the %varexist macro, takes two parameters

- *dsn* = the name of a data set
- *varlist* = a list of variable names

and creates a macro variable, &vars_in_dsn, which contains a list of all the variable names in *varlist* which are also variables in the *dsn* data set.

```
%MACRO vars_in(dsn, varlist);
  %GLOBAL vars_in_dsn;
  %LOCAL dsid i ok var num rc;
  %LET dsid=%SYSFUNC(OPEN(&dsn, i));
  %LET vars_in_dsn = ;
  %LET i=1;
  %LET var=%SCAN(&varlist, &i);
  %DO %WHILE(&var ne );
     %IF %varexist(&dsn, &var)=1
          %THEN %LET vars_in_dsn=&vars_in_dsn &var;
     %LET i=%EVAL(&i+1);
     %LET var=%SCAN(&varlist, &i);
  %END;
%MEND vars_in;
```

You can use the %vars_in macro, for example with PROC PRINT, to process as many variables as possible from a specified list, without causing errors if some of the variables are not present in the data set.

```
%vars_in(JES.Results, Vendor Month Week Resistance)
%PUT vars_in_dsn = [&vars_in_dsn];
PROC PRINT DATA=JES.Results;  ❶
   VAR Vendor Month Week Resistance;
RUN;
PROC PRINT DATA=JES.Results;  ❷
   VAR &vars_in_dsn;
RUN;
```

❶ The first PROC PRINT fails because Week is not a variable in JES.Results.

❷ The second PROC PRINT succeeds because &vars_in_dsn contains only variables in JES.Results.

```
39  %vars_in(JES.Results, Vendor Month Week Resistance)
40  %PUT vars_in_dsn = [&vars_in_dsn];
vars_in_dsn = [Vendor Month Resistance]
41  PROC PRINT DATA=JES.Results;
42     VAR Vendor Month Week Resistance;
ERROR: Variable WEEK not found.
43  RUN;
44  PROC PRINT DATA=JES.Results;
45     VAR &vars_in_dsn;
46  RUN;

NOTE: There were 754 observations read from the data set JES.RESULTS.
```

11.4.12 %DBMSlist, %RunQuery, and %MakeList

The %DBMSlist, %RunQuery, and %MakeList macros are taken from the paper

Helf, Garth W. 2002. "Can't Relate? A Primer on Using SAS with Your Relational Database." *Proceedings of the Twenty-Seventh Annual SAS Users Group International Conference.* Cary, NC: SAS Institute Inc. Paper 155-27. **[p155-27.pdf]**

This set of three macros provides a convenient method for extracting records from a relational database that match values in a list contained in a SAS data set. For example, find all of the records for which the value of Serial_Number is the same as any value of SN in a specified SAS data set. A very simple example of this idea was given in Section 4.3.7. These macros are a greatly enhanced version of that idea, including support for various data types, character, numeric, date, time, datetime, and a &bitesize parameter which controls the number of records sought in each SQL query. This is important because, in some cases, it is much more efficient to match, for example, 1000 values of SN in each of 10 queries than 10,000 values of SN in one query. Please consult the paper for instructions on how to use these macros.

```
%Macro DBMSlist(dsn, column, vtype, newdsn, dbname, query,
bitesize=200, test=no, dlm=%str(#) );
proc sql noprint; select count(*) into :gwhxxxx1 from &dsn; quit;
%if &gwhxxxx1=0 %then %do;
%put ====== WARNING: Input data set &dsn is empty, macro ends =======;
%goto exit;
%end;
%let totpass=%sysevalf(&gwhxxxx1/&bitesize,
ceil);
%if &test=no %then %do j=1 %to &gwhxxxx1 %by
&bitesize;
%let p=%sysevalf(&j/&bitesize, ceil);
%put ========== Starting pass &p of &totpass =============;
data gwhxxxx2; set &dsn (firstobs=&j obs=%eval(&j+&bitesize-1)); run;
%MakeList(mylist, gwhxxxx2, &column, &vtype);
%RunQuery(&dbname, gwhxxxx3, &query);
%if &j=1 %then %do;
data &newdsn;
set gwhxxxx3;
run;
%end;
%else %do;
Proc append base=&newdsn data=gwhxxxx3; run;
%end;
%end;
%else %do; %* Test=yes: do one query for timing;
data gwhxxxx2; set &dsn (firstobs=1 obs=&bitesize); Run;
%MakeList(mylist, gwhxxxx2, &column, &vtype);
%RunQuery(&dbname, gwhxxxx3, &query);
%end;
%exit: %mend DBMSlist;

%macro RunQuery(dbinfo, dsname, query);
%let DBMStype=DB2;
proc sql;
connect to &DBMStype (&dbinfo);
create table &dsname as select * from
connection to &DBMStype (
%unquote(&query) for fetch only);
```

```
%put &sqlxmsg;
disconnect from &DBMStype;
quit;
%mend RunQuery;

%macro MakeList(globname, dsn, varinfo, vartype);
%local i j;
%global &globname;
%let &globname=; /* return null if macro fails*/
%let numvars=1; %* find number of variables specified;
%do %while (%scan(&vartype,%eval(&numvars+1))
ne);
%let numvars=%eval(&numvars+1);
%end;
/*******************************************
Single variable entered
*******************************************/
%if &numvars=1 %then %do;
/*************** Character ********/
%if %upcase(&vartype)=C %then %do;
proc sql noprint;
Select
Distinct translate(quote(&varinfo),"'",'"')
into :&globname separated by ','
from &dsn;
quit;
%end;
/*************** Numeric ********/
%else %if %upcase(&vartype)=N %then %do;
proc sql noprint;
select distinct &varinfo
into :&globname separated by ','
from &dsn;
quit;
%end;
/*************** Date ********/
%else %if %upcase(&vartype)=D %then %do;
proc sql noprint;
select distinct "'"||
put(&varinfo, yymmdd10.)||"'"
into :&globname separated by ','
from &dsn;
quit;
%end;
/*************** Time ********/
%else %if %upcase(&vartype)=T %then %do;
proc sql noprint;
select distinct "'"||
translate(put(&varinfo, time.),
'.',':','0',' ')||"'"
into :&globname separated by ','
from &dsn;
quit;
%end;
                    (partial code listing)
```

11.4.13 %sas_papers

The %sas_papers macro creates a Web page that provides convenient access to most of the papers referenced in this book. It reads the ~JES\docs\SUGI_Papers.csv file (Display 11-1), which contains a list of all the paper titles and the names of the PDF files, and then creates the ~JES\docs\papers.html Web page, shown in Display 11-2. The left frame of the page allows you to select all the papers, or just the papers that are relevant to a particular chapter. The right frame lists the papers and allows you to open a paper by clicking on the name of the PDF, as shown in Display 11-3. You can add your own favorite papers to the SUGI_Papers.csv file and rerun the macro to customize the page for your own interests. Just be sure to put the new PDF files in the ~JES\docs\pdfs folder so the page can find them, and don't run the macro until you have closed the CSV file.

```
%MACRO SAS_papers;
PROC IMPORT DATAFILE="&JES.docs\SUGI_Papers.csv"
   OUT =Papers REPLACE ;
   GUESSINGROWS= 500 ;
RUN;
DATA Papers; SET Papers;
   ChapterTitle=TRANSLATE(ChapterTitle, ' ', ',');
RUN;
%MACRO Table_of_Papers(dsin, title, Chapter);
   ODS PROCLABEL="&title";
   TITLE2 H=5 "&Title";
   ODS NOPROCTITLE;
   PROC REPORT DATA=&dsin NOWINDOWS HEADLINE LS=256;
   COLUMN &Chapter Paper Author TITLE ;
   DEFINE &Chapter/ GROUP;
   DEFINE Paper/ DISPLAY;
   DEFINE Title / DISPLAY ;
   DEFINE Author / DISPLAY ;
   COMPUTE Paper;
       ref="./pdfs/"||TRIM(LEFT(Paper));
       CALL DEFINE (_COL_, 'URL', ref);
   ENDCOMP;
   BREAK AFTER &Chapter / SKIP;
   RUN;
%MEND Table_of_Papers;
ODS HTML PATH="&JES.docs" (URL=NONE)
   BODY ="papers_body.html" headtext="<base target=_blank>"
   CONTENTS="papers_toc.html"
   FRAME="papers.html" (TITLE="SAS Papers") STYLE=Minimal
       NEWFILE=PAGE;
   TITLE1 H=7  "Papers included with Just Enough SAS";
   FOOTNOTE;
   %Table_of_Papers(Papers, All Papers, Chapter);
   %DO i=1 %TO 12;
       DATA Chapter; SET Papers(where=(Chapter=&i)); RUN;
       %IF %obscnt(Chapter)>0 %THEN %DO;
           PROC SQL NOPRINT; SELECT ChapterTitle INTO :Title FROM Chapter; QUIT:
               %Table_of_Papers(Chapter, &Title, Chapter);
       %END;
   %END;
ODS HTML CLOSE;
%MEND sas_papers;
```

Display 11-1: The SUGI_Papers.csv File

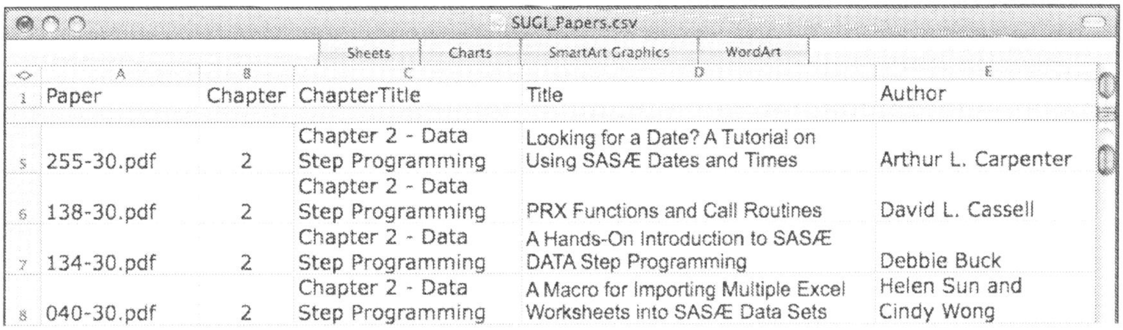

	A	B	C	D	E
1	Paper	Chapter	ChapterTitle	Title	Author
5	255-30.pdf	2	Chapter 2 - Data Step Programming	Looking for a Date? A Tutorial on Using SAS® Dates and Times	Arthur L. Carpenter
6	138-30.pdf	2	Chapter 2 - Data Step Programming	PRX Functions and Call Routines	David L. Cassell
7	134-30.pdf	2	Chapter 2 - Data Step Programming	A Hands-On Introduction to SAS® DATA Step Programming	Debbie Buck
8	040-30.pdf	2	Chapter 2 - Data Step Programming	A Macro for Importing Multiple Excel Worksheets into SAS® Data Sets	Helen Sun and Cindy Wong

Display 11-2: The JES/docs/papers.html Web Page

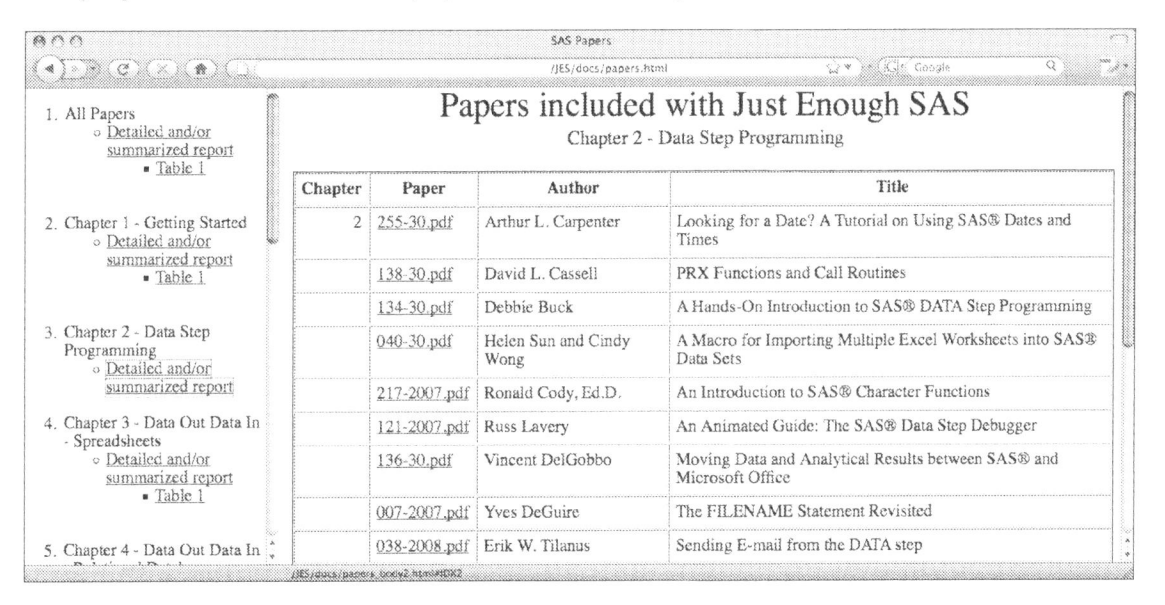

Papers included with Just Enough SAS
Chapter 2 - Data Step Programming

Chapter	Paper	Author	Title
2	255-30.pdf	Arthur L. Carpenter	Looking for a Date? A Tutorial on Using SAS® Dates and Times
	138-30.pdf	David L. Cassell	PRX Functions and Call Routines
	134-30.pdf	Debbie Buck	A Hands-On Introduction to SAS® DATA Step Programming
	040-30.pdf	Helen Sun and Cindy Wong	A Macro for Importing Multiple Excel Worksheets into SAS® Data Sets
	217-2007.pdf	Ronald Cody, Ed.D.	An Introduction to SAS® Character Functions
	121-2007.pdf	Russ Lavery	An Animated Guide: The SAS® Data Step Debugger
	136-30.pdf	Vincent DelGobbo	Moving Data and Analytical Results between SAS® and Microsoft Office
	007-2007.pdf	Yves DeGuire	The FILENAME Statement Revisited
	038-2008.pdf	Erik W. Tilanus	Sending E-mail from the DATA step

Sidebar navigation:
1. All Papers
   - Detailed and/or summarized report
     - Table 1
2. Chapter 1 - Getting Started
   - Detailed and/or summarized report
     - Table 1
3. Chapter 2 - Data Step Programming
   - Detailed and/or summarized report
4. Chapter 3 - Data Out Data In - Spreadsheets
   - Detailed and/or summarized report
     - Table 1
5. Chapter 4 - Data Out Data In

Display 11-3: Paper 255-30 Opened in Adobe Reader

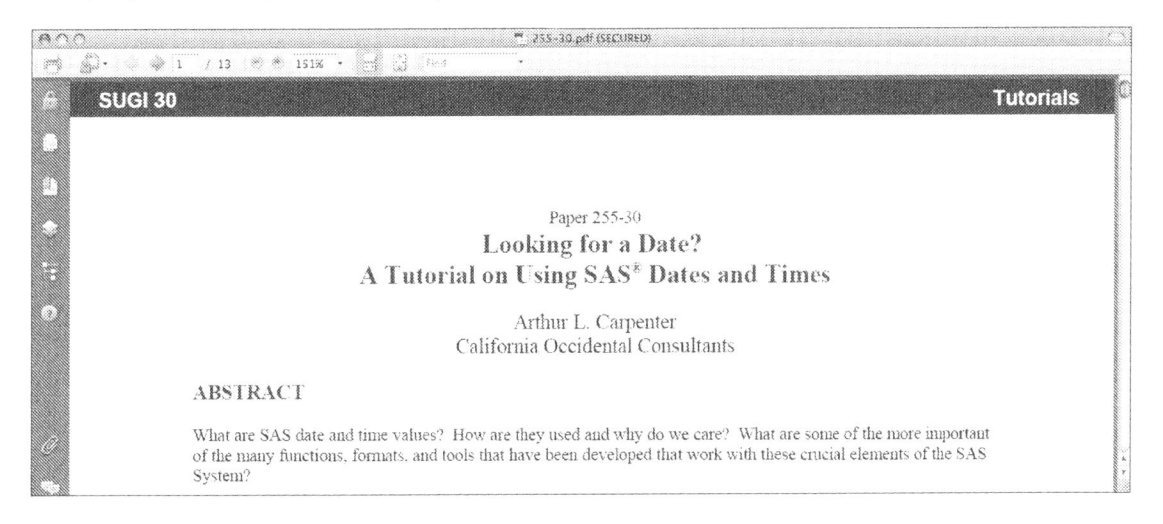

SUGI 30 — Tutorials

Paper 255-30
Looking for a Date?
A Tutorial on Using SAS® Dates and Times

Arthur L. Carpenter
California Occidental Consultants

ABSTRACT

What are SAS date and time values? How are they used and why do we care? What are some of the more important of the many functions, formats, and tools that have been developed that work with these crucial elements of the SAS System?

11.5 More Than Enough

This code in this section is written in Perl, not SAS. While you can do almost everything in SAS, a bit of Perl code is very useful for tasks such as controlling SAS batch jobs and, as illustrated here, collecting a group of SAS files into a single file. This book cannot teach you how to install or use Perl, but if you are a Perl user, or are willing to learn (it's not difficult), the code described here could be very useful to you. The Perl code, allcode.pl, is in your ~**JES**\utility_macros folder. To run the code

1. Open a terminal window (Command Prompt in Windows).

2. Navigate to the ~JES\utility_macros folder.

3. Type "perl allcode.pl" (without the quotes) and press the Return key.

The code creates a new file, all_code_*yyyy_mm_dd*.txt, shown below, with a list of all files in the directory that have a .sas extension, followed by a code listing for each file, with a header that shows the date when the code was most recently modified. The all_code_*yyyy_mm_dd*.txt file provides a time-stamped backup of all your macros. You can use an %INCLUDE statement with this file to define all of your macros at once, or attach this file to an e-mail to share all of your macros with a colleague. You can also copy allcode.pl to the folder containing your production code, and run it there to create a single file containing all your code. You can search this file to find all the places where particular macros, variables, data sets, or text phrases are used, which can be an invaluable aid in maintaining and debugging your code.

```
/*>==============================================
        all_code_2008_11_9.txt
        Last Modified Sun Nov  9 08:52:28 2008
------------------------------------------------
-----------  check_autocall.sas
-----------  DBMSlist.sas
-----------  delvars.sas
-----------  dups.sas
-----------  foot_note.sas
-----------  makedir.sas
-----------  MakeList.sas
-----------  my_symbols.sas
-----------  obscnt.sas
-----------  RunQuery.sas
-----------  sasver_os.sas
-----------  sas_papers.sas
-----------  thedate.sas
-----------  varexist.sas
-----------  vars_in.sas
-----------  xlxp2sas.sas
==============================================>*/
/*>==============================================
        check_autocall.sas
        Last Modified Sat Feb  9 19:42:58 2008
==============================================>*/
/*======================================
Macro: %check_autocall;
Action: Writes a message to the Log window
======================================*/
%MACRO check_autocall;
   %PUT Your Autocall Library is Working!;
%MEND check_autocall;
/*>==============================================
        END    check_autocall.sas
```

```
                    =============================================>*/

            ---(listing of all other .sas files in ~JES/utility_macros)----
```

This is the listing of the Perl program allcode.pl.

```perl
#!/usr/bin/perl -w
use strict;
use Cwd;
my $Day = (localtime)[3]; my $Mon = 1+(localtime)[4];
my $Yr  = 1900+(localtime)[5];
my $Dat = "$Yr"."_"."$Mon"."_$Day";
my( @FILES, $folder, $file, $afile, $line, $filedat);
my ($dev, $ino, $mode, $nlink, $uid, $gid, $rdev, $size,
        $atime, $mtime, $ctime, $blksize, $blocks);
my $sascode = "all_code"."_$Dat";
$folder=cwd;
open(OUT,">$folder"."/$sascode".".txt");
$filedat=scalar localtime;
print OUT ("/*>=============================================\n");
print OUT ("         $sascode \n");
print OUT ("         Last Modified $filedat\n");
print OUT ("---------------------------------------------\n");
# Writelist of sas files
opendir(TEMP, "$folder") || die;
while ($file=readdir TEMP){ chomp($file);
if($file=~m/\.sas$/){  print OUT ("------------ $file\n"); }  }
close(TEMP);
print OUT ("=============================================>*/\n\n");
# Write sas code
opendir(TEMP, "$folder") || die;
while ($file=readdir TEMP){ print("file=$file\n"); chomp($file);
if($file=~m/\.sas$/){
  print "$folder"."/"."$file\n";
        ($dev, $ino, $mode, $nlink, $uid, $gid, $rdev, $size,
         $atime, $mtime, $ctime, $blksize, $blocks)=stat("$file");
         $filedat=scalar localtime($mtime);
  print OUT ("/*>=============================================\n");
  print OUT ("         $file \n");
  print OUT ("         Last Modified $filedat\n");
  print OUT ("=============================================>*/\n");
              open(SAS,"<$folder"."/"."$file")    ||die "$!";
              while(<SAS>)
              { #s/\/\*.*\*\//  /;
                 $line="$_";
                 chomp($line);
                    print OUT ("$line\n");}
              close(SAS);
  print OUT ("/*>=============================================\n");
  print OUT ("         END    $file\n");
  print OUT ("=============================================>*/\n\n");
             }
} closedir(TEMP);
close(OUT);
```

11.6 Chapter Summary

11.6.1 Recap

After finishing this chapter you should know how to

- Create macro variables
    - using the %LET function
    - using CALL SYMPUT
    - using PROC SQL
- View all GLOBAL macro variables.
- Delete macro variables
    - using %SYMDEL
    - using %delvars
- Use macro variables and macro functions in your code.
- Write macro programs using positional and keyword parameters.
- Control the execution of the lines in your macros using these kinds of expressions
    - %DO I = 1 %TO &N
    - %IF logical expression %THEN ....;
    - %IF *logical expression* %THEN %DO; ... %END;
    - %GOTO *line*
- Define and use GLOBAL and LOCAL macro variables.
- Use the MPRINT, MLOGIC and SYMBOLGEN options to understand and debug your macros.
- Specify an autocall library that makes all of your macros accessible.
- Use the macros provided in the ~JES\utility_macros folder.

11.6.2 For More Information

Books

Aster, Rick. 2004. *Professional SAS Programmer's Pocket Reference, Fifth Edition*. Paoli PA: Breakfast Communications Corp. Chapter 8.

Aster, Rick, 2005. *Professional SAS Programming Shortcuts, Second Edition*. Paoli, PA: Breakfast Communications Corp. Chapters 61-63.

Burlew, Michele M. 2006. *SAS Macro Programming Made Easy, Second Edition*, Cary NC: SAS Institute Inc.

Carpenter, Art, 2004. *Carpenter's Complete Guide to the SAS Macro Language, Second Edition*. Cary NC: SAS Institute Inc.

Delwiche, Lora D., and Susan J. Slaughter. 2003. *The Little SAS Book: A Primer, Third Edition*. Cary, NC: SAS Institute Inc. Chapter 7.

SAS Institute Inc. 1995. *Combining and Modifying SAS Data Sets: Examples, Version 6, First Edition*. Cary, NC: SAS Institute Inc.

SAS Conference Papers

Carpenter, Arthur L. 1999. "Macro Quoting Functions, Other Special Character Masking Tools, and How To Use Them." *Proceedings of the Twenty-Fourth Annual SAS Users Group International Conference.* Cary, NC: SAS Institute Inc. Paper 38-24. **[p38-24.pdf]**

Carpenter, Arthur L., 2000. "Placing Dates in Your Titles: Do It Dynamically." *Proceedings of the Twenty-Fifth Annual SAS Users Group International Conference.* Cary, NC: SAS Institute Inc. Paper 93-25. **[25p093.pdf]**

Carpenter, Arthur L. 2002. "Building and Using Macro Libraries." *Proceedings of the Twenty-Seventh Annual SAS Users Group International Conference.* Cary, NC: SAS Institute Inc. Paper 17-27. **[p017-27.pdf]**

Carpenter, Arthur L. 2002. "Macro Functions: How to Make Them—How to Use Them." *Proceedings of the Twenty-Seventh Annual SAS Users Group International Conference.* Cary, NC: SAS Institute Inc. Paper 100-27. **[p100-27.pdf]**

Crawford, Peter. 2008. "Add SAS® Macros to Your Programming Skills: Achieve More, Write Less Code." *Proceedings of the SAS Global Forum 2008 Conference.* Cary, NC: SAS Institute Inc. Paper 175-2008. **[175-2008.pdf]**

First, Steven, and Katie Ronk. 2005. "SAS® Macro Variables and Simple Macro Programs." *Proceedings of the Thirtieth Annual SAS Users Group International Conference.* Cary, NC: SAS Institute Inc. Paper 130-30.

Larsen, Erik S. 2008. "Creating a Stored Macro Facility in Ten Minutes." *Proceedings of the SAS Global Forum 2008 Conference.* Cary, NC: SAS Institute Inc. Paper 101-2008. **[101-2008.pdf]**

Murphy, William C. 2007. "Changing Data Set Variables into Macro Variables." *Proceedings of the SAS Global Forum 2007 Conference.* Cary, NC: SAS Institute Inc. Paper 050-2007. **[050-2007.pdf]**

Philp, Stephen. 2008. "SAS® MACRO: Beyond the Basics." *Proceedings of the SAS Global Forum 2008 Conference.* Cary, NC: SAS Institute Inc. Paper 045-2008. **[045-2008.pdf]**

Slaughter, Susan J., and Lora D. Delwiche. 2004. "SAS® Macro Programming for Beginners." *Proceedings of the Twenty-Ninth Annual SAS Users Group International Conference.* Cary, NC: SAS Institute Inc. Paper 243-29. **[243-29.pdf]**

11.6.3 Exercises

Simplify the %Report6 macro (Section 11.3.3.5) by using the %obscnt macro (Section 11.4.8) to

- check for the existence of the JES.Rates_&Year data set
- compute the upper limit for the %DO loop
- replace the &Num macro variable in the %IF statement

Create a set of macros that can produce a report like the sample report in Section 7.1.1.

- Create a macro program from vendors_1.sas, with parameters Month_From and Month_To, that can be used to create a table similar to the first table in Display 7-2 for any range of months.
- Create similar macros from vendors_2.sas and vendors_3.sas.

- Create macros from chitronix_1.sas and chitronix_2.sas, each with three parameters: Vendor, Month_From, and Month_To, that control the vendor and range of months to be used for the plots.
- Create a macro with parameters Month_From and Month_To that calls the macros created above to produce the complete report.
    - Use the methods shown in Section 11.3.3.2 to loop through the vendor names in JES.Vendor_List.
    - Use the %obscnt macro as shown in the example in Section 11.4.8 to check that each data set exists and has at least one row before using it.

Bibliography

Books

Allison, Paul D. 1995. *Survival Analysis Using the SAS® System: A Practical Guide*. Cary, NC: SAS Institute Inc.

Aster, Rick. 2004. *Professional SAS® Programmer's Pocket Reference, Fifth Edition*. Paoli PA: Breakfast Communications Corp.

Aster, Rick. 2005. *Professional SAS® Programming Shortcuts, Second Edition*. Paoli, PA: Breakfast Communications Corp.

Bilenas, Jonas V. 2005. *The Power of PROC FORMAT*. Cary, NC: SAS Institute Inc.

Burlew, Michele M. 2006. *SAS® Macro Programming Made Easy, Second Edition*. Cary NC: SAS Institute Inc.

Carpenter, Art. 1999. *Annotate: Simply the Basics*. Cary, NC: SAS Institute Inc.

Carpenter, Art. 2004. *Carpenter's Complete Guide to the SAS® Macro Language, Second Edition*. Cary NC: SAS Institute Inc.

Carpenter, Art. 2007. *Carpenter's Complete Guide to the SAS® REPORT Procedure*. Cary NC: SAS Institute Inc.

Carpenter, Arthur L., and Charles E. Shipp. 1995. *Quick Results with SAS/GRAPH® Software*. Cary NC: SAS Institute Inc.

Cody, Ronald P., and Jeffrey K. Smith. 2006. *Applied Statistics and the SAS® Programming Language*. Upper Saddle River, NJ: Pearson Prentice Hall.

Delwiche, Lora D., and Susan J. Slaughter. 2003. *The Little SAS® Book: A Primer, Third Edition*. Cary, NC: SAS Institute Inc.

Duncan, Acheson J. 1986. *Quality Control and Industrial Statistics, Fifth Edition*. Homewood, IL: Irwin.

Grant, Eugene L., and Richard S. Leavenworth. 1988. *Statistical Quality Control, Sixth Edition*. New York: McGraw-Hill.

Gupta, Sunil. 2003. *Quick Results with the Output Delivery System*. Cary, NC: SAS Institute Inc.

Haworth, Lauren E. 1999. *PROC TABULATE by Example*. Cary, NC: SAS Institute Inc.

Haworth, Lauren E. 2001. *Output Delivery System: The Basics*. Cary, NC: SAS Institute Inc.

Henderson, Don. 2007. *Building Web Applications with SAS/IntrNet®: A Guide to the Application Dispatcher*. Cary NC: SAS Institute Inc.

Lafler, Kirk Paul. 2004. *PROC SQL: Beyond the Basics Using SAS®*. Cary, NC: SAS Institute Inc.

Meeker, William Q., and Luis A. Escobar. 1998. *Statistical Methods for Reliability Data*. New York, NY: John Wiley & Sons.

Miron, Thomas. 1995. *The How-To Book for SAS/GRAPH® Software*. Cary, NC: SAS Institute Inc.

Montgomery, Douglas C. 1996. *Introduction to Statistical Quality Control, Third Edition*. New York, NY: John Wiley & Sons.

Morgan, Derek P. 2006. *The Essential Guide to SAS® Dates and Times*. Cary, NC: SAS Institute Inc.

Nelson, Wayne. 1982. *Applied Life Data Analysis*. New York, NY: John Wiley & Sons.

Pratter, Frederick E. 2006. *Web Development with SAS® by Example, Second Edition*. Cary NC: SAS Institute Inc.

SAS Institute Inc. 1995. *Combining and Modifying SAS® Data Sets: Examples, Version 6, First Edition*. Cary, NC: SAS Institute Inc.

SAS Institute Inc. 1995. *Logistic Regression Examples Using the SAS® System, Version 6, First Edition*. Cary, NC: SAS Institute Inc.

Stokes, Maura E., Charles S. Davis, and Gary G. Koch. 2000. *Categorical Data Analysis Using the SAS® System, Second Edition*. Cary, NC: SAS Institute Inc.

Tobias, Paul A., and David C. Trindade. 1995. *Applied Reliability, Second Edition*. New York, NY: Van Nostrand Reinhold.

van Dobben de Bruyn, Cornelis S. 1968. *Cumulative Sum Tests: Theory and Practice*. London: Griffin.

Zdeb, Mike. 2002. *Maps Made Easy Using SAS®*. Cary, NC: SAS Institute Inc.

Papers

Brown, David. 2005. "%sas2xl: A Flexible SAS® Macro That Uses Tagsets to Produce Complex, Multi-Tab Excel Spreadsheets with Custom Formatting." *Proceedings of the Thirtieth Annual SAS Users Group International Conference*. Cary, NC: SAS Institute Inc. Paper 092-30. **[092-30.pdf]**

Buck, Debbie. 2005. "A Hands-On Introduction to SAS® DATA Step Programming." *Proceedings of the Thirtieth Annual SAS Users Group International Conference*. Cary, NC: SAS Institute Inc. Paper 134-30. **[134-30.pdf]**

Carpenter, Arthur L. 1999. "Macro Quoting Functions, Other Special Character Masking Tools, and How to Use Them." *Proceedings of the Twenty-Fourth Annual SAS Users Group International Conference*. Cary, NC: SAS Institute Inc. Paper 38-24. **[p38-24.pdf]**

Carpenter, Arthur L. 2000. "Placing Dates in Your Titles: Do It Dynamically." *Proceedings of the Twenty-Fifth Annual SAS Users Group International*. Cary, NC: SAS Institute Inc. Paper 93-25. **[25p093.pdf]**

Carpenter, Arthur L. 2002. "Building and Using Macro Libraries." *Proceedings of the Twenty-Seventh Annual SAS Users Group International Conference*. Cary, NC: SAS Institute Inc. Paper 17-27. **[p17-27.pdf]**

Carpenter, Arthur L. 2002. "Macro Functions: How to Make Them—How to Use Them." *Proceedings of the Twenty-Seventh Annual SAS Users Group International Conference*. Cary, NC: SAS Institute Inc. Paper 100-27. **[p100-27.pdf]**

Carpenter, Arthur L. 2005. "Looking for a Date? A Tutorial on Using SAS® Dates and Times." *Proceedings of the Thirtieth Annual SAS Users Group International Conference*. Cary, NC: SAS Institute Inc. Paper 255-30. **[255-30.pdf]**

Carpenter, Arthur L. 2005. "PROC REPORT Basics: Getting Started with the Primary Statements." *Proceedings of the Western Users of SAS Software Thirteenth Annual Conference*. San Jose, CA. **[how_proc_report_basics.pdf]**

Carpenter, Arthur L. 2006. "Data Driven Annotations: An Introduction to SAS/GRAPH's® Annotate Facility." *Proceedings of the Thirty-First Annual SAS Users Group International Conference*. Cary, NC: SAS Institute Inc. Paper 108-31. **[108-31.pdf]**

Carpenter, Arthur L. 2008. "The Path, the Whole Path, and Nothing But the Path, So Help Me Windows." *Proceedings of the SAS Global Forum 2008 Conference*. Cary, NC: SAS Institute Inc. Paper 023-2008. **[023-2008.pdf]**

Carpenter, Arthur L. 2008. "PROC REPORT: Compute Block Basics—Part I Tutorial." *Proceedings of the SAS Global Forum 2008 Conference*. Cary, NC: SAS Institute Inc. Paper 031-2008. **[031-2008.pdf]**

Carr, David W. 2008. "When PROC APPEND May Make More Sense Than the DATA STEP." *Proceedings of the SAS Global Forum 2008 Conference*. Cary, NC: SAS Institute Inc. Paper 085-2008. **[085-2008.pdf]**

Cartier, Jeff. 2006. "A Programmer's Introduction to the Graphics Template Language." *Proceedings of the Thirty-First Annual SAS Users Group International Conference*. Cary, NC: SAS Institute Inc. Paper 262-31. **[262-31.pdf]**

Cartier, Jeff, and Dan Heath. 2007. "Using ODS Styles with SAS/GRAPH®." *Proceedings of the SAS Global Forum 2007 Conference*. Cary, NC: SAS Institute Inc. Paper 088-2007. **[088-2007.pdf]**

Cheng, Wei. 2006. "ODS Statistical Graphics for Clinical Research." *Proceedings of the Thirty-First Annual SAS Users Group International Conference*. Cary, NC: SAS Institute Inc. Paper 095-31. **[095-31.pdf]**

Cisternas, Miriam, and Art Carpenter. 2005. "Extreme Graphics Make Over: Using SAS/GRAPH® to Get the Graphical Output You Need." *Proceedings of the Thirtieth Annual SAS Users Group International Conference*. Cary, NC: SAS Institute Inc. Paper 133-30. **[133-30.pdf]**

Cody, Ronald. 2005. "An Introduction to SAS® Character Functions (Including Some New SAS®9 Functions)." *Proceedings of the Thirtieth Annual SAS Users Group International Conference*. Cary, NC: SAS Institute Inc. Paper 233-30. **[233-30.pdf]**

Cody, Ronald. 2007. "An Introduction to SAS® Character Functions." *Proceedings of the SAS Global Forum 2007 Conference*. Cary, NC: SAS Institute Inc. Paper 217-2007. **[217-2007.pdf]**

Crawford, Peter. 2008. "Add SAS® Macros to Your Programming Skills: Achieve More, Write Less Code." *Proceedings of the SAS Global Forum 2008 Conference.* Cary, NC: SAS Institute Inc. Paper 175-2008. **[175-2008.pdf]**

Delaney, Kevin P. 2003. "Multiple Graphs on One Page, the Easy Way (PDF) and the Hard Way (RTF)." *Proceedings of the Twenty-Eighth Annual SAS Users Group International Conference.* Cary, NC: SAS Institute Inc. Paper 94-28. **[094-28.pdf]**

DelGobbo, Vincent. 2003. "A Beginner's Guide to Incorporating SAS® Output in Microsoft Office Applications." *Proceedings of the Twenty-Eighth Annual SAS Users Group International Conference.* Cary, NC: SAS Institute Inc. Paper 52-28. **[p52-28.pdf]**

DelGobbo, Vincent. 2004. "From SAS® to Excel via XML." *Proceedings of the Seventeenth Annual NorthEast SAS Users Group Conference.* Baltimore, MD. **[DelGobbo_ExcelXML.pdf]**

DelGobbo, Vincent. 2005. "Moving Data and Analytical Results between SAS® and Microsoft Office." *Proceedings of the Thirtieth Annual SAS Users Group International Conference.* Cary, NC: SAS Institute Inc. Paper 136-30. **[136-30.pdf]**

DelGobbo, Vincent. 2008. "Tips and Tricks for Creating Multi-Sheet Microsoft Excel Workbooks the Easy way with SAS®." *Proceedings of the SAS Global Forum 2008 Conference.* Cary, NC: SAS Institute Inc. Paper 192-2008. **[192-2008.pdf]**

Droogendyk, Harry, and Faisal Dosani. 2008. "Joining Data: Data Step Merge or SQL?" *Proceedings of the SAS Global Forum 2008 Conference.* Cary, NC: SAS Institute Inc. Paper 178-2008. **[178-2008.pdf]**

Erdman, Donald, Laura Jackson, and Arthur Sinko. 2008. "Zero-Inflated Poisson and Zero-Inflated Negative Binomial Models Using the COUNTREG Procedure." *Proceedings of the SAS Global Forum 2008 Conference.* Cary, NC: SAS Institute Inc. Paper 322-2008. **[322-2008.pdf]**

Feng, Ying. 2006. "The SQL Procedure: When and How to Use It?" *Proceedings of the Thirty-First Annual SAS Users Group International Conference.* Cary, NC: SAS Institute Inc. Paper 044-31. **[044-31.pdf]**

First, Steven. 2008. "The SAS INFILE and FILE Statements." *Proceedings of the SAS Global Forum 2008 Conference.* Cary, NC: SAS Institute Inc. Paper 166-2008.

First, Steven, and Katie Ronk. 2005. "SAS® Macro Variables and Simple Macro Programs." *Proceedings of the Thirtieth Annual SAS Users Group International Conference.* Cary, NC: SAS Institute Inc. Paper 130-30.

Gebhart, Eric A. 2005. "Tagset Spelunking and Cartography: Debugging and Exploring Tagsets with Battery-Powered Headlamps." *Proceedings of the Thirtieth Annual SAS Users Group International Conference.* Cary, NC: SAS Institute Inc. Paper 014-30. **[014-30.pdf]**

Gebhart, Eric. 2007. "ODS Markup, Tagsets, and Styles! Taming ODS Styles and Tagsets." *Proceedings of the SAS Global Forum 2007 Conference.* Cary, NC: SAS Institute Inc. Paper 225-2007. **[225-2007.pdf]**

Gebhart, Eric. 2008. "The Devil Is in the Details: Styles, Tips, and Tricks That Make Your Microsoft Excel Output Look Great!" *Proceedings of the SAS Global Forum 2008 Conference.* Cary, NC: SAS Institute Inc. Paper 036-2008. **[036-2008.pdf]**

Gupta, Sunil K. 2001. "Using Styles and Templates to Customize SAS® ODS Output." *Proceedings of the Twenty-Sixth Annual SAS Users Group International Conference.* Cary, NC: SAS Institute Inc. Paper 1-26. **[p001-26.pdf]**

Gupta, Sunil. 2008. "SAS® ODS Technology for Today's Decision Makers." *Proceedings of the SAS Global Forum 2008 Conference.* Cary, NC: SAS Institute Inc. Paper 193-2008. **[193-2008.pdf]**

Hadden, Louise. 2005. "PROC TABULATE and ODS RTF: The Perfect Fit for Complex Tables." *Proceedings of the Thirtieth Annual SAS Users Group International Conference.* Cary, NC: SAS Institute Inc. Paper 091-30. **[091-30.pdf]**

Hadden, Louise. 2006. "STOP! WAIT! GO! See What Traffic-Lighting Can Do For You!" *Proceedings of the Thirty-First Annual SAS Users Group International Conference.* Cary, NC: SAS Institute Inc. Paper 142-31. **[142-31.pdf]**

Hadden, Louise. 2009. "Turn the Tables on Boring Reports with SAS 9.2 and RTF Tagset Options." *Proceedings of the SAS Global Forum 2009 Conference.* Cary, NC: SAS Institute Inc. Paper 223-2009. **[223-2009.pdf]**

Haworth, Lauren. 2001. "ODS for PRINT, REPORT, and TABULATE." *Proceedings of the Twenty-Sixth Annual SAS Users Group International Conference.* Cary, NC: SAS Institute Inc. Paper 3-26. **[p003-26.pdf]**

Haworth, Lauren. 2003. "SAS Reporting 101: REPORT, TABULATE, ODS and Microsoft Office." *Proceedings of the Twenty-Eighth Annual SAS Users Group International Conference.* Cary, NC: SAS Institute Inc. Paper 71-28. **[p071-28.pdf]**

Haworth, Lauren. 2003. "SAS with Style: Creating Your Own ODS Style Template." *Proceedings of the Twenty-Eighth Annual SAS Users Group International Conference.* Cary, NC: SAS Institute Inc. Paper 195-28. **[195-28.pdf]**

Haworth, Lauren. 2004. "Introduction to ODS." *Proceedings of the Twenty-Ninth Annual SAS Users Group International Conference.* Cary, NC: SAS Institute Inc. Paper 245-29. **[245-29.pdf]**

Haworth, Lauren. 2004. "SAS with Style: Creating Your Own ODS Style Template for RTF Output." *Proceedings of the Twenty-Ninth Annual SAS Users Group International Conference.* Cary, NC: SAS Institute Inc. Paper 125-29. **[125-29.pdf]**

Haworth, Lauren. 2005. "SAS with Style: Creating Your Own ODS Style Template for PDF Output." *Proceedings of the Thirtieth Annual SAS Users Group International Conference.* Cary, NC: SAS Institute Inc. Paper 132-30. **[132-30.pdf]**

Haworth, Lauren. 2006. "PROC TEMPLATE: The Basics." *Proceedings of the Thirty-First Annual SAS Users Group International Conference.* Cary, NC: SAS Institute Inc. Paper 112-31. **[112-31.pdf]**

Heath, Dan. 2007. "SAS/GRAPH® Procedures for Creating Statistical Graphics in Data Analysis." *Proceedings of the SAS Global Forum 2007 Conference.* Cary, NC: SAS Institute Inc. Paper 193-2007. **[193-2007.pdf]**

Heath, Dan. 2008. "Effective Graphics Made Simple Using SAS/GRAPH® SG Procedures." *Proceedings of the SAS Global Forum 2008 Conference.* Cary, NC: SAS Institute Inc. Paper 255-2008. **[255-2008.pdf]**

Heaton, Ed. 2008. "Many-to-Many Merges in the DATA Step." *Proceedings of the SAS Global Forum 2008 Conference.* Cary, NC: SAS Institute Inc. Paper 81-2008. **[081-2008.pdf]**

Helf, Garth W. 2001. "Joining SAS® and DBMS Tables Efficiently." *Proceedings of the Twenty-Sixth Annual SAS Users Group International.* Cary, NC: SAS Institute Inc. Paper 127-26. **[p127-26.pdf]**

Helf, Garth W. 2002. "Can't Relate? A Primer on Using SAS® with Your Relational Database." *Proceedings of the Twenty-Seventh Annual SAS Users Group International Conference.* Cary, NC: SAS Institute Inc. Paper 155-27. **[p155-27.pdf]**

Kuiper, Daniel, and Koen Vyverman. 2008. "Put Your Customers on the Map: Integrating SAS/GRAPH® and Google Earth." *Proceedings of the SAS Global Forum 2008 Conference.* Cary, NC: SAS Institute Inc. Paper 252-2008. **[252-2008.pdf]**

Lafler, Kirk Paul. 2006. "A Hands-On Tour Inside the World of PROC SQL." *Proceedings of the Thirty-First Annual SAS Users Group International Conference.* Cary, NC: SAS Institute Inc. Paper 114-31. **[114-31.pdf]**

Larsen, Erik S. 2008. "Creating a Stored Macro Facility in Ten Minutes." *Proceedings of the SAS Global Forum 2008 Conference.* Cary, NC: SAS Institute Inc. Paper 101-2008. **[101-2008.pdf]**

Levine, Fred. 2001. "Using the SAS/ACCESS® LIBNAME Technology to Get Improvements in Performance and Optimizations in SAS/SQL® Queries." *Proceedings of the Twenty-Sixth Annual SAS Users Group International Conference.* Cary, NC: SAS Institute Inc. Paper 110-26. **[p110-26.pdf]**

Lund, Pete. 2008. "PDF Can Be Pretty Darn Fancy: Tips and Tricks for the ODS PDF Destination." *Proceedings of the SAS Global Forum 2008 Conference.* Cary, NC: SAS Institute Inc. Paper 033-2008. **[033-2008.pdf]**

Malby, Ann, and Sally Williams. 2008. "Save Time! Merge SAS® Files to Themselves." *Proceedings of the SAS Global Forum 2008 Conference.* Cary, NC: SAS Institute Inc. Paper 234-2008. **[234-2008.pdf]**

Matange, Sanjay. 2008. "Introduction to the Graph Template Language." *Proceedings of the SAS Global Forum 2008 Conference.* Cary, NC: SAS Institute Inc. Paper 313-2008. **[gtl.pdf]**

Matange, Sanjay. 2008. "ODS Graphics Editor." *Proceedings of the SAS Global Forum 2008 Conference.* Cary, NC: SAS Institute Inc. Paper 235-2008. **[235-2008.pdf]**

Massengill, A. Darrell. 2005. "Tips and Tricks: Using SAS/GRAPH® Effectively." *Proceedings of the Thirtieth Annual SAS Users Group International Conference.* Cary, NC: SAS Institute Inc. Paper **090**-30. **[090-30.pdf]**

Matthews, JoAnn, and Doug Zirbel. 2003. "SAS-L—A Very Powerful Resource for SAS Users Worldwide." *Proceedings of the Twenty-Eighth Annual SAS Users Group International Conference.* Cary, NC: SAS Institute Inc. Paper 247-28. **[247-28.pdf]**

McGill, Robert, John W. Tukey, and Wayne A. Larsen. 1978. "Variations of Box Plots." *The American Statistician,* February 1978, Vol. 32, No. 1. pp. 12–16.

McNeill, Sandy. 2001. "Changes & Enhancements for ODS by Example (through Version 8.2)." *Proceedings of the Twenty-Sixth Annual SAS Users Group International Conference.* Cary, NC: SAS Institute Inc. Paper 2-26. **[p002-26.pdf]**

Mink, David, and David J. Pasta. 2006. "Improving Your Graphics Using SAS/GRAPH® Annotate Facility." *Proceedings of the Thirty-First Annual SAS Users Group International Conference.* Cary, NC: SAS Institute Inc. Paper 085-31. **[085-31.pdf]**

Morgan, Derek. 2008. "The Essentials of SAS® Dates and Times." *Proceedings of the SAS Global Forum 2008 Conference.* Cary, NC: SAS Institute Inc. Paper 168-2008. **[168-2008.pdf]**

Murphy, William C. 2007. "Changing Data Set Variables into Macro Variables." *Proceedings of the SAS Global Forum 2007 Conference.* Cary, NC: SAS Institute Inc. Paper 050-2007. **[050-2007.pdf]**

Nelson, Lloyd S. 1984. "The Shewhart Control Chart—Tests for Special Causes." *Journal of Quality Technology,* 16: 237–239.

Nelson, Lloyd S. 1985. "Interpreting Shewhart $\overline{X}$ Control Charts." *Journal of Quality Technology,* 17:114–116.

Nelson, Wayne. 1988. "Graphical Analysis of System Repair Data." *Journal of Quality Technology* 20 (1): 24–35.

Nelson, Wayne. 1995. "Confidence Limits for Recurrence Data —Applied to Cost or Number of Product Repairs." *Technometrics* 37:147–157.

Olson, Diane. 2008. "New in SAS® 9.2: It's the Little Things That Count." *Proceedings of the SAS Global Forum 2008 Conference.* Cary, NC: SAS Institute Inc. Paper 176-2008. **[176-2008.pdf]**

Parker, Chevell. 2003. "Generating Custom Excel Spreadsheets Using ODS." *Proceedings of the Twenty-Eighth Annual SAS Users Group International Conference.* Cary, NC: SAS Institute Inc. Paper 12-28. **[012-28.pdf]**

Parker, Chevell. 2004. "SAS 9.1 MS OFFICE Integration." **[office91.pdf]**

Pass, Ray, and Daphne Ewing. 2006. "So You're Still Not Using PROC REPORT. Why Not?" *Proceedings of the Thirty-First Annual SAS Users Group International Conference.* Cary, NC: SAS Institute Inc. Paper 235-31. **[235-31.pdf]**

Philp, Stephen. 2008. "SAS® MACRO: Beyond the Basics." *Proceedings of the SAS Global Forum 2008 Conference.* Cary, NC: SAS Institute Inc. Paper 045-2008. **[045-2008.pdf]**

Pratter, Frederick E. 2007. "Using the SAS® Output Delivery System and PROC TEMPLATE to Create XHTML Files." *Proceedings of the SAS Global Forum 2007 Conference.* Cary, NC: SAS Institute Inc. Paper 118-2007. **[118-2007.pdf]**

Repole, Warren. 2007. "Exporting SAS/GRAPH® Output for Inclusion in Web Pages and Other Software Applications." *Proceedings of the SAS Global Forum 2007 Conference.* Cary, NC: SAS Institute Inc. Paper 269-2007. **[sgf2007-gsf.pdf]**

Rhodes, Dianne Louise. 2005. "Pretty Dates All in a Row." *Proceedings of the Thirtieth Annual SAS Users Group International Conference.* Cary, NC: SAS Institute Inc. Paper 055-30. **[055-30.pdf]**

Rodriguez, Robert N. 2008. "Getting Started with ODS Statistical Graphics in SAS® 9.2." *Proceedings of the SAS Global Forum 2008 Conference.* Cary, NC: SAS Institute Inc. Paper 305-2008. **[305-2008.pdf]**

Rodriguez, Robert N., and Tonya E. Balan. 2006. "Creating Statistical Graphics in SAS 9.2: What Every Statistical User Should Know." *Proceedings of the Thirty-First Annual SAS Users Group International Conference.* Cary, NC: SAS Institute Inc. Paper 192-31. **[192-31.pdf]**

SAS Institute Inc. 2003. SAS Technical Support. "SAS® Dates, Times, and Interval Functions." Cary, NC: SAS Institute Inc. TS-668. **[ts668.pdf]**

Schenker, Nathaniel, and Jane F. Gentleman. 2001. "On Judging the Significance of Differences by Examining the Overlap Between Confidence Intervals." *The American Statistician,* August 2001, Vol. 55, No. 3. pp. 182-186.

Slaughter, Susan J., and Lora D. Delwiche. 2004. "SAS® Macro Programming for Beginners." *Proceedings of the Twenty-Ninth Annual SAS Users Group International Conference.* Cary, NC: SAS Institute Inc. Paper 243-29. **[243-29.pdf]**

Smith, Kevin D. 2007. "The Output Delivery System (ODS) from Scratch." *Proceedings of the SAS Global Forum 2007 Conference.* Cary, NC: SAS Institute Inc. Paper 219-2007. **[219-2007.pdf]**

Smith, Kevin D. 2007. "PROC TEMPLATE Tables from Scratch." *Proceedings of the SAS Global Forum 2007 Conference.* Cary, NC: SAS Institute Inc. Paper 221-2007. **[221-2007.pdf]**

Tabladillo, Mark. 2008. "Return of the Codes: SAS'®, Windows'®, and Yours." *Proceedings of the SAS Global Forum 2008 Conference.* Cary, NC: SAS Institute Inc. Paper 004-2008. **[004-2008.pdf]**

Tilanus, Erik W. 2008. "Poor Man's Parallel Processing Using the DATA Step View." *Proceedings of the SAS Global Forum 2008 Conference.* Cary, NC: SAS Institute Inc. Paper 096-2008. **[096-2008.pdf]**

Tilanus, Erik W. 2008. "Sending E-mail from the DATA Step." *Proceedings of the SAS Global Forum 2008 Conference.* Cary, NC: SAS Institute Inc. Paper 038-2008. **[038-2008.pdf]**

Tilanus, Erik W. 2008. "SET, MERGE and Beyond." *Proceedings of the SAS Global Forum 2008 Conference.* Cary, NC: SAS Institute Inc. Paper 167-2008. **[167-2008.pdf]**

Tong, Cindy. 2003. "ODS RTF: Practical Tips." *Proceedings of the Sixteenth Annual NorthEast SAS Users Group Conference.* Washington, DC. Paper at007. **[at007.pdf]**

Whitlock, Ian. 2008. "The Art of Debugging." *Proceedings of the SAS Global Forum 2008 Conference.* Cary, NC: SAS Institute Inc. Paper 165-2008. **[165-2008.pdf]**

Williams, Christianna S. 2008. "PROC SQL for DATA Step Die-hards." *Proceedings of the SAS Global Forum 2008 Conference.* Cary, NC: SAS Institute Inc. Paper 185-2008. **[185-2008.pdf]**

Zdeb, Mike, and Robert Allison. 2006. "SAS/GRAPH® 101." *Proceedings of the Thirty-First Annual SAS Users Group International Conference.* Cary, NC: SAS Institute Inc. Paper 239-31. **[239-31.pdf]**

Zender, Cynthia. 2004. "Markup 101: Markup Basics." *Proceedings of the Twenty-Ninth Annual SAS Users Group International Conference.* Cary, NC: SAS Institute Inc. Paper 2603-29. **[p2603-29.pdf]**

Zender, Cynthia, and Catherine Truxillo. 2005. "Customizing ODS Statistical Graphs." *Proceedings of the Thirtieth Annual SAS Users Group International Conference*. Cary, NC: SAS Institute Inc. Paper 239-30. **[239-30.pdf]**

Index

A

ABS function 33
acceleration models
    Arrhenius 344–345, 348–349
    defined 335
    fitting multiple Weibull distribution 340–341
    fitting single Weibull distribution 338–339
    Weibull 342–343
ACROSS option
    KEYLEGEND statement (SGPLOT) 276
    LEGEND statement 160–161
Adobe Reader 218–219
AFTER option
    BREAK statement (REPORT) 115
    RBREAK statement (REPORT) 115
aliases 91
ALL option, TABLES statement (TABULATE) 113
ALPHA= option
    COMPARE statement (SGSCATTER) 298
    CUSUM procedure 326–327
    DOT statement (SGPLOT) 256–257
    HBAR statement (SGPLOT) 256
    HLINE statement (SGPLOT) 256–257, 288
    MATRIX statement (SGSCATTER) 298
    MEANS procedure 108
    PLOT statement (SGSCATTER) 298
    SGPLOT procedure 270
    SHEWHART procedure 323, 325
    VBAR statement (SGPLOT) 256, 258
    VLINE statement (SGPLOT) 256
    XCHART statement (ANOM) 120
ampersand (&) 6
analysis
    See data analysis
ANALYSIS option, DEFINE statement (REPORT) 115
Anderson-Darling test for normality 126–127
ANGLE option
    FOOTNOTE statement 156
    TITLE statement 156
ANNOTATE option
    GMAP procedure 203
    GPLOT procedure 176–177
ANOM procedure
    BOXCHART statement 120
    functionality 100, 120, 311, 316
    PCHART statement 120, 123
    UCHART statement 120, 122
    XCHART statement 120–121

ANSI standard 79
ANYALNUM function 39
ANYALPHA function 39
ANYCNTRL function 39
ANYDIGIT function 39
ANYFIRST function 39
ANYGRAPH function 39
ANYLOWER function 39
ANYNAME function 39
ANYPRINT function 39
ANYPUNCT function 39
ANYSPACE function 39
ANYUPPER function 39
ANYXDIGIT function 39
ARRAY statement 53
arrays 53, 248
Arrhenius acceleration model 344–345, 348–349
ATRISK option, LIFETEST procedure 352
attributes, data set variables 26–29
autocall programs 376
autoexec.sas 5, 8
AUTOHREF option, GPLOT procedure 162
automatic variables
    FIRST.var 45–46
    IN 48
    LAST.var 45–46
    viewing 368
AUTOVREF option, GPLOT procedure 162
axes
    multiple variables on different 151
    multiple variables on same 150
    second X or Y 268–269
AXIS option, REFLINE statement (SGPLOT) 280
AXIS statement
    example 182
    functionality 158–159
    LABEL option 158–159
    LABEL=NONE option 158
    MINOR option 158–159
    OFFSET option 158
    ORDER option 158–159, 170
    REFLABEL option 163
    VALUE option 158

B

backward slashes (\) 3
BAND statement, SGPLOT procedure 253, 263–264

bar charts
    character variables 184–185
    controlling statistics display 192–193
    dividing into segments 194–195
    GCHART procedure 182–191
    horizontal 182–189, 258–259
    numeric variables 184–185
    SGPLOT procedure 258–259
    stacked 196–197
    vertical 190–191, 258–259
BAR statement, GBARLINE procedure 139
Base SAS 12–13, 249
BASELINE statement, PHREG procedure 354
BAYES statement, LIFEREG procedure 360–361
Bayesian analysis 360–361
BETAINV function 32–33, 110, 292
binomial distribution
    adding confidence limits 110–111
    inverse probability functions 32–33
    MEANS procedure and 108, 110–111
bitmap formats 138, 144–145
block charts 139, 180, 198–199
BLOCK statement
    GCHART procedure 139, 180, 198–199
    GMAP procedure 139, 198, 203–204
BODY= option, ODS HTML statement 220, 222, 239
BORDER option
    GOPTIONS statement 142, 154–155
    INSET statement (SGPLOT) 282
    KEYLEGEND statement (SGPLOT) 276
BOUNDARY option, HISTOGRAM statement (SGPLOT) 284–285
box-and-whisker plots
    See box plots
BOX= option, TABLES statement (TABULATE) 113
box plots
    creating 316
    GPLOT procedure 170–171
    schematic 116, 118–119
    skeletal 116–117
BOXCHART statement
    ANOM procedure 120
    SHEWHART procedure 316
BOXCONNECT option
    PLOT statement (BOXPLOT) 116
    SHEWHART procedure 316
BOXPLOT procedure
    functionality 100, 116, 316
    ID statement 118
    PLOT statement 116
    schematic box plots 118–119
    skeletal box plots 116–117
BOXSTYLE option
    PLOT statement (BOXPLOT) 116, 118
    SHEWHART procedure 316
BOXWIDTH option
    HBOX statement (SGPLOT) 286
    PLOT statement (BOXPLOT) 116, 118
    VBOX statement (SGPLOT) 286
BOXWIDTHSCALE= option, PLOT statement (BOXPLOT) 118
branching macro programs 382–385
BREAK statement, REPORT procedure
    AFTER option 115
    functionality 115
    OL option 115
    SKIP option 115
    SUMMARIZE option 115
bubble plots
    BUBBLE statement (GPLOT) 139, 146, 198, 201
    BUBBLE2 statement (GPLOT) 139, 146, 198
BUBBLE statement, GPLOT procedure 139, 146, 198
BUBBLE2 statement, GPLOT procedure 139, 146, 198, 201
BWIDTH option, SYMBOL statement 170
BWSLEGEND option, PLOT statement (BOXPLOT) 118
BY statement
    GPLOT procedure 148
    SGPLOT procedure 264–265, 296
    SORT procedure 45–46
    TRANSPOSE procedure 52

C

CALL DEFINE routine 226–227
CALL RANNOR statement 31, 33
CALL routines 31–33, 235
CALL SYMPUT routine 235, 370–371
CAPABILITY procedure
    CFILL option 318
    CLEFT option 318
    CRIGHT option 318
    defined 311
    functionality 318–319
    HISTOGRAM statement 318
    INSET statement 318
    NORMAL option 318
    OUTTABLE option 318
    SPEC statement 318
    VAR statement 318
CAT function 37

CATEGORY option
    HBOX statement (SGPLOT) 286
    VBOX statement (SGPLOT) 286
CATS function 39
CATX function 37
CBOXES option, PLOT statement (BOXPLOT) 118
CBOXFILL option, PLOT statement (BOXPLOT) 118
CDF (Cumulative Distribution Function)
    functionality 32–33, 332–333
    Poisson distribution example 24
CEIL function 33
CFILL option, CAPABILITY procedure 318
CFREQ keyword, GCHART procedure 180, 182, 192
character variables
    character functions 36–39
    converting to numeric 37
    defined 22
    midpoint values 180
    on bar charts 184–185
    setting length 26
Characteristic Life 334
CHART statement, GRADAR procedure 139
chart variables 180
%check_autocall utility macro 390
CHISQ option, TABLES statement (FREQ) 130
CHORO statement, GMAP procedure 139, 198, 203
CHREF option, GPLOT procedure 162
CHTML format 213
CI option, SYMBOL statement 170–171
CINV function 32–33, 110
CLASS statement
    MEANS procedure 106–107
    TABULATE procedure 112
    UNIVARIATE procedure 128–129
classification variables 335
CLEFT option, CAPABILITY procedure 318
CLI option
    COMPARE statement (SGSCATTER) 298
    MATRIX statement (SGSCATTER) 298
    PBSPLINE statement (SGPLOT) 262
    PLOT statement (SGSCATTER) 298
    REG statement (SGPLOT) 262
CLM keyword 105
CLM option
    COMPARE statement (SGSCATTER) 298
    LOESS statement (SGPLOT) 262
    MATRIX statement (SGSCATTER) 298
    PBSPLINE statement (SGPLOT) 262
    PLOT statement (SGSCATTER) 298
    REG statement (SGPLOT) 262

CMYK color-naming scheme 168–169
CO option, SYMBOL statement 170
COALESCE function, SQL procedure 83
code
    See SAS code
color-naming schemes 168–169
COLOR option
    FOOTNOTE statement 156–157
    PATTERN statement 186
    SYMBOL statement 164
    TITLE statement 156
COLORS option, GOPTIONS statement 203
COLUMN statement, REPORT procedure 114, 227, 231
columns
    See also variables
    equivalent terms 20, 78
    extracting in data sets 80
    selecting 132
    selecting subsets 42
COLUMNS option, SGPANEL procedure 296
COMPARE procedure 51
COMPARE statement, SGSCATTER procedure
    ALPHA option 298
    CLI option 298
    CLM option 298
    DATALABEL option 298
    DEGREE option 298
    ELLIPSE option 298
    functionality 298–301
    GROUP option 298
    LOESS option 298
    NOGROUP option 298
    PBSPLINE option 298
    REG option 298
COMPBL function 36–37
COMPRESS function 37, 68–69
COMPUTE blocks
    adding hyperlinks to Web pages 226–227
    adding traffic lighting to reports 230–231
concatenating
    character strings 37
    data sets 133
concatenation operator (||) 36
conditional logic
    See also DO loops
    creating data sets 24
    %GOTO statement 385
    IF-THEN-ELSE statements 24
    IF-THEN-OUTPUT statements 43
    %IF...%THEN statements 383–384
    selecting records 46
    selecting subsets of rows 43

confidence intervals/limits
    adding normal limits on the means 108–109
    binomial distribution 110–111
    inverse probability functions 32
    MEANS procedure 105
    plotting group limits 172–179, 288–295
    Poisson distribution 110–111
    SGPLOT procedure 262
CONNECT TO statement, SQL procedure 86
CONTENTS option, ODS HTML statement 222
CONTENTS procedure 241
CONTENTS statement, DATASETS procedure 49
continuous variables
    acceleration models 335
    midpoint values 180
    XCHART statement (ANOM) 120–121
contour plots 139, 198, 206
control characters 68
control charts
    See SHEWHART procedure
converting
    lowercase/uppercase strings 36–37
    numeric/character strings 37
    phone numbers 36
COUNT function 372–373
COVARIATES option, BASELINE statement (PHREG) 354
CPCT keyword, GCHART procedure 180, 182, 192
CREATE TABLE statement, SQL procedure
    DBKEY option 92–93
    functionality 80, 86
    ON clause 82
CRIGHT option, CAPABILITY procedure 318
CSS keyword 105
CSV format
    exporting data sets to spreadsheets 58–61
    importing spreadsheet data 62–69
    ODS support 11
CSVALL format 60, 213
CTEXT option
    GFONT procedure 167
    GOPTIONS statement 154–156
Cumulative Distribution Function
    See CDF
cumulative distribution functions 32–33
Cumulative Hazard Function 333
CURVELABEL option, SGPLOT procedure 270, 278–279
CURVELABELLOC option, SGPLOT procedure 270, 278
CURVELABELPOS option, SGPLOT procedure 270, 278–279

CUSUM procedure 311, 326–327
CV keyword 105
CV option, SYMBOL statement 170
CVREF option, GPLOT procedure 162
CYCLEATTRS option, SGPLOT procedure 270, 274

D

D3D style 243
data analysis
    See also summarizing data for analysis
    acceleration models 335
    Arrhenius acceleration models 344–345, 348–349
    determining process control 314–317
    fitting multiple Weibull distribution 340–341
    fitting single Weibull distribution 338–339, 346–347
    fitting Weibull acceleration models 342–343
    LIFEREG procedure 346–351
    LIFETEST procedure 352–353
    measuring process capability 310, 318–319
    monitoring processes 310–311
    non-repairable units 332
    P-Charts for fraction failing 322–323
    parametric lifetime distributions 334
    PHREG procedure 354–357
    probability plots for temperatures 350–351
    proportional hazards models 335
    reliability of repairable units 358–359
    RELIABILITY procedure 338–339
    repairable units 332
    selecting variables for 103
    statistical quality improvement and 310–311
    U-Charts for defects per unit 324–325
data fit plots 262–263
DATA _NULL_ step 371
DATA= option
    EXPORT procedure 58
    GMAP procedure 203
    SORT procedure 44
data sets
    See also tables
    attributes of variables 26–29
    computing statistics for subsets 106–107
    concatenating 133
    creating 94
    creating by entering lines of data 22–23
    creating from existing data sets 25
    creating variables 31
    creating with DATALINES statement 23
    creating with INPUT statement 23
    DATA step programming support 10, 20–29

deleting 50
equivalent terms 20, 78
exporting to DBMS with LIBNAME statement 85
exporting to spreadsheets 58–61, 94–95
extracting rows/columns 80
importing data to 62–69, 86–87
inner joins 81, 133
joining rows 80–81
merging 47–49, 133
moving data between DBMS and 84
naming 9
outer joins 82–83, 133
output objects as 232
saving 30
saving output to 234–235
selecting records by 48–49
selecting subsets 42–43
sorting 44–46
SQL procedure 79–83
viewing in Explorer window 20–21
viewing in Output window 21, 29
DATA step programming
ARRAY statement 53
CALL routines 31–40, 235, 370–371
conditional logic 24
correcting problems 14
creating macro variables 235
data fit plots 263
DATALINES statement 23
DO loops 24
DROP statement 42
FIRSTOBS= option 43
FORMAT statement 27, 34, 65
functions 24, 31–40
INFORMAT statement 27, 65
INPUT statement 23, 65
KEEP statement 42
LABEL statement 28–29, 224
LENGTH statement 26–27
manipulating data sets 10, 20–29
MERGE statement 47–48, 83, 196–197
merging data sets 47–49
OBS= option 43
RETAIN statement 41
saving data sets 30
selecting data set subsets 42–43
SET statement 25, 42, 173
sorting data sets 44–46
stacked bar charts 196
TARGET= statement 225
TYPE statement 26
data summarization
See summarizing data for analysis

Database Administrator (DBA) 84
Database Management System
See DBMS
DATAFILE statement, IMPORT procedure 62
DATALABEL option
COMPARE statement (SGSCATTER) 298
HBOX statement (SGPLOT) 286
HLINE statement (SGPLOT) 288–289
MATRIX statement (SGSCATTER) 298
PLOT statement (SGSCATTER) 298
SGPLOT procedure 270, 290
VBOX statement (SGPLOT) 286
DATALINES statement 23
DATA_NULL_ statement 239
DATASETS procedure
CONTENTS statement 49
deleting graphs 220
KILL option 50
QUIT statement 50
%DATATYP autocall program 376
date functions 34–35
DATE7. format 34
DATE9. format 34, 63, 371
DATEPART function 35, 87, 91
DATETIME function 34–35
DATETIME20. format 63
DAY function 35
DBA (Database Administrator) 84
DBKEY option, CREATE TABLE statement (SQL) 92–93
DBMS (Database Management System)
accessing 84
equivalent term 78
exploring contents 88–89
exporting data sets with LIBNAME statement 85
extracting records with matching values 92–93
importing data from, using LIBNAME statement 90–91
importing data from, using SQL Pass-Through Facility 86–87
moving data between data sets and 84
predicted needed to access 12
SAS/ACCESS 84
treating as library 84
%DBMSlist utility macro 390, 400–401
debugging
control character problems 68
correcting problems 14–15
data import problems 64–67
detecting problems 14
escaping from runaway programs 15
macro programs 388

defects per unit 324–325
DEFINE statement, REPORT procedure
    ANALYSIS option 115
    functionality 114
    GROUP option 114
    NOPRINT option 227, 231
    WIDTH option 115
DEFINE statement, TEMPLATE procedure 304
DEFINE STYLE statement, TEMPLATE procedure 242
DEGREE option
    COMPARE statement (SGSCATTER) 298
    MATRIX statement (SGSCATTER) 298
    PBSPLINE statement (SGPLOT) 262
    PLOT statement (SGSCATTER) 298
    REG statement (SGPLOT) 262
DELTA option, CUSUM procedure 326–327
%delvars utility macro 390, 393
DENSITY statement, SGPLOT procedure
    functionality 253, 260–261, 264, 284–285
    HISTOGRAM statement support 266–267
    LEGENDLABEL option 285
    LINEATTRS option 284–285
    SCALE option 284
    TYPE=KERNEL option 260, 284–285
DEQUOTE function 37
DESCENDING option, BY statement (SORT) 45–46
DEVICE= option, GOPTIONS statement 154
DEVICE=GIF option, GOPTIONS statement 144, 155
DEVICE=PNG option, GOPTIONS statement 220
DEVICE=PSLEPSFC option, GOPTIONS statement 144, 155
DEVICE=SASEEMF option, GOPTIONS statement 216
DEVICE=SASPRTC option, GOPTIONS statement 218
DIAGONAL option, MATRIX statement (SGSCATTER) 302–303
DIFn function 40
DISCRETE option, HBAR statement (GCHART) 184–185, 188
DISPLAY option
    X2AXIS statement 272
    XAXIS statement 272
    Y2AXIS statement 272
    YAXIS statement 272
distribution percentiles 105
distribution plots 260–261
DISTRIBUTION statement
    LIFEREG procedure 346
    RELIABILITY procedure 338

DO loops
    ARRAY statement and 53
    creating data sets 24
    %DO statement 382
%DO statement 382, 384
DOCUMENT destination 213
DOL option, RBREAK statement (REPORT) 115
DONUT statement, GCHART procedure 139, 180
DOT statement, SGPLOT procedure
    ALPHA= option 256–257
    functionality 253, 256–257
    GROUP option 264
    HBAR/HLINE statement support 266
    LIMITS option 256–257
    LIMITSTAT option 256–257
    NUMSTD option 256
    RESPONSE option 256
    STAT option 256
DOWN option, KEYLEGEND statement (SGPLOT) 276
DROP data set option 43
DROP statement 42
duplicate records, selecting 46
%dups utility macro 390, 394–395

E

EDF (empirical distribution function) 127
ELLIPSE option
    COMPARE statement (SGSCATTER) 298
    MATRIX statement (SGSCATTER) 298, 302–303
    PLOT statement (SGSCATTER) 298, 300
ELLIPSE statement, SGPLOT procedure
    FILL option 262
    FILLATTRS option 262
    functionality 253, 262–264
    TYPE= option 262
EMBEDDED_FOOTNOTES tagset option 72
EMBEDDED_TITLES tagset option 72
END statement, DEFINE STYLE statement (TEMPLATE) 242
EPS format 138, 155
ERROR messages 14
%EVAL function 376–377, 383
ExcelXP format 70–72, 213
EXP function 33
explanatory variables 332, 335
Explorer window 4, 20–21
EXPORT procedure
    DATA= option 58
    OUTFILE statement 58
    REPLACE option 58–59

exporting data sets
    ExcelXP tagset 70–72
    to DBMS with LIBNAME statement 85
    to spreadsheets 58–61, 94–95
extracting records
    rows/columns in data sets 80
    that match values in a list 92–93
EXTREME option
    HBOX statement (SGPLOT) 286–287
    VBOX statement (SGPLOT) 286
extreme value distributions 334

F

F= option, TABLES statement (TABULATE) 113
FACTEX procedure 311
failure rate 333
fields
    See also columns
    equivalent terms 78
    exploring contents 89
FILE= option
    ODS PDF statement 218
    ODS RTF statement 216
FILENAME statement
    customizing report pages 239
    functionality 144
    MOD option 239
files, collecting 404–405
FILL option
    HISTOGRAM statement (SGPLOT) 284–285
    SGPLOT procedure 262, 270
FILLATTRS option
    HISTOGRAM statement (SGPLOT) 284–285
    SGPLOT procedure 262, 270
FIRSTOBS= option 43
FIRST.var 45–46
Fishbone diagrams 311
FISHER option, TABLES statement (FREQ) 130
Fisher's exact test 130
FIT= option, PROBPLOT statement (RELIABILITY) 342
FITPOLICY option
    X2AXIS statement 272
    XAXIS statement 272–273
    Y2AXIS statement 272
    YAXIS statement 272–273
FLOOR function 33
FONT option
    FOOTNOTE statement 156
    SYMBOL statement 164–167, 170
    TITLE statement 156
fonts 166–167

FOOTNOTE statement
    ANGLE option 156
    COLOR option 156–157
    customizing GPLOT output 156–157
    FONT option 156
    functionality 156–157
    HEIGHT option 156
    JUSTIFY option 156–157
    LINK option 156
    ROTATE option 156
    SGPLOT procedure and 271
    UNDERLIN option 156–157
%foot_note utility macro 390, 392
FORMAT option, INSET statement (UNIVARIATE) 126
FORMAT statement
    creating data sets 27
    fixing data import problems 65
    formatting date/time 34, 39
    GPLOT procedure 159
    MEANS procedure 103–104
formats
    bitmap 138, 144–145
    converting lowercase/uppercase strings 37
    date/time 34
    defining with FORMAT statement 27, 34
    defining with INFORMAT statement 27
    vector 144
forward slashes (/) 3
fraction failing 322–323
FRAME option
    LEGEND statement 160–161
    ODS HTML statement 222–223
Freeman-Hillman extension (Fisher's exact test) 130
FREQ keyword, GCHART procedure 180, 182–183, 192
FREQ procedure
    functionality 100, 130–131
    TABLES statement 130
FREQLABEL option, HBAR statement (GCHART) 192
FROM clause, SELECT statement (SQL) 80–82
FTEXT option, GOPTIONS statement 154–156
full outer join 82
functions
    creating data sets 24
    cumulative distribution functions 32
    date functions 34–35
    datetime functions 34–35
    defined 31
    inverse probability functions 32–33
    macro functions 6, 376–377

functions (*continued*)
    math functions 32–33
    numeric functions 32–33
    probability functions 32–33
    reliability functions 332–333
    rounding functions 33
    summary functions 132, 372–373
    time functions 34–35
    truncation functions 33

G

G3D procedure
    PLOT statement 139, 198, 207
    SCATTER statement 139, 198, 207
GAREABAR procedure 139
GBARLINE procedure 139
GCHART procedure
    block charts 199
    BLOCK statement 139, 180, 198–199
    chart types 180
    DONUT statement 139, 180
    functionality 180–181
    HBAR statement 138–139, 180–190
    HBAR3D statement 139, 180
    PATTERN statement 186–187
    PIE statement 139, 180, 198
    PIE3D statement 139, 180
    STAR statement 139, 180
    TYPE= option 180
    VBAR statement 138–139, 180
    VBAR3D statement 139, 180
GCONTOUR procedure
    PLOT statement 139, 198, 206
    plotting data 206
GFONT procedure 167
GIF bitmap format 138, 144–145
%GLOBAL statement 386–387
global variables 368, 374, 386–387
GMAP procedure
    ANNOTATE option 203
    BLOCK statement 139, 198, 203–204
    CHORO statement 139, 198, 203
    DATA= option 203
    MAP= option 203
    plotting variables on maps 202–205
    PRISM statement 139, 198, 203, 205
    SURFACE statement 139, 198, 203, 205
goodness of fit test 127
GOPTIONS statement
    BORDER option 142, 154–155
    COLORS option 203
    CTEXT option 154–156
    default values for graphic options 142
    DEVICE= option 154
    DEVICE=GIF option 144, 155
    DEVICE=PNG option 220
    DEVICE=PSLEPSFC option 144, 155
    DEVICE=SASEMF option 216
    DEVICE=SASPRTC option 218
    example 116
    FTEXT option 154–156
    functionality 154–155
    global options 154–155
    GSFMODE= option 154
    GSFMODE=REPLACE option 144, 155
    GSFNAME= option 144, 154–155
    GUNIT option 154–155
    HORIGIN option 218
    HSIZE option 218
    HTEXT option 154–156, 192
    image file characteristics 144
    INTERPOL option 154–155
    ORIENTATION=LANDSCAPE option 218
    ORIENTATION=PORTRAIT option 218
    RESET option 154–155
    setting graphic options for plots 116
    VORIGIN option 218
    VSIZE option 218
    XMAX option 144, 216
    YMAX option 144, 216
%GOTO statement 385
GPATH option, ODS LISTING statement 250
GPLOT procedure
    See also SYMBOL statement, GPLOT procedure
    adding traffic lighting to reports 231
    ANNOTATE option 176–177
    AUTOHREF option 162
    AUTOVREF option 162
    box plot support 170–171
    BUBBLE statement 139, 146, 198, 201
    BUBBLE2 statement 139, 146, 198, 201
    BY statement 148
    CHREF option 162
    customizing output 154–171
    CVREF option 162
    default symbols 166–167
    FORMAT statement 159
    functionality 146, 316
    HREF option 162
    HTML option 224, 227
    HTML_LEGEND option 224, 227
    LHREF option 162
    LVREF option 162
    PLOT statement 138–139, 143–144, 146–153, 161, 174–175
    PLOT2 statement 138–139, 146–153, 161
    plotting group confidence limits 172–179

Index 427

sample report 215
   VREF= option 162
   WHERE clause 150
GRADAR procedure 139
Graph Template Language (GTL) 249, 304
Graph window 142–143
graphic output
   LIFETEST procedure example 250–251
   naming 143–144
   saving 144–145
   templates and 249
   to RTF format 216
   viewing 138, 142–143
gray-scale color names 168–169
GRID option
   X2AXIS statement 272
   XAXIS statement 272
   Y2AXIS statement 272–273
   YAXIS statement 272
GROUP option
   BLOCK statement (GCHART) 199
   COMPARE statement (SGSCATTER) 298
   DEFINE statement (REPORT) 114
   HBAR statement (GCHART) 188–189
   MATRIX statement (SGSCATTER) 298
   PLOT statement (SGSCATTER) 298
   SGPANEL procedure 296
   SGPLOT procedure 264–265
   VBAR statement (GCHART) 190
GROUPN= option
   PCHART statement (ANOM) 123
   UCHART statement (ANOM) 122
GSFMODE= option, GOPTIONS statement 154
GSFMODE=REPLACE option, GOPTIONS statement 144, 155
GSFNAME= option, GOPTIONS statement 144, 154–155
GSPACE option, VBAR statement (GCHART) 190–191
GTL (Graph Template Language) 249, 304
GUESSINGROWS statement, IMPORT procedure 67, 69
GUNIT option, GOPTIONS statement 154–155

H

HAXIS option, PLOT statement (GPLOT) 159
hazard function 333, 335
HBAR statement, GAREABAR procedure 139
HBAR statement, GCHART procedure
   controlling display of statistics 192–193
   DISCRETE option 184–185, 188
   FREQLABEL option 192
   functionality 138–139, 180

GROUP option 188–189
MAXIS option 182
MEANLABEL option 192
MIDPOINTS option 184–185
PATTERNID option 186–188
RAXIS option 182
SUBGROUP option 194–195
SUMLABEL option 192
SUMVAR option 180, 182, 194
TYPE= option 180, 182, 194–195
HBAR statement, SGPLOT procedure
   ALPHA= option 256
   functionality 253, 256, 258–259
   GROUP option 264
   HLINE/DOT statement support 266
   LIMITS option 256
   LIMITSTAT option 256
   NUMSTD option 256
   RESPONSE option 256
   STAT option 256
   URL option 294
HBAR3D statement
   GAREABAR procedure 139
   GCHART procedure 139, 180
HBOX statement, SGPLOT procedure
   BOXWIDTH option 286
   CATEGORY option 286
   DATALABEL option 286
   EXTREME option 286–287
   functionality 253, 260–261, 286–287
   GROUP option restrictions 264
   SPREAD option 286
   VBOX statement support 266
HEADLINE option, REPORT procedure 114
HEADTEXT option, ODS HTML statement 231
HEIGHT option
   FOOTNOTE statement 156
   GFONT procedure 167
   INSET statement (UNIVARIATE) 126
   SYMBOL statement 164, 170
   TITLE statement 156
Help and Documentation
   Contents tab 12
   Documentation tab 12
   Index tab 12
   opening 12
   Search tab 12
HISTOGRAM statement, CAPABILITY procedure 318
HISTOGRAM statement, SGPLOT procedure
   BOUNDARY option 284–285
   DENSITY statement support 266–267
   FILL option 284–285
   FILLATTRS option 284–285

HISTOGRAM statement, SGPLOT procedure (*continued*)
    functionality 253, 260–261, 284–285
    GROUP option restrictions 264
    NOOUTLINE option 284
    OUTLINE option 284–285
    SCALE option 284–285
HISTOGRAM statement, UNIVARIATE procedure 126, 128–129
HISTORY= option, SHEWHART procedure 320–321
HLINE statement, SGPLOT procedure
    ALPHA= option 256–257, 288
    DATALABEL option 288–289
    functionality 253, 256–259, 288
    GROUP option 264
    HBAR/DOT statement support 266
    LIMITS option 256–257, 288
    LIMITSTAT option 256–257, 288
    NUMSTD option 256
    RESPONSE option 256
    STAT option 256, 288
HORIGIN option, GOPTIONS statement 218
horizontal bar charts
    GCHART procedure 182–189
    SGPLOT procedure 258–259
HREF option, GPLOT procedure 162
HSIZE option, GOPTIONS statement 218
HTEXT option, GOPTIONS statement 154–156, 192
HTML format
    adding hyperlinks to Web pages 224–229
    adding traffic lighting to reports 231
    bitmap graphics support 145
    customizing reports 236–237
    description 213, 252
    graphic output 138
    ODS support 11
    publishing report in 220–221
    putting output in frames 222–223
    selecting output objects 232
    SGPLOT procedure 252
HTML option, GPLOT procedure 224, 227
HTML3 format 213
HTMLCSS format 213
HTML_LEGEND option, GPLOT procedure 224, 227
hyperlinks
    adding to Web pages 224–229
    adding with SGPLOT procedure 294–295
hypothesis tests 105
HZERO option, PLOT statement (GPLOT) 158

I

ID option, SHEWHART procedure 316
ID statement
    BOXPLOT procedure 118
    TRANSPOSE procedure 52
IDCOLOR option, PLOT statement (BOXPLOT) 118
IDSYMBOL option, PLOT statement (BOXPLOT) 118
%IF statement 384
IF-THEN-ELSE statements 24
IF-THEN-OUTPUT statements 43
%IF...%THEN statements 383–384
IMAGE_DPI option, ODS LISTING statement 250
IMAGEFMT option, ODS GRAPHICS statement 250
IMAGEMAP option, ODS GRAPHICS statement 294
IMAGENAME option, ODS GRAPHICS statement 250, 253
IMODE format 213
IMPORT procedure
    DATAFILE statement 62
    fixing import problems 64–67
    GUESSINGROWS statement 67, 69
    OUT= option 62
    REPLACE option 62
importing data
    fixing problems 64–67
    from DBMS using LIBNAME statement 90–91
    from DBMS using SQL Pass-Through Facility 86–87
    from spreadsheets 62–69
    %xlxp2sas macro 73
IN automatic variable 48–49
IN clause, SELECT statement (SQL) 92
%INCLUDE function
    example 50
    functionality 6, 8, 389
    running SAS code 7
INDEX function 39
%INDEX function 376
INFORMAT statement
    creating data sets 27
    fixing data import problems 65
informats 27, 37
inner joins 81, 133
INPUT function 37–39, 69
INPUT statement
    creating data sets 23
    fixing data import problems 65

INSET statement
    CAPABILITY procedure   318
    LIFEREG procedure   346
    SGPLOT procedure   253, 282–283
    UNIVARIATE procedure   126
INSIDE option, VBAR statement (GCHART)   192–193, 197
INT function   33
INTCK function   35
Internet, publishing reports to   220–221
INTERPOL option
    GOPTIONS statement   154–155
    SYMBOL statement   164–165, 170–171, 174
INTERPOLATE= option, RELIABILITY procedure   358
INTERPOLATION= option, LOESS statement (SGPLOT)   262
INTNX function   35
INTO clause, SELECT statement (SQL)   372
'invalid data' messages   64
inverse probability functions   32–33
ISHIKAWA procedure   311

J

JES style   243
JPEG bitmap format   138, 144–145
JULDATE function   35
JUSTIFY option
    FOOTNOTE statement   156–157
    TITLE statement   156

K

Kaplan-Meier method   339, 348, 350, 352–353
KEEP data set option   42
KEEP statement   43
KEYLEGEND statement, SGPLOT procedure
    ACROSS option   276
    BORDER option   276
    DOWN option   276
    functionality   253, 276–277
    LOCATION option   276–277
    NOBORDER option   276
    POSITION option   276–277
    TITLE option   276–277
KILL option, DATASETS procedure   50
KURTOSIS keyword   105

L

LABEL option
    AXIS statement   158–159
    LEGEND statement   160
    ODS CSVALL statement   60
    PRINT procedure   29, 63, 226

REFLINE statement (SGPLOT)   280
X2AXIS statement   272
XAXIS statement   272
Y2AXIS statement   272
YAXIS statement   272–273
LABEL statement
    adding hyperlinks to Web pages   224
    creating data sets   28–29
LABELALIGN option, INSET statement (SGPLOT)   282
LABELATTRS option, REFLINE statement (SGPLOT)   280
LABELLOC option
    REFLINE statement (SGPLOT)   280
    SCATTER statement (SGPLOT)   290
LABEL=NONE option, AXIS statement   158
LABELPOS option, REFLINE statement (SGPLOT)   280
labels
    creating for variables   28–29
    suppressing default   158
LAGn function   40
LAST.var   45–46
LATEX destination   252
LAYOUT statement   304
LCLM keyword   105, 108
leading spaces   37
%LEFT autocall program   376
LEFT function   37
left outer join   82
LEGEND option, PLOT statement (GPLOT)   161
LEGEND statement
    ACROSS option   160–161
    FRAME option   160–161
    functionality   160
    LABEL option   160
    MODE option   160–161
    POSITION option   160
    VALUE option   160
LEGENDLABEL option
    SCATTER statement (SGPLOT)   290, 292
    SGPLOT procedure   270
LENGTH function   38–39
%LENGTH function   376
LENGTH statement   26–27
%LET function
    assigning values to macro variables   6
    creating macro variables   8, 366, 370
LEVELS option, CHORO statement (GMAP)   203
LHREF option, GPLOT procedure   162
LIBNAME statement
    exporting data sets with   85
    importing data from DBMS   90–91
    MULTI_DATASRC_OPT option   93

LIBNAME statement (*continued*)
    saving data sets 30
    SCHEMA option 90
libraries
    defined 21
    equivalent terms 78
    period in macro variables 375
    treating DBMS as 84
LIFEREG procedure 350–351
    Arrhenius acceleration models 348–349
    BAYES statement 360–361
    fitting single Weibull distribution 346–347
    functionality 332
    INSET statement 346
    MODEL statement 346, 348
    OUTPUT statement 346
    parametric lifetime distributions 334
    PREDICTED option 346
    probability plots for temperatures 350–351
    PROBPLOT statement 346–348
    QUANTILE option 346
    XDATA option 348
    XSDATA option 348
LIFETEST procedure
    ATRISK option 352
    functionality 332, 352–353
    graphic output example 250–251
    METHOD= option 352
    PLOTS option 352
    STRATA statement 352–353
    SURVIVAL statement 352
    TIME statement 352
lifetime distributions 334
limit plots 256–257
LIMITS option
    DOT statement (SGPLOT) 256–257
    HBAR statement (SGPLOT) 256
    HLINE statement (SGPLOT) 256–257, 288
    SHEWHART procedure 320–321
    VBAR statement (SGPLOT) 256, 258
    VLINE statement (SGPLOT) 256
LIMITSTAT option
    DOT statement (SGPLOT) 256–257
    HBAR statement (SGPLOT) 256
    HLINE statement (SGPLOT) 256–257, 288
    VBAR statement (SGPLOT) 256, 258
    VLINE statement (SGPLOT) 256
LINE option, SYMBOL statement 164, 171
line plots
    depicted 141
    GPLOT procedure 146, 149–153
    SGPLOT procedure 256–259
LINEAR option, RELIABILITY procedure 344

LINEATTRS option
    DENSITY statement (SGPLOT) 284–285
    SGPLOT procedure 270, 274–275
LINK option
    FOOTNOTE statement 156
    TITLE statement 156, 224, 226
LISTING destination
    description 213
    ODS LISTING statement 212, 219, 250
lists, extracting records with matching values 92–93
%LOCAL statement 386
local variables 386–387
LOCATION option, KEYLEGEND statement (SGPLOT) 276–277
location-scale distributions 334
LOESS option
    COMPARE statement (SGSCATTER) 298
    MATRIX statement (SGSCATTER) 298
    PLOT statement (SGSCATTER) 298
LOESS statement, SGPLOT procedure
    functionality 253, 262–263
    GROUP option 264
    INTERPOLATION= option 262
LOG function 33
Log-logistic distribution 334
Log-normal distribution 332, 334
Log window
    depicted 4
    ERROR messages 14
    'invalid data' messages 64
    %PUT macro function 6
    WARNING messages 14
LOG2 function 33
LOG10 function 33
logical expressions
    IF-THEN-ELSE statements 24
    IF-THEN-OUTPUT statements 43
    %IF...%THEN statements 383
LOGLOG method 353
logs 14
looping
    See DO loops
LOWCASE function 37
LOWER option, BASELINE statement (PHREG) 354
lowercase, converting to 37
LVREF option, GPLOT procedure 162

M

macro functions
    functionality 376–377
    prefixes 6

macro language statements 382
macro programs
    accessing macros 389
    branching 382–385
    conditional execution 382–385
    debugging 388
    %DO statement 382
    functionality 10, 366–367, 378
    global/local macro variables 386–387
    looping 382–385
    macro functions 6, 376–377
    macro language statements 382
    parameters in 380–381
    utility macros 390–403
    writing 379
%MACRO statement
    correcting problems 14
    functionality 10, 367
    macro parameters 380
    writing macro programs 379
macro variables
    appending characters 375
    creating 235, 370–373
    customizing TITLE statement 380
    deleting 374
    double quotes to resolve 375
    functionality 366, 368
    in %DO loops 382
    %LET macro function 6
    local 386–387
    periods to delimit names 375
    prefixes 6
    SQL procedure support 372–373
    viewing 368–369
    writing values of specific 368
MACRONTROL procedure 326
%makedir utility macro 390, 397
%MakeList utility macro 390, 400–401
MAP= option, GMAP procedure 203
MARKERATTRS option
    SCATTER statement (SGPLOT) 292
    SGPLOT procedure 270, 274–275
MARKERS option, SGPLOT procedure 270, 274–275
math functions 32–33
MATRIX statement, SGSCATTER procedure
    ALPHA= option 298
    CLI option 298
    CLM option 298
    DATALABEL option 298
    DEGREE option 298
    DIAGONAL option 302–303
    ELLIPSE option 298, 302–303
    functionality 298–300, 302–303
    GROUP option 298
    LOESS option 298
    NOGROUP option 298
    PBSPLINE option 298
    REG option 298
MAUTOSOURCE option, OPTIONS statement 389
MAX function 104–105, 372–373
MAX option
    X2AXIS statement 272
    XAXIS statement 272
    Y2AXIS statement 272–273
    YAXIS statement 272
MAXIS option, HBAR statement (GCHART) 182
MCF (Mean Cumulative Function) 333, 358–359
MCFPLOT statement, RELIABILITY procedure 358–359
MDY function 35
Mean Cumulative Function (MCF) 333, 358–359
MEAN function
    creating macro variables 373
    MEANS procedure 104–105
MEAN keyword
    DOT statement (SGPLOT) 257
    GCHART procedure 180, 182–183, 192
    HLINE statement (SGPLOT) 256
    TABULATE procedure 115
MEAN= option, TABLES statement (TABULATE) 113
MEANLABEL option, HBAR statement (GCHART) 192
MEANS procedure
    ALPHA= option 108
    binomial distribution and 108, 110–111
    CLASS statement 106–107
    FORMAT statement 103–104
    functionality 100
    NOPRINT option 102
    normal confidence limits on the means 108–109
    NWAY option 108
    OUTPUT statement 102–103
    Poisson distribution and 108, 110–111
    selecting statistics 104–105
    stacked bar charts 196
    summarizing numeric variables 102
    TYPES statement 196
    VAR statement 103
measuring process capability 310, 318–319
MEDIAN keyword 105
%MEND statement
    correcting problems 14
    functionality 10, 367
    writing macro programs 379

MERGE statement
    merging data sets   47–48
    outer joins   83
    RENAME option   48
    stacked bar charts   196–197
merging data sets   47–49, 133
METHOD= option, LIFETEST procedure   352
midpoint values   180
MIDPOINTS option, HBAR statement (GCHART)   184–185
MIN function   104–105, 373
MIN option
    X2AXIS statement   272
    XAXIS statement   272
    Y2AXIS statement   272–273
    YAXIS statement   272
MINOR option, AXIS statement   158–159
MISSTEXT= option, TABLES statement (TABULATE)   113
MLOGIC option, OPTIONS statement   388
MMDDYYY10. format   34
MOD function   33
MODE option, LEGEND statement   160–161
MODEL statement
    LIFEREG procedure   346, 348
    PHREG procedure   354
    RELIABILITY procedure   342, 344–345
monitoring processes
    SHEWHART procedure   320–321
    statistical quality improvement   310–311
MONTH function   35
MPRINT option, OPTIONS statement   388
MS Excel   61
MS Word   215, 217
MSOffice2K format   61, 213
MU0= option
    CUSUM procedure   326
    UNIVARIATE procedure   124
MULTI_DATASRC_OPT option, LIBNAME statement   93
MVAR statement, TEMPLATE procedure   304
%my_symbols utility macro   390, 392–393

N

N keyword   104–105
NAME= option
    GFONT procedure   167
    PLOT statement (BOXPLOT)   116
    PLOT statement (GPLOT)   143–144, 147
naming
    data sets   9
    graphic output   143–144
    macro variables   368
    variables   9
NEEDLE statement, SGPLOT procedure
    functionality   253–254
    GROUP option   264
    URL option   294
NEWFILE=PAGE option, ODS HTML statement   223
NHPP (Non-Homogeneous Poisson Process)   333
NMARKERS option, SHEWHART procedure   320
NMISS keyword   105
NOBORDER option
    INSET statement (SGPLOT)   282
    KEYLEGEND statement (SGPLOT)   276
NO_BOTTOM_MATTER option, ODS HTML statement   239
NOCENPRINT option, RELIABILITY procedure   358
NOCENTER option, OPTIONS statement   29
NODATE option, OPTIONS statement   29, 212
NOFILL option, SGPLOT procedure   270
NOGROUP option
    COMPARE statement (SGSCATTER)   298
    MATRIX statement (SGSCATTER)   298
    PLOT statement (SGSCATTER)   298
NOMARKERS option, SGPLOT procedure   270
Non-Homogeneous Poisson Process (NHPP)   333
non-repairable units
    defined   332
    reliability functions for   333
    sample data   336
NONUMBER option, OPTIONS statement   29
NOOBS option, PRINT procedure   29
NOOUTLINE option, SGPLOT procedure   270
NOPRINT option
    DEFINE statement (REPORT)   227, 231
    MEANS procedure   102
    SQL procedure   372
NORMAL option
    CAPABILITY procedure   318
    HISTOGRAM statement (UNIVARIATE)   126
    INSET statement (UNIVARIATE)   126
    PROBPLOT statement (UNIVARIATE)   127
NORMALTEST option, UNIVARIATE procedure   125
NOTALNUM function   39
NOTALPHA function   39
NOTCHES option, PLOT statement (BOXPLOT)   116
NOTCNTRL function   39
NOTDIGIT function   39
NOTFIRST function   39
NOTGRAPH function   39

NOTLOWER function 39
NOTNAME function 39
NOTOC option, ODS PDF statement 218
NO_TOP_MATTER option, ODS HTML
    statement 239
NOTPRINT function 39
NOTPUNCT function 39
NOTSPACE function 39
NOTUPPER function 39
NOTXDIGIT function 39
NOWINDOWS option, REPORT procedure 114
NROW= option, HISTOGRAM statement
    (UNIVARIATE) 128–129
numeric functions 32–33
numeric variables
    box plot support 116
    converting to character 37
    data fit plots 262–263
    date/time functions 34–35
    defined 22
    midpoint values 180
    numeric functions 32–33
    on bar charts 184–185
    summarizing 102
NUMSTD option
    DOT statement (SGPLOT) 256
    HBAR statement (SGPLOT) 256
    HLINE statement (SGPLOT) 256
    VBAR statement (SGPLOT) 256
    VLINE statement (SGPLOT) 256
NWAY option, MEANS procedure 108

O

OBS= option 43
%obscnt utility macro 390, 396
observations 20, 78
    See also rows
ODS (Output Delivery System)
    adding hyperlinks to Web pages 224–229
    adding traffic lighting to reports 230–231
    creating output styles 242–243
    functionality 11, 212–213
    publishing reports in PDF format 218–219
    publishing reports in RTF format 216–217
    publishing reports to the Web 220–221
    putting output in frames 222–223
    sample report 214–215
    saving procedure output 232–235
    selecting output style 240–241
    selecting procedure output 236–237
ODS CSVALL statement 60
ODS EXCLUDE statement 236–237
ODS GRAPHICS statement
    functionality 250
    IMAGEFMT option 250
    IMAGEMAP option 294
    IMAGENAME option 250, 253
    OFF action 250
    ON action 250
    RESET option 250
ODS HTML statement
    adding hyperlinks to Web pages 224–229
    BODY= option 220, 222, 239
    CLOSE action 220
    CONTENTS option 222
    EXCLUDE action 236–237
    FRAME option 222–223
    functionality 252
    HEADTEXT option 231
    including customized code 238–239
    NEWFILE=PAGE option 223
    NO_BOTTOM_MATTER option 239
    NO_TOP_MATTER option 239
    PATH= option 220
    publishing reports to the Web 220–221
    putting output in frames 222–223
    SELECT action 236–237
    STYLE=JES option 243
    STYLE=MINIMAL option 220, 239
    TITLE option 224
    URL option 220
ODS LATEX statement 252
ODS LISTING statement
    CLOSE action 212, 218
    functionality 212, 219
    GPATH option 250
    graphic output example 250
    IMAGE_DPI option 250
ODS OUTPUT statement 234–235, 314
ODS PDF statement
    CLOSE action 219
    FILE= option 218
    functionality 218
    NOTOC option 218
    PDFTOC option 218
    selecting output style 240
    STARTPAGE=NEVER option 218
    STARTPAGE=NOW option 219
    STARTPAGE=YES option 219
ODS PROCLABEL statement 219, 222
ODS RTF statement
    CLOSE action 212, 216
    EXCLUDE action 216
    FILE= option 216
    functionality 212, 216, 252
    STARTPAGE=NEVER option 216
ODS SELECT statement 236–237

ODS Statistical Graphics
    functionality 248–249
    plotting data 248–251
    SGPANEL procedure 296–297
    SGPLOT procedure 138, 252–295
    SGRENDER procedure 249, 304–305
    SGSCATTER procedure 298–303
ODS styles
    See styles
ODS TAGSETS.ExcelXP statement 70–72
ODS TAGSETS.MSOffice2K statement 61, 213
ODS TEXT= statement 217
ODS TRACE statement
    functionality 232–233
    OFF action 232
    ON action 232, 236
ODS_ALL_CLOSE statement 252–253
OFFSET option, AXIS statement 158
OL option, BREAK statement (REPORT) 115
ON clause, CREATE TABLE statement (SQL) 82
operators, concatenation 36
OPTEX procedure 311
OPTIONS statement
    debugging macros 388
    functionality 71
    MAUTOSOURCE option 389
    MLOGIC option 388
    MPRINT option 388
    NOCENTER option 29
    NODATE option 29, 212
    NONUMBER option 29
    SASAUTOS option 389
    SYMBOLGEN option 388
Oracle
    (+) notation 87
    fixing DateTime problem 87
    SAS/ACCESS support 84
    SELECT OWNER...FROM ALL_TABLES statement 88
    SELECT...FROM...WHERE statement 89
    WHERE ROWNUM statement 89
ORDER option, AXIS statement 158–159, 170
ORIENTATION=LANDSCAPE option, GOPTIONS statement 218
ORIENTATION=PORTRAIT option, GOPTIONS statement 218
OUT= option
    BASELINE statement (PHREG) 354
    IMPORT procedure 62
    SORT procedure 44
outer joins 82–83, 133
OUTFILE statement, EXPORT procedure 58
outliers 116

OUTLINE option
    HISTOGRAM statement (SGPLOT) 284–285
    SGPLOT procedure 270
Output Delivery System
    See ODS
OUTPUT destination 213
output objects
    defined 232
    identifying 232–233
    saving as data sets 232
    saving procedure output 232–235
    selecting procedure output 236–237
OUTPUT statement
    LIFEREG procedure 346
    MEANS procedure 102–103
Output window
    depicted 4
    sending results to 102–103
    viewing data sets 21, 29
    viewing graphic output 143
OUTSIDE option, VBAR statement (GCHART) 192–193, 197
OUTTABLE option, CAPABILITY procedure 318
OVERLAY option
    PLOT statement (GPLOT) 150, 153
    RELIABILITY procedure 358

P

P0= option, SHEWHART procedure 323
P1 keyword 105
P5 keyword 105
P10 keyword 105
P25 keyword 105
P50 keyword 105
P75 keyword 105
P90 keyword 105
P95 keyword 105
P99 keyword 105
P-Charts 322–323
    See also binomial distribution
page breaks 218–219
PANELBY statement, SGPANEL procedure 296
parameters, macro program 380–381
PARENT=STYLES option, DEFINE STYLE statement (TEMPLATE) 242
PARETO procedure 311
PATH= option, ODS HTML statement 220
PATTERN statement
    COLOR option 186
    controlling statistics display 192
    functionality 186–187
    VALUE option 186

PATTERNID option, HBAR statement
  (GCHART) 186–188
PBSPLINE option
  COMPARE statement (SGSCATTER) 298
  MATRIX statement (SGSCATTER) 298
  PLOT statement (SGSCATTER) 298
PBSPLINE statement, SGPLOT procedure
  CLI option 262
  CLM option 262
  DEGREE= option 262
  functionality 253–254, 262–263
  GROUP option 264
PCHART statement
  ANOM procedure 120, 123
  SHEWHART procedure 323
PCL format 213
PCT keyword, GCHART procedure 180, 182–183, 192
PCTLMINOR option, PROBPLOT statement (UNIVARIATE) 127
PDF (Probability Density Function)
  functionality 33, 333
  Poisson distribution example 24
PDF format
  bitmap graphics support 145
  description 213
  graphic output 138
  ODS support 11
  publishing reports in 218–219
  selecting output objects 232
PDFTOC option, ODS PDF statement 218
percent (%) 6
period (.) 375
Perl language 404–405
phone numbers 36
PHREG procedure
  BASELINE statement 354
  functionality 332, 354–357
  MODEL statement 354
  PLOT option 354–355
PHTML format 213
pie charts
  DONUT statement (GCHART) 139, 180
  PIE statement (GCHART) 139, 180, 198, 200
  PIE3D statement (GCHART) 139, 180
PIE statement, GCHART procedure
  functionality 139, 180, 198
  SUBGROUP option 200
PIE3D statement, GCHART procedure 139, 180
PLOT option, PHREG procedure 354–355
PLOT statement, BOXPLOT procedure
  BOXCONNECT option 116
  BOXSTYLE option 116
  BOXSTYLE=SCHEMATIC option 118

BOXSTYLE=SCHEMATICID option 118
BOXWIDTH option 116, 118
BOXWIDTHSCALE= option 118
BWSLEGEND option 118
CBOXES option 118
CBOXFILL option 118
functionality 116
IDCOLOR option 118
IDSYMBOL option 118
LEGEND option 161
NAME option 116
NOTCHES option 116
PLOT statement, G3D procedure 139, 198, 207
PLOT statement, GBARLINE procedure 139
PLOT statement, GCONTOUR procedure 139, 198, 206
PLOT statement, GPLOT procedure
  basic forms 146–153
  functionality 138–139
  HAXIS option 159
  HZERO option 158
  NAME= option 143–144, 147
  OVERLAY option 150, 153
  SKIPMISS option 174–175
  VAXIS option 159
  VZERO option 158
PLOT statement, SGSCATTER procedure
  ALPHA= option 298
  CLI option 298
  CLM option 298
  DATALABEL option 298
  DEGREE option 298
  ELLIPSE option 298, 300
  functionality 298–300
  GROUP option 298
  LOESS option 298
  NOGROUP option 298
  PBSPLINE option 298
  REG option 298
PLOT2 statement, GPLOT procedure
  AXIS statement and 159
  basic forms 146–153
  functionality 138–139
  LEGEND statement and 161
PLOTS option, LIFETEST procedure 352
plotting data
  adding footnotes to plots 156–157
  box plots 116–119, 170–171
  bubble plots 139, 146, 198, 201
  contour plots 139, 198, 206
  customizing GPLOT output 154–171
  data fit plots 262–263
  data for examples 140–141
  default values for graphic options 142

436 *Index*

plotting data (*continued*)
    distribution plots   260–261
    group confidence limits   172–179, 288–295
    limit plots   256–257
    line plots   141, 146, 149–153, 256–259
    multiple lines on same graph   264–269
    multiple variables on different axes   151
    multiple variables on same axis   150
    naming graphic output   143–144
    probability plots for temperatures   350–351
    same graph for each variable value   149
    SAS/GRAPH support   138–139
    saving graphic output   144–145
    scatter plots   138, 146–148, 248, 296–297
    schematic box plots   116, 118–119
    separate graphs for each variable value   148
    setting graphic options   116
    SGRENDER procedure   249, 304–305
    simple plots   147
    skeletal box plots   116–117
    trend plots   141, 178–179
    variables on maps   202–205
    viewing graphic output   142–143
    with G3D procedure   207
    with GCHART procedure   180–200
    with GCONTOUR procedure   206
    with GMAP procedure   202–205
    with GPLOT procedure   146–179, 199, 201
    with multiple plot statements   266–267
    with ODS Statistical Graphics   248–251
    with SGPANEL procedure   296–297
    with SGPLOT procedure   138, 252–295
    with SGSCATTER procedure   298–303
    X-Y plots   254–256
PNG bitmap format   138, 144–145, 250
Poisson distribution
    adding confidence limits   110–111
    MEANS procedure and   108, 110–111
    numeric functions   32–33
    probabilities table example   24
POSITION option
    INSET statement (SGPLOT)   282
    INSET statement (UNIVARIATE)   126
    KEYLEGEND statement (SGPLOT)   276–277
    LEGEND statement   160
PREDICTED option, LIFEREG procedure   346
prediction limits   262
PRINT procedure
    LABEL option   29, 63, 226
    NOOBS option   29
    sample report   214
    VAR statement   29, 63, 226
    viewing data sets in Output window   21, 29

PRINTER style   242
PRISM statement, GMAP procedure   139, 198, 203, 205
Probability Density Function
    See PDF
probability functions   32–33
probability limits   314
probability plots for temperatures   350–351
PROBIT function   32–33
PROBPLOT statement
    LIFEREG procedure   346–348
    RELIABILITY procedure   338, 342
    UNIVARIATE procedure   127
PROBT keyword   105
PROC SQL Pass-Through Facility
    functionality   84
    importing data from DBMS   86–87
procedures
    See also specific procedures
    Base SAS   12–13
    functionality   10
    SAS/GRAPH   12–13
    SAS/QC   12–13
    SAS/STAT   12–13
    saving output   232–235
    selecting output   236–237
processes
    determining control   314–317
    measuring capability   310, 318–319
    monitoring   310–311, 320–321
    statistical quality improvement   310–311
Program Editor window
    clearing code in   7
    depicted   4
    SAS Help and Documentation   12
    saving code   7
    writing and submitting code   6–7
programming
    See SAS programming
proportional hazards models   335
PS format   213
PSLEPS vector format   144
PSLEPSFC vector format   144
publishing reports
    adding traffic lighting   230–231
    customizing   238–239
    in PDF format   218–219
    in RTF format   216–217
    sample report   214–215
    to the Web   220–221
%PUT _ALL_ statement   368
PUT function   37–38, 371
%PUT function
    generating warning messages   374

writing text strings 6, 8, 371
%PUT _GLOBAL_ statement 368

Q

QRANGE keyword 105
QTR function 35
quality data
    defined 310
    determining process control 314–317
    for examples 312–313
    measuring process capability 310, 318–319
    monitoring processes 310–311, 320–321
    P-Charts for fraction failing 322–323
    statistical quality improvement 310–311
    U-Charts for defects per unit 324–325
quality improvement
    See statistical quality improvement
QUANTILE option, LIFEREG procedure 346
QUIT statement
    DATASETS procedure 50
    SQL procedure 80, 372
QUOTE function 37

R

random number generation 31, 33
RANGE function 105, 373
RANNOR function 31, 33
RAXIS option, HBAR statement (GCHART) 182
RBREAK statement, REPORT procedure 115
records
    See also rows
    equivalent terms 78
    exploring contents 89
    extracting 80, 92–93
    selecting by data sets 48–49
    selecting duplicate 46
    selecting unique 46
    selecting with conditional logic 46
reference lines 162–163
REFLABEL option, AXIS statement 163
REFLINE statement, SGPLOT procedure
    AXIS option 280
    functionality 253, 280–281, 292
    LABEL option 280
    LABELATTRS option 280
    LABELLOC option 280
    LABELPOS option 280
REFTICKS option
    X2AXIS statement 272
    XAXIS statement 272–273
    Y2AXIS statement 272
    YAXIS statement 272

REG option
    COMPARE statement (SGSCATTER) 298
    MATRIX statement (SGSCATTER) 298
    PLOT statement (SGSCATTER) 298
REG statement, SGPLOT procedure
    CLM option 262
    DEGREE= option 262
    functionality 253, 262–263
    GROUP option 264
REGISTRY procedure 168
RELATION= option, MODEL statement (RELIABILITY) 344–345
relational databases 78
    See also DBMS (Database Management System)
reliability data
    acceleration models 335
    Arrhenius acceleration models 344–345, 348–349
    defined 310
    fitting multiple Weibull distribution 340–341
    fitting single Weibull distribution 338–339, 346–347
    fitting Weibull acceleration models 342–343
    LIFEREG procedure 346–351
    LIFETEST procedure 352–353
    non-repairable units 332
    parametric lifetime distributions 334
    PHREG procedure 354–357
    probability plots for temperatures 350–351
    proportional hazards models 335
    reliability of repairable units 358–359
    RELIABILITY procedure 338–339
    repairable units 332
reliability functions 332–333
RELIABILITY procedure
    Arrhenius acceleration models 344–345
    DISTRIBUTION statement 338
    fitting multiple Weibull distribution 340–341
    fitting single Weibull distribution 338–339
    fitting Weibull acceleration models 342–343
    functionality 311, 332
    INTERPOLATE= option 358
    LINEAR option 344
    MCFPLOT statement 358–359
    MODEL statement 342, 344–345
    NOCENPRINT option 358
    OVERLAY option 358
    parametric lifetime distributions 334
    PROBPLOT statement 338, 342
    reliability of repairable units 358–359
    UNITID statement 358
RENAME data set option 42
RENAME option, MERGE statement 48

repairable units
    defined 332
    reliability of 333, 358–359
    sample data 337
REPLACE option
    EXPORT procedure 58–59
    IMPORT procedure 62
REPORT procedure
    adding traffic lighting 230–231
    BREAK statement 115
    COLUMN statement 114, 227, 231
    DEFINE statement 114–115, 227, 231
    functionality 100, 114
    HEADLINE option 114
    NOWINDOWS option 114
    RBREAK statement 115
    sample report 214
    TARGET= statement 229
reports
    See publishing reports
RESET option
    GOPTIONS statement 154–155
    ODS GRAPHICS statement 250
RESPONSE option
    DOT statement (SGPLOT) 256
    HBAR statement (SGPLOT) 256
    HLINE statement (SGPLOT) 256
    VBAR statement (SGPLOT) 256, 258
    VLINE statement (SGPLOT) 256
Results windows
    depicted 4
    viewing data sets 21
    viewing graphic output 142–143
RETAIN statement
    functionality 41
    probability plots for temperatures 350
REVERSE function 37
RGB color-naming scheme 168–169
RIGHT function 37
right outer join 82
ROTATE option
    FOOTNOTE statement 156
    TITLE statement 156
ROUND function 33
rounding functions 33
rows
    equivalent terms 20, 78
    extracting in data sets 80
    IN automatic variable 48
    inner joins 81
    joining, from multiple data sets 80–81
    outer joins 82
    selecting 132

    selecting subsets 43
    selecting with earliest install date 132
ROWS option, SGPANEL procedure 296
RTF format
    bitmap graphics support 145
    description 213, 252
    graphic output 138
    ODS LISTING statement 212
    ODS support 11
    publishing reports in 216–217
    sample report 214–215
    selecting output objects 232
    SGPLOT procedure 252
runaway programs 15
%RunQuery utility macro 390, 400–401

S

SAS/ACCESS
    DBMS support 84
    LIBNAME statement 85, 90–91, 93
SAS ADX Interface for Design of Experiments 311
SAS code
    customizing report pages 238–239
    debugging basics 14–15
    installing sample code 2–3
    macro functions 6
    macro variables 6
    running with %INCLUDE function 7
    saving 7
    submitting 6–7
    writing 6–7
SAS data sets
    See data sets
SAS for UNIX programming interface
    autoexec.sas 5, 8
    depicted 4–5
    escaping from runaway programs 15
    saving code 7
    shutting down SAS sessions 8
    site license 5
    starting SAS sessions 4–5
    submitting code 6
    viewing data sets 20–21
SAS for Windows programming interface
    autoexec.sas 5, 8
    depicted 4
    escaping from runaway programs 15
    saving code 7
    shutting down SAS sessions 8
    site license 5
    starting SAS sessions 4
    submitting code 6–7

viewing data sets   20–21
SAS formats
   See formats
SAS functions
   See functions
SAS/GRAPH
   color-naming schemes   168–169
   data for examples   140–141
   GCHART procedure   180–197
   GPLOT procedure   146–179
   naming graphic output   143–144
   procedures supported   12–13, 138–139
   saving graphic output   144–145
   symbols and fonts   166–167
   viewing graphic output   142–143
SAS Help and Documentation
   Contents tab   12
   Documentation tab   12
   Index tab   12
   opening   12
   Search tab   12
SAS informats   27, 37
SAS-L Web site   16
SAS logs   14
SAS macro programs
   See macro programs
SAS procedures
   See procedures
SAS programming
   DATA step programming   10
   debugging basics   14–15
   escaping from runaway programs   15
   key elements   10–11
   macro programs   10
   Output Delivery System   11
   procedures   10
SAS/QC
   ODS Graphics support   249
   procedures supported   12–13, 120, 311, 326
SAS sessions
   shutting down   8
   starting   4–5
   WORK library and   30
SAS/STAT
   ODS Graphics support   249
   procedures supported   12–13, 116
SAS statements
   See statements
SAS styles
   See styles
SASAUTOS option, OPTIONS statement   389
%sas_papers utility macro   390, 402–403
%sasver_os utility macro   390–391

saving
   data sets   30, 234–235
   graphic output   144–145
   procedure output   232–235
   SAS code   7
SCALE option, HISTOGRAM statement (SGPLOT)   284–285
SCAN function   36–37, 39
%SCAN function   376
scatter plots
   depicted   138
   GPLOT procedure   146–148
   SGPANEL procedure   296–297
   SGSCATTER procedure   248
SCATTER statement, G3D procedure   139, 198, 207
SCATTER statement, SGPANEL procedure   296
SCATTER statement, SGPLOT procedure
   functionality   253–254, 263, 290–292
   GROUP option   264
   LABELLOC option   290
   LEGENDLABEL option   290, 292
   MARKERATTRS option   292
   URL option   294
   XERRORLOWER option   290
   XERRORUPPER option   290
   YERRORLOWER option   292
SCHEMA option, LIBNAME statement   90
schematic box plots   116, 118–119
SEED option, BAYES statement (LIFEREG)   360
SELECT statement, SQL procedure
   creating macro variables   235
   FROM clause   80–82
   IN clause   92
   INTO clause   372
   manipulating data sets   80–81, 83, 372
   SEPARATED BY option   373
   WHERE clause   80–82, 372
SEPARATED BY option, SELECT statement (SQL)   373
SERIES statement, SGPLOT procedure
   CURVELABELPOS option   279
   functionality   253–254, 290–292
   GROUP option   264–265
   LEGENDLABEL option   276
   LINEATTRS option   274
   MARKERATTRS option   274
   MARKERS option   274
   multiple plot statement support   266–267
   NAME option   276
   URL option   294
   Y2AXIS option   268–269
sessions
   See SAS sessions

440 *Index*

SET statement
    creating data sets from existing data sets 25
    KEEP data set option 42
    plotting group confidence limits 173
    WHERE clause 43
SETINIT procedure 12, 84
SGPANEL procedure
    COLUMNS option 296
    functionality 248, 296–297
    GROUP option 296
    PANELBY statement 296
    ROWS option 296
    SCATTER statement 296
    UNISCALE= option 296
SGPLOT procedure
    adding hyperlinks 294–295
    ALPHA= option 270
    BAND statement 253, 263–264
    bar charts 258–259
    basic X-Y plots 254–256
    BY statement 264–265, 296
    CURVELABEL option 270, 278–279
    CURVELABELLOC option 270, 278
    CURVELABELPOS option 270, 278–279
    CYCLEATTRS option 270, 274
    data fit plots 262–263
    DATALABEL option 270
    DENSITY statement 253, 260–261, 264, 266–267, 284–285
    distribution plots 260–261
    DOT statement 253, 256–257, 264, 266
    ELLIPSE statement 253, 262–264
    FILL option 270
    FOOTNOTE statement and 271
    functionality 138, 248, 252–253, 258–259, 316
    GROUP option 264–265
    HBAR statement 253, 256, 258–259, 264, 266, 294
    HBOX statement 253, 260–261, 264, 266, 286–287
    HISTOGRAM statement 253, 260–261, 264, 266–267, 284–285
    HLINE statement 253, 256–259, 264, 266, 288
    INSET statement 253, 282–283
    KEYLEGEND statement 253, 276–277
    LEGENDLABEL option 270
    limit plots 256–257
    LINEATTRS option 270, 274–275
    LOESS statement 253, 262–264
    MARKERATTRS option 270, 274–275
    MARKERS option 270, 274
    multiple lines on same graph 264–269
    multiple plot statements 266–267
    NEEDLE statement 253–254, 264, 294
    NOFILL option 270
    NOMARKERS option 270
    NOOUTLINE option 270
    OUTLINE option 270
    PBSPLINE statement 253–254, 262–264
    plotting data 138, 252–295
    plotting group confidence limits 288–295
    probability plots for temperatures 350–351
    REFLINE statement 253, 280–281, 292
    REG statement 253, 262–264
    SCATTER statement 253–254, 263–264, 290–292, 294
    SERIES statement 253–254, 264, 266–269, 290–292, 294
    STEP statement 253–254, 264, 294
    TITLE statement and 271
    TRANSPARENCY option 270
    VBAR statement 253, 256, 258–259, 264, 266, 294
    VBOX statement 253, 260–261, 264, 266, 286–287
    VLINE statement 253, 256, 258–259, 264, 266
    X2AXIS statement 253, 272–273
    XAXIS statement 253, 272–273
    Y2AXIS statement 253, 272–273
    YAXIS statement 253, 272–273
SGRENDER procedure 249, 304–305
SGSCATTER procedure
    COMPARE statement 298, 300–301
    functionality 248, 298–303
    MATRIX statement 298, 302–303
    PLOT statement 298–300
SHEET_NAME tagset option 72
SHEWHART procedure
    ALPHA= option 323, 325
    BOXCHART statement 316
    BOXCONNECT option 316
    determining process control 314–317
    functionality 311, 313
    HISTORY= option 320–321
    ID option 316
    LIMITS option 320–321
    monitoring processes 320–321
    NMARKERS option 320
    P-Charts for fraction failing 322–323
    P0= option 323
    PCHART statement 323
    SIGMAS= option 325
    STDDEVIATIONS option 314, 316
    SUBGROUPN option 323, 325

TABLEALL option 314–315
TESTS= option 314
U0= option 325
UCHART statement 325
USYMBOL option 325
VREF option 316
VREFLABELS option 316
XCHART statement 314–315
XSCHART statement 316–317
ZONELABELS option 314
shutting down SAS sessions 8
sigma limits 314
SIGMA0 option, CUSUM procedure 326
SIGMAS= option, SHEWHART procedure 325
site licenses 5, 12
skeletal box plots 116–117
SKEWNESS keyword 105
SKIP option, BREAK statement (REPORT) 115
SKIPMISS option, PLOT statement (GPLOT) 174–175
SORT procedure
　BY statement 45–47
　DATA= option 44
　merging data sets 46
　OUT= option 44
　plotting group confidence limits 173
　sorting data sets 44–45
sorting data sets 44–46
SPACE option, VBAR statement (GCHART) 190–191
SPEC statement, CAPABILITY procedure 318
SPREAD option
　HBOX statement (SGPLOT) 286
　VBOX statement (SGPLOT) 286
spreadsheets
　customizing workbooks 95
　customizing worksheet names 72
　exporting data sets to 58–61, 70–72, 94–95
　importing data from 62–69
SQL (Structured Query Language) 79
SQL Pass-Through Facility
　functionality 84
　importing data from DBMS 86–87
SQL procedure
　alias support 91
　ANSI standard support 79
　COALESCE function 83
　CONNECT TO statement 86
　CREATE TABLE statement 80, 82, 86, 92–93
　creating macro variables 235
　data summarization 132–133
　extracting rows/columns 80
　in macro programs 372–373

inner joins 81
joining rows from multiple data sets 80–81
NOPRINT option 372
ON clause 82
outer joins 82–83
QUIT statement 80, 372
SELECT statement 80–83, 235, 372–373
with data sets 79–83
SQRT function 33
stacked bar charts 196–197
STAR statement, GCHART procedure 139, 180
starting SAS sessions 4–5
STARTPAGE=NEVER option
　ODS PDF statement 218
　ODS RTF statement 216
STARTPAGE=NOW option, ODS PDF statement 219
STARTPAGE=YES option, ODS PDF statement 219
STAT option
　DOT statement (SGPLOT) 256
　HBAR statement (SGPLOT) 256
　HLINE statement (SGPLOT) 256, 288
　VBAR statement (SGPLOT) 256, 258
　VLINE statement (SGPLOT) 256
statements
　See also specific statements
　basic syntax 9
　correcting problems 14
　macro variables in 375
STATGRAPH template 304
statistical graphics
　See ODS Statistical Graphics
statistical quality improvement
　determining process control 314–317
　measuring process capability 318–319
　monitoring processes 320–321
　P-Charts for fraction failing 322–323
　phases of 310–311
STD keyword 104
STDDEV keyword 105
STDDEVIATIONS option, SHEWHART procedure 314, 316
STDERR keyword 105
STEP statement, SGPLOT procedure
　functionality 253–254
　GROUP option 264
　URL option 294
STRATA statement, LIFETEST procedure 352–353
strings, concatenating 37
STRIP function 37
Structured Query Language (SQL) 79
　See also SQL procedure

442 *Index*

Student's t test 124
STYLE= option
    ODS tagsets 61, 70–72
        selecting for ODS output 240
STYLE statement, DEFINE STYLE statement (TEMPLATE) 242
STYLE=JES option, ODS HTML statement 243
STYLE=MINIMAL option
    ODS HTML statement 220, 239
    selecting for ODS output 240
styles
    creating 242–243
    SAS/GRAPH support 249
    selecting for output 240
    TEMPLATE procedure support 61, 71, 242
SUBGROUP option
    BLOCK statement (GCHART) 199
    HBAR statement (GCHART) 194–195
    PIE statement (GCHART) 200
    VBAR statement (GCHART) 197
subgroup variables 194
SUBGROUPN option, SHEWHART procedure 323, 325
submitting SAS code 6–7
subsets of data sets
    computing statistics 106–107
    selecting 42–43
SUBSTR function 36–37
%SUBSTR function 376
SUM keyword
    GCHART procedure 180, 182–183, 192
    MEANS procedure 104–105
SUMLABEL option, HBAR statement (GCHART) 192
SUMMARIZE option
    BREAK statement (REPORT) 115
    RBREAK statement (REPORT) 115
summarizing data for analysis
    ANOM procedure 100, 120–123
    BOXPLOT procedure 100, 116–119
    FREQ procedure 100, 130–131
    MEANS procedure 100, 102–111
    REPORT procedure 100, 114–115
    sample component data 100–101
    SQL procedure 132–133
    TABULATE procedure 100, 112–113
    UNIVARIATE procedure 10, 100, 124–129
summary functions 132, 372–373
SUMVAR option
    HBAR statement (GCHART) 180, 182, 194
    VBAR statement (GCHART) 197
SUMWGT keyword 105

SURFACE statement, GMAP procedure 139, 198, 203, 205
survival function 333
SURVIVAL option, BASELINE statement (PHREG) 354
SURVIVAL statement, LIFETEST procedure 352
SYMBOL statement, GPLOT procedure
    BWIDTH option 170
    CI option 170–171
    CO option 170
    COLOR option 164
    CV option 170
    FONT option 164–167, 170
    functionality 146, 149
    HEIGHT option 164, 170
    INTERPOL option 164–165, 170–171, 174
    LINE option 164, 171
    VALUE option 164, 166, 170
    WIDTH option 164, 170
SYMBOLGEN option, OPTIONS statement 388
%SYMDEL statement 374
syntax for statements 9
%SYSEVALF function 376–377, 383
%SYSFUNC statement 385

T

T keyword 105
TABLEALL option, SHEWHART procedure 314–315
tables
    See also data sets
    adding traffic lighting 230–231
    contents frames 222–223
    equivalent terms 20, 78
    exploring contents 88
    extracting rows/columns 80
    virtual 88
TABLES statement
    FREQ procedure 130
    TABULATE procedure 112–113
TABULATE procedure
    CLASS statement 112
    functionality 100, 112–113
    TABLES statement 112–113
    VAR statement 112–113
TARGET= statement
    adding hyperlinks to Web pages 225
    REPORT procedure and 229
temperatures, probability plots for 350–351
TEMPLATE procedure
    DEFINE statement 304
    DEFINE STYLE statement 242
    manipulating styles 61, 71, 242

MVAR statement   304
templates
    graphic   249
    STATGRAPH   304
TESTS= option, SHEWHART procedure   314
%theDate utility macro   390–391
TIME function   34
TIME statement, LIFETEST procedure   352
time-to-failure distributions   334
TIMEPART function   35
TITLE option
    INSET statement (SGPLOT)   282
    KEYLEGEND statement (SGPLOT)   276–277
    ODS HTML statement   224
TITLE statement
    ANGLE option   156
    COLOR option   156
    customizing data set view   29
    customizing GPLOT output   156–157
    customizing with macro variables   380
    example   116, 182
    FONT option   156
    functionality   156–157
    GOPTIONS statement and   154
    HEIGHT option   156
    JUSTIFY option   156
    LINK option   156, 224, 226
    ODS CSVALL statement   60
    ROTATE option   156
    setting graphic options for plots   116
    SGPLOT procedure and   271
    UNDERLIN option   156
TODAY function   34–35
Toolbox feature   4
traffic lighting   230–231
trailing spaces   37
TRANSLATE function   36–37
TRANSPARENCY option, SGPLOT procedure   270
TRANSPOSE procedure
    BY statement   52
    ID statement   52
    VAR statement   52
trend plots   141, 178–179
TRIM function   37
%TRIM function   388
troubleshooting
    See debugging
truncation functions   33
Tukey, John   116
TYPE= option
    DENSITY statement (SGPLOT)   260, 284–285

ELLIPSE statement (SGPLOT)   263
HBAR statement (GCHART)   180, 182, 194–195
VBAR statement (GCHART)   197
X2AXIS statement   272
XAXIS statement   272
Y2AXIS statement   272
YAXIS statement   272
TYPE statement   26
TYPES statement, MEANS procedure   196

U

U0= option, SHEWHART procedure   325
U-Charts   324–325
    See also Poisson distribution
UCHART statement
    ANOM procedure   120, 122
    SHEWHART procedure   325
UCLM keyword   105, 108
UNDERLIN option
    FOOTNOTE statement   156–157
    TITLE statement   156
unique records, selecting   46
UNISCALE= option, SGPANEL procedure   296
UNITID statement, RELIABILITY procedure   358
UNIVARIATE procedure
    CLASS statement   128–129
    creating output objects   232
    customizing reports   236–237
    example   10
    functionality   100, 124–125
    HISTOGRAM statement   126
    INSET statement   126
    MU0= option   124
    NORMALTEST option   125
    ODS RTF EXCLUDE statement and   216
    PROBPLOT statement   127
    sample report   215
    VAR statement   124
UNIX operating environment
    See SAS for UNIX programming interface
UPCASE function   36–37
%UPCASE function   376
UPPER option, BASELINE statement (PHREG)   354
uppercase, converting to   36
URL option
    ODS HTML statement   220
    SGPLOT procedure   294
USS keyword   105
USYMBOL option, SHEWHART procedure   325

utility macros
    %check_autocall 390
    %DBMSlist 390, 400–401
    %delvars 390, 393
    %dups 390, 394–395
    %foot_note 390, 392
    %makedir 390, 397
    %MakeList 390, 400–401
    %my_symbols 390, 392–393
    %obscnt 390, 396
    %RunQuery 390, 400–401
    %sas_papers 390, 402–403
    %sasver_os 390–391
    %theDate 390–391
    %varexist 390, 398
    %vars_in 390, 399

V

VALUE option
    AXIS statement 158
    LEGEND statement 160
    PATTERN statement 186
    SYMBOL statement 164, 166, 170
VALUEALIGN option, INSET statement (SGPLOT) 282
VALUES option
    X2AXIS statement 272
    XAXIS statement 272
    Y2AXIS statement 272
    YAXIS statement 272–273
VALUESHINT option
    X2AXIS statement 272
    XAXIS statement 272
    Y2AXIS statement 272
    YAXIS statement 272
VAR keyword 105
VAR statement
    CAPABILITY procedure 318
    IMPORT procedure 63
    MEANS procedure 103
    ODS CSVALL statement 60
    PRINT procedure 29, 63, 226
    TABULATE procedure 112–113
    TRANSPOSE procedure 52
    UNIVARIATE procedure 124
%varexist utility macro 390, 398
variables
    See also character variables
    See also columns
    See also macro variables
    See also numeric variables
    attributes 26–29
    automatic 45–46

chart 180
classification 335
continuous 120–121, 180
creating 132
creating in data sets 31
creating labels 28–29
equivalent terms 20, 78
explanatory 332, 335
formatting output 27
global 368, 374, 386–387
naming 9
plotting on maps 202–205
retaining values 41
selecting distinct values 132
selecting for analysis 103
subgroup 194
%vars_in utility macro 390, 399
VAXIS option
    CUSUM procedure 326
    PLOT statement (GPLOT) 159
VBAR statement, GAREABAR procedure 139
VBAR statement, GCHART procedure
    controlling display of statistics 192–193
    functionality 138–139, 180
    GSPACE option 190–191
    INSIDE option 192–193, 197
    OUTSIDE option 192–193, 197
    SPACE option 190–191
    stacked bar charts 196–197
    SUBGROUP option 197
    SUMVAR option 197
    TYPE= option 197
    WIDTH option 190–191
VBAR statement, SGPLOT procedure
    ALPHA= option 256, 258
    functionality 253, 256, 258–259
    GROUP option 264
    LIMITS option 256, 258
    LIMITSTAT option 256, 258
    NUMSTD option 256
    RESPONSE option 256, 258
    STAT option 256, 258
    URL option 294
    VLINE statement support 266
VBAR3D statement
    GAREABAR procedure 139
    GCHART procedure 139, 180
VBOX statement, SGPLOT procedure
    BOXWIDTH option 286
    CATEGORY option 286
    DATALABEL option 286
    EXTREME option 286
    functionality 253, 260–261, 286–287
    GROUP option restrictions 264

HBOX statement support 266
SPREAD option 286
vector formats 144
VERIFY function 38–39
vertical bar charts
GCHART procedure 190–191
SGPLOT procedure 258–259
viewing
automatic variables 368
data sets in Explorer window 20–21
data sets in Output window 21, 29
global variables 368
graphic output 138, 142–143
macro variables 368–369
views 88
virtual tables 88
VLINE statement, SGPLOT procedure
ALPHA= option 256
functionality 253, 256, 258
GROUP option 264
LIMITS option 256
LIMITSTAT option 256
NUMSTD option 256
RESPONSE option 256
STAT option 256
VBAR statement support 266
VORIGIN option, GOPTIONS statement 218
VREF= option
BOXCHART statement (ANOM) 120
GPLOT procedure 162
SHEWHART procedure 316
VREFLABELS option, SHEWHART procedure 316
VSIZE option, GOPTIONS statement 218
VZERO option, PLOT statement (GPLOT) 158

W

WARNING messages
OVERLAY option 153
writing to Log window 14
Web, publishing reports to 220–221
Web pages, adding hyperlinks 224–229
WEEKDATE15. format 34
WEEKDAY function 35
Weibull distribution
acceleration models 335
fitting acceleration models 342–343
fitting multiple 340–341
fitting single 338–339, 346–347
location-scale equivalent 334
non-repairable units 332
probability plots for temperatures 350
WEIBULLSHAPEPRIOR option, BAYES statement (LIFEREG) 360
WHERE clause
GPLOT procedure 150
SELECT statement (SQL) 80–82, 372
SET statement 43
WIDTH option
DEFINE statement (REPORT) 115
SYMBOL statement 164, 170
VBAR statement (GCHART) 190–191
Wilcoxon Signed Rank test 124
Windows operating environment
See SAS for Windows programming interface
WORK library 30
writing
macro programs 379
SAS code 6–7
text strings 6
values of specific macro variables 368
WARNING messages to Log window 14

X

X-Y plots 254–256
XAXIS statement, SGPLOT procedure
DISPLAY option 272
FITPOLICY option 272–273
functionality 253, 272–273
GRID option 272
LABEL option 272
MAX option 272
MIN option 272
REFTICKS option 272–273
TYPE= option 272
VALUES option 272
VALUESHINT option 272
X2AXIS statement, SGPLOT procedure
DISPLAY option 272
FITPOLICY option 272
functionality 253, 272–273
GRID option 272
LABEL option 272
MAX option 272
MIN option 272
REFTICKS option 272
TYPE= option 272
VALUES option 272
VALUESHINT option 272
XCHART statement
ANOM procedure 120–121
CUSUM procedure 326
SHEWHART procedure 314–315
XDATA option, LIFEREG procedure 348
XERRORLOWER option, SCATTER statement (SGPLOT) 290

XERRORUPPER option, SCATTER statement (SGPLOT) 290
XHTML format 213
%xlxp2sas macro 73
XMAX option, GOPTIONS statement 144, 216
XML format
    formatting output 71
    ODS support 11
    %XLXP2SAS macro 73
XSCHART statement, SHEWHART procedure 316–317
XSDATA option, LIFEREG procedure 348

Y

YAXIS statement, SGPLOT procedure
    DISPLAY option 272
    FITPOLICY option 272–273
    functionality 253, 272–273
    GRID option 272–273
    LABEL option 272–273
    MAX option 272–273
    MIN option 272–273
    REFTICKS option 272
    TYPE= option 272
    VALUES option 272–273
    VALUESHINT option 272
Y2AXIS option, SERIES statement (SGPLOT) 268–269
Y2AXIS statement, SGPLOT procedure
    DISPLAY option 272
    FITPOLICY option 272
    functionality 253, 272–273
    GRID option 272
    LABEL option 272
    MAX option 272
    MIN option 272
    REFTICKS option 272
    TYPE= option 272
    VALUES option 272
    VALUESHINT option 272
YEAR function 35
YERRORLOWER option, SCATTER statement (SGPLOT) 292
YERRORUPPER option, SCATTER statement (SGPLOT) 292
YMAX option, GOPTIONS statement 144, 216

Z

ZONELABELS option, SHEWHART procedure 314

Special Characters

& (ampersand) 6
\ (backward slashes) 3
|| (concatenation operator) 36
/ (forward slashes) 3
(+) notation 87
% (percent) 6
. (period) 375